An Illustrated Guide to
VETERINARY MEDICAL TERMINOLOGY

FOURTH EDITION

Janet Amundson Romich, DVM, MS

CENGAGE
Learning®

Australia • Brazil • Mexico • Singapore • United Kingdom • United States

An Illustrated Guide to Veterinary Medical Terminology, Fourth Edition

Janet Amundson Romich, DVM, MS

General Manager: Dawn Gerrain

Associate Product Manager: Dan Johnson

Senior Director, Development:
   Marah Bellegarde

Senior Product Development Manager:
   Larry Main

Senior Content Developer: Darcy M. Scelsi

Product Assistant: Scott Royael

Executive Director of Marketing:
   Michelle McTighe

Senior Production Director: Wendy A. Troeger

Production Manager: Mark Bernard

Senior Content Project Manager:
   Elizabeth C. Hough

Senior Art Director: David Arsenault

Media Editor: Deborah Bordeaux

Cover image: Horse: Volodymyr Burdiak/
   Shutterstock.com

Cover Design: Dave Gink/Shadowland Studios

© 2015, 2009, 2006, 2000 Cengage Learning

WCN: 01-100-101

For product information and technology assistance, contact us at
Cengage Learning Customer & Sales Support, 1-800-354-9706

For permission to use material from this text or product,
submit all requests online at **www.cengage.com/permissions**.
Further permissions questions can be e-mailed to
permissionrequest@cengage.com

Library of Congress Control Number: 2013956042

ISBN-13: 978-1-133-12576-1

Cengage Learning
200 First Stamford Place, 4th Floor
Stamford, CT 06902
USA

Cengage Learning is a leading provider of customized learning solutions with office locations around the globe, including Singapore, the United Kingdom, Australia, Mexico, Brazil, and Japan. Locate your local office at:
www.cengage.com/global

Cengage Learning products are represented in Canada by Nelson Education, Ltd.

To learn more about Cengage Learning, visit **www.cengage.com**

Purchase any of our products at your local college store or at our preferred online store **www.cengagebrain.com**

Notice to the Reader

Printed in the United States of America
2 3 4 5 6 7 18 17 16 15

An Illustrated Guide to

VETERINARY MEDICAL TERMINOLOGY

CONTENTS

PREFACE

TO THE STUDENT

Medical terminology may seem like a foreign language because many of the terms are unfamiliar, seem strange, or do not make sense. However, to communicate in the medical world, you need a thorough understanding of the language.

Most medical terms are based on word parts that already may be familiar. You may have heard words such as *appendicitis*, *gastritis*, and *tonsillectomy* or used them in the past. You may not realize how many medical terms you already know. Building on this foundation, learning new word parts will make medical terminology seem more logical.

This text and the accompanying materials simplify the process of learning medical terminology. Review the introductory sections so you are familiar with the organizational scheme of the textbook and CourseMate™. Once you become comfortable with the materials, you will find yourself learning medical terms faster than you ever imagined possible.

Chapter Organization

The chapters in *An Illustrated Guide to Veterinary Medical Terminology, Fourth Edition*, are organized in the following fashion:

- Introduction to medical terms
- Anatomical foundations
- Body systems
- Species-specific chapters

Chapter 1 provides the basics of how medical terms are formed, analyzed, and defined. Chapter 2 provides terms used in everyday dialogue regarding positioning of animals and relationships between body parts. Chapters 3 and 4 discuss anatomical landmarks both internally (musculoskeletal system) and externally (common terms for landmarks on an animal's body).

Chapter 5 consists of terms used in the animal industry to describe males and females of selected species and terms for their young and for groups of their species. Chapters 6 through 15 are organized by body systems. These chapters describe the anatomy of the body system; include clinical terms used in reference to it; and conclude with diagnostic tests, pathology, and procedures for the body system. Chapters 16 and 17 relate tests, procedures, and treatments used in the care of animals in the veterinary medical field. Chapters 18 through 23 are species-specific chapters that you can study independently to enhance your knowledge of a particular species or that your instructor may incorporate into other chapters to assess your progress. Appendix A consists of tables of abbreviations, and Appendix B contains plural forms of medical terms. Appendix C lists prefixes, combining forms, and suffixes.

COURSEMATE

CourseMate complements your textbook with several robust and noteworthy components:

- An interactive eBook, with highlighting, note taking, and search capabilities.
- Interactive and engaging learning tools, including flashcards quizzes, games, PowerPoint® presentations, and much more!
- Engagement Tracker, a first-of-its-kind tool that monitors student participation and retention in the course.

To access CourseMate content:

- Go to www.cengagebrain.com.
- For an Internet access code (Order # 978-1-1331-2582-2)
- For a Print access code (Order # 978-1-1331-2583-9)

TO THE INSTRUCTOR

An Instructor Companion Website is available to facilitate classroom preparation, presentation, and testing. This includes an instructor's manual that provides answer keys for all exercises in the text, teaching tips, and activities to enhance your teaching of medical terminology. A test bank contains 1,000 questions in the following formats: multiple choice, short answer, and matching. An image library containing the images from the text can be used to create PowerPoint® slides, transparencies, or handouts for students. PowerPoint® presentations can be used to deliver lectures or to provide as handouts to students.

This content can be accessed through your Instructor SSO account.

To set up your account:

- Go to www.cengagebrain.com/login.
- Choose **Create a New Faculty Account.**
- Next you will need to select your **Institution.**
- Complete your personal **Account Information**.
- Accept the **License Agreement**.
- Choose **Register.**
- Your account will be pending validation; you will receive an e-mail notification when the validation process is complete.
- If you are unable to find your Institution; complete an **Account Request Form.**

Once your account is set up or if you already have an account:

- Go to www.cengagebrain.com/login.
- Enter your e-mail address and password and select **Sign In.**
- Search for your book by author, title, or ISBN.
- Select the book and click **Continue.**
- You will receive a list of available resources for the title you selected.
- Choose the resources you would like and click **Add to My Bookshelf.**

ACKNOWLEDGMENTS

Special thanks to the following people who helped review this text and answered many questions regarding medical terminology throughout its development. Without their expertise, the text would not have been as complete.

Kevin R. Berry, CVT
Gaska Dairy Health Services, Columbus, WI

Kay Bradley, BS, CVT
Madison Area Technical College, Madison, WI

Kenneth Brooks, DVM, Diplomate ABVP
Lodi Veterinary Hospital, SC, Lodi, WI

Eric Burrough, DVM
Kirkwood Community College, Cedar Rapids, IA

Stephen J. Carleton, DVM
Quinnipiac University, Connecticut

Anne E. Chauvet, DVM, Diplomate ACVIM—Neurology
University of Wisconsin Veterinary Medical Teaching Hospital, Madison, WI

Jane Clark, DVM
Madison Area Technical College, Madison, WI

Michael T. Collins, DVM, PhD
University of Wisconsin School of Veterinary Medicine, Madison, WI

Thomas Curro, DVM, MS
Henry Doorly Zoo, Omaha, Nebraska

Deb Donohoe, LATG
Medical College of Wisconsin, Milwaukee, WI

Wendy Eubanks, CVT
Chickasaw Trail Animal Hospital, Orlando, FL

Ron Fabrizius, DVM, Diplomate ACT
Poynette Veterinary Service, Inc., Poynette, WI

Kelly Gilligan, DVM
Four Paws Veterinary Clinic, LLC, Prairie du Sac, WI

Carmen M. Gorder
Science Student, Waterloo, WI

John H. Greve, DVM, PhD
Iowa State University, Ames, IA

Gerald Hackett, DVM
California Polytechnic University, Pomona, CA

Brain J. Heim, DVM
Cedar Valley College, Lancaster, Texas

Mark Jackson, DVM, PhD, Diplomate ACVIM, MRCVS
Glasgow University, Scotland

Linda Kratochwill, DVM
Crow-Goebel Veterinary Clinic, Scanlon, MN

Amy Lang, RTR
University of Wisconsin Veterinary Medical Teaching Hospital, Madison, WI

Laura L. Lien, CVT, BS, MS, VTS (LAIM)
Madison Area Technical College, Madison, WI

Carole Maltby, DVM
Maple Woods Community College, Kansas City, MO

A. Edward Marshall, DVM, PhD
Auburn University, Auburn, AL

Sheila McGuirk, DVM, PhD, Diplomate ACVIM
University of Wisconsin Veterinary Medical Teaching Hospital, Madison, WI

James T. Meronek, DVM, MPH
ABS Global Inc, DeForest, WI

David Morales, DVM
Oklahoma State University, Oklahoma City, OK

Karl Peter, DVM
Foothill College, Los Altos Hills, CA

Kathrine Polzin, BA, CVT
University of Wisconsin Veterinary Medical Teaching Hospital, Madison, WI

Stuart Porter, VMD
Blue Ridge Community College, Weyers Cave, VA

Teri Raffel, CVT
Madison Area Technical College, Madison, WI

Linda Sullivan, DVM
University of Wisconsin Veterinary Medical Teaching Hospital, Madison, WI

Laurie Thomas, BA, MA
Clinicians Publishing Group/Partners in Medical Communications, Clifton, NJ

Beth Uldal Thompson, VMD
Veterinary Technician/Veterinary Learning Systems, Trenton, NJ

Animart
Madison, WI

I also would like to express my gratitude to Beth Thompson, VMD, and Laurie Thomas, BA, MA, of Veterinary Learning Systems for their determination in advancing my writing skills through the publication of journal articles for *Veterinary Technician Journal*. Without their guidance I would not have honed my writing skills. I also would like to thank the many veterinary technician and laboratory animal technician students at Madison Area Technical College for their support and continued critique of the veterinary terminology course. A special thank-you goes to the 1998 veterinary technician and laboratory animal technician students at Madison Area Technical College, who learned terminology through my rough draft of the original text. Finally, I would like to thank the excellent staff at Cengage Learning and my family for their continued support.

ABOUT THE AUTHOR

Dr. Janet Romich received her Bachelor of Science degree in Animal Science from the University of Wisconsin–River Falls and her Doctor of Veterinary Medicine and Master of Science degree from the University of Wisconsin–Madison. Currently, Dr. Romich teaches at Madison Area Technical College in Madison, Wisconsin, where she has taught and continues to teach a variety of science-based courses. Dr. Romich was honored with the Distinguished Teacher Award in 2004 for use of technology in the classroom, advisory and professional activities, publication list, and fundraising efforts. She received the Wisconsin Veterinary Technician Association's Veterinarian of the Year Award in 2007 for her contributions in educating veterinary technician students and promoting the use of veterinary technicians in the workplace. Dr. Romich authored the textbooks *Fundamentals of Pharmacology for Veterinary Technicians* and *Understanding Zoonotic Diseases*, as well as served as a coauthor on *Delmar's Veterinary Technician Dictionary*. Dr. Romich remains active in veterinary practice through her relief practice, where she works in both small- and mixed-animal practices.

How to Use This Text

An Illustrated Guide to Veterinary Medical Terminology, Fourth Edition, helps you learn and retain medical terminology using a logical approach to medical word parts and associations. Following are the keys to learning from this text.

Illustrations

Complete with detailed labeling, the text's line drawings clarify key concepts and contain important information of their own. In addition to line drawings, photos are included to enhance the visual perception of medical terms and improve retention of medical terms and use of these terms in the real world. Review each illustration and photo carefully for easy and effective learning.

Charts and Tables

Charts and tables condense material in a visually appealing and organized fashion to ensure rapid learning. Some tables include terms organized by opposites or body systems to facilitate relating the information to various situations.

New Terms

New terms appear in bold type, followed by the pronunciation and definition.

Pronunciation System

The pronunciation system is an easy approach to learning the sounds of medical terms. This system is not laden with linguistic marks and variables, ensuring that students do not get bogged down in understanding the key. Once students become familiar with the key, it is very easy for them to progress in speaking the medical language.

Pronunciation Key

Pronunciation Guide

- Pronunciation guides for common words are omitted.
- Any vowel that has a dash above it represents the long sound, as in ā *hay*, ē *we*, ī *ice*, ō *toe*, and ū *unicorn*.
- Any vowel followed by an "h" represents the short sound, as in ah *apple*, eh *egg*, ih *igloo*, oh *pot*, and uh *cut*.
- Unique letter combinations are as follows: oo *boot*, ər *higher*, oy *boy*, aw *caught*, and ow *ouch*.

Other Pronunciation Guidelines

Word parts are represented in the text as prefixes, combining forms, and suffixes. The notation for a prefix is a word part followed by a hyphen. The notation for a combining form (word root and its vowel to ease pronunciation) is the root followed by a slash and its vowel, as in *nephr/o*. The notation for a suffix is a hyphen followed by the word part. The terms *prefix, combining form,* and *suffix* do not appear in the definitions.

Learning Objectives

The beginning of each chapter lists learning objectives so that students know what is expected of them as they read the text and complete the exercises.

Review Exercises

Exercises at the end of each chapter help you interact with and review the chapter's content. The exercises include several formats: multiple choice, matching, case studies, word building, diagram labeling, crossword puzzles, and critical thinking. The answers to the exercises are found in the Instructor's Manual.

NEW TO THIS EDITION

Features

- An overview has been added to the beginning of each chapter. This serves as a road map to the chapter content.
- Biology Bonus—a boxed feature that correlates the terminology to biological concepts. These will be found throughout the reading where appropriate.
- A section on relevant abbreviations has been added at the end of most chapters.
- Additional Review Questions have been added to chapters including True/False and Critical Thinking questions.
- Additional artwork has been added to most chapters to further engage students and enhance student comprehension.

Chapter 1

- Added a section on root words
- Added some additional suffixes (-ia, -ion, -ive, -us)
- Added examples related to spelling and pronunciation of terms
- Added box emphasizing the importance of proper communication both verbal and written to avoid medical errors

Chapter 2

- Added discussion of the suffix -logist

Chapter 3

- Added terms: fissure, densitometer, goniometer, epiphysitis, spur, crepitus, displaced fracture, pathologic fracture, stress fracture, external fixation, atrophy, leiomyoma, myoma, and rhabdomyoma
- New table distinguishing between bone depressions and extensions
- New section on chiropractic care and related terminology

Chapter 6

- New terms added: prandial, coprophagic therapy, reflux, inflammatory bowel disease, malnutrition, total parenteral nutrition, and transfaunation
- New word parts: cirrh/o and dips/o
- Expanded definition of colic as presented in the horse
- Expanded definition of dehydration and diarrhea

Chapter 7
- Added the following word parts: home/o and azot/o
- Added the following terms: nephroplasty, nephrostomy, nephrotomy, ureterotomy, ureteroplasty, urethrolithiasis urethrolith, and urethrotomy
- Expanded definition of urinalysis

Chapter 8
- Added terms: atrial septal defect, mitral stenosis, and thromboembolism
- Added section on abbreviations related to the cardiovascular system

Chapter 9
- Added terms: arterial blood gases, consolidation, and sputum culture
- Added the following word parts: -plegia

Chapter 10
- Added terms: root, caseous, excoriation, pododermatitis, bumblefoot, and sore hocks
- Added the word parts: -grade, alopec/o, carcin/o, fistul/o, papill/o, and papill/i

Chapter 11
- More descriptive definition of the secretions of the pancreas
- Expanded definition of hypoglycemia
- Added the word parts: mellit/o, ket/o, gynec/o, and mast/o

Chapter 12
- Added definition of artificial insemination
- Added the word parts: copulat/o, coit/o, abort/o, crypt/o, -tocia, terat/o, mut/a, phim/o, fet/o, and vas/o
- Added term: ovoviviparous

Chapter 13
- Added definition of the blood brain barrier
- Added these word parts: lept/o, home/o, -plegia, opisth/o, and bifid/o

Chapter 14
- Added the word parts: extrins/o
- Added the following terms: episclera, fundus, funduscopy, infectious keratoconjunctivitis, keratoconjunctivitis sicca, dry eye, nictitating gland prolapse, and cherry eye or third eyelid prolapse

Chapter 15
- Updates and clarifications of definitions throughout the chapter; includes differentiation between mammals and non-mammals
- Expanded discussion of immunity

- Includes a list of common terminology used to describe tumors
- Added discussion of various types of biopsies
- Added the terms: hematology, colloid, reticulocytosis, segmented neutrophil, heterophils, ultrasound, disseminated intravascular coagulation, modified transudate, herd immunity, innate immunity, inherent or genetic immunity, titer, apoptosis, and encapsulated, leiomyosarcoma, rhabdomyosarcoma, fulguration, protocol, rads, radiosensitive, radioresistant, and leukotrichia
- Added word parts: poikilo, mut/a, and blast/o

Chapter 16
- Expanded description of pH
- Expanded discussions of the various imaging techniques
- Added term: enzyme linked immunosorbent assay
- Added word parts: staphyl/o and strept/o

Chapter 17
- Expanded discussion of surgery
- Added the terms: pharmacogenomics, vaccine, antacid, preventative, manipulative, diagnostic, minimally invasive, reconstructive, cryogenic, and cauterizing
- Added word parts: thec/o

Chapter 18
- Expanded discussion of variations in personalities of cats and dogs as well as the roll of the American Kennel Club and the Cat Fancier's Association
- Added terms: brindle, *ad libitum* feeding, allogrooming, Association of American Feed Control Officials, bloom, body condition, cattery, champion, choke chain, collar, fancier, food hopper, guaranteed analysis, harness, head halter, brush, digest, kennel, kneading, leash, libido, limit feeding, litter pan (box), mask, natural breed, net quality statement, nutritional adequacy statement, obligate carnivore, pedigreed cat, pet quality, ruff, sex-linked trait, sexing, socialization, spraying, steatitis, stray, disposition, tie, topknot, undercoat, and wirecoat

Chapter 19
- Expanded introduction to the chapter
- Added terms: barrel racing, English, gait, gymkhana, racing, showing, bay, buckskin, overo, pinto, roan, stock type, tobiano, curb bit, snaffle bit, reins, covered by a stallion, groom, grooming equipment, in foal, saddle pad, and stock

Chapter 20
- Expanded introduction to cattle, sheep, goats, and camelids

- Added terms: auction, backgrounding, beef checkoff, breed character, bulk tank, California mastitis test, colostrometer, composite breed, condition, Confined Animal Feeding Operations, dairy character, downer, estrus synchronization, finish, frame, grade, gutter, gutter cleaner, alley scraper, herd health program, implant, market animal, muscling, nose tongs, preconditioning, range, registered, tilt table, milk grade, milk letdown, milking parlor, energy feed, energy efficiency, grass, handfed, milk replacer, total mixed ration, accelerated lambing, broken mouth, browse, early spring lambs, fall lambs, farm flock, fed lambs, feeder lambs, flocking instinct, lambing pen, late spring lamb, mixed grazing, open face, yearling, dust pile, humming, and kush

Chapter 21
- Expanded introduction to swine
- Added the following terms: dressing, free access gestation housing, and tail docking

Chapter 22
- Added the following terms: down feather, preening, average daily gain, battery cage, bleaching, breed, brooding, cannibalism, class, confinement, contract growers, coop, dressed bird, feed efficiency, force-feeding, free range, hatchability, hatchery, hen-day production, hierarchy, pecking order, incubation, litter, variety, aviary, aviculture, cage, and self-mutilation
- Discussion of feather patterns and comb types

Chapter 23
- Added the following terms: altricial, arboreal, autonomy, bolt hole, Bruce effect, Harderian gland, dwarf, ectotherm, endotherm, fixed formula, free choice, fuzzy, genome, homeotherm, hoppers, hutch, hygrometer, parthenogenesis, pinky, poikilotherm, precocial, whorl, scruff, urethral cone, urostyle, warren, Whitten effect, Coprophagic therapy, urolithiasis, polyphyodontic, hemipenes, metamorphosis, amplexus, spawn, and poikilotherms
- Expanded discussion of mice, rats, guinea pigs, hamsters, gerbils, chinchillas, ferrets, rabbits, and amphibians

Chapter 1
READY, SET, GO

Objectives

Upon completion of this chapter, the reader should be able to:

- Identify and recognize the types of word parts that make up a medical term
- Define commonly used prefixes, roots, combining forms, and suffixes presented in this chapter
- Analyze and understand basic medical terms
- Recognize the importance of spelling medical terms correctly
- Practice proper pronunciation of medical terms using the pronunciation guide
- Recognize the importance of and practice medical dictionary use
- Recognize, define, spell, and pronounce the medical terms in this chapter

INTRODUCTION TO MEDICAL TERMINOLOGY

Medical terms are used every day in medical offices, newspapers, television, and conversational settings. Most people are familiar with many medical terms; however, other medical terms seem complicated and foreign. Learning and understanding how medical terminology developed can help in mastering these terms.

Current medical vocabulary is based on terms of Greek and Latin origin, **eponyms** (words formed from a person's name), and modern language terms. The majority of medical terms are derived from word parts based on Greek and Latin words. Increasing familiarity with these Greek and Latin terms as well as the ability to identify word parts aids in learning common medical terms and recognizing unfamiliar medical terms by word analysis. Medical terminology may seem daunting at first because of the length of medical words and the seemingly complex spelling rules, but once the basic rules of breaking down a word into its constituents are mastered, the words become easier to read and understand.

ANATOMY OF A MEDICAL TERM

Many medical terms are composed of word part combinations. Recognizing these word parts and their meanings simplifies learning medical terminology. These word parts are as follows:

- **prefix:** word part found at the beginning of a word. Usually indicates number, location, time, or status.
- **root:** word part that gives the essential meaning of the word. A root cannot stand alone; a suffix must be added to complete the term.
- **combining vowel:** single vowel, usually an *o*, that is added to the end of a root to make the word easier to pronounce. A combining vowel is used when a suffix begins with a consonant or when two or more roots are joined.
- **combining form:** combination of the root and combining vowel.
- **suffix:** word part found at the end of a word. Usually indicates procedure, condition, disease, or disorder.

Understanding the meaning of the word parts allows the dissection of medical terms in a logical way. By breaking down unfamiliar terms into recognizable word parts, the veterinary professional can greatly increase his or her medical vocabulary.

PREFIXES

Prefixes are added to the beginning of a word or root to modify its meaning. For example, the term *operative* can be modified using various prefixes.

- The prefix **pre-** means before. **Preoperative** means before or preceding an operation.
- The prefix **peri-** (pehr-ē) means around. **Perioperative** means pertaining to the period around an operation or the period before, during, and after an operation.
- The prefix **post-** means after. **Postoperative** means after an operation.

Many prefixes have another prefix whose meaning is opposite of its own. Initially, when learning prefixes, it is helpful to learn them in these pairs or in similar groups (Table 1–1, Table 1–2, and Figure 1–1).

ROOTS

Roots are the foundation of most medical terms and provide the essential meaning of the word. Roots usually, but not always, describe the part of the body that is involved. Examples of roots are found distributed throughout this chapter and an extensive root list categorized by body system will be presented in Chapter 2 (Table 2–2).

COMBINING VOWELS

A combining vowel sometimes is used to make the medical term easier to pronounce. The combining vowel is used when the suffix begins with a consonant, as in the suffix **-scope**. An **arthroscope** is an instrument used to visually examine the joint. Because the suffix *-scope* begins with a consonant, the combining vowel *o* is used. *O* is the most commonly used combining vowel; however, *i* and *e* may be used as well. A combining vowel is not used when the suffix begins with a vowel (a, e, i, o, or u), as in the suffix **-itis**. **Gastritis** is inflammation of the stomach. Because the suffix *-itis* begins with a vowel, the combining vowel *o* is not used.

A combining vowel is always used when two or more root words are joined. For example, when **gastr/o** (stomach) is joined with **enter/o** (small intestine), the combining vowel is used with **gastr/o,** as in the term **gastroenteritis.** A combining vowel is not used between a prefix and the root word.

TABLE 1–1 Contrasting Prefixes

Without a prefix, the root **traumatic** means pertaining to injury.	**A-** (ah or ā) means without or no. **Atraumatic** means without injury.
Without a prefix, the root **uria** means urination.	**An-** (ahn) means without or no. **Anuria** means absence of urine.
Ab- (ahb) means away from. **Abduction** means to take away from the midline.	**Ad-** (ahd) means toward. **Adduction** means move toward the midline.
Without a prefix, the root **emetic** means pertaining to vomiting.	**Anti-** (ahn-tī or ahn-tih) means against. **Antiemetics** work against or prevent vomiting.
Dys- (dihs) means difficult, painful, or bad. **Dysphagia** means difficulty eating or swallowing.	**Eu-** (yoo) means good, easy, or normal. **Euthyroid** means having a normally functioning thyroid gland.
Endo- (ehn-dō) means within or inside. **Endocrine** means to secrete internally.	**Ex-** (ehcks) or **exo-** (ehcks-ō) means without, out of, outside, or away from. **Exocrine** means to secrete externally (via a duct).
Endo- means within or inside. **Endoparasite** is an organism that lives within the body of the host.	**Ecto-** (ehck-tō) means outside. **Ectoparasite** is an organism that lives on the outer surface of the host.
Hyper- (hī-pər) means elevated, higher, or more than normal. **Hyperglycemia** means elevated amounts of blood glucose.	**Hypo-** (hī-pō) means depressed, lower, or less than normal. **Hypoglycemia** means depressed amounts of blood glucose.
Inter- (ihn-tər) means between. **Intercostal** means between the ribs.	**Intra-** (ihn-trah) means within. **Intramuscular** means within the muscle.
Poly- (pohl-ē) means many or excessive. **Polyuria** means excessive amount or frequency of urination.	**Oligo-** (ohl-ih-gō) means scant or little. **Oliguria** means scant amount or frequency of urination.
Pre- (prē) means before. **Preanesthetic** means pertaining to before anesthesia.	**Post-** (pōst) means after. **Postanesthetic** means pertaining to after anesthesia.
Sub- (suhb) means below, under, or less. **Sublingual** means under the tongue.	**Super-** (soo-pər) and **supra-** (soo-prah) mean above, beyond, or excessive. **Supernumerary** means more than the regular number. **Suprascapular** means above the shoulder blade.

COMBINING FORMS

The combining form is a word root plus a combining vowel. Combining forms usually describe a part of the body. New words are created when combining forms are added to prefixes, other combining forms, and suffixes. For example, the term *panleukopenia* is composed of the following word parts:

- **pan-** (pahn), a prefix meaning all
- **leuk/o** (loo-kō), a combining form meaning white
- **-penia** (pē-nē-ah), a suffix meaning deficiency or reduction in number

Panleukopenia is a deficiency of all types of white blood cells.

SUFFIXES

Suffixes are attached to the end of a word part to modify its meaning. For example, the combining form *gastr/o* means stomach and can be modified using various suffixes.

- The suffix **-tomy** means cutting into or incision. **Gastrotomy** is incision into the stomach.

TABLE 1–2 Directional Prefixes and Their Meanings

Prefix	Pronunciation	Definition
epi-	(eh-pē)	upper
extra-	(ehcks-trah)	outside
hyper-	(hī-pər)	above, increased, or more than normal
hypo-	(hī-pō)	below, under, or decreased
infra-	(ihn-frah)	below or beneath
inter-	(ihn-tər)	between
intra-	(ihn-trah)	within
meta-	(meht-ah)	beyond
per-	(pər)	throughout
sub-	(suhb)	below, under, or decreased
super-	(soo-pər)	above, increased, or more than normal
supra-	(soo-prah)	above, increased, or more than normal
trans-	(trahnz)	across
ultra-	(uhl-trah)	above, increased, or more than normal

Copyright © 2015 Cengage Learning®.

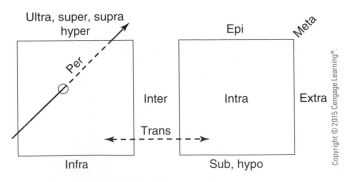

FIGURE 1–1 Directional prefixes.

- The suffix -**stomy** means a surgically created opening. **Gastrostomy** is a surgically created opening between the stomach and the body surface.
- The suffix -**ectomy** means surgical removal or excision. **Gastrectomy** is surgical removal of the stomach.

Many suffixes can be grouped together by meaning or by the category they modify. Initially, when learning suffixes, it is easiest if the learner groups them by meaning or category.

"Pertaining To" Suffixes

- -**ac** (ahck), as in **cardiac** (pertaining to the heart).
- -**al** (ahl), as in **renal** (pertaining to the kidney).
- -**an** (ahn), as in **ovarian** (pertaining to the ovary).
- -**ar** (ahr), as in **lumbar** (pertaining to the loin, lower back).
- -**ary** (ahr-ē), as in **alimentary** (pertaining to the gastrointestinal tract).
- -**eal** (ē-ahl), as in **laryngeal** (pertaining to the larynx).
- -**ic** (ihck), as in **enteric** (pertaining to the small intestine).
- -**ine** (ihn), as in **uterine** (pertaining to the uterus).
- -**ous** (uhs), as in **cutaneous** (pertaining to the skin).
- -**tic** (tihck), as in **nephrotic** (pertaining to the kidneys).

Surgical Suffixes

- -**ectomy** (ehck-tō-mē) = surgical removal, as in **mastectomy,** surgical removal of the breast or mammary glands.
- -**pexy** (pehck-sē) = suture to stabilize, as in **gastropexy,** surgically stabilizing the stomach to the abdominal wall.
- -**plasty** (plahs-tē) = surgical repair, as in **rhinoplasty,** surgical repair of the nose.
- -**stomy** (stō-mē) = surgically created opening, as in **colostomy,** a surgically created opening between the colon and body surface.
- -**tomy** (tō-mē) = cutting into, as in **laparotomy,** an incision into the abdomen.

Procedural Suffixes

- -**centesis** (sehn-tē-sihs) = surgical puncture to remove fluid or gas (for diagnosis or for treatment to remove excess fluid or gas), as in **cystocentesis,** a surgical puncture of the urinary bladder with a needle to remove fluid (urine).
- -**gram** (grahm) = record of, as in **electrocardiogram,** the electrocardiographic hard copy record.

- **-graph** (grahf) = instrument that records (or used as a record), as in **electrocardiograph,** the machine that records the electrical activity of the heart.
- **-graphy** (grahf-ē) = procedure that records, as in **electrocardiography,** the procedure used to record the electrical activity of the heart.
- **-lysis** (lī-sihs) = separation or breakdown, as in **urinalysis,** separation of the urine into its constituents.
- **-scope** (skōp) = instrument to visually examine, as in **endoscope,** an instrument used to visually examine inside the body.
- **-scopy** (skōp-ē) = procedure to visually examine, as in **endoscopy,** the procedure of visually examining inside the body.
- **-therapy** (thehr-ah-pē) = treatment, as in **chemotherapy,** treatment with chemical substances or drugs.

Double *R* Suffixes

- **-rrhagia** or **-rrhage** (rā-jē-ah or rihdj) = bursting forth, as in **hemorrhage,** bursting forth of blood from the vessels.
- **rrhaphy** (rahf-ē) = to suture, as in **enterorrhaphy,** suturing of the small intestine.
- **-rrhea** (rē-ah) = flow, discharge, as in **diarrhea,** complete discharge of the bowels.
- **-rrhexis** (rehck-sihs) = rupture, as in **myorrhexis,** rupture of the muscle.

What Is the Difference Between Human and Veterinary Medical Terminology?

Most times the medical terms used in human medical settings are identical to the ones used in veterinary medical settings. The greater number of species in veterinary medicine and the addition of terms used in animal production greatly expand the vocabulary of veterinary professionals. Species-specific anatomical differences also influence the terms used in a specific area. Do you know where the calf muscle is located on a person? Where is the calf muscle in a calf?

Conditional and Structural Suffixes

- **-algia** and **-dynia** (ahl-jē-ah and dihn-ē-ah) = pain, as in **arthralgia** and **arthrodynia,** joint pain.
- **-ia** (ē-ah) = state or condition, as in **hyperglycemia,** condition of elevated amounts of blood glucose.
- **-ion** (shuhn) = action, process, state, or condition, as in **infarction,** condition of blood flow obstruction.
- **-itis** (ī-tihs) = inflammation, as in **hepatitis,** inflammation of the liver.
- **-ive** (ihv) = performs or tends toward, as in **congestive,** tends toward accumulation of fluid.
- **-malacia** (mah-lā-shē-ah) = abnormal softening, as in **osteomalacia,** abnormal softening of bone.
- **-megaly** (mehg-ah-lē) = enlargement, as in **cardiomegaly,** enlargement of the heart.
- **-osis** (ō-sihs) = abnormal condition, as in **cardiosis,** an abnormal condition of the heart.
- **-pathy** (pahth-ē) = disease, as in **enteropathy,** a disease of the small intestine.
- **-sclerosis** (skleh-rō-sihs) = abnormal hardening, as in **arteriosclerosis,** abnormal hardening of the arteries.
- **-um** (uhm) = structure, as in **pericardium,** the structure surrounding the heart.
- **-us** (uhs) = thing, as in **tarsus,** the joint ("thing") between the tibia and fibula and metatarsal bones.

Suffixes may change a word's part of speech. Different suffixes may change the word from a noun (naming people, places, or things) to an adjective (descriptor) (Figure 1–2). Examples of this include the following terms:

- **Cyan*osis*** is a noun meaning condition of blue discoloration, whereas **cyano*tic*** is an adjective meaning pertaining to blue discoloration.
- **Anem*ia*** is a noun meaning a blood condition of deficient red blood cells and/or hemoglobin, whereas **anem*ic*** is an adjective meaning pertaining to a blood condition of deficient red blood cells and/or hemoglobin.
- **Muc*us*** is a noun meaning a slime-like substance that is composed of glandular secretion, salts, cells, and leukocytes (white blood cells), whereas **muc*ous*** is an adjective meaning pertaining to mucus.
- **Ili*um*** is a noun meaning a structure of the hip, whereas **ili*ac*** is an adjective meaning pertaining to a structure of the hip.
- **Condyl*e*** is a noun meaning a rounded projection on a bone, whereas **condyl*ar*** is an adjective meaning pertaining to a rounded projection on a bone.
- **Carp*us*** is a noun meaning the joint (thing) between the radius and ulna and metacarpal bones, whereas **carp*al*** is an adjective meaning pertaining to the joint between the radius and ulna and metacarpal bones.

Noun	Suffix	Adjective	Suffix
cyanosis	**-osis**	cyanotic	**-tic**
anemia	**-ia**	anemic	**-ic**
mucus	**-us**	mucous	**-ous**
ilium	**-um**	iliac	**-ac**
condyle	**-e**	condylar	**-ar**
carpus	**-us**	carpal	**-al**

FIGURE 1–2 Suffix variation depending upon usage.

ANALYZING MEDICAL TERMS

Medical terminology can be more easily understood when the following objectives are adhered to when a medical term is examined for the first time:

- *Dissect:* Analyze the word structurally by dividing it into its basic components.
- *Begin at the end:* After dividing the word into its basic parts, define the suffix first, the prefix second, and then the root. If there are two roots, divide each one and read them from left to right.
- *Anatomical order:* Where body systems are involved, the words usually are built in the order in which the organs occur in the body. For example, **gastroenteritis** is the proper term for inflammation of the stomach and small intestine. Because food passes from the stomach into the small intestine, the medical term for *stomach* appears before the medical term for *small intestine.* The order of word parts in a medical term may also represent the order of blood flow through organs. The exception to this involves some diagnostic procedures in which tools or substances are passed retrograde, or in the opposite direction of anatomical order. In these cases, the words are built in the order in which the equipment passes the body part.

Using these guidelines, analyze the term *ovariohysterectomy.* First, divide the term into its basic components: **ovari/o/hyster/ectomy.** Defining from back to front, the suffix -**ectomy** is surgical removal, one combining form **ovari/o** means ovary, and the other combining form **hysteri/o** means uterus. Together the term **ovariohysterectomy** means surgical removal of the ovaries and uterus. This term is based on the order in which the ovaries and uterus are found in the body.

WHAT DID YOU SAY?

Proper pronunciation of medical terms takes time and practice. Listening to how medical professionals pronounce words, using medical dictionaries and textbooks, and listening to prepared audio are the best ways to learn pronunciation. There are individual variations based on geographic location and personal preference. Medical dictionaries also vary in how they present pronunciation of medical terms. Some sources mark the syllable receiving the greatest emphasis with a primary accent (') and the syllable receiving the second most emphasis with a secondary accent ("). Other sources boldface and capitalize the syllable receiving the most emphasis, and other sources do not emphasize syllables. Consult with references before pronouncing a word. When starting to work with an unfamiliar dictionary (print or online version), spend a few minutes reviewing its user guide, table of contents, and appendices.

GENERAL PRONUNCIATION GUIDELINES

A medical term is easier to understand and remember when you learn to pronounce it properly. To help you master the pronunciation of new terms, a commonly accepted pronunciation of the word appears in parentheses next to the term. Vowels can be short or long (Table 1–3). Consonants are generally pronounced as in other English words.

DOES SPELLING COUNT?

Be aware of spelling when using medical terminology. Changing one or two letters can change the meaning of a word. **Hepatoma** is a liver mass, whereas **hematoma** is a mass or collection of blood (-**oma** is the suffix for mass). The **urethra** takes urine from the urinary bladder to the outside of the body, whereas **ureters** collect urine from the kidney and transport it to the urinary bladder.

Medical terms may be pronounced the same but have different meanings, so spelling is important. For example, ileum and ilium are pronounced the same. However, **ileum** is the distal part of the small intestine (*e* = enter/o or *e* = eating or small int*e*stin*e* has 2 e's in it), whereas **ilium** is part of the pelvic bone (pelv*i*c has an **i** in it). Some medical terms have the same spelling as terms used for other body parts. For example, the combining form **myel/o** represents the spinal cord and bone marrow. (It originates from the term meaning white substance.) Cervical means pertaining to the neck; however, it may pertain to the neck that joins the head to the body (cervical vertebrae) or the neck of an organ (the cervix of the uterus). Other terms have different spellings depending on how the term is used grammatically. For example, when used as a noun, *mucus* (the slimy stuff secreted from mucous membranes) is spelled differently than when it is used as an adjective (as in *mucous membrane*).

Some medical terms, if pronounced incorrectly, sound the same. For example, the term prostate (proh-stāt) is a gland that secretes a fluid to aide in sperm motility in males, whereas the term prostrate (proh-strāt) means exhausted or to lie flat on the ground. Train your ears to hear the differences in word pronunciation. Prostate is not prostrate; reflex (an automatic, involuntary response to change) is not reflux (backward flow).

Many medical terms form a verb, noun, plural, and adjective form. For example, you will need to know diagnose (verb), diagnosis (noun), diagnoses (plural), and diagnostic (adjective). Keep this in mind when you are learning new terms.

TABLE 1–3 Pronunciation Guide

Vowel	Sound	Example
"a" at the end of a word	ah	idea
"ae" followed by r or s	ah	aerobic
"i" at the end of a word	ī	bronchi
"oe"	eh	oestrogen (old English form)
"oi"	oy	sarcoid
"eu"	ū	euthanasia
"ei"	ī	Einstein
"ai"	ay	air
"au"	aw	auditory

Exceptions to Consonant Pronunciations

Consonant	Sound	Example
"c" before e, i, and y	s	cecum
"c" before a, o, and u	k	cancer
"g" before e, i, and y	j	genetic
"g" before a, o, and u	g	gall
"ps" at beginning of word	s	psychology
"pn" at beginning of word	n	pneumonia
"c" at end of word	k	anemic
"cc" followed by i or y	first c = k, second c = s	accident
"ch" at beginning of word	k	chemistry
"cn" in middle of word	both c (pronounce k) and n (pronounce ehn)	gastrocnemius
"mn" in middle of word	both m and n	amnesia
"pt" at beginning of word	t	pterodactyl
"pt" in middle of word	both p and t	optical
"rh"	r	rhinoceros
"x" at beginning of word	z	xylophone xenograph

Why Is Precision Needed in Veterinary Medical Communication?

Each year animals die because of drug reactions and medical errors. Many of these errors are due to inaccurate or imprecise written or verbal communication between members of the veterinary health care team. Many words, when written or pronounced, have a word part that if misspelled or mispronounced gives the intended word an entirely different meaning. A treatment based on the response to a medical term with a different meaning than the intended term could cause a medical error and perhaps the death of a patient. Precision in written and verbal communication is essential to prevent errors in patient care. Keep in mind that the medical record in which you document a patient's care and your actions is a legal document. It may be used in court as evidence in professional medical liability cases.

When looking up a medical term in the dictionary, spelling plays an important role. However, the term may not be spelled the way it sounds. The following guidelines can be used to find a word in the dictionary:

- If it sounds like *f*, it may begin with *f* or *ph*.
- If it sounds like *j*, it may begin with *g* or *j*.
- If it sounds like *k*, it may begin with *c*, *ch*, *k*, or *qu*.
- If it sounds like *s*, it may begin with *c*, *ps*, or *s*.
- If it sounds like *z*, it may begin with *x* or *z*.

REVIEW EXERCISES

Multiple Choice

Choose the correct answer.

1. The prefix _____ means away from.
   - a. ad-
   - b. ab-
   - c. ex-
   - d. endo-

2. The suffix _____ means an instrument to visually examine.
   - a. -ectomy
   - b. -scope
   - c. -scopy
   - d. -graphy

3. The prefix _____ means elevated, while the prefix _____ means depressed.
   - a. pre-, post-
   - b. endo-, exo-
   - c. hyper-, hypo-
   - d. inter-, intra-

4. The suffix _____ means pertaining to.
   - a. -al
   - b. -ary or -ar
   - c. -ic
   - d. all of the above

5. The suffix _____ means incision.
   - a. -ex
   - b. -tomy
   - c. -ectomy
   - d. -graphy

6. The suffix _____ means abnormal condition.
   - a. -osis
   - b. -rrhea
   - c. -rrhagia
   - d. -uria

7. The suffix _____ means separation or breaking into parts.
   - a. -gram
   - b. -pexy
   - c. -um
   - d. -lysis

8. The prefix _____ means below.
   - a. supra-
   - b. super-
   - c. inter-
   - d. sub-

9. The prefix(es) _____ mean(s) many.
   a. olig
   b. a-, an-
   c. poly-
   d. eu-

10. The prefix(es) _____ mean(s) without or no.
    a. a-, an-
    b. olig-
    c. dys-
    d. hyper-

11. The suffix -algia means
    a. pain
    b. excessive
    c. liver
    d. abnormal condition

12. The prefix pre- means
    a. after
    b. around
    c. before
    d. during

13. Which suffix may be part of the term meaning a procedure to visually examine?
    a. -lysis
    b. -scopy
    c. -rrhexis
    d. -scope

14. Which type of word part is always placed at the end of a term?
    a. combining form
    b. prefix
    c. suffix
    d. root

15. Which type of word part is always placed at the beginning of a term?
    a. combining form
    b. prefix
    c. suffix
    d. root

16. Which word part gives the essential meaning of a term?
    a. combining form
    b. prefix
    c. suffix
    d. root

17. Which word association is incorrect?
    a. inter- means between
    b. sub- means below, under, or less
    c. an- means without or no
    d. ad- means away from

18. Which suffix means to rupture?
    a. -rrhage
    b. -rrhaphy
    c. -rrhea
    d. -rrhexis

19. Which prefix means around?
    a. hyper-
    b. hypo-
    c. peri-
    d. supra-

20. In the term panleukopenia, the o between leuk and penia is called a
    a. combining form
    b. suffix
    c. combining vowel
    d. root

21. The term cardiomyopathy has a suffix meaning
    a. action
    b. condition
    c. disease
    d. pertaining to

22. Which term has the suffix meaning study of?
    a. myocardium
    b. cardiologist
    c. intravenous
    d. neurology

23. Which set contains prefixes that have opposite meaning?
    a. hyper- and hypo-
    b. micro- and macro-
    c. dys- and eu-
    d. all of the above

24. Which of the following terms is an adjective?
    a. mucus
    b. ilium
    c. anemic
    d. cyanosis

25. Which term does not contain a prefix?
    a. transfusion
    b. gastritis
    c. intercostal
    d. polyuria

Matching

Match the word parts in Column I with the definition in Column II.

Column I	Column II
1. _____ -itis	a. incision or cutting into
2. _____ -gram	b. before
3. _____ post-	c. surgical puncture to remove fluid or gas
4. _____ -tomy	d. difficult, painful, or bad
5. _____ pre-	e. enlargement
6. _____ -centesis	f. excision or surgical removal
7. _____ -therapy	g. liver
8. _____ dys-	h. kidney
9. _____ peri-	i. inflammation
10. _____ ren/o	j. record
11. _____ hepat/o	k. after
12. _____ -megaly	l. treatment
13. _____ -ectomy	m. around
14. _____ hyper-	n. pain
15. _____ -algia	o. excessive, increased

Match the definition in Column I to its word part in Column II.

Column I	Column II
1. _____ abnormal condition or disease	a. -lysis
2. _____ abnormal softening	b. -rrhea
3. _____ deficient, decreased	c. -plasty
4. _____ bursting forth	d. -rrhage
5. _____ creation of an artificial opening to the body surface	e. eu-
6. _____ surgical repair	f. -graphy
7. _____ all	g. epi-
8. _____ suture to stabilize	h. inter-
9. _____ procedure to visually examine	i. -scopy
10. _____ good, easy, or normal	j. -pexy
11. _____ upper	k. hypo-

12. _____ flow or discharge
13. _____ separation or breakdown
14. _____ procedure that records
15. _____ between

l. -malacia

m. -osis

n. -stomy

o. pan-

Fill in the Blanks

Write the medical terms that represent the following definitions.

1. Pertaining to the stomach _____

2. Inflammation of the liver _____

3. Abnormal softening of bone _____

4. Joint pain _____

5. Procedure to visually examine inside the body _____

6. Heart enlargement _____

7. Pertaining to the kidney _____

8. Bursting forth of blood from vessels _____

9. Suturing of stomach to body wall _____

10. Treatment with chemicals or drugs _____

True or False

If the statement is true, write T on the line. If the statement is false, write F on the line.

1. _____ An ectoparasite is an organism that lives within the body of the host.

2. _____ Osteomalacia is abnormal hardening of bone.

3. _____ A colostomy is the surgical creation of an opening between the colon and the body surface body surface.

4. _____ Mucus is a slimelike substance composed of glandular secretion, salts, cells, and leukocytes.

5. _____ An ovariohysterectomy is surgical removal of the ovaries and uterus.

Spelling

Cross out any misspelled words in the following sentences and replace them with the proper spelling.

1. Thick mucous was evident in the cat with upper respiratory disease. _____

2. Urine was collected via cistocentesis so that the urinanalysis could be performed to determine whether the dog had a urinary tract infection. _____

3. The horse's diarhea was caused by intestinal parasites. _____

4. The cutaneus lesion was not cancerous. _____

5. A local anestetic was used so that the surgery could be performed on the cow. _____

6. The medical term meaning disease of the small intestine is interopathy. _____

7. Hepititis is the medical term meaning inflammation of the liver. _____

8. A disease named from a person's name is an enaponym. _____

9. A hepatoma is a mass or collection of blood. _____

10. The ileum is a bone in the pelvis._____

Word Part Identification

Underline the word root(s) in the following terms.

1. hepat/it is
2. gastr/o/intestin/al
3. cardi/o/logy
4. intra/ven/ous
5. nephr/osis

Underline the suffix in the following terms.

6. hepat/itis
7. gastr/o/intestin/al
8. cardi/o/logy
9. intra/ven/ous
10. nephr/osis

Underline the prefix in the following terms.

11. hyper/secretion
12. peri/card/itis
13. endo/cardi/um
14. poly/uria
15. ur/o/lith

Crossword Puzzles

Prefixes

Supply the correct prefix in the appropriate space for the definition listed.

Across

1 beyond
3 across
6 between
8 throughout
9 above, beyond, excessive
13 inside
14 less than normal

Down

2 against
4 without
5 excessive
6 within
7 outside
8 after
10 before
11 many
12 outside

Suffixes

Supply the correct suffix in the appropriate space for the definition listed.

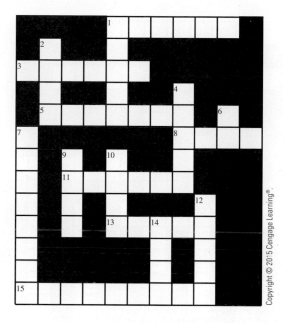

Across

1 bursting forth
3 procedure that records
5 abnormal softening
8 cutting into
11 surgical removal
13 surgically create new opening
15 abnormal hardening

Down

1 discharge
2 record of
4 disease
6 structure
7 surgical puncture to remove fluid
9 suture to stabilize
10 inflammation
12 breakdown
14 abnormal condition

Medical Terms

Supply the correct medical term in the appropriate space for the definition listed.

Across

1 enlargement of the heart

3 before an operation

8 pertaining to the skin

11 separation of urine into its components

13 inflammation of the liver

14 increased amount of blood glucose

Down

2 without injury

4 away from midline

5 infrequent urination

6 toward midline

7 frequent urination

9 organism that lives on the outer surface of the host

10 pertaining to the heart

12 surgical removal of the mammary glands

Medical Terms

Supply the correct medical term in the appropriate space for the definition listed.

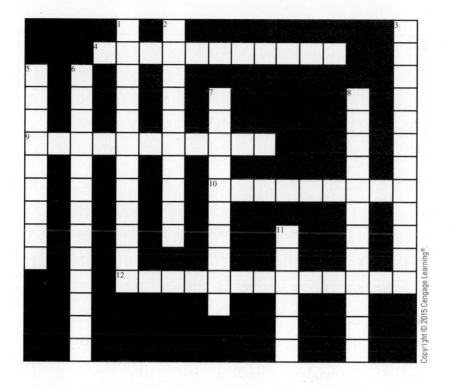

Across

4 disease of small intestine

9 structure surrounding the heart

10 instrument to visually examine inside the body

12 excess of regular number

Down

1 between the ribs (plural form)

2 bursting forth of blood from vessels

3 treatment with drugs

5 difficulty eating or swallowing

6 within muscle

7 pertaining to the gastrointestinal tract

8 abnormal softening of bone

11 pertaining to the loin

Word Search

Find the following medical terms or word parts in the puzzle. (Make sure you understand what the terms mean as you find them.)

```
E T I S A R A P O D N E A E S
R E N A L E S O C U L G Y Y G
E N T E R I C P E Y N C S G A
Y M O T C E O R N S O I Y O I
O P Y H Y A E O I P I T X L M
A T C L I P C N R E T A E O E
I E Y Y Y G O U E N A M P N C
Y O U H P A L N T I N U O I Y
H Y O U R S O C U A I A R M L
Y M O T E T S I X E R R T R G
Y L E E F R T A I I U T S E R
M E T U I I O T N K F G A T E
O U T G X T M I M H K F G F P
T K U U C I Y O O I T I U I Y
S O E O E S T N A I H P C S H
```

colostomy	penia
ectomy	prefix
endoparasite	pronunciation
enteric	renal
gastritis	stomy
gastropexy	suffix
glucose	terminology
hyper	tomy
hyperglycemia	traumatic
hypo	urination
leuko	uterine

Case Studies

Fill in the blanks to complete the case history.

A 5-year-old male neutered cat is presented to a veterinary clinic with _____ (painful urination) and _____ (scant urine production). Upon examination the abdomen is palpated and _____ (enlarged urinary bladder) is noted. After completing the examination, the veterinarian suspects an obstruction of the _____ (tube that carries urine from the urinary bladder to outside the body). Blood is taken for analysis, and the cat is admitted to the clinic. The cat is anesthetized, and a urinary catheter is passed. Urine is collected for _____ (breakdown of urine into its components). In addition to the obstruction, the cat is treated for _____ (inflammation of the urinary bladder).

In this case study, the meanings of some unfamiliar medical terms (underlined) cannot be understood by breaking up the term into its basic components. Using a print or online dictionary, define the following medical terms.

1. palpated _____

2. obstruction _____

3. catheter _____

Define the medical terms in the following case study. A space is provided after the terms that need a definition.

A 6-month-old female Golden Retriever presented to the clinic for a preoperative _____ examination prior to her scheduled ovariohysterectomy _____. On physical exam, her vital signs were normal. Blood was collected for a preanesthetic _____ analysis to check if the dog is anemic _____ or has metabolic disturbances such as hyperglycemia _____; a fecal sample was collected and set up to check for endoparasites _____; and urine was collected via cystocentesis _____ to assess whether or not she had a urinary tract infection. After her laboratory samples were analyzed, it was determined that she was healthy and able to have surgery.

In this case study, the meanings of some unfamiliar medical terms (underlined) cannot be understood by breaking up the term into its basic components. Using a print or online dictionary, define the following medical terms.

1. examination _____

2. vital _____

3. metabolic _____

Critical Thinking Exercise

The following story and questions are designed to stimulate critical thinking through class discussion or brief writing responses. There are many correct answers to these questions.

Stephanie Adams, a first-year veterinary technician student, sits on the couch with her veterinary medical terminology book open to Chapter 1. Her gray cat, Sooty, sits rolled up beside her. Stephanie is frustrated because she feels the words look too much alike yet have totally different meanings. "Why did I think I would be able to learn all of these words? There are just too many to memorize."

Her phone rings and it is her mother calling to see how she is doing. Stephanie tells her she has a lot of homework and does not know if she will be able to learn all of the new words for her medical terminology class. Her mom reminds her that she already knows plenty of medical terms. "Remember when I had my physical exam last year and I had to schedule an appointment with the cardiologist? You were the one who told me I would be seeing a heart specialist." Her mom made her realize she may be familiar with some of the words in Chapter 1.

Stephanie remembered that she had taken her pets to the veterinarian and understood many of the terms used by the veterinary staff. She also goes to the doctor for her annual physical exam and has had laboratory tests run such as a urinalysis and bloodwork, has had the Hepatitis B vaccine for a former job, and has arthritis in her ankle from a sports injury. She knows what these medical terms mean and she soon realizes she knows more than she gave herself credit.

She then thinks back to her veterinary medical terminology class and remembers she must complete her assignment before tomorrow's class. "Mucous and mucus; both sound the same. I know one is an adjective and one is a noun, but which is which?" she wonders aloud. Then she remembers what her teacher said in class: "The longer word is a type of membrane (mucous) and is an adjective because the terms *adjective* and *mucous* have more letters in them. The shorter word is the secretion (mucus) and is a noun because the terms *noun* and *mucus* have fewer letters in them." Stephanie is beginning to believe in her ability to learn medical terminology.

Suggested Discussion Topics

1. Stephanie needs to learn veterinary medical terminology to become a veterinary technician. What study habits would help her achieve her goal?

2. How can this textbook and other resource materials help her, and you, learn veterinary medical terminology?

3. What veterinary medical terms are you already familiar with?

Chapter 2
WHERE, WHY, AND WHAT?

Objectives

Upon completion of this chapter, the reader should be able to:

- Identify and recognize body planes, positional terms, directional terms, and body surface terms
- Define terms used to describe specific branches of study
- Identify terms used to describe the oral cavity and tooth surfaces
- Identify terms used to describe the structure of cells, tissues, and glands
- Define terms related to body cavities and structure
- Describe how an animal is positioned while lying down
- Recognize the parts of the cell and describe their functions
- Define terms related to genes or heredity
- List the four tissue types and describe terms associated with cell and tissue structure
- Recognize, define, spell, and pronounce medical terms related to pathology and procedures
- Identify body systems by their components, functions, and combining forms
- Identify prefixes that assign numeric value

IN POSITION

Positional terms are important for accurately and concisely describing body locations and relationships of one body structure to another. The terms *forward* and *backward*, *up* and *down*, *in* and *out*, and *side to side* are not clear enough descriptions by themselves to have universal understanding in the medical community. Therefore, very specific terms were developed so that there would be no confusion as to the meaning being conveyed. Listed in Table 2–1 and illustrated in Figures 2–1, 2–2, 2–3, and 2–4 are directional terms used in veterinary settings.

THE PLANE TRUTH

Planes are imaginary lines that are used descriptively to divide the body into sections.

- **Midsagittal** (mihd-sahdj-ih-tahl) **plane** is the plane that divides the body into *equal* right and

TABLE 2–1 Terms Used to Describe Direction and Surface

Ventral (vehn-trahl) refers to the belly or underside of a body or body part. (*Ventr/o* in Latin means belly.) (*Venture* means to undertake.) (A ventral fin is on the belly.)	**Dorsal** (dōr-sahl) refers to the back. (*Dors/o* in Latin means back.) (*Endorse* means sign on the back.) (A dorsal fin is on the back.) Also refers to the cranial surface of the manus (front paw) and pes (rear paw).
Cranial (krā-nē-ahl) means toward the head. (*Crani/o* in Latin means skull.)	**Caudal** (kaw-dahl) means toward the tail. (*Cauda* in Latin means tail.)
Anterior (ahn-tēr-ē-ər) means front of the body. (*Anteri/o* in Latin means before.) Used more in description of organs or body parts because front and rear are confusing terms in quadrupeds. A quadruped's belly is oriented downward, not forward as in humans.	**Posterior** (pohs-tēr-ē-ər) means rear of the body. (*Posteri/o* in Latin means behind.)
Rostral (rohs-trahl) means nose end of the head. (*Rostrum* in Latin means beak.) **Cephalic** (seh-fahl-ihck) means pertaining to the head. (*Kephale* in Greek means head.)	**Caudal** (kaw-dahl) means toward the tail. (*Cauda* in Latin means tail.)
Medial (mē-dē-ahl) means toward the midline. (*Medi/o* in Latin means middle.)	**Lateral** (laht-ər-ahl) means away from the midline. (*Later/o* in Latin means side.)
Superior (soo-pēr-ē-ər) means uppermost, above, or toward the head. Used more commonly in bipeds. (*Super* in Latin means above.)	**Inferior** (ihn-fēr-ē-ər) means lowermost, below, or toward the tail. Used more commonly in bipeds. (*Inferi* in Latin means lower.)
Proximal (prohck-sih-mahl) means nearest the midline or nearest to the beginning of a structure. (*Proxim/o* in Latin means next.)	**Distal** (dihs-tahl) means farthest from the midline or farthest from the beginning of a structure. (*Dist/o* in Latin means distant.)
Superficial (soop-ər-fihsh-ahl) means near the surface; also called external. (*Super* in Latin means above.)	**Deep** (dēp) means away from the surface; also called internal. (*Deep* means beneath the surface.)
Palmar (pahl-mahr) means the caudal surface of the manus (front paw) including the carpus (from the antebrachial joint distally). (*Palmar* in Latin means hollow of the hand.)	**Plantar** (plahn-tahr) means the caudal surface of the pes (rear paw) including the tarsus (from the tibiotarsal joint distally). (*Plantar* in Latin means sole of the foot.)

left halves. It also is called the **median** (mē-dē-ahn) plane and the midline (Figure 2–3).

- **Sagittal** (sahdj-ih-tahl) **plane** is the plane that divides the body into unequal right and left parts (Figure 2–4).
- **Dorsal** (dōr-sahl) **plane** is the plane that divides the body into dorsal (back) and ventral (belly) parts (Figure 2–4). It also is called the **frontal** (frohn-tahl) plane or coronal (kō-roh-nahl) plane. In humans, the frontal plane is a vertical plane because people stand erect.

- **Transverse** (trahnz-vərs) **plane** is the plane that divides the body into cranial and caudal parts (Figure 2–4). It also is called the **horizontal plane** or **cross-sectional plane.** The transverse plane also may be used to describe a perpendicular transection to the long axis of an appendage.

FIGURE 2–3 Planes of the body. The midsagittal, or median, plane divides the body into *equal* left and right portions

FIGURE 2–1 Directional and surface terms. The arrows on this Boston terrier represent the following directional terms: A = cranial, B = caudal, C = ventral, D = dorsal, E = rostral, F = proximal, G = distal, H = palmer, and I = plantar.

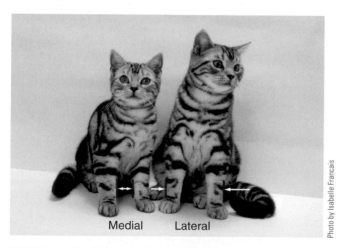

FIGURE 2–2 Medial versus lateral. The lines on these cats represent the directional terms *medial* and *lateral*.

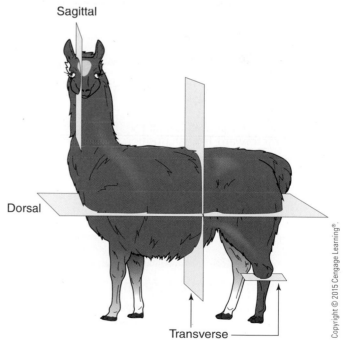

FIGURE 2–4 Planes of the body. The sagittal plane divides the body into *unequal* right and left parts, the dorsal plane divides the body into back and belly parts, and the transverse plane divides the body into cranial and caudal parts. The transverse plane also describes a perpendicular transection to the long axis of an appendage.

STUDYING

The suffix **-logy** means the study of. Specific terms are used to describe specific branches of study. The study of body structure is called **anatomy** (ah-naht-ō-mē). **Physiology** (fihz-ē-ohl-ō-jē) is the study of body function(s); **physi/o** means nature. **Pathology** (pahth-ohl-ō-jē) is the study of the nature, causes, and development of abnormal conditions; **path/o** means disease. Combining physiology and pathology results in the term **pathophysiology** (pahth-ō-fihz-ē-ohl-ō-jē), which is the study of changes in function caused by disease. The study of disease causes is **etiology** (ē-tē-ohl-ō-jē); **eti/o** means to cause.

The suffix **-logist** means specialist. A cardiologist is a veterinarian who specializes in diagnosing and treating disorders of the heart (**cardi/o** is the combining form for heart). A dermatologist is a veterinarian who specializes in diagnosing and treating disorders of the skin (**dermat/o** is the combining form for skin).

YOU HAVE SAID A MOUTHFUL

Describing positions in the mouth has become increasingly important with the rise of veterinary dentistry. The dental **arcade** (ahr-kād) is the term used to describe how teeth are arranged in the mouth. *Arcade*

Are Directional Terms the Same in Humans and Animals?

Original human anatomy drawings posed a man with his palms forward. Human positional terminology is still based on that pose. If this original drawing had had the palms turned away, human and veterinary anatomical terminology would have been identical.

Directional Term Confusion

The terms *anterior*, *posterior*, *superior*, and *inferior* can be confusing when used with quadrupeds. In quadrupeds, *ventral* is a better term for *anterior* and *dorsal* is a better term than *posterior*.

What does *anterior* mean in a quadruped (cranial or dorsal)?

What does *superior* mean in a quadruped (cranial, dorsal, superficial, or proximal)?

means a series of arches, which is how the teeth are arranged in the oral cavity. Surfaces of the teeth are named for the area in which they contact (Figure 2–5). The **lingual** (lihng-gwahl) **surface** is the aspect of the tooth that faces the tongue. Remember that linguistics is the study of language, and the tongue is used to make sounds. Some people use lingual surface to describe the tooth surface that faces the tongue on both the maxilla (upper jaw) and mandible (lower jaw). More correctly, the **palatal** (pahl-ah-tahl) **surface** is the tooth surface of the maxilla that faces the tongue, and the lingual surface is the tooth surface of the mandible that faces the tongue. The **buccal** (buhk-ahl or būk-ahl) **surface** is the aspect of the tooth that faces the cheek. *Bucca* is Latin for cheek. The buccal surface is sometimes called the **vestibular** (vehs-tih-buh-lahr) **surface.** *Vestibule* in Latin means space or cavity at an entrance. The **occlusal** (ō-klū-zahl) **surfaces** are the aspects of the teeth that meet when the animal chews. Think of the teeth occluding, or stopping, things from passing between them when the teeth are clenched. The **labial** (lā-bē-ahl) **surface** is the tooth surface facing the lips. *Labia* is the medical term for lips. **Contact** (kohn-tahckt) **surfaces** are the aspects of the tooth that touch other teeth. Contact surfaces are divided into **mesial** (mē-zē-ahl) and **distal** (dihs-tahl). The mesial contact surface is the one closest to the midline of the dental arcade or arch. The distal contact surface is the one furthest from the midline of the dental arcade (think distance). Each tooth has both contact surfaces, even the last molar, which touches only one tooth surface.

THE HOLE TRUTH

A body **cavity** (kahv-ih-tē) is a hole or hollow space in the body that contains and protects internal organs. The **cranial** (krā-nē-ahl) cavity is the hollow space that contains the brain in the skull. The **spinal** (spī-nahl) **cavity** is the hollow space that contains the spinal cord within the spinal column. The **thoracic** (thō-rahs-ihck) **cavity**, or **chest cavity**, is the hollow space that contains the heart and lungs within the ribs between the neck and diaphragm. The **abdominal** (ahb-dohm-ih-nahl) **cavity** is the hollow space that contains the major organs of digestion located between the diaphragm and pelvic cavity. The abdominal cavity is commonly called the **peritoneal** (pehr-ih-tohn-ē-ahl) **cavity**, but that is not quite accurate. The peritoneal cavity is the hollow space within the abdominal cavity between the parietal peritoneum and the visceral peritoneum. The **pelvic** (pehl-vihck) **cavity** is the hollow space that contains the reproductive and some

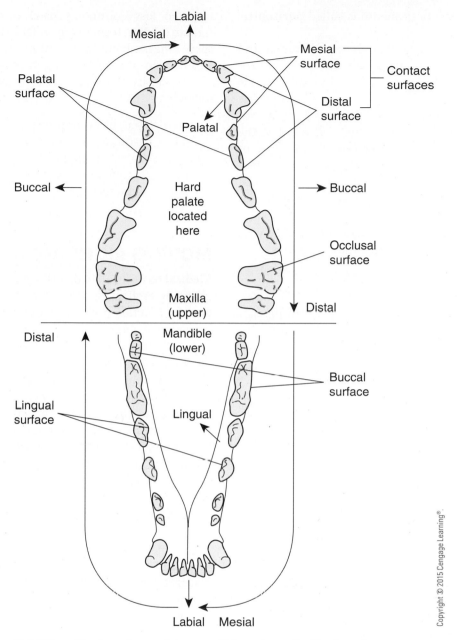

FIGURE 2–5 Teeth surfaces of the mandible and maxilla. Teeth surfaces are identified by the area they are near.

excretory systems' (urinary bladder and rectum) organs bounded by the pelvic bones.

Cavities are just one way to segregate the body. Regional terms are also used to describe areas of the body. The **abdomen** (ahb-dō-mehn) is the portion of the body between the thorax and the pelvis containing the abdominal cavity. The **thorax** (thaw-rahcks) is the chest region located between the neck and the diaphragm. The **groin** (groyn) is the caudal region of the abdomen adjacent to the thigh; it also is known as the **inguinal** (ihng-gwih-nahl) **area**.

Membranes (mehm-brānz) are thin layers of tissue that cover a surface, line a cavity, or divide a space or an organ. The **peritoneum** (pehr-ih-tō-nē-uhm) is the membrane lining the walls of the abdominal and pelvic cavities and it covers some organs in this area. The peritoneum may be further divided in reference to its location. The **parietal** (pah-rī-eh-tahl) **peritoneum** is the outer layer of the peritoneum that lines the abdominal and pelvic cavities, and the **visceral** (vihs-ər-ahl) **peritoneum** is the inner layer of the peritoneum that surrounds the abdominal organs.

Inflammation of the peritoneum is called **peritonitis** (pehr-ih-tō-nī-tihs).

Other terms associated with the abdomen and peritoneum include umbilicus, mesentery, and retroperitoneal. The **umbilicus** (uhm-bihl-ih-kuhs) is the pit in the abdominal wall marking the point where the umbilical cord entered the fetus (Figure 2–6). In veterinary terminology, the umbilicus is also called the **navel** (nā-vuhl). The **mesentery** (mehs-ehn-tehr-ē or mehz-ehn-tehr-ē) is the layer of the peritoneum that suspends parts of the intestine in the abdominal cavity. **Retroperitoneal** (reh-trō-pehr-ih-tō-nē-ahl) means superficial to the peritoneum.

Other membranes of the body are described with the specific body region in which they are found.

LYING AROUND

Lay, *lie*, *laid*, and *lying* are confusing words in English. However, the only medical term for lying down is **recumbent** (rē-kuhm-behnt). Recumbent is then modified depending on which side is facing down (Figure 2–7).

- **Dorsal recumbency** (dōr-sahl rē-kuhm-behn-sē) is lying on the back.
- **Ventral recumbency** (vehn-trahl rē-kuhm-behn-sē) is lying on the belly = **sternal** (stər-nahl) **recumbency.**
- **Left lateral recumbency** (laht-ər-ahl rē-kuhm-behn-sē) is lying on the left side.
- **Right lateral recumbency** is lying on the right side.

FIGURE 2–6 The umbilicus marks the point where the umbilical cord entered the fetus.

Umbilicus

© iStockphoto/Eric Isselée

Two less commonly used terms derived from human medical terminology refer to lying down. **Prone** (prōn) means lying in ventral or sternal recumbency; **supine** (soo-pīn) means lying in dorsal recumbency.

To clarify the recumbency terms, remember the following:

- lay = to put, place, or prepare
- laid = past tense of *lay*
- laying = present tense of *lay*
- lie = to recline or be situated
- lain = past tense of *lie*
- lying = present tense of *lie*

MOVING RIGHT ALONG

Medical terms used to describe movement may involve changing prefixes or suffixes to change direction. The terms *adduction* and *abduction* look very similar yet have opposite meanings (Figure 2–8).

Adduction (ahd-duhck-shuhn) means movement toward the midline (think addition to something), and **abduction** (ahb-duhck-shuhn) means movement away from the midline (think child abduction).

Flexion (flehck-shuhn) means closure of a joint angle, or reduction of the angle between two bones. Contracting the biceps involves flexing the elbow. **Extension** (ehcks-tehn-shuhn) means straightening of a joint angle or an increase in the angle between two bones (Figure 2–9). You extend your hand for a handshake. **Hyperflexion** (hī-pər-flehcks-shuhn) and **hyperextension** (hī-pər-ehcks-tehn-shuhn) occur when a joint is flexed or extended too far. Hyperflexion is the palmar or plantar movement of the joint angles. Hyperextension is the dorsal movement of the joints beyond the reference angle.

Supination and *pronation* are two less commonly used terms in veterinary settings. **Supination** (soo-pih-nā-shuhn) is the act of rotating the limb or body part so that the palmar surface is turned upward, and **pronation** (prō-nā-shuhn) is the act of rotating the limb or body part so that the palmar surface is turned downward. Think of supination as the movement involved with eating soup while cupping the hand. **Rotation** (rō-tā-shuhn) is another term of movement that means circular movement around an axis.

SETTING OUR CYTES AHEAD

Cells are the structural units of the body (Figure 2–10). The combining form for cell is **cyt/o** (sī-tō). Cells are specialized and grouped together to form tissues and organs. **Cytology** (sī-tohl-ō-jē) is the study

(a) Dorsal recumbency

X-ray cassette

(b) Ventral recumbency/sternal recumbency

(c) Lateral recumbency

FIGURE 2–7 Recumbency positions. The position in which an animal lies is important in veterinary medicine, especially in radiographing an animal. (a) This dog is in *dorsal recumbency*. (b) This dog is in *ventral*, or *sternal*, *recumbency*. (c) This dog is in *right lateral recumbency*.

Midline

Midline

Abduction

Adduction

Carpus (knee)

Flexion

Extension

FIGURE 2–8 *Adduction* versus *abduction*.

FIGURE 2–9 *Flexion* and *extension* of the carpus (knee) of a horse.

FIGURE 2–10 Parts of an animal cell.

of cells. The suffix *-logy* means the study of. Cytology involves studying cell origin, structure, function, and pathology.

The cell membrane, cytoplasm, and nucleus are collectively called the **protoplasm** (prō-tō-plahzm). The suffix *-plasm* (plahzm) means formative material of cells, and the combining form *prot/o* means first. The **cell membrane** (also called the plasma membrane) is the structure lining the cell that protects the cell's contents and regulates what goes in and out of the cell. **Cytoplasm** (sī-tō-plahzm) is the gelatinous

material located in the cell membrane that is not part of the nucleus. The cytoplasm is highly organized and contains organelles suspended in it that are the functional machinery of the cell. Each type of organelle has a specific role in the metabolic reactions that take place in the cytoplasm (Figure 2–10):

- nucleolus (produces RNA that forms ribosomes)
- centrioles (rod-shaped organelles that maintain cell shape and move chromosomes during cell replication)

- mitochondria (energy producers of the cell)
- Golgi apparatus (chemical processor of the cell)
- endoplasmic reticulum (collection of folded membranes that may contain ribosomes, known as rough endoplasmic reticulum, which synthesize protein, or may be void of ribosomes, known as smooth endoplasmic reticulum, which synthesizes lipids, and some carbohydrates)
- ribosomes (site of protein synthesis)
- vacuoles (small membrane-bound organelles containing water, food, or metabolic waste)
- lysosomes (digestive system of the cell)

The **nucleus** (nū-klē-uhs) is the structure in a cell that contains nucleoplasm, chromosomes, and the surrounding membrane. **Nucleoplasm** (nū-klē-ōplahzm) is the material in the nucleus, and **chromosomes** (krō-mō-sōmz) are the structures in the nucleus composed of DNA that transmits genetic information.

IT'S IN THE GENES

Genetic is a term used to denote something that pertains to genes or heredity. A **genetic** (jehn-eh-tihck) **disorder** is any inherited disease or condition caused by defective genes. This term is different from **congenital** (kohn-jehn-ih-tahl), which denotes something that is present at birth. A genetic defect may be congenital, but a congenital defect implies only that something faulty is present at birth. An **anomaly** (ah-nohm-ah-lē) is a deviation from what is regarded as normal. *Anomaly* may be used instead of *defect*.

GROUPING THINGS TOGETHER

A group of specialized cells that is similar in structure and function is a **tissue** (tihsh-yoo). The study of the structure, composition, and function of tissue is **histology** (hihs-tohl-ō-jē). **Hist/o** is the combining form for tissue.

There are four types of tissue: epithelial, connective, muscle, and nervous. **Epithelial tissue** (ehp-ih-thē-lē-ahl tihsh-yoo) or **epithelium** (ehp-ih-thē-lē-uhm) covers internal and external body surfaces and is made up of tightly packed cells in a variety of arrangements (Figure 2–11). *Epi-* is a prefix that means above, *thel/o* is a combining form that means nipple but is now used to denote any thin membrane, and *-um* is a suffix that means structure. Epithelial tissue is further divided into

FIGURE 2–11 Epithelial tissue. Some examples of epithelial tissue include (a) simple (single layer) squamous (cells are flattened) epithelial tissue, (b) simple cuboidal (cells are cube shaped) epithelial tissue, (c) simple columnar (cells are column shaped) epithelial tissue, and (d) stratified (multilayered) squamous epithelial tissue.

mesothelium and endothelium. **Endothelium** (ehn-dō-thē-lē-uhm) is the cellular covering that forms the lining of the internal organs, including the blood vessels. *Endo-* is a prefix meaning within. **Mesothelium** (mēs-ō-thē-lē-uhm) is the cellular covering that forms the lining of serous membranes such as the peritoneum. The prefix *meso-* means middle.

Connective tissue is another tissue type. Connective tissue adds support and structure to the body by holding the organs in place and binding body parts together (Figure 2–12). Bone, cartilage, dense connective tissue (found in tendons and ligaments), loose connective tissue, and blood are all types of connective tissue. **Adipose** (ahd-ih-pohs) tissue, another form of connective tissue, is also known as fat. **Adip/o** is the combining form for fat.

Muscle tissue is another tissue type that contains cell material with the specialized ability to contract and relax. Three muscle types exist in animals: skeletal, smooth, and cardiac (Figure 2–13). These muscle types are covered in Chapter 3.

Nervous tissue is the last tissue type (Figure 2–14). Nervous tissue contains cells with the specialized ability to react to stimuli and conduct electrical impulses. The nervous system is covered in greater depth in Chapter 13.

Tissue can form normally or abnormally. The suffix **-plasia** (plā-zē-ah) is used to describe formation, development, and growth in the *number* of cells in an organ or tissue. The suffix **-trophy** (trō-fē) means formation, development, and growth in the *size* of an organ or a tissue or individual cells. The use of different prefixes describes problems with tissue formation.

- **Anaplasia** (ahn-ah-plā-zē-ah) is a change in the structure of cells and their orientation to each other.
- **Aplasia** (ā-plā-zē-ah) is lack of development of an organ or a tissue or a cell.

(a)

(b)

(c)

(d)

Photomicrographs courtesy of William J. Bacha, PhD, and Linda M. Bacha, MS, VMD.

FIGURE 2–12 Connective tissue. Some examples of connective tissue include (a) loose connective tissue (typically found attached to abdominal organs), (b) dense connective tissue (found in tissues such as ligaments), (c) adipose tissue (lipid or fat tissue), and (d) cartilage (articular cartilage is found on the ends of bones).

FIGURE 2–13 Muscle tissue. The three types of muscle include (a) skeletal muscle, (b) cardiac muscle, and (c) smooth muscle.

FIGURE 2–14 Nervous tissue. Photomicrograph of a neuron from the spinal cord of a bovine.

- **Dysplasia** (dihs-plā-zē-ah) is abnormal growth or development of an organ or a tissue or a cell.
- **Hyperplasia** (hī-pər-plā-zē-ah) is an abnormal increase in the number of normal cells in normal arrangement in an organ or a tissue.

- **Hypoplasia** (hī-pō-plā-zē-ah) is incomplete or less than normal development of an organ or a tissue or a cell.
- **Neoplasia** (nē-ō-plā-zē-ah) is any abnormal new growth of tissue in which multiplication of cells is uncontrolled, more rapid than normal, and progressive. Neoplasms usually form a distinct mass of tissue called a **tumor** (too-mər). Tumors may be **benign** (beh-nīn), meaning not recurring, or **malignant** (mah-lihg-nahnt), meaning tending to spread and be life threatening. The suffix **-oma** (ō-mah) means tumor or neoplasm.
- **Atrophy** (ah-trō-fē) is decrease in size or complete wasting of an organ or tissue or cell.
- **Dystrophy** (dihs-trō-fē) is defective growth in the size of an organ or tissue or cell.
- **Hypertrophy** (hī-pər-tō-fē) is increase in the size of an organ or a tissue or cell.

The prefix *a-* means without, *hypo-* means less than normal, *hyper-* means more than normal, *dys-* means bad, *ana-* means without, and *neo-* means new.

BIOLOGY BONUS

As you learn about cells, remember different cells have different functions. Their shape influences their function. Epithelial cells that line the oral cavity are flat and tightly packed so they form a protective layer over underlying cells. Nerve cells typically have long, thin extensions that transmit nerve impulses over a distance. Muscle cells are slender rods that attach to the ends of structures they move. As cell types specialize further, their shape and function change to fit a specific need.

Glands (glahndz) are groups of specialized cells that secrete material used elsewhere in the body. *Aden/o* is the combining form for gland. Glands are divided into two categories: exocrine and endocrine (Figure 2–15). **Exocrine** (ehck-sō-krihn) **glands** are groups of cells that secrete their chemical substances into ducts that lead out of the body or to another organ. Examples of exocrine glands are sweat glands, sebaceous glands, and the portion of the pancreas that secretes digestive chemicals. **Endocrine** (ehn-dō-krihn) **glands** are groups of cells that secrete their chemical substances directly into the bloodstream, which transports them throughout the body. Endocrine glands are ductless. Examples of endocrine glands are the thyroid gland, the pituitary gland, and the portion of the pancreas that secretes insulin.

An **organ** (ohr-gahn) is a part of the body that performs a special function or functions. Each organ has its own combining form or forms, as listed in Table 2–2. The combining forms have either Latin or Greek origins. If a body part has two combining forms that are used to describe it, how do you know which form to use? In general, the Latin term is used to describe or modify something, as in *renal disease* and *renal tubule*. The Greek term generally is used to describe a pathological finding, as in *nephritis* and *nephropathy*.

1, 2, 3, GO

Medical terms can be further modified by the use of prefixes to assign number value (Table 2–3), numerical order, or proportions. The following prefixes are also used in everyday English, so some of them may be familiar. For example, unicorns are animals with one horn (*uni* = one, *corn* = horn). It would make sense then that a **bicornuate uterus** (*bi* = two, *corn* = horn) is a uterus with two horns. Knowing that *lateral*

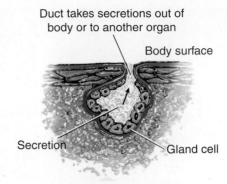

(a) Exocrine gland (has duct)

(b) Endocrine gland (ductless)

FIGURE 2–15 Types of glands. Exocrine glands secrete their chemical substances into ducts that lead out of the body or to another organ. Endocrine glands secrete their chemical substances (hormones) directly into the bloodstream.

means pertaining to the side, it would make sense that **unilateral** (yoo-nih-lah-tər-ahl) means pertaining to one side. **Bilateral** (bī-lah-tər-ahl) means pertaining to two sides.

TABLE 2–2 Combining Forms for Organs

Body System	Combining Form	Major Functions
Skeletal system	bones = **oste/o** (ohs-tē-ō), **oss/e** (ohs-ē), or **oss/i** (ohs-ih) joints = **arthr/o** (ahr-thrō) cartilage = **chondr/o** (kohn-drō)	Support and shape, protects internal organs, hematopoiesis, mineral storage
Muscular system	muscles = **my/o** (mī-ō) fascia = **fasc/i** (fahs-ē) or **fasci/o** (fahs-ē-ō) tendons = **ten/o** (tehn-ō), **tend/o** (tehn-dō), or **tendin/o** (tehn-dih-nō)	Locomotion, movement of body fluids, body heat generation
Cardiovascular system	heart = **cardi/o** (kahr-dē-ō) arteries = **arteri/o** (ahr-tē-rē-ō) veins = **ven/o** (vēn-ō) or **phleb/o** (fleh-bō) blood = **hem/o** (hē-mō) or **hemat/o** (hē-maht-ō)	Via pumping mechanism delivers oxygen, nutrients, electrolytes, and hormones to tissue, transports cellular waste from body, transports immune cells and antibodies
Lymphatic and Immune systems	lymph vessels, fluid, and nodes = **lymph/o** (lihm-fō) tonsils = **tonsill/o** (tohn-sih-lō) spleen = **splen/o** (spleh-nō) thymus = **thym/o** (thī-mō)	Provide nutrients to and remove waste from tissues, protect the body from harmful substances and invading pathogens
Respiratory system	nose or nares = **nas/o** (nā-zō) or **rhin/o** (rī-nō) pharynx = **pharyng/o** (fahr-ihn-gō) trachea = **trache/o** (trā-kē-ō) larynx = **laryng/o** (lahr-ihng-gō) lungs = **pneum/o** (nū-mō), **pneumon/o** (nū-mohn-ō), **pulm/o** (puhl-mō), or **pulmon/o** (puhl-mohn-ō)	Brings oxygen into the body for transportation to the cells, removes carbon dioxide and some water waste from the body
Digestive system	mouth = **or/o** (ōr-ō) or **stomat/o** (stō-maht-ō) esophagus = **esophag/o** (eh-sohf-ah-gō) stomach = **gastr/o** (gahs-trō) small intestine = **enter/o** (ehn-tər-ō) large intestine = **col/o** (kō-lō) or **colon/o** (kō-lohn-ō) liver = **hepat/o** (hehp-ah-tō) pancreas = **pancreat/o** (pahn-krē-ah-tō)	Digests ingested food, absorption of digested food, elimination of solid waste
Urinary system	kidneys = **ren/o** (rē-nō) or **nephr/o** (nehf-rō) ureters = **ureter/o** (yoo-rē-tər-ō) urinary bladder = **cyst/o** (sihs-tō) urethra = **urethr/o** (yoo-rē-thrō)	Filters blood to remove waste, maintains electrolyte balance, regulates fluid balance

(continued)

TABLE 2–2 Combining Forms for Organs (*Continued*)

Body System	Combining Form	Major Functions
Nervous system and Special Senses	nerves = **neur/o** (nū-rō) or **neur/i** (nū-rē) brain = **encephal/o** (ehn-sehf-ah-lō) spinal cord = **myel/o** (mī-eh-lō) eyes = **ophthalm/o** (ohf-thahl-mō), **ocul/o** (ohck-yoo-lō), **opt/o** (ohp-tō), or **opt/i** (ohp-tē) sight = **optic/o** (ohp-tih-kō) ears = **ot/o** (ō-tō), **audit/o** (aw-dih-tō), and **aud/i** (aw-dē) external ear = **aur/i** (awr-ih) and **aur/o** (awr-ō) sound = **acoust/o** (ah-koo-stō) and **acous/o** (ah-koo-sō)	Coordinating mechanism, reception of stimuli, transmission of messages
Integumentary system	skin = **dermat/o** (dər-mah-tō), **derm/o** (dər-mō), or **cutane/o** (kyoo-tā-nē-ō)	Protection of body, regulate body temperature and water content
Endocrine system	adrenals = **adren/o** (ahd-reh-nō) gonads = **gonad/o** (gō-nahd-ō) pineal = **pineal/o** (pī-nē-ahl-ō) pituitary = **pituit/o** (pih-too-ih-tō) thyroid = **thyroid/o** (thī-royd-ō) or **thyr/o** (thī-ro)	Integrates body functions, homeostasis, and growth with chemicals called hormones
Reproductive system	testes = **orch/o** (ōr-kō), **orchi/o** (ōr-kē-ō), **orchid/o** (ōr-kihd-ō), or **testicul/o** (tehst-tihck-yoo-lō) ovaries = **ovari/o** (ō-vā-rē-ō) or **oophor**/o (ō-ohf-ehr-ō) uterus = **hyster/o** (hihs-tehr-ō), **metr/o** (mē-trō), **metr/i** (mē-trē), **metri/o** (mē-trē-ō), or **uter/o** (yoo-tər-ō)	Production of new life

TABLE 2–3 Prefixes Assigning Number Value

Number Value	Latin Prefix	Greek Prefix	Examples
1	uni-	mono-	unicorn, unilateral monochromatic, monocyte
2	duo- or bi-	dyo-	duet, bilateral, dyad
3	tri-	tri-	trio, triceratops, triathlon
4	quadri- or quadro-	tetr- or tetra-	quadruplet, tetralogy, tetroxide
5	quinqu- or quint-	pent- or penta-	quintet, pentagon
6	sex-	hex- or hexa-	sexennial, hexose, hexagon
7	sept- or septi-	hept- or hepta-	septuple, heptarchy
8	octo-	oct-, octa-, or octo-	octave, octopus
9	novem- or nonus-	ennea-	nonuple, ennead
10	deca- or decem-	dek- or deka-	decade, dekanem

REVIEW EXERCISES

Multiple Choice

Choose the correct answer.

1. Lateral means
   a. near the beginning
   b. near the front
   c. toward the side
   d. toward midline

2. The sagittal plane divides the body into
   a. cranial and caudal portions
   b. left and right portions
   c. equal left and right halves
   d. dorsal and ventral portions

3. The paw is _____ to the shoulder.
   a. caudal
   b. cranial
   c. proximal
   d. distal

4. The transverse plane divides the body into
   a. cranial and caudal portions
   b. left and right portions
   c. equal left and right halves
   d. dorsal and ventral portions

5. The lining of the abdominal cavity and some of its organs is called the
   a. mesentery
   b. peritoneum
   c. thoracum
   d. membrane

6. The study of structure, composition, and function of tissues is called
   a. cytology
   b. histology
   c. pathology
   d. organology

7. The _____ plane divides the body into dorsal and ventral portions.
   a. sagittal
   b. midsagittal
   c. dorsal
   d. transverse

8. The medical term for lying down is
   a. lateral
   b. sternal
   c. recumbent
   d. surface

9. The medical term for increase in size of an organ, tissue, or cell is
   a. atrophy
   b. hypertrophy
   c. dystrophy
   d. hyperplasia

10. The medical term for the caudal surface of the rear paw, hoof, or foot is
    a. ventral
    b. dorsal
    c. palmar
    d. plantar

11. The term for toward the midline is
    a. medial
    b. lateral
    c. proximal
    d. distal

12. The term for nearest the midline or the beginning of a structure is
    a. medial
    b. lateral
    c. proximal
    d. distal

13. The term for away from the midline is
    a. medial
    b. lateral
    c. proximal
    d. distal

14. The term for farthest from the midline or beginning of a structure is
    a. medial
    b. lateral
    c. proximal
    d. distal

15. The term that refers to the back is
    a. ventral
    b. dorsal
    c. cranial
    d. caudal

16. The term that means toward the tail is
    a. ventral
    b. dorsal
    c. cranial
    d. caudal

17. The term that means toward the head is
    a. ventral
    b. dorsal
    c. cranial
    d. caudal

18. The term that refers to the belly or underside of a body is
    a. ventral
    b. dorsal
    c. cranial
    d. caudal

19. Which type of tissue covers internal and external body surfaces?
    a. adipose
    b. epithelial
    c. connective
    d. cytoplasm

20. The term for a hole or hollow space in the body that contains and protects internal organs is
    a. abdomen
    b. peritoneal
    c. cavity
    d. membrane

21. Which of the following is the only horizontal plane in quadrupeds?
    a. midsagittal
    b. dorsal
    c. sagittal
    d. transverse

22. Which of the following best describes *dermato* and *cutaneo*?
    a. They are both roots that mean gland.
    b. They are both roots but have different meanings.
    c. They are both combining forms that mean skin.
    d. One is a prefix and one is a suffix.

23. Which two terms have prefixes denoting numbers?
    a. hypoglycemia and hyperglycemia
    b. preanesthetic and postanesthetic
    c. bilateral and unilateral
    d. polyuria and oliguria

24. The term that is opposite of caudal is
    a. proximal
    b. cranial
    c. superficial
    d. plantar

25. The term that is opposite of medial is
    a. lateral
    b. cephalic
    c. rostral
    d. ventral

Matching

Match the number in Column I with its prefix in Column II. Each number may have more than one correct answer.

Column I	**Column II**
1. _____ one	a. hept-
2. _____ two	b. mono-
3. _____ three	c. tri-
4. _____ four	d. deka-
5. _____ five	e. duo-
6. _____ six	f. uni-
7. _____ seven	g. penta-
8. _____ eight	h. octo-
9. _____ nine	i. quadri-, quadro-
10. _____ ten	j. tetra-
	k. sex-
	l. nonus-
	m. deca-
	n. quinqu-

Match the pathology term in Column I with its definition in Column II.

Column I

11. _____ anaplasia

12. _____ aplasia

13. _____ dysplasia

14. _____ hyperplasia

15. _____ hypoplasia

Column II

a. abnormal growth or development of an organ or a tissue or a cell

b. a change in the structure of cells and their orientation to each other

c. an abnormal increase in the number of normal cells in normal arrangement in an organ, a tissue, or a cell

d. incomplete or less than normal development of an organ, a tissue, or a cell

e. lack of development of an organ, a tissue, or a cell

Match the definition in Column I with its word part in Column II

Column I	Column II
16. _____ fat	a. aden/o
17. _____ front	b. adip/o
18. _____ gland	c. anteri/o
19. _____ study of	d. caud/o
20. _____ cell	e. -logy
21. _____ head	f. cyt/o
22. _____ rear of the body	g. cephal/o
23. _____ toward the tail	h. hist/o
24. _____ formation of cell size	i. path/o
25. _____ mouth	j. -trophy
26. _____ formation of cell number	k. -oma
27. _____ first	l. or/o
28. _____ tumor	m. posteri/o
29. _____ disease	n. -plasia
30. _____ tissue	o. prot/o

Match the term in Column I with its definition in Column II

Column I

31. _____ anatomy

32. _____ pathophysiology

33. _____ etiology

34. _____ physiology

35. _____ pathology

Column II

a. branch of biological science that studies and describes how body parts work or function

b. branch of biological science that studies the structure of body parts

c. branch of biological science that studies how disease changes the function of the body

d. branch of biological science that studies the causes of disease

e. branch of biological science that studies the nature, causes, and development of abnormal conditions

Fill in the Blanks

Write the medical term or description to complete the following sentences.

1. The _____ is also known as the navel.

2. _____ glands secrete chemical substances directly into the bloodstream.

3. A(n) _____ is any new growth of tissue in which multiplication of cells is uncontrolled, more rapid than normal, and progressive.

4. A(n) _____ is a deviation from what is regarded as normal.

5. The _____ cavity contains the heart and lungs.

6. The caudal surface of the front paw, foot, or hoof is the _____ surface.

7. The shoulder is _____ to the pelvis.

8. A(n) _____ is the basic structural unit of the body.

9. The stomach is located _____ to the heart.

10. _____ is the palmar or plantar movement of joint angles.

11. Another term for groin is _____.

12. The _____ is a layer of the peritoneum that suspends parts of the intestine in the abdominal cavity.

13. _____ is the suffix for formative material of cells.

14. Not malignant is _____.

15. The five combining forms for uterus are _____, _____, _____, _____, and _____.

16. The plane that divides the animal into equal right and left halves is the _____.

17. The plane that divides the animal into cranial and caudal parts is the _____.

18. The study of body structure is called _____.

19. The study of body function is called _____.

20. The aspect of the tooth of the mandible that faces the tongue is called the _____.

21. The aspect of the tooth that faces the cheek is called the _____.

22. Movement toward the midline is known as _____.

23. Movement away from the midline is known as _____.

24. Groups of specialized cells that secrete material used elsewhere in the body are known as _____.

25. A part of the body that performs a special function or functions is known as a(n) _____.

True or False

If the statement is true, write T on the line. If the statement is false, write F on the line.

1. _____The abdominal cavity contains the major organs of digestion.

2. _____The term *lateral* means the direction toward or nearer the midline.

3. _____The parietal peritoneum is the outer layer of the peritoneum that lines the interior of the abdominal cavity.

4. _____The layer of the peritoneum that suspends parts of the intestine in the abdominal cavity is the umbilicus.

5. _____The term *inguinal* refers to the entire caudal region of the abdomen adjacent to the thigh.

Spelling

Cross out any misspelled words in the following sentences and replace them with the proper spelling.

1. The mesantry is a layer of the peritoneum that suspends parts of the intestine in the abdominal cavity. _____

2. Hypretrophy is a general increase in the bulk of a body part or organ due to an increase in the size, but not the number, of cells in tissue. _____

3. The protective covering for all of the internal and external surfaces of the body is formed by epithealial tissue. _____

4. An abnomolly is any deviation from what is regarded as normal. _____

5. Lack of development of cell numbers in an organ or a tissue is aplazia. _____

6. The term *retroperetonal* means superficial to the peretoneum. _____

7. The imbilikus is the pit in the abdominal wall marking the point where the imbilikal cord entered the fetus. _____

8. The caudal region of the abdomen adjacent to the thigh is the inguynal area. _____

9. The oclusal surfaces are the aspects of the teeth that meet when you chew._____

10. A body kavity is a hole or hollow space in the body that contains and protects internal organs._____

Crossword Puzzles

Directional Terms and Planes of the Body

Supply the correct term in the appropriate space for the definition listed.

Across

1 pertaining to the belly
5 to reduce the angle between two bones
6 situated nearest the midline
7 bottom of the rear foot, hoof, or paw
8 pertaining to toward the side
9 pertaining to toward the head
13 positioned away from the surface
15 to move toward midline
17 situated farthest from midline
18 plane dividing the body into cranial and caudal portions
19 pertaining to the nose end of the head
20 plane dividing the body into unequal left and right portions

Down

2 to increase the angle between two bones
3 pertaining to the head
4 below or lowermost
6 bottom of the front foot, hoof, or paw
9 pertaining to toward the tail
10 to move away from midline
12 pertaining to the back
16 pertaining to toward midline

Organ Combining Forms

Across

2 kidney

5 lungs

8 pharynx

10 spleen

12 blood

14 nose or nares

15 bone

19 vein

20 ear

22 spinal cord

23 eye

28 urethra

30 mouth

32 small intestine

33 lymph vessel, fluid, or node

34 uterus

35 cartilage

36 urinary bladder

Down

1 joint

2 nerves

3 brain

4 liver

6 fascia

7 ureters

9 stomach

11 esophagus

13 testes

16 thymus

17 tendon

18 skin

21 trachea

24 pancreas

25 tonsil

26 muscle

27 larynx

31 large intestine

Label the Diagrams

For Figures 2–16, 2–17, 2–18, 2–19, 2–20, and 2–21, follow the instructions provided in the captions.

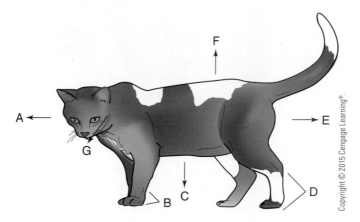

FIGURE 2–16 Label the arrows with the proper directional term.

FIGURE 2–17 Label the arrows with the proper directional term.

FIGURE 2–18 Label the sacs of the rumen. Through which plane is this goat sectioned?

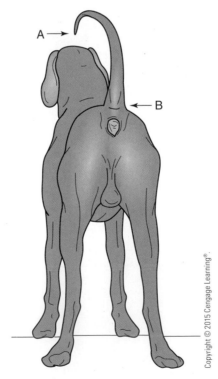

FIGURE 2–19 Is point A the more proximal or more distal end of the tail? Is point B the more proximal or more distal end of the tail?

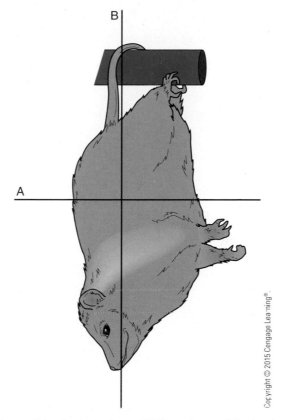

FIGURE 2–20 What plane of the body is plane A? What plane of the body is plane B?

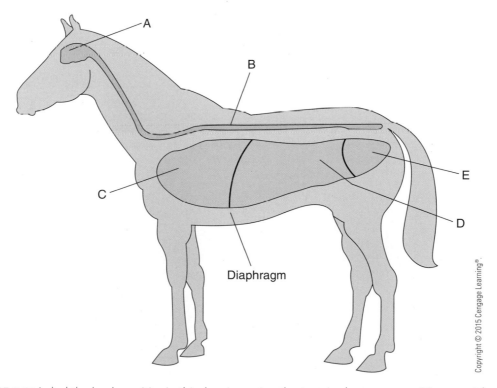

FIGURE 2–21 Label the body cavities in this drawing using the terminology on page 22 as a guide.

Case Studies

After reading each case study, define the list of terms.

A 7-year-old male Siberian husky presented to the clinic with a cough that has become more severe in the past few weeks. Today the dog collapsed while playing fetch and was rushed to the veterinary clinic. Once the dog was stabilized, thoracic radiographs were taken and a tumor was seen in the cranial thoracic area. There also was hypertrophy of the right side of the heart. The veterinarian was concerned that the dog may have a malignant tumor and requested more tests.

Define the terms using the word parts.

1. thoracic _____

2. tumor _____

3. cranial _____

4. hypertrophy _____

5. malignant _____

Charlie, a 10-year-old domestic shorthair cat, was diagnosed with diabetes mellitus three years previous to his presentation at the clinic. His owners noticed that his front leg looked abnormal, but were not aware of any injury to Charlie's leg. As you place Charlie on the exam table, you notice he is holding his front limb abnormally so that the palmar surface touches the top of the table (Figure 2–22). While waiting for the veterinarian to perform a physical exam, you decide to research complications of diabetes mellitus. You find information that states up to 50% of diabetic cats can develop peripheral neuropathy due to hyperglycemia, which causes glucose toxicity and cell starvation and has no effective therapy. Clinical signs of peripheral neuropathy include a plantigrade or palmigrade stance and posterior weakness. As you finish your research, the veterinarian asks you to assist in the exam room. After the initial physical exam, the veterinarian tells you to place the cat in left lateral recumbency so that he can do a neurologic exam. The veterinarian decides to refer Charlie to a veterinary neurologist for further assessment.

Define the terms using their word parts.

1. palmar _____

2. neuropathy _____

FIGURE 2–22 Palmigrade stance in a diabetic cat.

3. hyperglycemia _____

4. plantigrade (the suffix -grade means to go) _____

5. palmigrade _____

6. left lateral recumbency _____

Critical Thinking Exercise

The following story and questions are designed to stimulate critical thinking through class discussion or brief writing responses. There are many correct answers to these questions.

There was an overwhelming smell as Jason entered the classroom and he immediately knew why: on each table was sitting a preserved cat for dissection. "Do these cats have to stink?" he asked his instructor. "Well, if they didn't stink from the preservative, they would be rotting and could be spreading disease," his instructor said. "The next thing I'll hear is you complaining about having to learn planes of the body and directional terms. You probably think they have no significance in the real world." Jason smiled and believed he would not have to use these terms once he had a real job.

As the students began dissection, the instructor felt that Jason was not interested in learning the positional terms in his lab manual. "I know you don't think you will need to know these words in the 'real world,' but you'll be surprised when and how you use the knowledge you acquire in the classroom," his instructor told him. "Identifying fractures on X-rays, submitting biopsy samples, and describing the location of lesions are just a few examples of when you will use directional and positional terms on the job. Learning them well now will make it easier for you in the future." Once Jason understood the clinical application of these terms, he was more willing to spend time committing them to memory.

Suggested Discussion Topics

1. Describe how attitude plays a role in learning.

2. One way to master positional and directional terms is to learn them as oppositional pairs. Complete the following chart of terms and their opposite pairs:

Term	Definition	Opposite Term	Definition
Dorsal			
Caudal			
Rostral			
Lateral			
Superior			
Distal			
Deep			
Plantar			
Abduction			
Flexion			

Chapter 3
MEAT AND BONES

OVERVIEW

Functions of the Skeletal System	Introduction of the skeletal system's role in an animal's body
Structures of the Skeletal System	Description of bones, cartilage, joints, ligaments, and tendons
Boning Up	Differentiation of the axial and appendicular skeleton
Structural Support	Terms used to describe bone structures
Test Me: Skeletal System	Terms used to describe diagnostic procedures of the skeletal system
Pathology: Skeletal System	Terms used to describe pathological conditions of the skeletal system
Procedures: Skeletal System	Terms used to describe procedures of the skeletal system
Functions of the Muscular System	Introduction of the muscular system's role in an animal's body
Structures of the Muscular System	Description of the specialized cells of contractibility
Show Some Muscle	Terms used to describe movement
What's in a Name?	Terms used to name muscles
Test Me: Muscular System	Terms used to describe diagnostic procedures of the muscular system
Pathology: Muscular System	Terms used to describe pathological conditions of the muscular system
Procedures: Muscular System	Terms used to describe procedures of the muscular system
Abbreviations Related to the Skeletal and Muscular Systems	Lists abbreviations for this chapter

Objectives

Upon completion of this chapter, the reader should be able to:

- Identify and describe the major structures and functions of the musculoskeletal system
- Define terms used to describe the musculoskeletal system
- Describe bone anatomy terms
- Differentiate between the axial and appendicular skeletons
- Define terms used to describe the parts of the axial and appendicular skeletons
- Define terms used to describe muscle movement
- Describe how muscles are named
- Recognize, define, spell, and pronounce terms related to the diagnosis, pathology, and treatment of the musculoskeletal system
- Construct musculoskeletal terms from word parts
- Interpret the meaning of abbreviations of the musculoskeletal system

FUNCTIONS OF THE SKELETAL SYSTEM

The **musculoskeletal** (muhs-kyoo-lō-skehl-eh-tahl) **system** consists of two systems that work together to support the body and allow movement of the animal—the skeletal system and the muscular system. The skeletal system forms the framework that supports and protects an animal's body. Within bone is the red bone marrow, which functions to form red blood cells, white blood cells, and clotting cells. Joints aid in the movement of the body. Cartilage protects the ends of bones where they contact each other. Cartilage is also found in the ear and nose. Tendons connect muscle to bone allowing animals to move, while ligaments connect bone to bone which supports joints. The muscular system is covered later in this chapter.

STRUCTURES OF THE SKELETAL SYSTEM

The skeletal system consists of bones, cartilage, joints, ligaments, and tendons.

Make the Connection

The skeleton is made up of various forms of connective tissues. Connective tissue is a type of tissue in which the proportion of cells to extracellular matrix is small. Connective tissue binds together and supports various structures of the body. Bone, tendons, ligaments, and cartilage are all connective tissues associated with the skeletal system.

Bone

Bone, a form of connective tissue, is one of the hardest tissues in the body. Embryonically, the skeleton is made of cartilage and fibrous membranes that harden into bone before birth. **Ossification** (ohs-ih-fih-kā-shuhn), the formation of bone from fibrous tissue, continues until maturity, which varies with species. Normal bone goes through a continuous process of building up and breaking down throughout an animal's life. This process allows bone to heal and repair itself. Bone growth is balanced between the actions of **osteoblasts** (ohs-tē-ō-blahsts) and **osteoclasts** (ohs-tē-ō-klahsts). Osteoblasts (**oste/o** = bone, **-blasts** = immature) are immature bone cells that produce bony tissue, and osteoclasts (**oste/o** = bone, **-clasts** = break) are phagocytic cells that eat away bony tissue from the medullary cavity of long bone (Figure 3–1). When osteoblasts

FIGURE 3–1 Microscopic structure of compact bone, showing osteons with a central canal. The bone matrix is deposited in a circular arrangement. Darkly stained osteocytes are visible in the matrix.

Courtesy of William J. Bacha, PhD, and Linda M. Bacha, MS, VMD

mature, they become **osteocytes** (ohs-tē-ō-sītz). The combining forms for bone are **oste/o**, **oss/e**, and **oss/i**.

Red bone marrow, located in cancellous bone, is **hematopoietic** (hēm-ah-tō-poy-eht-ihck). The combining form **hemat/o** means blood, and the suffix **-poietic** means pertaining to formation. Thus, red bone marrow produces red blood cells, white blood cells, and clotting cells. The **medullary** (mehd-yoo-lahr-ē) **cavity** of long bone, or the inner space of bone, contains yellow bone marrow. In adult animals, yellow bone marrow replaces red bone marrow. Yellow bone marrow is composed mainly of fat cells and serves as a fat storage area.

Bone is divided into different categories based on bone types, bone shapes, and bone functions (Table 3–1 and Figure 3–2).

Cartilage

Cartilage (kahr-tih-lihdj) is another form of connective tissue that is more elastic than bone. The elasticity of cartilage makes it useful in the more flexible

Descriptive Word Parts for the Skeletal System

epi- = above, **physis** = growth, **dia-** = between, **peri-** = surrounding, **oste/o-** = bone, **-um** = structure, **endo-** = within or inner, **meta-** = beyond

TABLE 3–1 Terminology Applied to Bone

Types of Bone	
cortical bone (kōr-tih-kahl)	hard, dense, strong bone that forms the outer layer of bone; also called **compact bone** **cortex** = bark or shell in Latin
cancellous bone (kahn-sehl-uhs)	lighter, less strong bone that is found in the ends and inner portions of long bones; also called **spongy bone** **cancellous** = latticework in Latin

Bone Anatomy Terms	
epiphysis (eh-pihf-ih-sihs)	wide end of a long bone, which is covered with articular cartilage and is composed of cancellous bone **proximal epiphysis** = located nearest the midline of the body **distal epiphysis** = located farthest away from the midline of the body
diaphysis (dī-ahf-ih-sihs)	shaft of a long bone that is composed mainly of compact bone
physis (fī-sihs)	cartilage segment of long bone that involves growth of the bone; also called the **growth plate** or **epiphyseal cartilage** (Figure 3–3)
metaphysis (meh-tahf-ih-sihs)	wider part of long bone shaft located adjacent to the physis; in adult animals, it is considered part of the epiphysis
periosteum (pehr-ē-ohs-tē-uhm)	tough, fibrous tissue that forms the outer covering of bone
endosteum (ehn-dohs-tē-uhm)	tough, fibrous tissue that forms the lining of the medullary cavity

Bone Classification	
long bones	bones consisting of a shaft, two ends, and a marrow cavity (i.e., femur)
short bones	cube-shaped bones with no marrow cavity (i.e., carpal bones)
flat bones	thin, flat bones (i.e., pelvis)
pneumatic bones	sinus-containing bones (i.e., frontal bone)
irregular bones	unpaired bones (i.e., vertebrae)
sesamoid bones	small bones embedded in a tendon (i.e., patella) (the only exception is the distal sesamoid of the horse)

portions of the skeleton. **Articular** (ahr-tihck-yoo-lahr) **cartilage,** a specific type of cartilage, covers the joint surfaces of bone. The **meniscus** (meh-nihs-kuhs) is a curved fibrous cartilage found in some joints, such as the canine stifle, that cushions forces applied to the joint. The combining form for cartilage is **chondr/o.**

Joints

Joints or **articulations** (ahr-tihck-yoo-lā-shuhns) are connections between bones. *Articulate* means to join in a way that allows motion between the parts. The combining form for joint is **arthr/o.** The different types of joints are classified based on their function and degree of movement (Figure 3–4). **Synarthroses** (sihn-ahrth-rō-sēz) allow no movement, **amphiarthroses** (ahm-fih-ahrthr-ō-sēz) allow slight movement, and **diarthroses** (dī-ahrth-rō-sēz) allow free movement.

Synarthroses are immovable joints usually united with fibrous connective tissue. An example of a synarthrosis is a suture. A **suture** (soo-chuhr) is a jagged line where bones join and form a nonmovable joint. Sutures typically are found in the skull. A **fontanelle** (fohn-tah-nehl) is a soft spot remaining at the junction of sutures that usually closes after birth.

Amphiarthroses are semimovable joints. An example of an amphiarthrosis is a symphysis. A **symphysis** (sihm-fih-sihs) is a joint where two bones join and are held firmly together so that they function as one bone. Another term for *symphysis* is **cartilaginous joint.** The halves of the mandible fuse at a symphysis to form one bone. This fusion is the **mandibular symphysis.** The halves of the pelvis also fuse at a symphysis, which is called the **pubic symphysis.**

Diarthroses are freely movable joints. Examples of diarthroses are synovial joints. **Synovial** (sih-nō-vē-ahl) **joints** are further classified as

(a)

(c)

FIGURE 3–2 Anatomy of a long bone. (a) Sheep tibia cranial view. (b) Interior view. (c) Sagittal section.

Radius   Growth Plate

Metacarpals

Carpal Bones

Ulna

FIGURE 3–3 Radiograph of the radius and ulna of a young dog. Note the growth plate (physis) visible in the radius and ulna.

Soft versus Hard

Bone diseases can cause abnormal changes. Bones can become softer than normal or harder than normal. To describe these changes, the suffixes **-malacia** (abnormal softening) and **-sclerosis** (abnormal hardening) are used.

ball-and-socket joints (also called **enarthrosis** (ehn-ahr-thrō-sihs) or spheroid joints), arthrodial (ahr-thrō-dē-ahl) or condyloid (kohn-dih-loyd) joints, trochoid (trō-koyd) or pivot (pih-voht) joints, ginglymus (jihn-glih-muhs) or hinge joints, and gliding joints. **Ball-and-socket joints** allow a wide range of motion in many directions, such as the hip and shoulder joints. **Arthrodial** or **condyloid joints** are

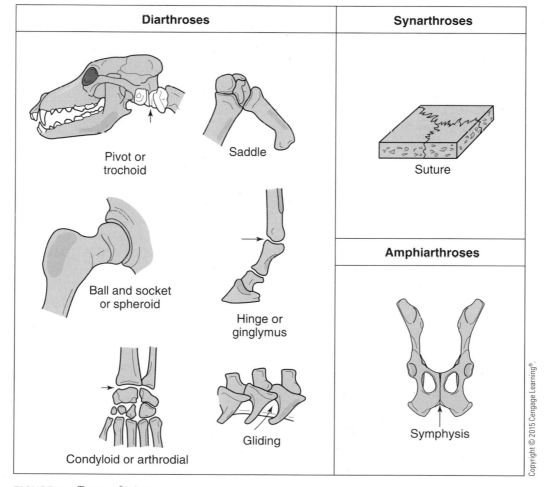

FIGURE 3–4 Types of joints.

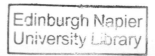

joints with oval projections that fit into a socket, such as the carpal joints (where the radius meets the carpus). **Trochoid joints** include pulley-shaped joints like the connection between the atlas to the axis. **Hinge joints** allow motion in one plane or direction, such as canine stifle and elbow joints. **Gliding joints** move or glide over each other, as in the radioulnar joint or the articulating process between successive vertebrae. Primates have an additional joint called the saddle joint. The only **saddle joint** is located in the carpometacarpal joint of the thumb. This saddle joint allows primates to flex, extend, abduct, adduct, and circumduct the thumb.

Ligaments and Tendons

A **ligament** (lihg-ah-mehnt) is a band of fibrous connective tissue that connects one bone to another bone to help stabilize joints. **Ligament/o** is the combining form for ligament. A ligament is different from a tendon. A **tendon** (tehn-dohn) is a band of fibrous connective tissue that connects muscle to bone to help the animal move. The combining forms for tendon are **ten/o, tend/o,** and **tendin/o.**

Bursa

A **bursa** (bər-sah) is a fibrous sac that acts as a cushion to ease movement in areas of friction. Within the shoulder joint is a bursa where a tendon passes over bone. The combining form for bursa is **burs/o.** More than one bursa is **bursae** (bər-sā).

Synovial Membrane and Fluid

Bursae and synovial joints have an inner lining called the **synovial** (sih-nō-vē-ahl) **membrane.** The synovial membrane secretes **synovial fluid,** which acts as a lubricant to make joint movement smooth. **Synovi/o** is the combining form for synovial membrane and synovial fluid.

A way to remember that a tendon connects a muscle to bone is that both tendon and muscle have the same number of letters or that the Achilles tendon attaches the calf muscle to a bone.

BONING UP

The skeleton is descriptively divided into two parts: the axial skeleton and the appendicular skeleton. The **axial** (ahcks-ē-ahl) **skeleton** is the framework of the body that includes the skull, auditory ossicles, hyoid bones, vertebral column, ribs, and sternum. The **appendicular** (ahp-ehn-dihck-yoo-lahr) **skeleton** is the framework of the body that consists of the extremities, shoulder, and pelvic girdle. *Append* means to add or hang, so think of the appendages or extremities as structures that hang from the axial skeleton.

The Axial Skeleton

The axial skeleton is composed of bones that lie around the body's center of gravity.

Take It from the Top

The **cranium** (krā-nē-uhm) is the portion of the skull that encloses the brain. The combining form **crani/o** means skull. The cranium consists of the following bones (Figure 3–5):

- **frontal** (frohn-tahl) = forms the roof of the cranial cavity or "front" or cranial portion of the skull. In some species, the horn, or cornual (kohrn-yoo-ahl) process, arises from the frontal bone (Figure 3–6).
- **parietal** (pah-rī-ih-tahl) = paired bones that form the roof of the caudal cranial cavity.
- **occipital** (ohck-sihp-ih-tahl) = forms the caudal aspect of the cranial cavity where the foramen magnum, or opening for the spinal cord, is located. **Foramen** (fō-rā-mehn) is an opening in bone through which tissue passes. **Magnum** (māg-nuhm) means large.
- **temporal** (tehm-pohr-ahl) = paired bones that form the sides and base of the cranium.
- **sphenoid** (sfeh-noyd) = paired bones that form part of the base of the skull and parts of the floor and sides of the bony eye socket.
- **ethmoid** (ehth-moyd) = forms the rostral part of the cranial cavity.
- **incisive** (ihn-sīs-ihv) = forms the rostral part of the hard palate and lower edge of nares.
- **pterygoid** (tahr-ih-goyd) = forms the lateral wall of the nasopharynx.

In addition to bones, the skull also has air- or fluid-filled spaces. These air- or fluid-filled spaces are called **sinuses** (sīn-uhs-ehz).

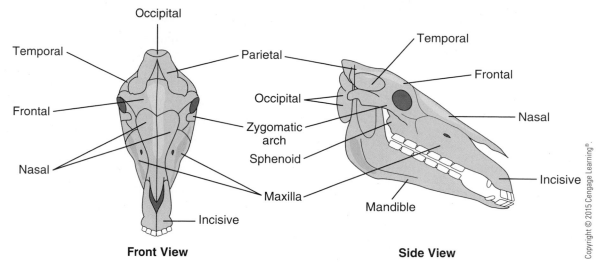

FIGURE 3–5 Selected bones of the skull and face.

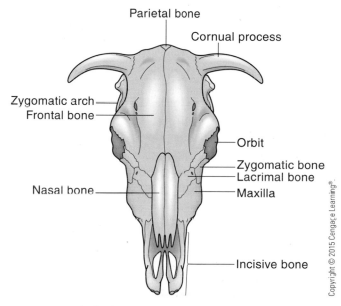

FIGURE 3–6 Skull of an ox, dorsal view.

Skull shapes in dogs can vary by breed. Examples of skull shapes in dogs include the following:

- **Brachycephalic** (brā-kē-seh-fahl-ihck) dogs have short, wide heads, as do pugs and Pekingese.
- **Dolichocephalic** (dō-lih-kō-seh-fahl-ihck) dogs have narrow, long heads, as do collies and greyhounds.
- **Mesocephalic** (mehs-ō-seh-fahl-ihck) dogs have average width to their heads, as do Labrador retrievers. Also called **mesaticephalic** (mehs-āt-ih-seh-fahl-ihck).

Let's Face It

The bones of the face consist of the following:

- **zygomatic** (zī-gō-mah-tihck) = forms the orbit and cheekbone; projections from the temporal and zygomatic bone form the zygomatic arch (cheeckbone)
- **maxilla** (mahck-sih-lah) = forms the upper jaw.
- **mandible** (mahn-dih-buhl) = forms the lower jaw.
- **palatine** (pahl-ah-tihn) = forms part of the hard palate.
- **lacrimal** (lahck-rih-mahl) = forms the medial part of the orbit.
- **incisive** (ihn-sī-sihv) = forms the rostral part of the hard palate and lower edge of nares.
- **nasal** (nā-sahl) = forms the bridge of the nose.
- **vomer** (vō-mər) = forms the base of the nasal septum. The **nasal septum** (nā-sahl sehp-tuhm) is the cartilaginous structure that divides the two nasal cavities.
- **hyoid** (hī-oyd) = bone suspended between the mandible and the laryngopharynx.

Back to Basics

The **vertebral** (vər-teh-brahl) **column** (also called the **spinal column** and **backbone**) supports the head and body and protects the spinal cord. The vertebral column consists of individual bones called **vertebra** (vər-teh-brah). The combining forms for vertebra are **spondyl/o** and **vertebr/o**. More than one vertebra are called **vertebrae** (vər-teh-brā).

Vertebrae are divided into parts, and the parts may vary depending on the location of the vertebra and its function (Figure 3–7). The **body** is the solid

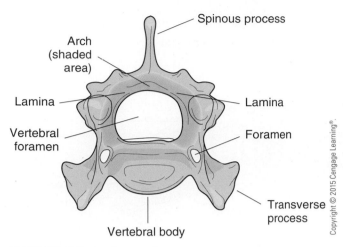

Spinous process

Arch (shaded area)

Lamina

Lamina

Vertebral foramen

Foramen

Transverse process

Vertebral body

Copyright © 2015 Cengage Learning®

FIGURE 3–7 Parts of a vertebra.

portion ventral to the spinal cord. The **arch** is the dorsal part of the vertebra that surrounds the spinal cord. The **lamina** (lahm-ih-nah) is the left or right dorsal half of the arch. Processes project from the vertebrae. The term *process* means projection. A **spinous process** is a single projection from the dorsal part of the vertebral arch. **Transverse processes** project laterally from the right and left sides of the vertebral arch. **Articular processes** are paired cranial and caudal projections located on the dorsum of the vertebral arch.

Foramen (fō-rā-mehn) means opening. The opening in the middle of the vertebra through which the spinal cord passes is the **vertebral foramen.** The vertebrae are separated and cushioned from each other by cartilage discs called **intervertebral discs.** The combining form for discs is **disc/o.**

Vertebrae are organized and named by region. The regions are identified in Table 3–2 and Figures 3–8 and 3–9.

In addition, the first two vertebrae have individual names. C1 (or cervical vertebra one) is called the **atlas,** and C2 (or cervical vertebra two) is called the **axis.** (Remember that they follow alphabetical order from cranial to caudal.)

Stick to Your Ribs

Ribs are paired bones that attach to thoracic vertebrae (Figure 3–10). The combining form for rib is **cost/o.** Ribs are sometimes called **costals.**

The **sternum** (stər-nuhm), or breastbone, forms the midline ventral portion of the rib cage. The sternum is divided into three parts: the manubrium, body, and xiphoid process. The **manubrium** (mah-nū-brē-uhm) is the cranial portion of the sternum. The **body** of the sternum is the middle portion. The caudal portion of the sternum is known as the **xiphoid** (zī-foyd) **process**.

The ribs, sternum, and thoracic vertebrae make up the boundaries of the thoracic cavity. The **thoracic cavity**, or **rib cage**, protects the heart and lungs (Figure 3–11).

The Appendicular Skeleton

The appendicular skeleton is composed of the bones of the limbs.

From the Front

The bones of the front limb from proximal to distal consist of the scapula, clavicle, humerus, radius, ulna, carpus, metacarpals, and phalanges.

The **scapula** (skahp-yoo-lah), or **shoulder blade**, is a large triangular bone on the side of the thorax (Figure 3–12). The **clavicle** (klahv-ih-kuhl), or **collarbone**, is a slender bone that connects the sternum to the scapula. Some animal species have only a **vestigial** (vehs-tihj-ē-ahl), or **rudimentary**, clavicle, whereas other animal species, such as swine, ruminants, and equine, do not have a clavicle.

The **humerus** (hū-mər-uhs) is the long bone of the proximal front limb. The humerus is sometimes called the **brachium** (brā-kē-uhm). The radius and ulna are the two bones of the forearm or distal front limb. This region is called the **antebrachium** (ahn-tih-brā-kē-uhm). **Ante-** means before. The **radius** (rā-dē-uhs) is the cranial bone of the front limb, and the **ulna**

TABLE 3–2 Vertebral Regions

Cervical (sihr-vih-kahl)	Thoracic (thō-rahs-ihck)	Lumbar (luhm-bahr)	Sacral (sā-krahl)	Coccygeal (kohck-sih-jē-ahl) (also called caudal)
Neck area "C"	Chest area "T"	Loin area "L"	Sacrum area "S"	Tail area "Cy" or "Cd"

Cranial articular process
Caudal articular process
Transverse process
Spinous process
Lamina
Vertebral canal
Body

(a) Cervical Vertebrae—Caudal View

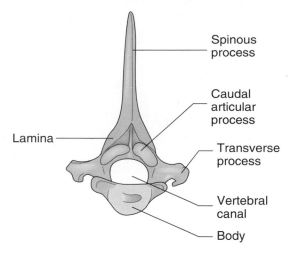

Spinous process
Caudal articular process
Transverse process
Vertebral canal
Body
Lamina

(b) Thoracic Vertebrae—Caudal View

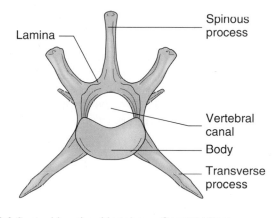

Lamina
Spinous process
Vertebral canal
Body
Transverse process

(c) Typical Lumbar Vertebra—Cranial View

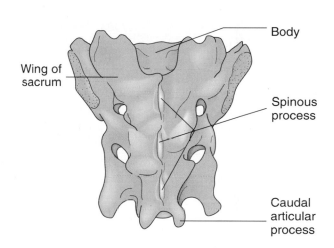

Body
Wing of sacrum
Spinous process
Caudal articular process

(d) Sacrum—Dorsal View

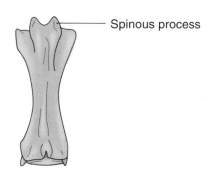

Spinous process

(e) Coccygeal Vertebrae—Dorsal View

FIGURE 3–8 Comparison of the structure of cervical, thoracic, lumbar, sacral, and coccygeal vertebrae: (a) caudal view of a cervical vertebra, (b) caudal view of a thoracic vertebra, (c) cranial view of a lumbar vertebra, (d) dorsal view of fused sacral vertebrae, and (e) dorsal view of a coccygeal vertebra.

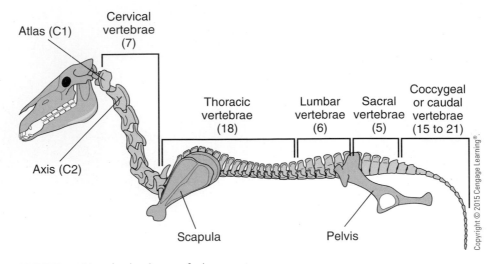

FIGURE 3–9 Vertebral column of a horse.

Atlas (C1)

Cervical vertebrae (7)

Axis (C2)

Thoracic vertebrae (18)

Lumbar vertebrae (6)

Sacral vertebrae (5)

Coccygeal or caudal vertebrae (15 to 21)

Scapula

Pelvis

The vertebral formulas for different species are as follows:

- dogs and cats: C = 7, T = 13, L = 7, S = 3, Cy = 6–23
- equine: C = 7, T = 18, L = 6 (or L = 5 in some Arabians), S = 5, Cy = 15–21
- bovine: C = 7, T = 13, L = 6, S = 5, Cy = 18–20
- pigs: C = 7, T = 14–15, L = 6–7, S = 4, Cy = 20–23
- sheep and goats: C = 7, T = 13, L = 6–7, S = 4, Cy = 16–18
- chicken: C = 14, T = 7, LS = 14, Cy = 6 (lumbar and sacral vertebrae are fused)

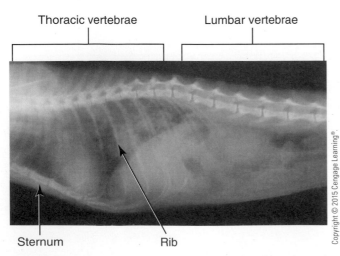

Thoracic vertebrae

Lumbar vertebrae

Sternum

Rib

FIGURE 3–10 Radiograph of a cat showing the thoracic and lumbar spine. Ribs and sternum also are visible.

(uhl-nah) is the caudal bone of the front limb. The ulna has a proximal projection called the **olecranon** (ō-lehck-rah-nohn) that forms the point of the elbow. Some species have a fused radius and ulna.

The **carpal** (kahr-pahl) **bones** are irregularly shaped bones in the area known as the wrist in people. In small animals, this joint is called the **carpus**, and in large animals, this joint is called the **knee**. The **metacarpals** (meht-ah-kahr-pahlz) are bones found distal to the carpus (**meta-** = beyond). The metacarpals are identified by numbers from medial to lateral.

In some species such as the horse, certain metacarpals do not articulate with the phalanges. In the horse, metacarpals (and metatarsals) II and IV do not articulate with the phalanges and are commonly called **splint bones.** Splint bones are attached by an **interosseous** (ihn-tər-ohs-ē-uhs) ligament to the large third metacarpal (or metatarsal) bone, which is commonly called the **cannon bone** (Figure 3–13). In ruminants, the cannon bones are metacarpal (or metatarsal) bones III and IV.

The **phalanges** (fā-lahn-jēz) are the bones of the digit. One bone of the digit is called a **phalanx** (Figure 3–14a and b). Phalanges are numbered from proximal to distal. Most digits have three phalanges, but the most medial phalanx (digit I) has only two phalanges. **Digits** are the bones analogous to the human finger and vary in number in animals (Figure 3–15). Digit I of dogs is commonly called the **dewclaw** and may be removed shortly after birth. **Ungulates** (uhng-yoo-lātz), or animals with hooves, also have digits that are numbered in the same fashion. Animals with a cloven hoof, or split hoof, have digits III and IV, and digits II and V are vestigial. The vestigial digits

FIGURE 3–11 Cat skeleton.

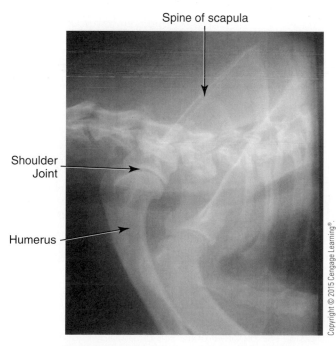

FIGURE 3–12 Radiograph of the scapula and shoulder (scapulohumeral) joint of a dog.

of cloven-hoofed animals are also called dewclaws. Cloven-hoofed animals, such as ruminants and swine, have three phalanges in their digits, with the distal-most phalanx (P3) encased in the hoof. Equine species have one digit (digit III). Within that digit are three phalanges. In livestock, the joints between the phalanges or between the phalanges and other bones have common names. The joint between metacarpal (metatarsal) III and the proximal phalanx is the **fetlock joint**. The joint between P1 and P2 is known as the **pastern joint**. The joint between P2 and P3 is known as the **coffin joint**. The phalangeal bones also have common names in livestock. P1 is the **long pastern bone**, P2 is the **short pastern bone**, and P3 is the **coffin bone**. (See Chapter 4 for an illustration.)

Phalanx 3 (P3) also may be called a **claw** in non-hooved animals. The combining form for claw is **onych/o**. In cats, a surgical procedure to remove the claws is commonly called a **declaw**; the medical term is **onychectomy** (ohn-ih-kehk-tō-mē).

Sesamoid (sehs-ah-moyd) **bones** are small nodular bones embedded in a tendon or joint capsule. There

FIGURE 3–13 Horse skeleton.

(a)

Carpal bones   Metacarpals

Phalanges

(b)

Proximal phalanx (P1)

Middle phalanx (P2)

Distal phalanx (P3)

FIGURE 3–14 (a) Radiograph of the carpal bones, metacarpals, and phalanges of a dog. (b) Radiograph of the foot of a horse. The phalanges of a horse's foot are numbered proximal to distal. The proximal phalanx is commonly known as P1, or the long pastern bone. The middle phalanx is commonly known as P2, or the short pastern bone. The distal phalanx is commonly known as P3, or the coffin bone.

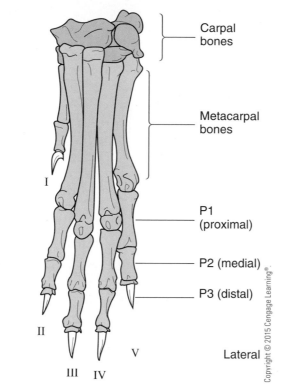

FIGURE 3–15 Digits versus phalanges. The digits of the hoof or paw are numbered medially to laterally. The phalanges are numbered from proximal to distal.

are multiple sesamoid bones in animals. Some sesamoid bones also have a common name. The **navicular bone** of horses is the common name for the sesamoid bone located inside the hoof on the palmar or plantar surface of P3.

To the Back

The bones of the rear limb include the pelvis, femur, tibia, fibula, tarsals, metatarsals, and phalanges.

The **pelvis** (pehl-vihs), or **hip**, consists of three pairs of bones: ilium, ischium, and pubis. The **ilium** (ihl-ē-uhm) is the largest pair and is blade-shaped. The ilium articulates with the sacrum to form the **sacroiliac** (sā-krō-ihl-ē-ahck) **joint**. The **ischium** (ihs-kē-uhm) is the caudal pair of bones. The **pubis** (pehw-bihs) is the ventral pair of bones that are fused on midline by a cartilaginous joint called the **pubic symphysis** (pehw-bihck sihm-fih-sihs). The **acetabulum** (ahs-eh-tahb-yoo-luhm) is the large socket of the pelvic bone that forms where the three bones meet. The acetabulum forms the ball-and-socket joint with the femur (Figure 3–16).

The **femur** (fē-muhr), or **thigh bone**, is the proximal long bone of the rear leg. The head of the femur articulates proximally with the acetabulum. The **femoral** (fehm-ohr-ahl) **head**, or head of the femur,

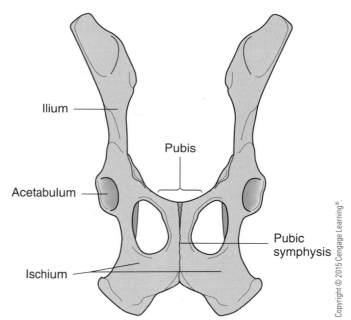

FIGURE 3–16 Parts of the pelvis.

One way to remember the order of the ilium and ischium is that they follow alphabetical order from cranial to caudal. In cattle, the points of the ilium and ischium are called **hooks** and **pins**, respectively. They too follow alphabetical order from cranial to caudal.

Sound Alikes

Ileum and *ilium* are pronounced the same, yet they have different meanings. Ileum is the distal or aboral (the end opposite the mouth) part of the small intestine, and ilium is part of the pelvic bone. One way to keep these spellings straight is to remember that ileum has an *e* in it, as in *eating* and *enter/o*, which involve the digestive tract. Ilium and pelvis both have an *i* in them.

is connected to a narrow area, which is called the **femoral neck**. Other structures found on the femur are the **trochanters** (trō-kahn-tehrs), which means large, flat, broad projections on a bone, and **condyles** (kohn-dīlz), which means rounded projection (Figure 3–17).

The **patella** (pah-tehl-ah) is a large sesamoid bone in the rear limb. In people, it is called the kneecap and

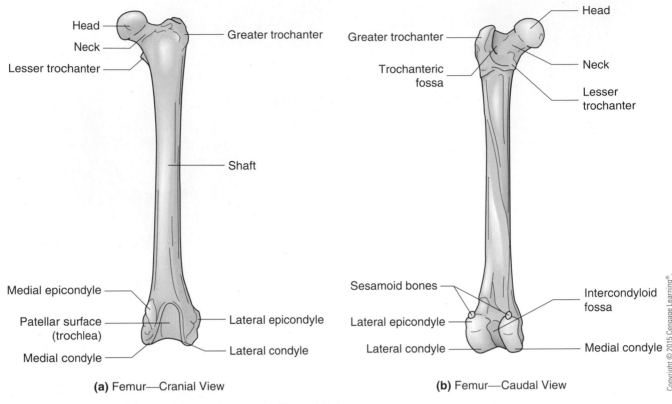

FIGURE 3–17 Femur of the cat: (a) cranial view and (b) caudal view.

Knee Deep in Trouble

The term *knee* can be a confusing term in veterinary medicine. Laypeople may use the term *knee* to refer to the stifle joint of dogs and cats; however, in large animals, *knee* is used to describe the carpal joint. Most people in veterinary medicine use the term *stifle* for the joint located in the rear leg between the femur and tibia/fibula and reserve *knee* for the carpal joint in large animals.

the joint is known as the knee. *Knee* is not a good term to use to describe the joint between the femur and tibia in animals because in large animals, *knee* is commonly used to describe the carpus. In animals, the joint that houses the patella is called the **stifle** (stī-fuhl) **joint.** Another sesamoid bone in the rear limb of some animals is the **popliteal** (pohp-liht-ē-ahl). The popliteal sesamoid is located on the caudal surface of the stifle.

The tibia and fibula are the distal long bones of the rear limb. The **tibia** (tihb-ē-ah) is the larger and more weight-bearing bone of the two. The **fibula** (fihb-yoo-lah) is a long, slender bone. Some animals do not have a fibula that extends to the distal end, whereas other animals have the tibia and fibula fused. The area of the rear limb between the stifle and hock is called the **crus** (kruhs).

The tarsal bones are irregularly shaped bones found in the area known as the ankle in people. In small animals, this joint is called the **tarsus** (tahr-suhs), and in large animals, it is called the **hock** (hohck). One of the tarsal bones is the **talus** (tahl-uhs). The talus is the shorter, medial tarsal bone located in the proximal row of tarsal bones. Talus and tarsus both begin with *t* and sound similar, which makes associating them together easy. The long, lateral tarsal bone located in the proximal row of tarsal bones is the **calcaneus** (kahl-k ā-nē-uhs). Calcaneus and carpus both begin with *c* and sound similar, but they are not located in the same area. Calcaneus was named because it reminded someone of a piece of chalk, which consists mainly of calcium. The metatarsals are bones found distal to the tarsus (**meta-** = beyond). The metatarsals are numbered and have similar names as the metacarpals.

The phalanges are the bones of the digit (both front and rear limbs). The terminology used for the phalanges in the front limb is also used for the rear limb.

STRUCTURAL SUPPORT

Bones are not structurally smooth and often have bumps or grooves or ridges (Figure 3–18a, b, and c). Bones also have depressions and extensions. These are described in Table 3–3. All of these structures have a medical term that describes them. Knowing what these descriptive terms mean can make learning bone parts easier.

aperture (ahp-ər-chər) = opening.
canal (kahn-ahl) = tunnel.
condyle (kohn-dīl) = rounded projection (that articulates with another bone).
crest (krehst) = high projection or border projection.
crista (krihs-tah) = ridge.
dens (dehnz) = toothlike structure.
eminence (ehm-ih-nehns) = surface projection.
facet (fahs-eht) = smooth area.
fissure (fihz-uhr) = deep cleft.
foramen (fō-rā-mehn) = hole.
fossa (fohs-ah) = trench or hollow depressed area.
fovea (fō-vē-ah) = small pit.

head = major protrusion.
lamina (lahm-ih-nah) = thin, flat plate.
line = low projection or ridge.
malleolus (mah-lē-ō-luhs) = rounded projection (distal end of tibia and fibula).
meatus (mē-ā-tuhs) = passage or opening.
process (proh-sehs) = projection.
protuberance (prō-too-bər-ahns) = projecting part.
ramus (rā-muhs) = branch or smaller structure given off by a larger structure.
sinus (sīn-uhs) = space or cavity.
spine (spīn) = sharp projection.
sulcus (suhl-kuhs) = groove.
suture (soo-chuhr) = seam.
trochanter (trō-kahn-tehr) = broad, flat projection (on femur).
trochlea (trōck-lē-ah) = pulley-shaped structure in which other structures pass or articulate.
tubercle (too-behr-kuhl) = small, rounded surface projection.
tuberosity (too-beh-rohs-ih-tē) = projecting part.

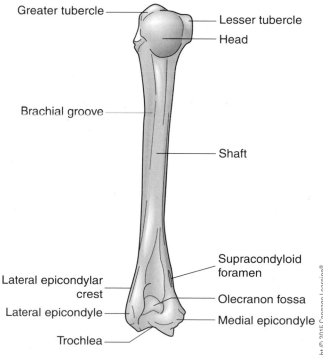

Humerus   Cranial View

Lesser tubercle
Greater tubercle
Teres minor tuberosity
Crest of the greater tubercle
Tricipital line
Deltoid tuberosity
Supracondyloid foramen
Coronoid fossa
Radial fossa
Medial epicondyle
Lateral epicondyle
Trochlea

Humerus   Caudal View

Greater tubercle
Lesser tubercle
Head
Brachial groove
Shaft
Supracondyloid foramen
Lateral epicondylar crest
Olecranon fossa
Lateral epicondyle
Medial epicondyle
Trochlea

(a)

FIGURE 3–18 (a) Cranial and caudal view of the humerus of the cat; (b) radius and ulna of the cat; and (c) ventral view of the tibia and fibula of the cat.

Radius

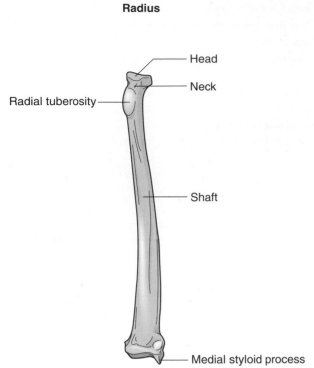

- Head
- Neck
- Radial tuberosity
- Shaft
- Medial styloid process

Ulna

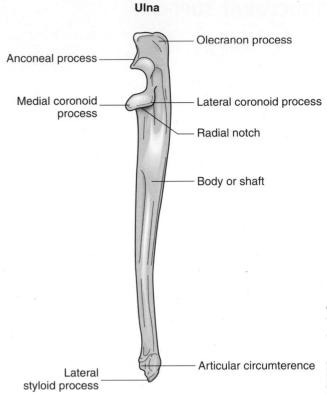

- Olecranon process
- Anconeal process
- Medial coronoid process
- Lateral coronoid process
- Radial notch
- Body or shaft
- Lateral styloid process
- Articular circumterence

(b)

Tibia—Ventral View

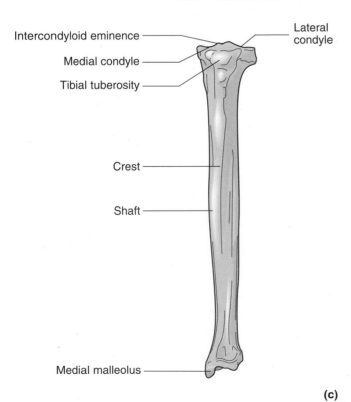

- Intercondyloid eminence
- Lateral condyle
- Medial condyle
- Tibial tuberosity
- Crest
- Shaft
- Medial malleolus

Fibula—Ventral View

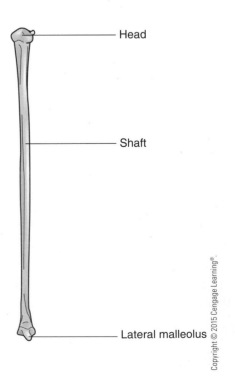

- Head
- Shaft
- Lateral malleolus

(c)

FIGURE 3–18 (continued)

TABLE 3–3 Bone Extensions versus Depressions

Bone Extensions	
• enlargements, usually at the ends of bones, where muscles, tendons, and other bones are attached	condyle crest crista dens eminence facet head lamina line malleolus process protuberance ramus spine suture trochanter trochlea tubercle tuberosity

Bone Depressions	
• depressions in bone allow bones to attach to each other or serve as passageways for blood vessels and nerves	aperture canal fissure foramen fossa fovea meatus sinus sulcus

TEST ME: SKELETAL SYSTEM

Diagnostic procedures performed on the skeletal system include the following:

- **arthrocentesis** (ahr-thrō-sehn-tē-sihs) = surgical puncture of a joint to remove fluid for analysis.
- **arthrography** (ahr-throhg-rah-fē) = injection of a joint with contrast material for radiographic examination.
- **arthroscopy** (ahr-throhs-kō-pē) = visual examination of the joint using a fiberoptic scope (called an arthroscope when used in the joint).
- **densitometer** (dehn-sih-tohm-eh-ər) = device that measures bone density using light and X-rays.

- **goniometer** (gō-nē-ohm-eh-ər) = instrument that measures angles or range of motion in a joint (**goni/o** means angle or seed).
- **radiology** (rā-dē-ohl-ō-jē) = study of internal body structures after exposure to ionizing radiation; used to detect fractures and diseases of bones (Figure 3–19).

PATHOLOGY: SKELETAL SYSTEM

Pathologic conditions of the skeletal system include the following:

- **ankylosis** (ahng-kih-lō-sihs) = loss of joint mobility caused by disease, injury, or surgery; **ankyl/o** means bent or stiff.
- **arthralgia** (ahr-thrahl-jē-ah) = joint pain.

Acetabulum

FIGURE 3–19 Radiograph of the canine hip with hip dysplasia. Note the shallow acetabulum present in this dog's pelvis.

- **arthritis** (ahr-thrī-tihs) = inflammatory condition of joints.
- **arthrodynia** (ahr-thrō-dihn-ē-ah) = joint pain.
- **arthropathy** (ahr-throhp-ah-thē) = joint disease.
- **bursitis** (bər-sī-tihs) = inflammation of the bursa.
- **chondromalacia** (kohn-drō-mah-lā-shē-ah) = abnormal cartilage softening.
- **chondropathy** (kohn-drohp-ah-thē) = cartilage disease.
- **discospondylitis** (dihs-kō-spohn-dih-lī-tihs) = inflammation of the intervertebral disc and vertebrae.
- **epiphysitis** (eh-pihf-ih-sī-tihs) = inflammation of the growth plate.
- **exostosis** (ehck-sohs-tō-sihs) = benign growth on the bone surface.
- **gouty arthritis** (gow-tē ahr-thrī-tihs) or **gout** = joint inflammation associated with the formation of uric acid crystals in the joint (seen more commonly in birds).
- **hip dysplasia** (dihs-plā-zē-ah) = abnormal development of the pelvic joint causing the head of the femur and the acetabulum not to be aligned properly; most commonly seen in large breed dogs (Figure 3–19).

- **intervertebral disc disease** (ihn-tər-vər-tē-brahl dihsk dih-zēz) = rupture or protrusion of the cushioning disc found between the vertebrae that results in pressure on the spinal cord or spinal nerve roots; also called **herniated disc**, **ruptured disc**, or **IVDD** (Figure 3–20).
- **kyphosis** (kī-fō-sihs) = dorsal curvature of the spine; also called hunchback.
- **Legg-Calvé-Perthes disease** (lehg cah-veh pər-thehz dih-zēz) = idiopathic necrosis of the femoral head and neck of small breed dogs; also called avascular necrosis of the femoral head and neck.
- **lordosis** (lōr-dō-sihs) = position in which the vertebral column is abnormally curved ventrally; seen in cats in heat; commonly called **swayback**.
- **luxation** (luhck-sā-shuhn) = dislocation or displacement of a bone from its joint.
- **myeloma** (mī-eh-lō-mah) = tumor composed of cells derived from hematopoietic tissues of bone marrow.
- **ostealgia** (ohs-tē-ahl-jē-ah) = bone pain.
- **osteitis** (ohs-tē-ī-tihs) = inflammation of bone.
- **osteoarthritis** (ohs-tē-ō-ahr-thrī-tihs) = degenerative joint disease commonly associated with aging or wear and tear on the joints; also called **degenerative joint disease**, or **DJD**.
- **osteochondrosis** (ohs-tē-ō-kohn-drō-sihs) = degeneration or necrosis of bone and cartilage followed by regeneration or recalcification.
- **osteochondrosis dissecans** (ohs-tē-ō-kohn-drō-sihs dehs-ih-kahns) = degeneration or necrosis of bone and cartilage followed by regeneration or recalcification with dissecting flap of articular cartilage and some inflammatory joint changes; detached

FIGURE 3–20 Myelogram showing intervertebral disc disease. Dye injected into the spinal column shows compression of the spinal cord.

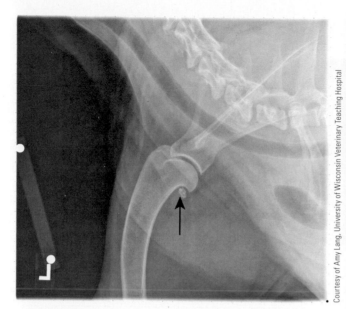

Courtesy of Amy Lang, University of Wisconsin Veterinary Teaching Hospital

FIGURE 3–21 Radiograph of dog's shoulder with an osteophyte at the proximal end of the humerus.

Courtesy of Amy Lang, University of Wisconsin Veterinary Teaching Hospital

FIGURE 3–22 Radiograph of horse's leg with a sequestrum.

pieces of articular cartilage are called **joint mice** or **osteophytes** (ohs-tē-ō-fītz) (Figure 3–21).

- **osteomalacia** (ohs-tē-ō-mah-lā-shē-ah) = abnormal softening of bone.
- **osteomyelitis** (ohs-tē-ō-mī-eh-lī-tihs) = inflammation of bone and bone marrow.
- **osteonecrosis** (ohs-te-o-neh-kro-sihs) = death of bone tissue.
- **osteoporosis** (ohs-tē-ō-neh-krō-sihs) = abnormal condition of marked loss of bone density and an increase in bone porosity.
- **osteosclerosis** (ohs-tē-ō-skleh-rō-sihs) = abnormal hardening of bone.
- **periostitis** (pehr-ē-ohs-tī-tihs) = inflammation of the fibrous tissue that forms the outermost covering of bone.
- **rheumatoid arthritis** (roo-mah-toyd ahr-thrī-tihs) = autoimmune disorder of the connective tissues and joints; abbreviated RA.
- **sequestrum** (sē-kwehs-truhm) = piece of dead bone that is partially or fully detached from the adjacent healthy bone (Figure 3–22).
- **spondylitis** (spohn-dih-lī-tihs) = inflammation of the vertebrae.
- **spondylosis** (spohn-dih-lō-sihs) = any degenerative disorder of the vertebrae.
- **spondylosis deformans** (spohn-dih-lō-sihs dē-fōr- mahnz) = chronic degeneration of the articular processes and the development of bony outgrowths around the ventral edge of the vertebrae (Figure 3–23).

- **spur** (spuhr) = bony projection growing out of a bone.
- **subluxation** (suhb-luhck-sā-shuhn) = partial dislocation or displacement of a bone from its joint (Figure 3–24).
- **synovitis** (sihn-o-vī-tihs) = inflammation of the synovial membrane of joints.

Fracture Terminology

See Figure 3–25.

- **avulsion** (ā-vuhl-shuhn) **fracture** = broken bone in which the site of muscle, tendon, or ligament insertion is detached by a forceful pull.
- **callus** (kahl-uhs) = bulging deposit around the area of a bone fracture that may eventually become bone.
- **closed fracture** = broken bone in which there is no open wound in the skin; also known as a **simple fracture**.
- **comminuted** (kohm-ih-noot-ehd) **fracture** = broken bone that is splintered or crushed into multiple pieces (Figure 3–26).

Lumbar vertebrae

Pelvis

Bony outgrowth

Courtesy of Eli Larson, DVM

FIGURE 3–23 Radiograph of the lumbar spine of a dog. This dog has an age-related change called spondylosis deformans. In this disease, bone spurs (bony outgrowths) are formed that eventually can bridge between vertebrae.

Courtesy of Anne E. Chauvet, DVM, Diplomate ACVIM–Neurology, University of Wisconsin School of Veterinary Medicine.

FIGURE 3–24 Radiograph of atlas-axis subluxation in a dog.

- **compression** (kohm-prehs-shuhn) **fracture** = broken bone produced when the bones are pressed together.
- **crepitation** (krehp-ih-tā-shuhn) = cracking sensation that is felt and heard when broken bones move together; also referred to as **crepitus** (krehp-ih-tuhs).

- **displaced** (dihs-plāsd) **fracture** = bone fracture parts are out of line.
- **fracture** (frahck-shər) = broken bone.
- **greenstick fracture** = bone that is broken only on one side and the other side is bent; also called **incomplete fracture.**
- **immobilization** (ihm-mō-bihl-ih-zā-shuhn) = act of holding, suturing, or fastening a bone in a fixed position, usually with a bandage or cast.
- **manipulation** (mahn-ihp-yoo-lā-shuhn) = attempted realignment of the bone involved in a fracture or dislocation; also known as **reduction.**
- **oblique** (ō-blēck) **fracture** = broken bone that has an angular break diagonal to the long axis.
- **open fracture** = broken bone in which there is an open wound in the skin; also known as a **compound fracture.**
- **pathologic fracture** = broken bone in an area of bone weakened by disease.

BIOLOGY BONUS

When a bone is fractured, blood vessels bleed into the fracture site and form a hematoma. After a few days, osteoblasts move into the area, start to produce new bone matrix, and develop into osteocytes. Eventually the new bone fuses together the segments of the fracture. Sometimes fractures do not heal properly, resulting in **malunion** (mahl-yoon-uhn) in which two bony ends of the fracture fail to heal together correctly or **nonunion** in which there is total failure of fracture healing.

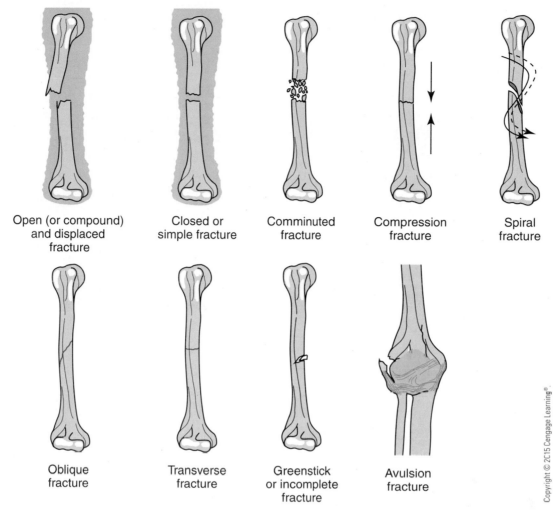

Open (or compound) and displaced fracture

Closed or simple fracture

Comminuted fracture

Compression fracture

Spiral fracture

Oblique fracture

Transverse fracture

Greenstick or incomplete fracture

Avulsion fracture

FIGURE 3–25 Fracture types.

FIGURE 3–26 Comminuted fracture of the radius and ulna in a cat.

In or Out

Bones may abnormally bend in or bend out. Medical terms for this condition in bones are *varus* (vahr-uhs) and *valgus* (vahl-guhs). *Valgus* means bend out (think bend laterally; both *valgus* and *lateral* have an *l*), and *varus* means bend in.

- **physeal** (fī-sē-ahl) **fracture** = bone that is broken at the epiphyseal line or growth plate; these fractures are further categorized as **Salter-Harris I–V fractures**.
- **spiral** (spī-rahl) **fracture** = broken bone in which the bone is twisted apart or spiraled apart.

- **stress fracture** = broken bone caused by repetitive, local stress on a bone.
- **transverse** (trahnz-vərs) **fracture** = broken bone that is broken at right angles to its axis or straight across the bone.

PROCEDURES: SKELETAL SYSTEM

Procedures performed on the skeletal system (Figure 3–27) include the following:

- **amputation** (ahmp-yoo-tā-shuhn) = removal of some or all of a body part.

- **arthrodesis** (ahr-thrō-dē-sihs) = fusion of a joint or the spinal vertebrae by surgical means; **-desis** means to bind or surgical fixation of a bone or joint.
- **chemonucleolysis** (kē-mō-nū-klē-ō-lī-sihs) = process of dissolving part of the center of an intervertebral disc by injecting a foreign substance.
- **craniotomy** (krā-nē-oht-ō-mē) = surgical incision or opening into the skull.
- **external fixation** = alignment of bone maintained by immobilizing the bone near the fracture through the use of casts, splints, or external fixators (rods or pins).

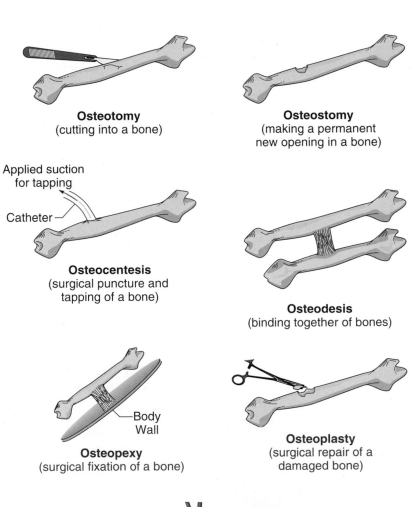

Osteotomy
(cutting into a bone)

Osteostomy
(making a permanent new opening in a bone)

Applied suction for tapping

Catheter

Osteocentesis
(surgical puncture and tapping of a bone)

Osteodesis
(binding together of bones)

Body Wall

Osteopexy
(surgical fixation of a bone)

Osteoplasty
(surgical repair of a damaged bone)

Ostectomy
(removal of a bone)

FIGURE 3–27 Surgical procedures of bone.

BIOLOGY BONUS

Body systems of animals do not work in isolation from other body systems. For an animal to walk, the skeletal, muscular, and nervous systems must all work together to coordinate the movement of an individual limb with the others so that a horse can gallop or a gerbil can run around in an exercise wheel.

- **internal fixation** = alignment of bone maintained by immobilizing the bone directly at the fracture site through the use of wires, screws, pins, or plates.
- **laminectomy** (lahm-ih-nehck-tō-mē) = surgical removal of the dorsal arch of a vertebra.
- **onychectomy** (ohn-ih-kehk-tō-mē) = surgical removal of a claw.
- **ostectomy** (ohs-tehck-tō-mē) = surgical removal of bone.
- **osteocentesis** (ohs-tē-ō-sehn-tē-sihs) = surgical puncture of a bone.
- **osteodesis** (ohs-tē-ō-dē-sihs) = fusion of bones.
- **osteopexy** (ohs-tē-ō-pehck-sē) = surgical fixation of a bone to the body wall.
- **osteoplasty** (ohs-tē-ō-plahs-tē) = surgical repair of bone.
- **osteostomy** (ohs-tē-ohs-tō-mē) = surgical creation of a permanent new opening in bone.
- **osteotomy** (ohs-tē-oht-ō-mē) – surgical incision or sectioning of bone.
- **trephination** (treh-fih-nā-shuhn) – process of cutting a hole into a bone using a **trephine** (trē-fīn) (circular sawlike instrument used to remove bone or tissue).

Chiropractic Care

Sometimes standard veterinary care can be supplemented with complementary care to help diagnose and treat diseases. Chiropractic care is an example of complementary care that can help manage musculoskeletal and neurologic components of certain injuries and disease conditions in animals. Chiropractic is derived from the Greek word parts *cheir* (hand) and *praktike* (business or practice). It is an emerging subspecialization in veterinary medicine that focuses on the structure and function of the musculoskeletal and nervous systems and how their relationship affects health. Spinal manipulation, manual therapy, and other holistic and conservative techniques are all parts of chiropractic care that address the relationship between the spinal column and nervous system and the crucial role of this relationship to the maintenance of overall animal health. The goals of chiropractic treatment are to restore normal joint motion, stimulate nerve reflexes, and reduce pain and abnormal muscle tone. Terms used in veterinary chiropractic include the following:

- **adjustment** = the application of a controlled, rapid thrust of precise direction and depth delivered to a specific contact point; also called **manipulation**.
- **basic movements of the vertebral column**
  - **extension** = ventral arching or straightening of the spine
  - **flexion** = dorsal arching of the spine
  - **lateral flexion** = bowing or bending to each side (right or left)
  - **rotation** = twisting to the right or left
- **fixation** = when a joint becomes immobilized in a fixed position that may occur at rest or during a normal movement.
- **malarticulations** = the connections between bones do not line up correctly.
- **misalignment** = a motion segment in which alignment, movement integrity, and/or physiological function are altered, although contact between joint surfaces remains intact; chiropractic use of this term typically implies vertebral misalignment; also called **subluxation.** Misalignment of the spine allegedly interferes with nerve signals from the brain.
- **motion segment** = basic functional unit of the spine; consists of two adjacent articulating surfaces and the connecting tissues binding them to each other. A motion segment includes two vertebrae, the intervertebral disc, nerves, arteries, and soft tissue structures (muscles, ligaments, tendons), which participate in movement.
- **range of motion** = amount of movement that occurs at a particular joint or region of the body; abbreviated ROM.
- **spinal manipulation** = manipulation of the spine to restore its normal motion and alignment. Spinal manipulation relieves any restrictions on the nerves or spinal cord that could impair their function and function of the organs they innervate; typically used to treat animals with sore necks or backs.

FUNCTIONS OF THE MUSCULAR SYSTEM

Muscles are organs that contract to produce movement. Muscles make movement possible. One type of movement is **ambulation** (ahm-bū-lā-shuhn), or walking, running, or otherwise moving from one place to another. Another type of movement is contraction of organs or tissues that result in normal functioning of the body. For example, contraction of sections of the gastrointestinal tract allows food to move through the digestive system, and contraction of vessels allows movement of fluids such as blood. Movement also results in heat generation to keep the body warm.

STRUCTURES OF THE MUSCULAR SYSTEM

The muscular system is composed of specialized cells called muscle fibers whose predominant function is contractibility.

Muscle Fibers

Muscles are made up of long, slender cells called muscle fibers. Each muscle consists of a group of muscle fibers encased in a fibrous sheath. The combining forms for muscle are **my/o** and less commonly **myos/o;** the combining forms for fibrous tissue are **fibr/o** and **fibros/o**.

Three types of muscle cells are based on their appearance and function. The three types are skeletal, smooth, and cardiac (Table 3–4 and Figure 3–28).

Making Another Connection

Like the skeletal system, the muscular system also contains various forms of connective tissue. The connective tissues that support the muscular system are the fascia, tendons, and aponeuroses.

Fascia

Fascia (fahsh-ē-ah) is a sheet of fibrous connective tissue that covers, supports, and separates muscles. The combining forms for fascia are **fasci/o** and **fasc/i.**

Tendons

A **tendon** (tehn-dohn) is a narrow band of connective tissue that attaches muscle to bone. The combining forms for tendon are **tend/o, tendin/o,** and **ten/o.** Remember to make the distinction between a tendon and a ligament (Figure 3–29). Tendons connect muscles to bones or other structures. One example is the linea alba. *Linea alba* means white line in Latin. The **linea alba** is a fibrous band of connective tissue on the ventral abdominal wall that is the median attachment of the abdominal muscles (Figure 3–30).

Aponeurosis

An **aponeurosis** (ahp-ō-nū-rō-sihs) is a fibrous sheet that provides attachment to muscular fibers and is a means of origin or insertion of a flat muscle. The combining form for aponeurosis is **aponeur/o,** and the plural form of aponeurosis is **aponeuroses.**

TABLE 3–4 Muscle Types

Muscle Type	Description	Microscopic Appearance	Function
Skeletal	**striated** (strī-āt-ehd) **voluntary**	long, cylindrical, multinucleated cells with dark and light bands to create a striated or striped look; the combining form for striated muscle is **rhabdomy/o** (rahb-dō-mī-ō)	attach bones to the body and make motion possible
Smooth	**nonstriated** (also called **unstriated**) **involuntary** **visceral**	spindle-shaped without stripes or striations; the combining form for smooth muscle is **leiomy/o** (lī-ō-mī-ō)	produce slow contractions to allow unconscious functioning of internal organs
Cardiac	**striated** **involuntary**	elongated, branched cells that lie parallel to each other and have dark and light bands; connected by intercalated disks	involuntary contraction of heart muscle

FIGURE 3–28 The three types of muscle tissues. (a) Skeletal muscle from the tongue of a cat. The photomicrograph shows cells cut along their length and in cross section. (b) Cardiac muscle from the heart of a goat. Note the intercalated disks that connect the cells. These structures allow the cells to act together with an organized contraction. (c) Smooth muscle from the colon of a horse. Note that smooth muscle lacks the striations found in cardiac and skeletal muscle.

Courtesy of William J. Bacha, PhD, and Linda M. Bacha, MS, VMD

Copyright © 2015 Cengage Learning®.

FIGURE 3–29 Tendons versus ligaments. Tendons are dense fibrous connective tissues that connect muscle to bone; ligaments are dense fibrous connective tissues that connect bone to bone.

FIGURE 3–30 The linea alba of a cat.

SHOW SOME MUSCLE

One of the functions of muscle is to allow movement. The combining form **kinesi/o** and the suffix **-kinesis** mean movement. **Kinesiology** (kih-nē-sē-ohl-ō-jē) is the study of movement.

Some muscles are arranged in pairs that work opposite or against each other. One muscle pair may produce movement in one direction, whereas another muscle pair produces movement in the opposite direction. Muscles that work against or opposite each other are called **antagonistic** (ahn-tā-gohn-ihs-tihck).

Anti- means against; **agon** means struggle. Other muscles of the body are arranged to work with another muscle. **Synergists** (sihn-ər-jihsts) are muscles that contract at the same time as another muscle to help movement or support movement (synergists are also called **agonistic** (ā-gohn-ihs-tihck)). **Syn-** means together; **erg** means work. Antagonistic muscles work by producing contraction of one pair of muscles while the other pair relaxes. **Contraction** (kohn-trahck-shuhn) means tightening. **Relaxation** (rē-lahk-sā-shuhn) means lessening of tension. During contraction, the muscle becomes shorter and thicker. During relaxation, the muscle returns to its original shape. Muscles are signaled to contract or relax by nerve impulses. A **neuromuscular** (nū-rō-muhs-kū-lahr) **junction** is the point at which nerve endings come in contact with the muscle cells.

Tonus (tō-nuhs), or muscle tone, is balanced muscle tension. The combining form for tone, tension, or stretching is **ton/o**.

WHAT'S IN A NAME?

At first glance, the names of muscles and the task of learning them may seem impossible. Dividing muscle names into their basic components or taking a closer look at how the names were derived may help in learning their names and functions.

Beginning and Ending

Muscles are formed by where they begin and where they end. Terms used to denote these two locations are **muscle origin** (ōr-ih-jihn) and **muscle insertion** (ihn-sihr-shuhn), respectively. Muscle origin is the place where a muscle begins, or originates, and is the more fixed attachment or the end of the muscle closest to the midline. Muscle insertion is the place where a muscle ends, or inserts, and is the more movable end or portion of the muscle farthest from the midline. Muscles may be named according to where they originate and end. Brachioradialis muscles are connected to the brachium (humerus) and to the radius.

How Do They Move?

Muscles move in a variety of ranges. **Range of motion** is a term used to describe the types of muscle movements. Range of motion is sometimes abbreviated ROM. Muscles may be named for the manner in which they move, as follows:

- **abductor** (ahb-duhck-tər) = muscle that moves a part away from the midline.

- **adductor** (ahd-duhck-tər) = muscle that moves a part toward the midline.
- **flexor** (flehck-sər) = muscle that bends a limb at its joint or decreases the joint angle.
- **extensor** (ehcks-tehn-sər) = muscle that straightens a limb or increases the joint angle.
- **levator** (lē-vā-tər) = muscle that raises or elevates a part.
- **depressor** (dē-prehs-sər) = muscle that lowers or depresses a part.
- **rotator** (rō-tā-tər) = muscle that turns a body part on its axis.
- **supinator** (soo-pih-nā-tər) = muscle that rotates the palmar or plantar surface upward.
- **pronator** (prō-nā-tər) = muscle that rotates the palmar or plantar surface downward.

Where Are They?

Muscles also are named for their location on the body or organ they are near. **Pectoral muscles** are located on the chest (**pector** = chest). Muscles may also be named for their location in relation to something else. **Epaxial** (ehp-ahcks-ē-ahl) **muscles** are located above the pelvic axis (**epi-** = above, **axis** = line about which rotation occurs), **intercostal muscles** are located between the ribs (**inter-** = between, **cost/o** = rib), **infraspinatus muscles** are located beneath the spine of the scapula (**infra-** = beneath or below), and **supraspinatus muscles** are located above the spine of the scapula (**supra-** = above). Muscle names also may indicate their location within a group, such as **inferior** (below or deep), **medius** (middle), and **superior** (above). Other terms indicating depth of muscles are **externus** (outer) and **internus** (inner). **Orbicularis** are muscles surrounding another structure.

Which Way Do They Go?

Muscles may also be named according to the direction of the muscle fibers.

- **Rectus** (rehck-tuhs) means straight. Rectus muscles align with the vertical axis of the body.
- **Oblique** (ō-blēck) means slanted. Oblique muscles slant outward away from the midline.
- **Transverse** means crosswise. Transverse muscles form crosswise to the midline.
- **Sphincter** means tight band. Sphincter muscles are ringlike and constrict the opening of a passageway.

How Many Parts Are There?

Some muscles are named for the number of divisions they have.

- **Biceps** (bī-sehpz) generally have two divisions (heads); **bi-** means two.
- **Triceps** (trī-sehpz) generally have three divisions (heads); **tri-** means three.
- **Quadriceps** (kwohd-rih-sehps) generally have four divisions (heads); **quadri-** means four.

Some muscles are not paired or divided. **Azygous** (ah-zī-guhs) means not paired (**a-** means without; **zygot/o** means joined).

How Big Is It?

Muscles may also be named for their size. Muscles may be small (**minimus**) or large (**maximus** or **vastus**), broad (**latissimus**) or narrow (**longissimus** or **gracilis**). **Major** and **minor** also are used to describe larger and smaller parts, respectively.

How Is It Shaped?

Some muscles are shaped like familiar objects and have been named accordingly.

- **Deltoid** (dehl-toyd) muscles look like the Greek letter delta (Δ).
- **Quadratus** (kwohd-rā-tuhs) muscles are square or four-sided.
- **Rhomboideus** (rohm-boy-dē-uhs) muscles are diamond-shaped. (Rhomboid is a four-sided figure that may have unequal adjoining sides but equal opposite sides.)
- **Scalenus** (skā-lehn-uhs) muscles are unequally three-sided. (*Skalenos* is Greek for uneven.)
- **Serratus** (sihr-ā-tuhs) muscles are saw-toothed. (*Serratus* is Latin for notched.)
- **Teres** (tər-ēz) muscles are cylindrical. (*Teres* is Latin for smooth and round or cylindrical.)

No Rules

Sometimes muscles are named for what they look like or how they relate to something else. **Sartorius** (sahr-tōr-ē-uhs) **muscle** (one muscle of the thigh area) is named because this muscle flexes and adducts the leg of a human to that position assumed by a tailor sitting cross-legged at work (*sartorius* means tailor). The **gemellus** (jeh-mehl-uhs) is named because it is a twinned muscle (*gemellus* means twin). The **gastrocnemius** (gahs-trohck-nē-mē-uhs) **muscle** is the leg muscle that resembles the shape of the stomach (**gastr/o** means stomach, **kneme** means leg).

TEST ME: MUSCULAR SYSTEM

A diagnostic procedure performed on the muscular system is as follows:

- **electromyography** (ē-lehck-trō-mī-ohg-rah-fē) = process of recording the electrical activity of the muscle cells near the recording electrodes; abbreviated EMG. An **electromyogram** (ē-lehck-trō-mī-ō-grahm) is the record of the strength of muscle contraction caused by electrical stimulation.

PATHOLOGY: MUSCULAR SYSTEM

Pathologic conditions of the muscular system include the following:

- **adhesion** (ahd-hē-shuhn) = band of fibers that hold structures together in an abnormal fashion.
- **ataxia** (ā-tahck-sē-ah) = lack of voluntary control of muscle movement; "wobbliness"; **tax/o** is the combining form for coordination or order.
- **atonic** (ā-tohn-ihck) = lacking muscle tone.
- **atrophy** (ah-trō-fē) = decrease in size or complete wasting of an organ or tissue or cell.
- **dystrophy** (dihs-trō-fē) = defective growth.
- **fasciitis** (fahs-ē-ī-tihs) = inflammation of the sheet of fibrous connective tissue that covers, supports, and separates muscles (fascia).
- **fibroma** (fī-brō-mah) = tumor composed of fully developed connective tissue; also called **fibroid** (fī-broyd).
- **hernia** (hər-nē-ah) = protrusion of a body part through tissues that normally contain it.

- **laxity** (lahcks-ih-tē) = looseness.
- **leiomyoma** (lī-ō-mī-ō-mah) = benign tumor of smooth muscle.
- **leiomyositis** (lī-ō-mī-ō-sī-tihs) = inflammation of smooth muscle.
- **myasthenia** (mī-ahs-thē-nē-ah) = muscle weakness; **-asthenia** means weakness.
- **myoclonus** (mī-ō-klō-nuhs) = spasm of muscle; **clon/o** means violent action (spasm).
- **myoma** (mī-ō-mah) = benign tumor of muscle.
- **myopathy** (mī-ohp-ah-thē) = abnormal condition or disease of muscle.
- **myositis** (mī-ō-sī-tihs) = inflammation of voluntary muscles.
- **myotonia** (mī-ō-tō-nē-ah) = delayed relaxation of a muscle after contraction.
- **rhabdomyoma** (rahb-dō-mī -ō) = benign tumor of striated muscle.
- **tendinitis** (tehn-dih-nī-tihs) = inflammation of the band of fibrous connective tissue that connects muscle to bone.
- **tetany** (teht-ahn-ē) = muscle spasms or twitching.

PROCEDURES: MUSCULAR SYSTEM

Procedures performed on the muscular system include the following:

- **myectomy** (mī-ehck-tō-mē) = surgical removal of muscle or part of a muscle.
- **myoplasty** (mī-ō-plahs-tē) = surgical repair of muscle.
- **myotomy** (mī-oht-ō-mē) = surgical incision into a muscle.
- **tenectomy** (teh-nehck-tō-mē) = surgical removal of a part of a tendon.
- **tenotomy** (teh-noht-ō-mē) = surgical division of a tendon.

BIOLOGY BONUS

Normal muscles contract and extend when an animal moves. In disease conditions or lack of use due to pain or paralysis, muscles can atrophy. On the other hand, overuse of muscles due to excessive exercise can cause hypertrophy, an increase in the size of an organ or tissue or cell. Some medications, such as anabolic steroids, have been used to build muscle and due to their abuse by humans are now controlled substances.

ABBREVIATIONS RELATED TO THE SKELETAL AND MUSCULAR SYSTEMS

- DJD = degenerative joint disease
- EMG = electromyography
- fx = fracture
- IVDD = intervertebral disc disease
- P1 = phalanx 1
- P2 = phalanx 2
- P3 = phalanx 3
- RA = rheumatoid arthritis
- ROM = range of motion
- TPO = triple pelvic osteotomy

REVIEW EXERCISES

Multiple Choice

Choose the correct answer.

1. A common name for the tarsus is the
   a. elbow
   b. calcaneus
   c. hock
   d. wrist

2. The _____ joints are the freely movable joints of the body.
   a. suture
   b. synovial
   c. symphysis
   d. cartilaginous

3. The correct order of the vertebral segments is
   a. cervical, thoracic, lumbar, sacral, and coccygeal
   b. cervical, lumbar, thoracic, coccygeal, and sacral
   c. thoracic, lumbar, cervical, sacral, and coccygeal
   d. thoracic, cervical, lumbar, sacral, and coccygeal

4. A _____ is a fibrous band of connective tissue that connects bone to bone.
   a. fascia
   b. tendon
   c. synovial membrane
   d. ligament

5. The acetabulum is the
   a. patella
   b. cannon bone
   c. large socket in the pelvic bone
   d. crest of the scapula

6. The three parts of the pelvis are
   a. ileum, pubis, and acetabulum
   b. ilium, pubis, and sacrum
   c. ilium, sacrum, and coccyx
   d. ilium, ischium, and pubis

7. The digits contain bones that are called
   a. carpals
   b. phalanges
   c. tarsals
   d. tarsus

8. Components of the axial skeleton include
   a. scapula, humerus, radius, ulna, and carpus
   b. skull, auditory ossicles, hyoid, vertebrae, ribs, and sternum
   c. pelvic girdle, femur, tibia, fibula, and tarsus
   d. scapula, pelvis, humerus, femur, tibia, fibula, radius, and ulna

9. Another term for growth plate is
   a. physis
   b. shaft
   c. diaphysis
   d. trophic

10. Bones of the front limb include
    a. humerus, tibia, fibula, tarsal, metatarsal, and phalanges
    b. humerus, radius, ulna, carpal, metacarpal, and phalanges
    c. femur, tibia, fibula, tarsal, metatarsal, and phalanges
    d. radius, humerus, ulna, carpal, metatarsal, and phalanges

11. *Rectus* means
    a. ringlike
    b. straight
    c. angled
    d. rotating

12. Muscles may be classified as
    a. voluntary
    b. involuntary
    c. cardiac
    d. all of the above

13. A term for when a muscle becomes shorter and thicker is
    a. relaxation
    b. contraction
    c. rotation
    d. depression

14. Levator muscles _____ a body part.
    a. decrease the angle of
    b. increase the angle of
    c. raise
    d. depress

15. A fibrous band of connective tissue that connects muscle to bone is
    a. cartilage
    b. tendon
    c. ligament
    d. aponeurosis

16. Looseness is called
    a. laxity
    b. rigidity
    c. spasm
    d. tonus

17. Protrusion of a body part through tissues that normally contain it is called a
    a. projection
    b. hernia
    c. prominence
    d. myotonia

18. A muscle that forms a tight band is called a(n)
    a. purse-string
    b. sartorius
    c. sphincter
    d. oblique

19. Surgical removal of a muscle or part of a muscle is called
    a. myositis
    b. myotomy
    c. myectomy
    d. myostomy

20. Abnormal condition or disease of muscle is called
    a. myodynia
    b. myography
    c. myasthenia
    d. myopathy

21. Which term describes the shaft of a long bone?
    a. diaphysis
    b. epiphysis
    c. endosteum
    d. periosteum

22. The tarsal bones are found in the
    a. digits
    b. wrist
    c. stifle
    d. hock

23. The manubrium is the
    a. lower jaw
    b. cranial portion of the sternum
    c. upper jaw
    d. caudal portion of the sternum

24. Which term describes the freely movable joints of the body?
    a. synarthroses
    b. amphiarthroses
    c. diarthroses
    d. articulations

25. The opening in a bone through which blood vessels, nerves, and ligaments pass is a(n)
    a. fontanel
    b. foramen
    c. meatus
    d. lamina

26. The acetabulum is commonly called the
    a. collar bone
    b. patella
    c. hip socket
    d. knee

27. Muscles located above the pelvic axis are called
    a. spinatus muscles
    b. orbicularis muscles
    c. epaxial muscles
    d. inferior muscles

28. Muscles under voluntary control are known as
    a. involuntary
    b. nonstriated
    c. skeletal
    d. visceral

29. Minimus is to maximus as
    a. longissimus is to gracilis
    b. biceps is to triceps
    c. oblique is to sphincter
    d. minor is to major

30. A muscle that bends a limb at its joint or decreases the joint angle is called a(n)
    a. flexor
    b. extensor
    c. supinator
    d. pronator

Matching

Match the bone or joint in Column I with its common name in Column II.

Column I

1. _____ P1

2. _____ P2

3. _____ P3

4. _____ tarsus

5. _____ splint bone

6. _____ fetlock joint

7. _____ pastern joint

8. _____ coffin joint

9. _____ knee

10. _____ stifle

11. _____ clavicle

12. _____ cannon bone

13. _____ dewclaw

14. _____ sternum

Column II

a. carpus in large animals

b. hock

c. coffin bone

d. short pastern

e. long pastern

f. metacarpal/metatarsal III in equine and metacarpal/metatarsal III and IV in ruminants

g. collarbone

h. metacarpo-/metatarsophalangeal joint of equine and ruminants

i. metacarpal/metatarsal II and IV in equine

j. connection between phalanx I and II in equine and ruminants

k. distal interphalangeal joint of phalanx II and III in equine and ruminants

l. variable digit depending on species; digit I in dogs, digits II and V in ruminants

m. synovial joint located between the femur and tibia; also known as the femorotibial joint

n. breastbone

Match the bone in Column I with the area it is located in Column II.

Column I	Column II
15. _____ humerus	a. distal front limb
16. _____ fibula	b. proximal front limb
17. _____ tibia	c. proximal hind limb
18. _____ ulna	d. distal hind limb
19. _____ femur	e. joint in front limb
20. _____ tarsus	f. joint in hind limb
21. _____ radius	g. distal part of front and hind limbs
22. _____ carpus	
23. _____ metacarpal	
24. _____ metatarsal	
25. _____ phalanx	

Fill in the Blanks

Write the correct term in the space provided.

1. _____ and _____ are terms used for displacement of a bone from its joint.

2. The _____ is the tough, fibrous tissue that forms the outermost covering of bone.

3. A(n) _____ is a curved fibrous cartilage found in some synovial joints.

4. Connections between two bones are called _____ or _____.

5. The caudal portion of the sternum is called the _____.

6. A(n) _____ is removal of all or part of a limb or body part.

7. A(n) _____ is a piece of dead bone that is partially or fully detached from the surrounding healthy bone.

8. Inward curvature of a bone is called _____.

9. Visual examination of the internal structure of a joint using a fiberoptic instrument is _____.

10. _____ is loss of mobility of a joint.

11. _____ is abnormal softening of cartilage.

12. A muscle that straightens a limb at a joint is called a(n) _____.

13. Straightening of a limb beyond its normal limits is called _____.

14. A(n) _____ is a band of fibers that holds structures together in an abnormal fashion.

15. Dogs with short, wide skulls are said to be _____.

16. Involuntary muscle is also called _____, _____, or
_____ .

17. Surgical removal of a claw is _____.

18. A(n) _____ is a broken bone in which there is an open wound in the skin.

19. The _____ is the fibrous band of connective tissue on the ventral abdominal wall
that is the median attachment of the abdominal muscles.

20. A(n) _____ is the place where muscle ends that is the more movable end or portion
away from the midline.

21. Inflammation of a tendon is called _____.

22. The opposite of extension is _____.

23. The opposite of contraction is _____.

24. A muscle that lowers or depresses a part is called a(n) _____.

25. The opposite of inferior is _____.

26. The term for crosswise is _____.

27. The term for slanted is _____.

28. The crackling sensation that is felt and heard when broken bones move together is called
_____.

29. Broken bones that are splintered or crushed into multiple pieces are called _____.

30. Abnormal development of the pelvic joint causing the head of the femur and the acetabulum not to be
aligned properly is called _____.

True or False

If the statement is true, write T on the line. If the statement is false, write F on the line.

1. _____Arthrodesis is fusion of a joint or the spinal vertebrae by surgical means.

2. _____A craniotomy is a surgical incision into a joint.

3. _____An osteotomy is the surgical removal of a bone.

4. _____Ataxia is lack of voluntary control of muscle movement.

5. _____An adhesion is a band of fibers that holds structures together in an abnormal fashion.

6. _____The femur is the proximal bone in the front limb.

7. _____Cervical vertebrae attach to the ribs.

8. _____The radius is a bone in the rear limb.

9. _____The clavicle is the caudal end of the shoulder bone.

10. _____The coccygeal vertebrae are found in the tail.

Spelling

Cross out any misspelled words in the following sentences and replace them with the proper spelling.

1. An obligue fracture is also known as a compound fracture. _____

2. Hip displasia is common in German Shepherd dogs._____

3. Osteomyilitis is inflammation of the bone and bone marrow. _____

4. The metitarsal bones are distal to the tarsal bones._____

5. The large socket of the pelvis is the asetabulum. _____

6. The navacular bone is the sesamoid bone located inside the hoof on the palmar or plantar surface of P3.
   _____

7. P3 is the distal phalnx. _____

8. Animals with hooves are called ungoolates. _____

9. Surgical removal of claws is an onchectomy. _____

10. Another name for the sternim is the breastbone. _____

Crossword Puzzle

Bone Extensions and Depressions

Supply the correct term in the appropriate space for the definition listed.

Across

3 projection

6 sharp projection

11 low projection or ridge

13 space or cavity

15 projecting part

16 rounded projection (that articulates with another bone)

18 passage or opening

19 tunnel

20 seam

22 toothlike structure

23 smooth area

Down

1 trench or hollow depressed

2 small pit

4 surface projection

5 rounded projection (distal end of tibia and fibula)

7 projecting part

8 opening

9 branch or smaller structure given off by a larger structure

10 high projection or border projection

12 broad, flat projection (on femur)

14 groove

17 thin, flat plate

21 major protrusion

Word Search

Define the following terms; then find each term in the puzzle.

```
G S N E S I T I S O Y M X O N
P T S I G R E N Y S I O I E O
S S E C O R P D I O H P I X I
I E L T S I N O G A T N A D T
Y S S C T M I A N I G I R O A
H A Y G O L O E T S O I O O T
P M S T B O E T A L U G N U O
O O G T N O I D S O O I H T R
R I Q U A D R U P E D I P G G
T D O G M Y O T O M Y I D O X
S T E O I T C I N O T A N O H
Y M N U U M T M A N D I B L E
D M V A L G U S M H H P O S A
M A X I L L A I H O I G O O A
D O L I C H O C E P H A L I C
```

defective growth _____

bent outward _____

things that work together _____

place where muscle begins _____

narrow, long skull _____

surgical incision into a muscle _____

inflammation of voluntary muscle _____

lack of muscle tone _____

twisting to the right or left _____

small bone embedded in tendon _____

four-footed animal _____

study of bone _____

upper bone of jaw _____

lower bone of jaw _____

animal with hooves _____

caudal part of sternum _____

things that work opposite each other _____

Label the Diagrams

Label the diagrams in Figures 3–31 and 3–32.

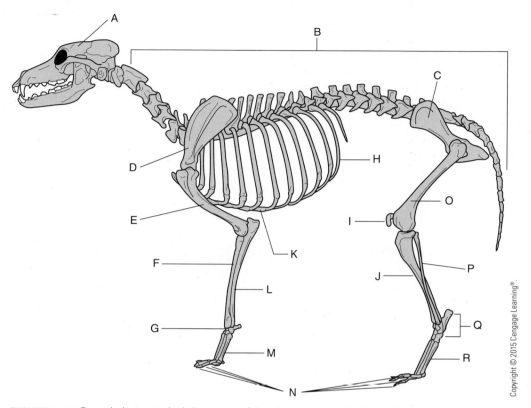

FIGURE 3–31 Dog skeleton. Label the parts of the dog skeleton indicated by the letters.

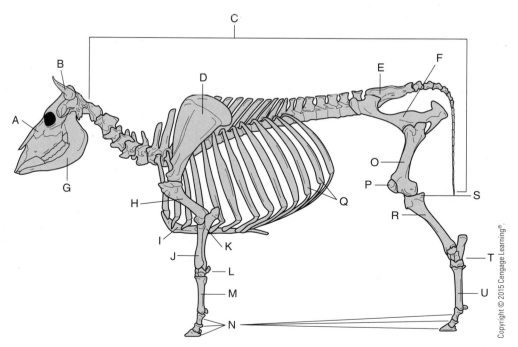

FIGURE 3–32 Bovine skeleton. Label the parts of the bovine skeleton indicated by the letters.

Case Study

After reading each case study, define the list of terms. Abbreviations can be found in Appendix A.

Blitzen, a 4-year-old M/N Whippet, presented to the veterinary clinic with left hind limb lameness after he injured tissues in his left stifle while running in the back yard. On PE, the dog's gait (the way the animal moves) consisted of short strides on the affected limb while the stifle remained in a fixed, partially flexed position. Crepitus was palpated during flexion and extension of the stifle. Blitzen was placed in right lateral recumbency so the veterinarian could assess laxity in his stifle. The cranial drawer test (the femur is stabilized with one hand and the proximal tibia manipulated with the other hand in a cranial-caudal direction; (Figure 3–33)) was positive (the tibia moved forward in relation to the femur) and it was determined that Blitzen had torn his cranial cruciate ligament (this ligament is the main ligament in the stifle that connects the distal femur to the proximal tibia). Blitzen had surgery to stabilize his stifle and remove his torn meniscus. After surgery, the veterinarian told Blitzen's owners to focus on strengthening the muscles around the stifle and regaining joint mobility and stability by taking their dog on controlled leash walks and performing flexion/extension exercises with him. A recheck appointment was scheduled for one week after he was discharged.

Copyright © 2015 Cengage Learning®

FIGURE 3–33 A cranial drawer test is used to assess cranial cruciate ligament rupture in the stifle of dogs. The arrows indicate the direction of force being applied to the femur and tibia.

1. M/N   _____

2. stifle   _____

3. PE   _____

4. flexed   _____

5. crepitus   _____

6. flexion   _____

7. extension   _____

8. right lateral recumbency   _____

9. laxity   _____

10. cranial _____

11. femur _____

12. tibia _____

13. ligament _____

14. meniscus _____

Critical Thinking Exercise

Blitzen's case study and the following questions are designed to stimulate critical thinking through class discussion, brief Internet research, or writing responses.

15. What is the function of a ligament?

16. A cruciate ligament forms what shape?

17. What two parts of the patient's stifle were repaired?

18. What is the function of a meniscus?

19. Describe "joint mobility and stability."

20. Why and how do dogs tear their cranial cruciate ligaments?

Critical Thinking Exercise

Chapter 4
HEAD TO TOE (AND ALL PARTS IN BETWEEN)

Objectives

Upon completion of this chapter, the reader should be able to:

- Identify and describe the body parts of various species

- Recognize, define, spell, and pronounce anatomical terms

TWO WORDS, SAME MEANING

Medical terminology is a specific language that healthcare professionals (both human and veterinary) use to describe conditions in a concise manner. Laypeople also have a language that is used to describe anatomy and medical conditions in a concise manner. In veterinary medicine, many different terms are used to describe the anatomy and diseases of many different species. In this chapter, the anatomical lay terms that many people use in the veterinary community are described. The lay terms for diseases are covered in the chapters on individual species.

COMMON ANATOMICAL TERMS FOR EQUINE SPECIES

See Figure 4–1a and b.

barrel (bah-rəl): capacity of the chest or trunk.

bars (bahrz): raised V-shaped structure on distal surface of hoof.

cannon (kahn-nohn) **bone:** third metacarpal (metatarsal) of the horse; also called the **shin bone**.

cheek (chēk): fleshy portion of either side of the face; forms the sides of the mouth and continues rostrally to the lips.

chest (chehst): part of the body between the neck and abdomen; the thorax.

chestnuts (chehs-nuhtz): horny, irregular growths on the medial surface of the equine leg; in the front legs, the chestnuts are just above the knee; in the rear legs, the chestnuts are near the hock.

coffin (kawf-ihn) **joint:** distal interphalangeal joint (joint between the short pastern and coffin bones (phalanges II and III, respectively)) in ungulates (Figure 4–2a, b, and c).

corners (kawr-nrz): third incisors of equine.

(a)

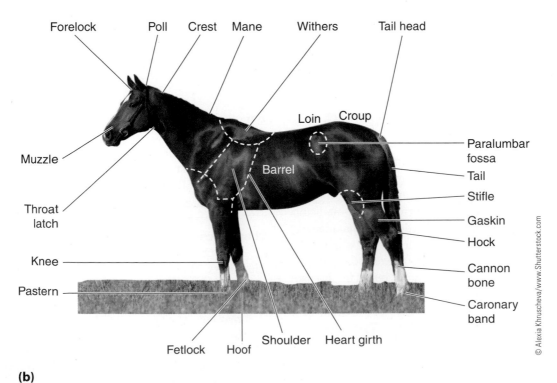

(b)

FIGURE 4–1 Anatomical parts of a horse.

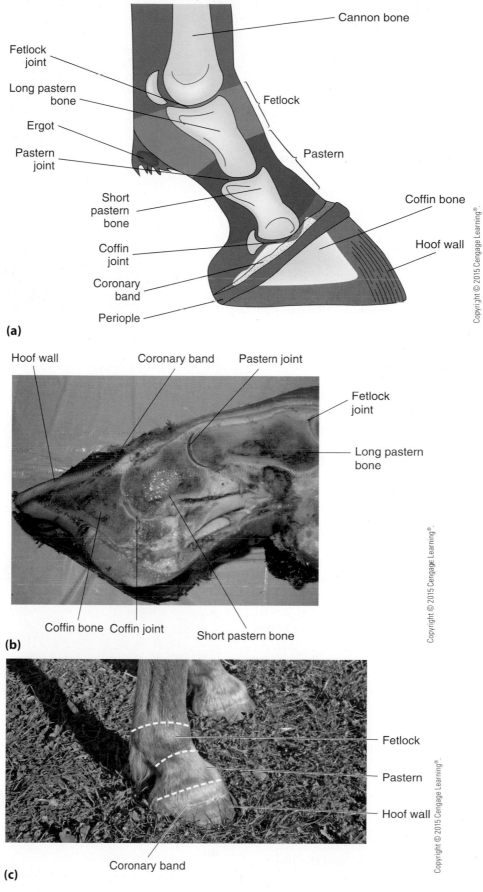

FIGURE 4–2 Anatomical parts of a horse's foot.

coronary (kohr-ō-nār-ē) **band:** junction between the skin and the horn of the hoof; also called the **coronet** (kŏr-oh-neht).

crest (krehst): root of the mane.

croup (kroop): muscular area around and above the tail base.

cutters (kuht-ərz): second incisors of equine.

dock (dohck): solid part of the equine tail.

elbow (ehl-bō): forelimb joint formed by distal humerus, proximal radius, and proximal ulna.

ergot (ahr-goht or ər-goht): small keratinized (kehr-ə-tə-nīzd) mass of horn in a small bunch of hair on the palmar or plantar aspects of the equine fetlock.

fetlock (feht-lohck): area of the limb between the pastern and the cannon.

fetlock (feht-lohck) **joint:** metacarpophalangeal and metatarsophalangeal joint (joint between the cannon bone and long pastern bone (phalanx I)) in ungulates.

flank (flānk): side of the body between the ribs and ilium.

forelock (fōr-lohck): in maned animals, the most cranial part of the mane hanging down between the ears and onto the forehead.

frog (frohg): V-shaped pad of soft horn between the bars on the sole of the equine hoof.

gaskin (gahs-kihn): muscular portion of the hindlimb between the stifle and hock; also called the **crus.**

heart girth (hahrt gərth): circumference of the chest just caudal to the shoulders and cranial to the back.

heel: caudal region of the hoof that has an area of soft tissue called the bulb.

hock (hohck): tarsal joint; also called the **tarsus.**

hoof (hoof): hard covering of the digit in ungulates (Figure 4–3a and b).

hoof wall (hoof wahl): hard, horny outer layer of the covering of the digit in ungulates.

knee (nē): carpus in ungulates (an ungulate is an animal with hooves).

loin (loyn): lumbar region of the back between the thorax and pelvis.

mane (mān): region of long, coarse hair at the dorsal border of the neck and terminating at the poll.

muzzle (muh-zuhl): two nostrils (including the skin and fascia) and the muscles of the upper and lower lip.

nippers (nihp-pərz): central incisors of equine.

paralumbar fossa (pahr-ah-luhm-bahr fohs-ah): hollow area of the flank whose boundaries are the transverse processes of the lumbar vertebrae (dorsally), the last rib (cranially), and the thigh muscles (caudally).

(a)

(b)

FIGURE 4–3 Anatomical parts of a horse's hoof.

pastern (pahs-tərn): area of the limb between the fetlock and hoof.

pastern (pahs-tərn) **joint:** proximal interphalangeal joint (joint between the long and short pastern bones (phalanges I and II, respectively)) in ungulates.

poll (pōl): top of the head; **occiput; nuchal crest.**

quarter: lateral or medial side of the hoof.

shoulder (shōl-dər): region around the large joint between the humerus and scapula.

sole (sōl): palmar or plantar surface of the hoof; irregular crescent-shaped bottom of hoof.

stifle (stī-fuhl) **joint:** femorotibial and femoropatellar joint in quadrupeds.

tail (tāl): caudal part of the vertebral column extending beyond the trunk.

tail head (tāl hehd): base of the tail where it connects to the body.

teat (tēt): nipple of mammary gland.

toe: cranial side of the hoof.

udder (uh-dər): mammary gland.

white line: fusion between the wall and sole of the hoof.

withers (wih-thərz): region over the dorsum where the neck joins the thorax and where the dorsal margins of the scapula lie just below the skin.

COMMON ANATOMICAL TERMS FOR CATTLE

See Figure 4–4a, b, and c.

barrel (bah-rəl): capacity of the chest or trunk.

brisket (brihs-kiht): mass of connective tissue, muscle, and fat covering the cranioventral part of the ruminant chest between the forelegs.

cannon (kahn-nohn) **bone:** third and fourth metacarpal (metatarsal) of the ruminant (not commonly used); also called the **shin bone.**

coffin (kawf-ihn) **joint:** distal interphalangeal joint (joint between the short pastern and coffin bones (phalanges II and III, respectively)) in ungulates.

crest (krehst): dorsal margin of the neck.

dewclaw (doo-klaw): accessory claw of the ruminant foot that projects caudally from the fetlock.

dewlap (doo-lahp): loose skin under the throat and neck, which may become pendulous in some breeds.

dock (dohck): solid part of the tail.

elbow (ehl-bō): forelimb joint formed by distal humerus, proximal radius, and proximal ulna.

fetlock (feht-lohck) **joint:** metacarpophalangeal and metatarsophalangeal joint (joint between the cannon bone and the long pastern bone (phalanx I)) in ungulates.

flank (flānk): side of the body between the ribs and ilium.

forearm (fōr-ahrm): part of the foreleg supported by the radius and ulna, between the elbow and knee.

forehead (fōr-hehd): region of the head between the eyes and ears.

heart girth (hahrt gərth): circumference of the chest just caudal to the shoulders and cranial to the back.

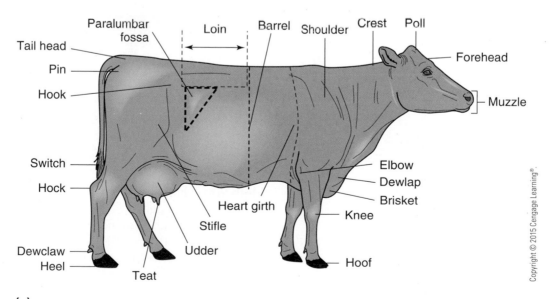

FIGURE 4–4 Anatomical parts of a cow.

(a)

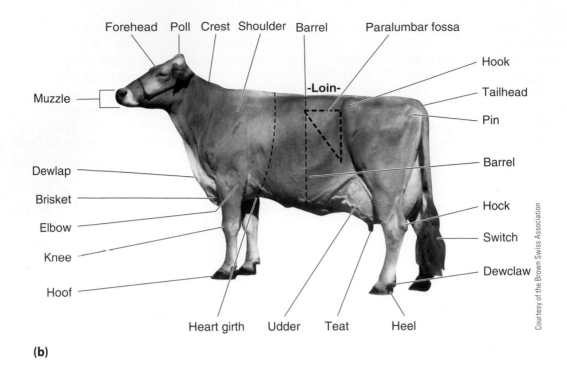

(b)

Courtesy of the Brown Swiss Association

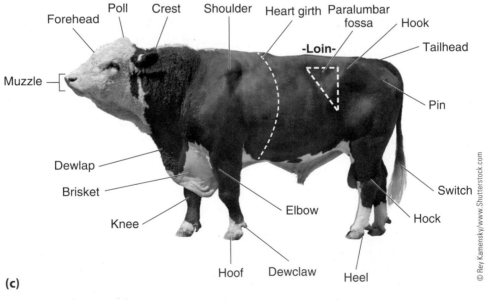

(c)

FIGURE 4–4 *(continued)*

heel (hēl): caudal region of the hoof that has an area of soft tissue called the bulb.

hock (hohck): tarsal joint; also called **tarsus**.

hoof (hoof): hard covering of the digit in ungulates.

hoof wall (hoof wahl): hard, horny outer layer of the covering of the digit in ungulates.

hooks (hookz): protrusion of the wing of the ilium on the dorsolateral area of ruminants.

knee (nē): carpus in ungulates.

loin (loyn): lumbar region of the back, between the thorax and pelvis.

muzzle (muh-zuhl): two nostrils (including the skin and fascia) and the muscles of the upper and lower lip.

paralumbar fossa (pahr-ah-luhm-bahr fohs-ah): hollow area of the flank whose boundaries are the transverse processes of the lumbar vertebrae (dorsally), the last rib (cranially), and the thigh muscles (caudally).

pastern (pahs-tərn) **joint:** proximal interphalangeal joint (joint between the long and short pastern bones (phalanges I and II, respectively)) in ungulates.

pedal (pē-dahl): pertaining to the foot.

pins (pihnz): protrusion of the ischium bones just lateral to the base of the tail in ruminants.

poll (pōl): top of the head; occiput; nuchal crest.

quarter: one of the four glands in the cow's udder.

shoulder (shōl-dər): region around the large joint between the humerus and scapula.

sole (sōl): palmar or plantar surface of the hoof; bottom of hoof.

stifle (stī-fuhl) **joint:** femorotibial and femoropatellar joint in quadrupeds.

switch (swihtch): tuft of hair at the end of the tail.

tail (tāl): caudal part of the vertebral column extending beyond the trunk.

tail head (tāl hehd): base of the tail where it connects to the body.

teat (tēt): nipple of mammary gland.

toe (tō): cranial end of the hoof.

udder (uh-dər): mammary gland.

COMMON ANATOMICAL TERMS FOR GOATS

See Figure 4–5a and b.

brisket (brihs-kiht): mass of connective tissue, muscle, and fat covering the cranioventral part of the ruminant chest between the forelegs.

cannon (kahn-nohn) **bone:** third and fourth metacarpal (metatarsal) of the ruminant (not commonly used); also called the **shin bone.**

chine (chīn): thoracic region of the back.

coffin (kawf-ihn) **joint:** distal interphalangeal joint (joint between the short pastern and coffin bones (phalanges II and III, respectively)) in ungulates.

crest (krehst): dorsal margin of the neck.

dewclaw (doo-klaw): accessory claw of the ruminant foot that projects caudally from the fetlock.

elbow (ehl-bō): forelimb joint formed by distal humerus, proximal radius, and proximal ulna.

fetlock (feht-lohck) **joint:** metacarpophalangeal and metatarsophalangeal joint (joint between the cannon bone and the long pastern bone (phalanx I)) in ungulates.

flank (flānk): side of the body between the ribs and ilium.

forearm (fōr-ahrm): part of the foreleg supported by the radius and ulna, between the elbow and knee.

forehead (fōr-hehd): region of the head between the eyes and ears.

heart girth (hahrt gərth): circumference of the chest just caudal to the shoulders and cranial to the back.

heel (hēl): caudal region of the hoof that has an area of soft tissue called the bulb.

hock (hohck): tarsal joint; also called **tarsus.**

hoof (hoof): hard covering of the digit in ungulates.

hoof wall (hoof wahl): hard, horny outer layer of the covering of the digit in ungulates.

hooks (hookz): protrusion of the wing of the ilium on the dorsolateral area of ruminants.

horn butt (hōrn buht): poll region between the eyes and ears of previous horn growth.

knee (nē): carpus in ungulates.

loin (loyn): lumbar region of the back, between the thorax and pelvis.

muzzle (muh-zuhl): two nostrils (including the skin and fascia) and the muscles of the upper and lower lip.

paralumbar fossa (pahr-ah-luhm-bahr fohs-ah): hollow area of the flank whose boundaries are the transverse processes of the lumbar vertebrae (dorsally), the last rib (cranially), and the thigh muscles (caudally).

pastern (pahs-tərn) **joint:** proximal interphalangeal joint (joint between the long and short pastern bones [phalanges I and II, respectively]) in ungulates.

pedal (pē-dahl): pertaining to the foot.

pins (pihnz): protrusion of the ischium bones just lateral to the base of the tail in ruminants.

poll (pōl): top of the head; occiput; nuchal crest.

rump (ruhmp): sacral to tailhead region of the back.

shoulder (shōl-dər): region around the large joint between the humerus and scapula.

sole (sōl): palmar or plantar surface of the hoof; bottom of hoof.

stifle (stī-fuhl) **joint:** femorotibial and femoropatellar joint in quadrupeds.

tail (tāl): caudal part of the vertebral column extending beyond the trunk.

tail head (tāl hehd): base of the tail where it connects to the body.

teat (tēt): nipple of mammary gland.

toe (tō): cranial end of the hoof.

udder (uh-dər): mammary gland.

wattle (waht-tuhl): appendages suspended from the head (usually under the chin).

withers (wih-thərz): region over the dorsum where the neck joins the thorax and where the dorsal margins of the scapula lie just below the skin.

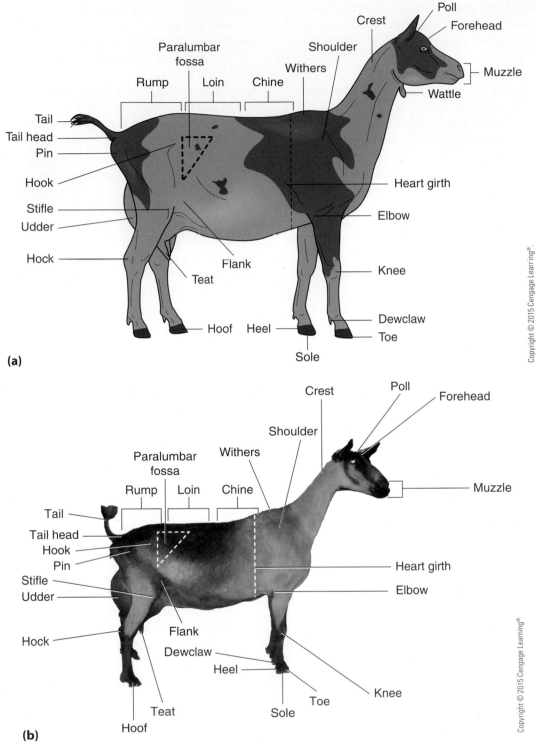

FIGURE 4–5 Anatomical parts of a goat.

COMMON ANATOMICAL TERMS FOR SHEEP

See Figure 4–6a and b.

brisket (brihs-kiht): mass of connective tissue, muscle, and fat covering the cranioventral part of the ruminant chest between the forelegs.

cannon (kahn-nohn) **bone**: third and fourth metacarpal (metatarsal) of the ruminant (not commonly used); also called the **shin bone**.

coffin (kawf-ihn) **joint**. distal interphalangeal joint (joint between the short pastern and coffin bones (phalanges II and III, respectively)) in ungulates.

crest (krehst): dorsal margin of the neck.

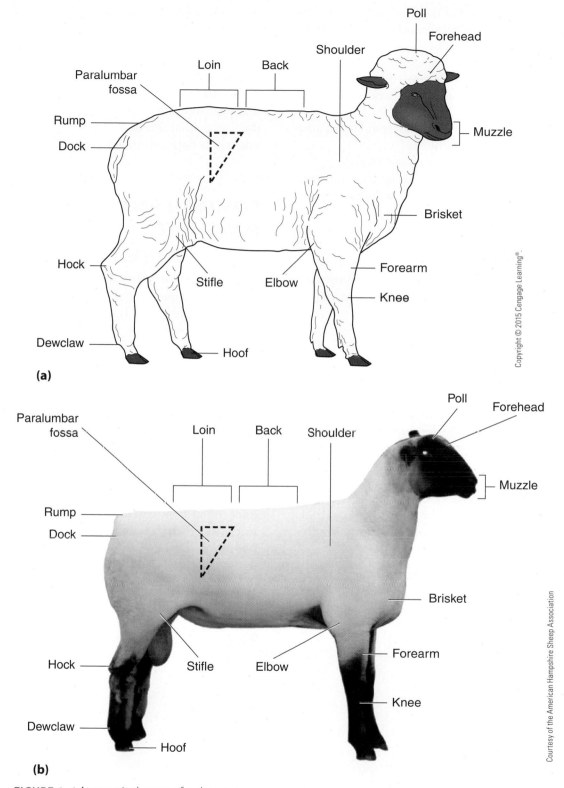

FIGURE 4–6 Anatomical parts of a sheep.

dewclaw (doo-klaw): accessory claw of the ruminant foot that projects caudally from the fetlock.

dock (dohck): solid part of the tail.

elbow (ehl-bō): forelimb joint formed by distal humerus, proximal radius, and proximal ulna.

fetlock (feht-lohck) **joint**: metacarpophalangeal and metatarsophalangeal joint (joint between the cannon bone and the long pastern bone (phalanx I)) in ungulates.

flank (flānk): side of the body between the ribs and ilium.

forearm (fōr-ahrm): part of the foreleg supported by the radius and ulna, between the elbow and knee.

forehead (fōr-hehd): region of the head between the eyes and ears.

heart girth (hahrt gərth): circumference of the chest just caudal to the shoulders and cranial to the back.

heel (hēl): caudal region of the hoof that has an area of soft tissue called the bulb.

hock (hohck): tarsal joint; also called **tarsus**.

hoof (hoof): hard covering of the digit in ungulates.

hoof wall (hoof wahl): hard, horny outer layer of the covering of the digit in ungulates.

knee (nē): carpus in ungulates.

loin (loyn): lumbar region of the back, between the thorax and pelvis.

muzzle (muh-zuhl): two nostrils (including the skin and fascia) and the muscles of the upper and lower lip.

paralumbar fossa (pahr-ah-luhm-bahr fohs-ah): hollow area of the flank whose boundaries are the transverse processes of the lumbar vertebrae (dorsally), the last rib (cranially), and the thigh muscles (caudally).

pastern (pahs-tərn) **joint**: proximal interphalangeal joint (joint between the long and short pastern bones (phalanges I and II, respectively)) in ungulates.

pedal (pē-dahl): pertaining to the foot.

poll (pōl): top of the head; occiput; nuchal crest.

rump (ruhmp): sacral to tailhead region of the back.

shoulder (shōl-dər): region around the large joint between the humerus and scapula.

sole (sōl): palmar or plantar surface of the hoof; bottom of hoof.

stifle (stī-fuhl) **joint**: femorotibial and femoropatellar joint in quadrupeds.

tail head (tāl hehd): base of the tail where it connects to the body.

teat (tēt): nipple of mammary gland.

toe (tō): cranial end of the hoof.

udder (uh-dər): mammary gland.

COMMON ANATOMICAL TERMS FOR SWINE

See Figure 4–7a and b.

coffin (kawf-ihn) **joint**: distal interphalangeal joint (joint between the short pastern and coffin bones (phalanges II and III, respectively)) in ungulates.

dewclaw (doo-klaw): accessory claw of the porcine foot that projects caudally from the fetlock.

elbow (ehl-bō): forelimb joint formed by distal humerus, proximal radius, and proximal ulna.

fetlock (feht-lohck) **joint**: metacarpophalangeal and metatarsophalangeal joint (joint between the cannon bone and long pastern bone (phalanx I)) in ungulates.

flank (flānk): side of the body between the ribs and ilium.

ham (hahm): musculature of the upper thigh.

hock (hohck): tarsal joint; also called **tarsus**.

hoof (hoof): hard covering of the digit in ungulates.

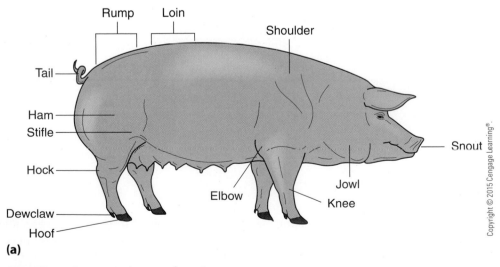

(a)

FIGURE 4–7 Anatomical parts of a swine.

(b)

FIGURE 4–7 *(continued)*

hoof wall (hoof wahl): hard, horny outer layer of the covering of the digit in ungulates.

jowl (jowl): external throat, especially when fat or loose skin is present.

knee (nē): carpus in ungulates.

loin (loyn): lumbar region of the back, between the thorax and pelvis.

pastern (pahs-tərn) **joint**: proximal interphalangeal joint (joint between the long and short pastern bones (phalanges I and II, respectively)) in ungulates.

rump (ruhmp): sacral to tailhead region of the back.

shoulder (shōl-dər): region around the large joint between the humerus and scapula.

snout (snowt): upper lip and apex of the nose of swine.

stifle (stī-fuhl) **joint**: femorotibial and femoropatellar joint in quadrupeds.

tail (tāl): caudal part of the vertebral column extending beyond the trunk.

COMMON ANATOMICAL TERMS FOR DOGS AND CATS

See Figures 4–8a and b and 4–9a and b.

cheek (chēk): fleshy portion of either side of the face; forms the sides of the mouth and continues rostrally to the lips.

chest (chehst): part of the body between the neck and abdomen; the thorax.

chin (chihn): rostroventral protrusion of the mandible.

dewclaw (doo-klaw): rudimentary first digit of dogs and cats.

elbow (ehl-bō): forelimb joint formed by distal humerus, proximal radius, and proximal ulna.

flank (flānk): side of the body between the ribs and ilium.

forearm (fōr-ahrm): part of the foreleg supported by the radius and ulna, between the elbow and carpus.

forehead (fōr-hehd): region of the head between the eyes and ears.

hock (hohck): tarsal joint; also called **tarsus**.

muzzle (muh-zuhl): two nostrils (including the skin and fascia) and the muscles of the upper and lower lip.

pinna (pihn-ah): projecting part of the ear lying outside the head; the **auricle**.

rump (ruhmp): sacral to tailhead region of the back; also called the **croup**.

shoulder (shōl-dər): region around the large joint between the humerus and scapula.

stifle (stī-fuhl) **joint**: femorotibial and femoropatellar joint in quadrupeds.

tail (tāl): caudal part of the vertebral column extending beyond the trunk.

(a)

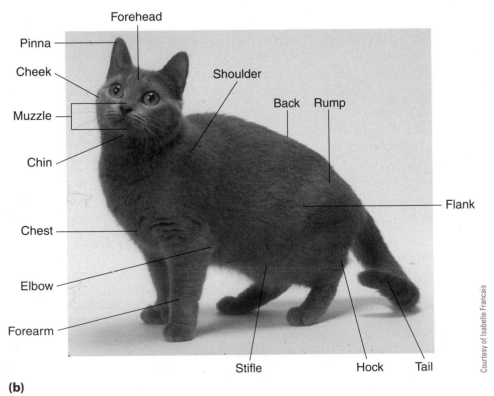

(b)

FIGURE 4–8 Anatomical parts of a cat.

(a)

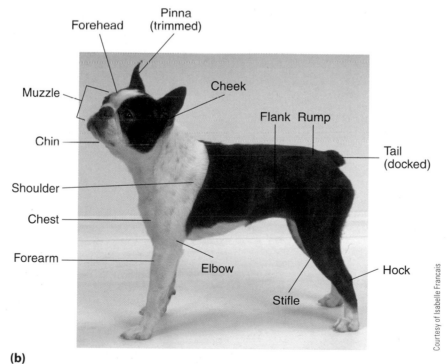

(b)

FIGURE 4–9 Anatomical parts of a dog.

REVIEW EXERCISES

Multiple Choice

Choose the correct answer.

1. Another term for the distal interphalangeal joint in ungulates is the
   a. coffin joint
   b. fetlock joint
   c. pastern joint
   d. coronary band

2. The V-shaped pad of soft horn between the bars on the sole of the equine hoof is known as the
   a. sole
   b. white line
   c. ergot
   d. frog

3. The anatomical term for the top of the head is the
   a. crest
   b. poll
   c. sole
   d. forehead

4. The common name for the tarsal joint in animals is the
   a. loin
   b. stifle
   c. hock
   d. wrist

5. The side of the body between the ribs and ilium is called the
   a. rump
   b. flank
   c. loin
   d. ham

6. The upper lip and apex of the nose of swine is called the
   a. jowl
   b. ham
   c. snout
   d. pinna

7. The two nostrils and the muscles of the upper and lower lip are called the
   a. snout
   b. muzzle
   c. cheek
   d. crest

8. The proximal interphalangeal joint in ungulates is called the
   a. pastern joint
   b. coffin joint
   c. fetlock joint
   d. stifle joint

9. The protrusions of the ischium bones just lateral to the base of the tail in ruminants are known as
   a. tailheads
   b. hooks
   c. pins
   d. docks

10. The protrusions of the wing of the ilium on the dorsolateral area of ruminants are known as
    a. tailheads
    b. hooks
    c. pins
    d. docks

11. The hollow area of the flank is called the
    a. pedal fossa
    b. paralumbar fossa
    c. rump fossa
    d. tail fossa

12. The mass of connective tissue, muscle, and fat covering the cranioventral part of the ruminant chest is the
    a. brisket
    b. crest
    c. flank
    d. loin

13. The rudimentary first digit of dogs and cats is the
    a. claw
    b. digit
    c. dewclaw
    d. declaw

14. The "knee" in people is known as what in animals?
    a. hock joint
    b. pastern joint
    c. coffin joint
    d. stifle joint

15. The auricle is also known as the
    a. cheek
    b. forehead
    c. chin
    d. pinna

16. In swine, the external throat, especially when fat or loose skin is present, is called the
    a. nout
    b. ham
    c. jowl
    d. loin

17. In equine, the region over the dorsum where the neck joins the thorax and where the dorsal margins of the scapula lie is called the
    a. shoulder
    b. withers
    c. flank
    d. loin

18. Ilium is to ischium as
    a. hooks is to pins
    b. pins is to hooks
    c. pastern is to coffin
    d. coffin is to pastern

19. The lumbar region of the back is called the
    a. rump
    b. tailhead
    c. loin
    d. flank

20. The lateral or medial side of the hoof is the
    a. frog
    b. coronet
    c. heel
    d. quarter

21. The cranial side of the equine hoof is the
    a. quarter
    b. gaskin
    c. heel
    d. toe

22. The mammary gland in cattle is the
    a. wattle
    b. poll
    c. udder
    d. teat

23. The musculature of the upper thigh in swine is the
    a. ham
    b. flank
    c. jowl
    d. rump

24. Which of the following is found in the chest or trunk area of a horse?
    a. crest
    b. barrel
    c. chestnut
    d. gaskin

25. Which of the following is the correct order of equine joints from proximal to distal?
    a. fetlock, pastern, coffin
    b. long pastern, short pastern, coffin
    c. fetlock, cannon, coffin
    d. pastern, cannon, coffin

Matching

Match the anatomical term in Column I with its definition in Column II.

Column I

1. _____ mane
2. _____ cannon bone
3. _____ knee
4. _____ stifle joint
5. _____ sole
6. _____ cutters
7. _____ corners
8. _____ dock
9. _____ dewclaw
10. _____ barrel
11. _____ crest
12. _____ chestnuts
13. _____ pinna
14. _____ croup
15. _____ loin
16. _____ jowl
17. _____ heart girth
18. _____ teat
19. _____ poll
20. _____ flank

Column II

a. carpus in ungulates

b. capacity of the chest or trunk

c. amputation of the tail; solid part of the equine tail

d. region of long, coarse hair at the dorsal border of the neck

e. rudimentary first digit of dogs and cats

f. third metacarpal or metatarsal bone of the horse

g. femorotibial and femoropatellar joint in quadrupeds

h. bottom of the hoof

i. second incisors of equine

j. third incisors of equine

k. side of body between ribs and ilium

l. nipple of mammary gland

m. top of head

n. dorsal margin of neck of cattle, sheep, and goats

o. external throat of swine

p. rump in dogs and cats

q. lumbar region of back

r. auricle

s. circumference of chest

t. horny growths on medial surface of equine leg

Fill in the Blanks

Write the medical term or description to complete the following sentences.

1. The poll region between the eyes and ears of previous horn growth in goats is the _____ _____.

2. The palmar or plantar surface of the hoof in livestock is the _____.

3. The hock is the _____ joint.

4. The pins are the protrusion of the _____ bones just lateral to the base of the tail in ruminants.

5. The hooks are the protrusion of the wing of the _____ on the dorsolateral area of ruminants.

6. The tail is the _____ part of the vertebral column.

7. The rostroventral protrusion of the mandible in dogs and cats is the _____.

8. The snout is the upper lip and apex of the nose in _____.

9. The term *pedal* pertains to the _____.

10. The dock is the solid part of the _____.

Spelling

Cross out any misspelled words in the following sentences and replace them with the proper spelling.

1. The heal is the caudal region of the hoof. _____

2. The soul is the bottom of the hoof in livestock._____

3. The lumber region of the back is the loin._____

4. The femorotibal and femoropatellar joint in quadrupeds is known as the styfle._____

5. The outer ear in dogs and cats is the pina._____

6. The area of the limb between the fetlock and the hoof is the pasturn._____

7. The cranial part of the mane in horses is the fourlock._____

8. The third metacarpal and metatarsal in horses is the canyon bone._____

9. The base of the tail is known as the tailhed._____

10. The fehtlock is the area of the limb between the pastern and cannon. _____

Crossword Puzzle

Common Anatomical Terms

Supply the correct anatomical term in the appropriate space for the definition listed.

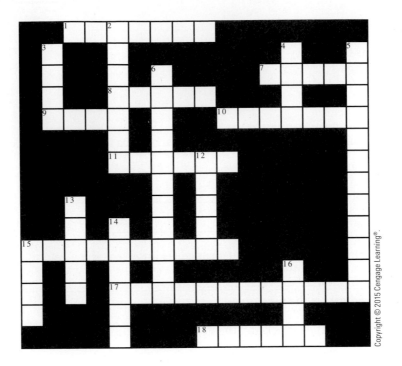

Copyright © 2015 Cengage Learning®.

Across

1 Rudimentary first digit of dogs and cats

7 Projecting part of the ear lying outside the head; the auricle

8 Protrusions of the wing of the ilium on the dorsolateral area of ruminants

9 The carpus in ungulates (an ungulate is an animal with hooves)

10 Mass of connective tissue, muscle, and fat covering the cranioventral part of the ruminant chest between the forelegs

11 Tuft of hair at the end of the tail

15 Circumference of the chest just caudal to the shoulders and cranial to the back

17 Metacarpophalangeal and metatarsophalangeal joint in ungulates (2 wds.)

18 The two nostrils (including the skin and fasica) and the muscles of the upper and lower lip

Down

2 Region in equine over the dorsum where the neck joins the thorax and where the dorsal margins of the scapula lie just below the skin

3 Amputation of the tail

4 Protrusions of the ischium bones just lateral to the base of the tail in ruminants

5 Joint between the long and short pastern bones (phalanx I and II, respectively) in ungulates (2 wds.)

6 Joint between the short pastern and coffin bones (phalanx II and III, respectively) in ungulates (2 wds.)

12 The root of the mane

13 Side of the body between the ribs and ilium

14 Femorotibial and femoropatellar joint in quadrupeds

15 Tarsal joint

16 Top of the head; the occiput

Word Searches

Find the following anatomical terms in the puzzle below. (Make sure you understand what the terms mean as you find them.)

Dog and Cat Word Search

```
W L L E C L N W I C E A I L M
L N W L E C C I L L W W U W L
E K A Z W O T C Z C L L L H H
A L T W N S L Z E H I C Z T L
T N I A I H U L L M F W I W E
L I W L L M U I L W L L T W S
O L U L C Z S U I M I U L L W
T N O L L U L C F I Z L Z C A
W Z L W N N I H C Z T I M K U
A I K S T I F L E J O I N T W
L L N T I T A C A T L T A W Z
C K C O H A N A N E W I L N W
W Z L C N I N O I L I W C P U
E E Z H T L I U C N F D H L W
D L H T K C P N A T Z F W S C
```

hock

stifle joint

pinna

muzzle

chin

dewclaw

tail

Ruminant Word Search

```
T A W A T T L E R T L O T U L
R D R O A D C H K P T L U W K
O N I D N R P T N L T R L T S
T P T L R A B S A N S D L K L
L T P E D A L P L A O A R O O
E N D P S P T D F D A T O L I
N P D T N A L W O G E W T R T
O U I N A C I O A E L W T D K
B I C N N P C R K O L C L L C
N U L U S T D L I I A R T A L
O D T R L T N N W R S P R N P
N D A E P O U P T E K S I R B
N E S P C U O I L C R B G A S
A R T I C E S O C S K O O H T
C H T R I G T R A E H W L T L
```

cannon bone   loin

udder   flank

wattle   pedal

heart girth   brisket

pins   dewlap

hooks

Equine Word Search

```
O T O T E N A H E E S U E N L
G N A N T S N G E N T N N N A
A S R E H U N G G E U T H T S
I U L E N G A T O E N N P N O
L L O P H S S O R P T I K I A
U R E H T T U G F T S O S O N
T G A S K I N R S U E J U J W
T T S M S R H E N I H N G N I
N O E O H E E E E C I N R T
A G U A U A C N E N A F N E H
C P E G T G H U P E K F M T E
E E N T E U O S O U N O A S R
L A R S H E U T P E E C N A S
H N I R R I G R E G E S E P A
G E S K N N S O L E E A E K H
```

knee

poll

mane

pastern joint

coffin joint

sole

frog

gaskin

withers

chestnuts

ergot

Swine Word Search

```
O O L P O O O L N T O S O O H
L R H L O F M L O R M T O O O
H O R T S F R O L H H O M L O
A A L H H H A O O N M P T F
M M O O N M L O O H R O U A U
T T O T M O B W S R T U O N S
O P T A L M L H O U S H L L S
S H S H M R L T O M H L O A O
O O P O O U U L O L M S O F R
A O N O R U F H M R L L O S O
O L A O J O L O S L E L H U O
J J S P O P H S U S O L O H L
O J S O O M S M H O O U B S S
W O L S H U M J A M P O H O O
L M F U O R O O T H H S H O W
```

rump	snout
ham	hoof
jowl	elbow

Label the Diagrams

Label diagrams (Figures 4–10, 4–11, 4–12, 4–13, 4–14, and 4–15).

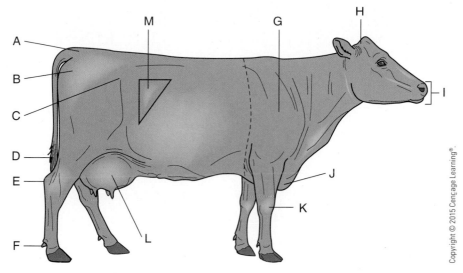

FIGURE 4–10 Identify the parts of this cow.

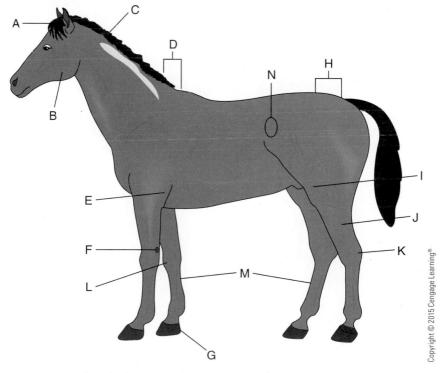

FIGURE 4–11 Identify the parts of this horse.

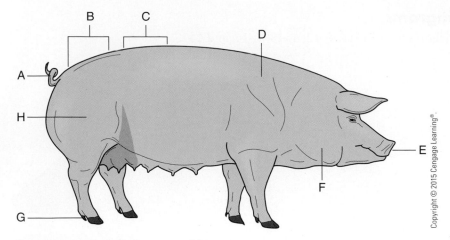

FIGURE 4–12 Identify the parts of this swine.

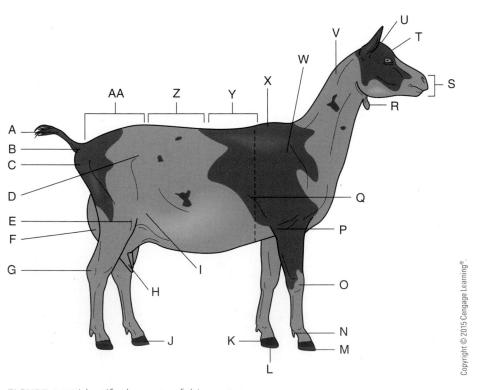

FIGURE 4–13 Identify the parts of this goat.

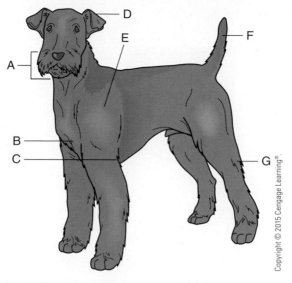

FIGURE 4–14 Identify the parts of this dog.

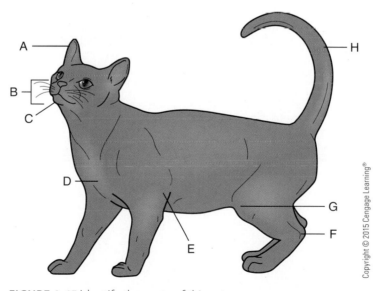

FIGURE 4–15 Identify the parts of this cat.

Label the anatomical parts of a horse's foot.

1. cannon bone
2. coffin bone
3. coffin joint
4. coronary band
5. ergot
6. fetlock

7. fetlock joint
8. hoof wall
9. long pastern bone
10. pastern
11. pastern joint
12. short pastern bone

Case Studies

Fill in the blanks to complete the following case history.

A 2-month-old female foal was presented to the clinic with a history of left front limb lameness and swelling near the _____ joint (joint between the cannon and long pastern bone). The foal has been eating and drinking normally and her attitude was good. On _____ (abbreviation for physical examination), the veterinarian noted that the foal had knuckling of the metacarpophalangeal joint and abnormal contraction of the _____ tendon (closure of a joint angle). The veterinarian asked the owner if the foal may have injured herself while on pasture causing the joint to fill with _____ (lubricant secreted by the synovial membrane); however, the owner was not aware of any history of trauma. The veterinarian asked what type of feed the foal ate and the owner stated she was feeding the foal lots of calorie-dense, high energy food to help her grow more quickly. Based on the animal's age and history of lameness and overfeeding, the veterinarian suspected that the foal had _____ (inflammation of the growth plate) and recommended rest and dietary change.

Define the following anatomical terms in this case study.

Becky is a veterinary technician who is working with local FFA members to help them learn cattle judging. The students give her a diagram of a cow and the following information:

Cattle generally deposit subcutaneous fat in the following places: tail head, loin, brisket, barrel, and udder.

Becky finds a line drawing of a cow and believes the students would learn a lot of cattle terminology if they label the drawing with the terms described above. Define the following terms and identify which number correlates with the term:

tail head:

loin:

brisket:

barrel:

udder:

What body parts do the other numbers represent?

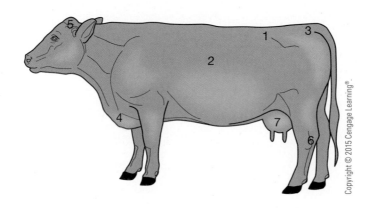

Critical Thinking Exercise

The following story and questions are designed to stimulate critical thinking through class discussion or brief writing responses. There are many correct answers to these questions.

Justine Penn is a veterinary technician who worked in a small animal practice for 4 years. She was just hired to work at a new mixed animal veterinary clinic and is concerned because she has not worked with large animals since she graduated from college. The first day at her new job proved to be challenging because she had trouble understanding the clients and what they wanted for their animals. One client asked for a teat dip to prevent mastitis; another told her that the pins and hooks on his cattle were more prominent due to lack of food from the previous summer's drought; and yet another said his pigs have deformed snouts likely due to atrophic rhinitis. Justine was used to working with dogs and cats and was familiar with the terms the clients used; however, she felt out of her league with livestock and could not remember the terms quickly enough to understand what clients wanted. She wondered if she should have stayed at her former job in the small animal clinic.

1. What are some things Justine could do to quickly learn veterinary medical terminology relating to large animals?

2. What are some resources she could use to overcome her apprehension about working in a mixed animal practice?

3. What obstacles may Justine encounter if she does not learn the terms commonly used by the clients in this practice?

Chapter 5
WHAT IS IN A NAME?

Objectives

Upon completion of this chapter, the reader should be able to:

- Identify and recognize common terms used for animals

- Define common terms used to denote sex and age of animals

- Define common terms used to denote birthing and grouping of animals

WHAT IS YOUR NAME?

Laypeople and professionals use terms to describe in one word the status of an animal. The term may relate to the sexual status of an animal (**intact** meaning sexually functional or **altered** meaning sexually nonfunctional) or the age status of an animal. Terms have also been derived to denote the process of giving birth and the grouping of animals. The following lists provide the terms used to describe animals.

canine (kā-nīn) = dog

dog/stud = intact (not sexually altered) male dog.
bitch = intact female dog.
whelp (wehlp) or **pup** = young dog.
whelping (wehl-pihng) = giving birth to whelps.
pack = group of dogs.
litter = multiple offspring born during same labor.

feline (fē-līn) = cat

tom = intact male cat.
queen = intact female cat.
kitten = young cat.
queening = giving birth to kittens.

lagomorph (lāg-ō-mōrf) = rabbit

buck = intact male rabbit.
doe = intact female rabbit.
lapin (lahp-ihn) = neutered male rabbit.
kit = young (blind, deaf) rabbit.
kindling (kihnd-lihng) = giving birth to rabbits.
herd = group of rabbits.

ferret (fehr-reht)

hob = intact male ferret.
jill = intact female ferret.
gib (gihb) = neutered male ferret.
sprite (sprīt) = spayed female ferret.
kit = young ferret.
kindling = giving birth to ferrets.

cavy (kā-vē) = guinea pig

boar = intact male guinea pig.
sow = intact female guinea pig.
pup = young guinea pig.

murine (moo-rēn) = mouse or rat

sire (sī-ər) = intact male mouse or rat.
dam (dahm) = intact female mouse or rat.
pup = young mouse or rat.
gang = group of mice.

psittacine (siht-ah-sēn) = parrot includes other birds with bills for cracking seeds

cock = intact male parrot.
hen = intact female parrot.
chick = young parrot.
flock = group of parrots (also called a **company**).

turkey = one kind of poultry

tom = intact male turkey.
hen = intact female turkey.
poult (pōlt) = young turkey.
flock = group of turkeys.
clutch (kluhtch) = group of eggs.

chicken = one kind of poultry

rooster = sexually mature male chicken; also called **cock**.
hen = intact female chicken.
capon (kā-pohn) = young castrated male chicken or domestic fowl.
cockerel (kohck-ər-ehl) = immature male chicken.
pullet (puhl-eht) = immature female chicken.
poult = young chicken.
chick = very young chicken.
flock = group of chickens.

Gender versus Sex

Veterinary professionals talk only about the sex of an animal. Gender is used to denote whether words pertaining to a noun are masculine, feminine, or neutral. Gender also can be used to denote social constructs, such as the gender or social roles of men and women.

Parentage

Dam and *sire* are terms used to denote female parent and male parent, respectively, for many species. When animals are bred, these terms may be used instead of the ones in the lists here. Used correctly, these mean that the male and female have mated and produced an embryo or fetus. **Dam** is a female parent; **sire** is a male parent.

goose = one kind of anserine (ahn-sehr-ihn)

gander = intact male goose.
goose = intact female goose.
gosling = young goose.
gaggle = group of geese.

duck = one type of anserine

drake = intact male duck.
duck = intact female duck.
duckling = young duck.
flock = group of ducks.

ratite (rah-tīt) = large, flightless bird (ostrich, emu, and rhea)

rooster = intact male ratite.
hen = intact female ratite.
hatchling = ostrich up to 2 days old.
chick = young ratite (usually less than 6 months of age).
yearling = ostrich from 6 months to 12 months of age; rhea from 6 months to 18 months of age.
flock = group of ratites.

porcine (poor-sīn) = pig = swine

boar (bōr) = intact male pig.
sow = intact female pig.
barrow (bār-ō) = male pig castrated when young.
stag = male pig castrated after maturity.
gilt (gihlt) = young female pig that has not farrowed.
pig or **piglet** = young pig; old term is *shoat*.
farrowing (far-ō-ihng) = giving birth to pigs.
herd = group of pigs.

Avian Terms

There are unique names for groups of birds that are particular to one type of bird. Examples include the following:

- flock, flight, pod, or volery of birds
- pandemonium of parrots
- murder of crows
- dole or flight of doves
- brace, bunch, flock, or team of ducks
- aerie or convocation of eagles
- gaggle of geese
- colony of gulls
- cast of hawks
- charm of hummingbirds
- band of blue jays
- watch of nightingales
- ostentation of peacocks
- bevy or covey of quails

equine (ē-kwīn) = horse, pony, donkey, and mule

stallion (stahl-yuhn) = intact male equine >4 years old.
colt (kōlt) = intact male equine <4 years old.
mare (mār) = intact female equine >4 years old.
filly (fihl-ē) = intact female equine <4 years old.
gelding (gehld-ihng) = castrated male equine.
ridgeling (rihdj-lihng) or **rig** = cryptorchid equine (one or both testicles have not descended from the abdomen).
foal = young equine of either sex.
weanling = young equine <1 year old that has been removed (weaned) from its mother.
suckling = a still-nursing foal.
yearling = young equine between 1 and 2 years old; a short yearling is closer to 1 year old while a long yearling is closer to 2 years old.
foaling = giving birth to equine.
herd = group of equine.
band = group of horses consisting of one mature stallion, his breeding mares, and the immature male and female offspring of his mares.
brood mare (bruhd mār) = breeding female equine.
maiden mare (mā-dehn mār) = female equine never bred.

What Is a Mule?

Mule is a general term that applies to the hybrid crossing of equines. *Mule* also is used to denote the offspring of a jack (male donkey) and a mare (female horse). **Hinny** (hihn-ē) is used to denote the offspring of a stallion (male horse) and a jenny (female donkey). Think mule:mare to remember the lineage of this hybrid. Both mules and hinnies are sterile.

barren mare (bār-ehn mār) = intact female horse that was not bred or did not conceive the previous season; also called **open mare**.

wet mare = intact female horse that has foaled during the current breeding season.

agalactic mare (ā-gahl-ahck-tihck mār) = intact female horse not producing milk.

pony = equine between 8.2 and 14.2 hands when mature (not a young horse).

donkey = ass = burro

jack or **jack ass** = intact male donkey.
jenny = intact female donkey.

ovine (ō-vīn) = sheep

ram = intact male sheep.
ewe (yoo) = intact female sheep.
wether (weh-thər) = castrated male sheep.
lamb = young sheep.
hothouse lamb = young sheep less than 3 months of age.
spring lamb = young sheep 3–7 months of age.
yearling = sheep 1–2 years of age (used when marketing sheep).
lambing = giving birth to sheep.
flock = group of sheep.

What Is a Bellwether?

A *wether* is a neutered sheep or goat. *Bell* is a ringing device. Originally, *bellwether* was used to describe the practice of putting a bell on the lead wether of a flock or herd.

Where Are Animals Bred?

The term **stud** is commonly used in reference to a male animal used for breeding. However, the term *stud* actually refers to the facility or farm where breeding animals are kept.

caprine (kahp-rīn) = goat

buck = intact male goat.
doe = intact female goat.
wether = castrated male goat.
kid = young goat.
kidding = giving birth to goats.
freshening = giving birth to dairy animals; also called **freshen**.
herd = group of goats.

camelid = llama, alpaca, guanaco, vicuna

bull = intact male llama; also called a **stallion**.
cow = intact female llama.
gelding = castrated male llama.
cria (krē-ah) = young llama.

bovine = cattle

bull = intact male bovine.
jumper bull = intact male bovine that has just reached maturity and is used for breeding.
cow = intact female bovine that has given birth.
steer = male bovine castrated when young.
stag = male bovine castrated after maturity.
heifer (hehf-ər) = young female bovine that has not given birth.
calf = young bovine.
calving = giving birth to cattle.
freshening = giving birth to dairy animals; also called **freshen**.
herd = group of cattle.
springing heifer = young female pregnant with her first calf; also called **first calf heifer**.
freemartin (frē-mahr-tihn) = sexually imperfect, usually sterile female calf born as a twin with a male calf.
gomer bull = bull used to detect female bovines in heat; bull may have penis surgically deviated to the side, may be treated with androgens, or may be vasectomized so as not to impregnate female; also called **teaser bull**.

cervidae = deer, elk, moose, caribou

buck = intact male deer.
bull = intact male elk, moose, caribou.
doe = intact female deer.
cow = intact female elk, moose, caribou.
fawn = young deer.
heifer = young female elk, moose, caribou.
herd = group of cervidae.

Symbols Denoting Gender

The symbols denoting male ♂ and female ♀ originally stood for Mars (the god of war) and Venus (the goddess of love), respectively. *Mars* and *male* both begin with *m*.

REVIEW EXERCISES

Multiple Choice

Choose the correct answer.

1. A neutered male sheep or goat is a(n)
   a. bull
   b. ovine
   c. wether
   d. caprine

2. A sexually imperfect, usually sterile female calf born as a twin with a male calf is a
   a. heifer
   b. freemartin
   c. gilt
   d. filly

3. A cross between a stallion and a jenny is a
   a. donkey
   b. mule
   c. jenny
   d. hinny

4. Cow is to mare as steer is to
   a. stallion
   b. ridgling
   c. colt
   d. gelding

5. Parrots are in a group of birds called
   a. amazon
   b. psittacine
   c. lagomorph
   d. murine

6. In canines, the act of giving birth is
   a. whelping
   b. pupping
   c. packing
   d. gestation

7. Intact male and female ferrets are called
   a. jacks and jills, respectively
   b. kits and jills, respectively
   c. hobs and jills, respectively
   d. jacks and kits, respectively

8. A young llama is called a
   a. calf
   b. cria
   c. clutch
   d. colt

9. *Freshening* is a term that means
   a. cleaning an animal to make it smell fresh
   b. giving birth to a dairy animal
   c. the act of mating in cattle
   d. removing the horns of a bovine to enhance mating

10. Giving birth to swine is called
    a. barrowing
    b. queening
    c. farrowing
    d. tupping

11. Tom is to queen as hob is to
    a. doe
    b. bitch
    c. jill
    d. cow

12. An intact male rabbit is known as a
    a. stud
    b. stallion
    c. buck
    d. gander

13. A young female bovine that has not given birth is called a
    a. cow
    b. stag
    c. calf
    d. heifer

14. A female equine that has never been bred is known as a
    a. ridgeling mare
    b. brood mare
    c. maiden mare
    d. barren mare

15. A group of eggs is known as a
    a. flock
    b. clutch
    c. poult
    d. herd

16. The symbol for female is
    a. f
    b. I
    c. ♂
    d. ♀

17. A ram is to a wether as a bull is to a
    a. barrow
    b. steer
    c. gelding
    d. gib

18. A mule is an offspring of a
    a. male donkey and a female pony
    b. male pony and a female horse
    c. male horse and a female donkey
    d. male donkey and a female horse

19. Giving birth to rabbits is known as
    a. whelping
    b. farrowing
    c. freshening
    d. kindling

20. A group of dogs is known as a
    a. pack
    b. herd
    c. flock
    d. gaggle

21. An intact female llama is a
    a. doe
    b. sow
    c. cow
    d. ewe

22. A bitch is an intact
    a. female cat
    b. female dog
    c. female ferret
    d. female rat

23. A group of horses is called a
    a. herd
    b. pack
    c. litter
    d. clutch

24. An intact female is a hen in all of the following species except:
    a. duck
    b. turkey
    c. parrot
    d. chicken

25. An intact female sheep is a/an
    a. lamb
    b. ewe
    c. doe
    d. cow

Matching

Match the species common name in Column I with its taxonomic name in Column II.

Column I   **Column II**

1. _____ cattle   a. canine

2. _____ cat   b. bovine

3. _____ pig   c. equine

4. _____ parrot   d. feline

5. _____ rat   e. caprine

6. _____ dog   f. ovine

7. _____ sheep   g. porcine

8. _____ mouse   h. psittacine

9. _____ donkey   i. murine

10. _____ horse

11. _____ goat

Match the female name in Column I with its species in Column II.

Column I   **Column II**

12. _____ bitch   a. horse

13. _____ queen   b. bovine

14. _____ dam   c. goat

15. _____ hen   d. dog

16. _____ ewe   e. mouse or rat

17. _____ doe   f. cat

18. _____ goose   g. sheep

19. _____ jill   h. goose

20. _____ mare   i. chicken

21. _____ cow   j. ferret

Fill in the Blanks

Write the correct term in the space provided.

1. A young dog is called a(n) _____ or _____.

2. A young cat is called a(n) _____.

3. A young horse is called a(n) _____.

4. A young bovine is called a(n) _____.

5. A young goat is called a(n) _____.

6. A young sheep is called a(n) _____.

7. A young swine is called a(n) _____ or _____.

8. Young rabbits or ferrets are called _____.

9. Young mice or rats are called _____.

10. A young llama is called a(n) _____.

11. A group of rabbits is called a(n) _____.

12. A group of cattle is called a(n) _____.

13. A group of turkeys is called a(n) _____.

14. A group of sheep is called a(n) _____.

15. A group of goats is called a(n) _____.

16. An ostrich up to 2 days old is called a(n) _____.

17. A ratite less than 6 months of age is called a(n) _____.

18. A young deer is called a(n) _____.

19. A young castrated male chicken is called a(n) _____.

20. A group of eggs is called a(n) _____.

Spelling

Cross out any misspelled words in the following sentences and replace them with the proper spelling.

1. An intact female horse that was not bred is a barrin mare. _____

2. A young equine between 1 and 2 years of age is a yearing. _____

3. Multiple offspring born during the same labor is a liter. _____

4. An intact male guinea pig is a bore. _____

5. A male pig castrated when young is a barow. _____

6. A young castrated male chicken or domestic fowl is a capron. _____

7. A hurd is a group of horses, pigs, rabbits, and cattle. _____

8. A hiefer is a young female bovine that has not given birth. _____

9. A young sheep is a lam. _____

10. A young lama is a crea. _____

Crossword Puzzle

Animal Status Terms

Supply the correct term in the appropriate space for the definition listed.

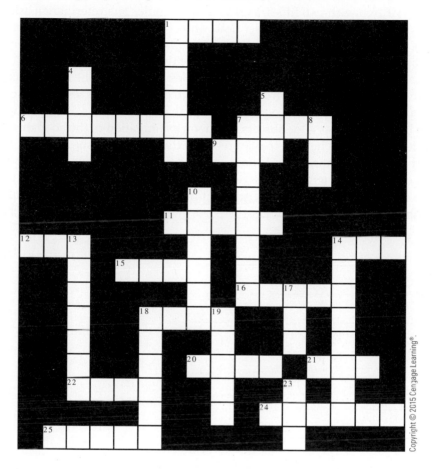

Across

1 male rabbit or goat

6 intact male horse over 4 years old

7 intact male horse less than 4 years old

9 female rabbit or goat

11 female cat

12 canine

14 young goat

15 young horse

16 castrated rabbit

18 feline (pl.)

20 group of bovine

21 young ferret or rabbit

22 caprine

24 group of geese

25 female dog

Down

1 male swine castrated when young

4 intact male swine

5 female caprine

7 immature male chicken

8 male turkey

10 immature female chicken

13 castrated llama

14 giving birth to a ferret or rabbit

17 young mouse or rat

18 group of eggs

19 ovine

23 intact male ovine

Case Study

After reading the case study, define the list of terms.

A 6-year-old quarter horse mare presented for a slight change in her regular gait (the way the animal moves). She is a barren mare who is pastured with two fillies, a gelding, and a mule. The veterinarian asks the owner to trot the horse to assess the lameness. The horse bobs her head while trotting, and after the owner dismounts, the horse stands resting the right front limb while pointing the hoof. On PE, the veterinarian feels for a digital pulse in the right front limb by placing his thumb and forefinger on either side of the fetlock joint. He then compares the digital pulse from the right front limb to the one from the left front limb. The right front limb digital pulse is stronger and faster than the one from the left front limb. The veterinarian believes the horse has a hoof abscess (localized collection of pus) and needs to rule out a toe, sole, bar, or heel abscess. Because the horse points with her front right hoof, he believes she has a toe abscess and upon examination of the hoof sees the hoof wall has a vertical crack that is filled with dirt. The veterinarian removes the dirt and flushes the hoof to drain the abscess. The veterinarian tells the owner the horse should be monitored for the appearance of a small horizontal crack in the hoof wall near the site of the original abscess, which may develop a few months after toe abscesses are drained. He explains to the owner that as the hoof grows, it may get contaminated with more dirt and another abscess may develop. The owner agreed to monitor the horse and contact a farrier to fit the horse with proper shoes.

1. mare _____

2. barren mare _____

3. fillies _____

4. gelding _____

5. mule _____

6. hoof _____

7. PE _____

8. digital _____

9. fetlock joint _____

10. toe _____

11. sole _____

12. bar _____

13. heel _____

14. hoof wall _____

Critical Thinking Exercise

The following story and questions are designed to stimulate critical thinking through class discussion or brief writing responses. There are many correct answers to these questions.

Justine Penn has been working as a veterinary technician in a new mixed animal veterinary clinic for 1 month and is beginning to become comfortable using large animal terminology after working the previous 4 years in a small animal practice. There are days she still struggles with the terms clients use to describe their animals. Sometimes taking a thorough history on livestock can be difficult when owners use words such as *farrowing, poult, steer,* and *gelding.* Justine is familiar with terms such as *whelping* and *stud*; however, large animal terms are difficult to remember quickly because there are so many terms and different species. She wonders how she can keep them all straight.

1. What are some things Justine could do to quickly learn species-specific terminology?

2. What are some resources she could use to master these terms?

3. What obstacles may Justine encounter if she does not learn the terms commonly used by the clients in this practice?

Chapter 6
GUT INSTINCTS

Objectives

Upon completion of this chapter, the reader should be able to:

- Identify and describe the major structures and functions of the digestive tract
- Distinguish between monogastric and ruminant digestive system anatomy and physiology
- Describe the processes of digestion, absorption, and metabolism
- Recognize, define, spell, and pronounce terms related to the diagnosis, pathology, and treatment of the digestive system
- Interpret the meaning of abbreviations of the digestive system

FUNCTIONS OF THE DIGESTIVE SYSTEM

The **digestive** (dī-jehs-tihv) **system, alimentary** (āl-ih-mehn-tahr-ē) **system, gastrointestinal** (gahs-trō-ihn-tehst-ihn-ahl) **system**, and **GI system** are terms used to describe the body system that is a long, muscular tube beginning at the mouth and ending at the anus. The digestive system is responsible for the intake and digestion of food and water, the absorption of nutrients, and the elimination of solid waste products. The combining form for nourishment is **aliment/o**.

STRUCTURES OF THE DIGESTIVE SYSTEM

The major structures of the digestive system include the oral cavity, pharynx, esophagus, stomach, and small and large intestines. The liver, gallbladder, and pancreas are organs associated with the digestive system.

Mouth

The mechanical and chemical process of digestion begins in the mouth. The mouth contains the major structures of the oral cavity. The **oral** (ōr-ahl) **cavity**

127

contains the lips and cheeks, hard and soft palates, salivary glands, tongue, teeth, and periodontium. The combining forms **or/o** and **stomat/o** mean mouth.

The maxilla and mandible are bones that are the boundaries of the oral cavity. The combining form for jaw is **gnath/o**. An animal that has **prognathia** (prohg-nah-thē-ah) has an elongated mandible, or a mandible that is overshot. Prognathia is sometimes called sow mouth. An animal that has **brachygnathia** (brahck-ē-nah-thē-ah) has a shortened mandible, or a mandible that is undershot. Brachygnathia is sometimes called parrot mouth.

The lips form the opening to the oral cavity. The term **labia** (lā-bē-ah) is the medical term for lips. A single lip is a **labium** (lā-bē-uhm). The combining forms for lips are **cheil/o** and **labi/o**. The cheeks form the walls of the oral cavity. The combining form for cheek is **bucc/o**. The term **buccal** (buhk-ahl or būk-ahl) means pertaining to or directed toward the cheek.

The **palate** (pahl-aht) forms the roof of the mouth. The palate consists of two parts: the hard and soft palates. The **hard palate** forms the bony rostral portion of the palate that is covered with specialized mucous membrane. This specialized mucous membrane contains irregular folds called **rugae** (roo-gā) (Figure 6–1). Rugae also are found in the stomach. **Rug/o** is the combining form for wrinkle or fold. The **soft palate** forms the flexible caudal portion of the palate. The soft palate is involved in closing off the nasal passage during swallowing so food does not move into the nostrils. The combining form **palat/o** means palate.

The **tongue** (tuhng) is a movable muscular organ in the oral cavity used for tasting and processing food, grooming, and articulating sound. The tongue moves food during chewing and swallowing. The dorsum of the tongue has **papillae** (pah-pihl-ā), or elevations, and the ventral surface of the tongue is highly vascular (Figure 6–2). The types of papillae located on the dorsum of the tongue may appear threadlike, or **filiform** (fihl-ih-fōrm); mushroomlike, or **fungiform** (fuhn-jih-fōrm); or cup-shaped, or **vallate** (vahl-āt). Taste buds are located in the fungiform and vallate papillae. The tongue is connected to the ventral surface of the oral cavity by a band of connective tissue called the **frenulum** (frehn-yoo-luhm). The combining forms for tongue are **gloss/o** and **lingu/o**. The **lingual surface** of the cheek is the side adjacent to the tongue.

The combining forms **dent/o**, **dent/i**, and **odont/o** are used to refer to teeth. **Dentition** (dehn-tihsh-uhn) refers to the teeth as a whole, that is, the teeth arranged in the maxillary (upper) and mandibular (lower) arcades. The primary dentition or **deciduous** (deh-sihd-yoo-uhs) **dentition** is the temporary set of teeth that erupt in young animals and are replaced at

FIGURE 6–1 Rugae of the stomach are visible on this preserved cat.

FIGURE 6–2 Papillae are found on the dorsum of the tongue.

or near maturity. **Decidu/o** is the combining form for shedding. The **permanent dentition** is the set of teeth designed to last the lifetime of an animal. Sometimes a deciduous tooth of brachydontic animals is not shed at the appropriate time, and both the deciduous and permanent teeth are situated beside each other. The period in which both deciduous and permanent teeth are present is the mixed dentition. The deciduous tooth that has not been shed is called a **retained deciduous tooth** and may be extracted (removed) professionally.

The four types of teeth have different functions (Figure 6–3):

- **incisor** (ihn-sīz-ōr) = front tooth used for cutting; an incision is a cut; abbreviated I.
- **canine** (kā-nīn) = long, pointed bonelike tooth located between the incisors and premolars; also called **fang** and **cuspid** (*cusp* means having a tapering projection; *cuspid* means having one point); abbreviated C.
- **premolar** (prē-mō-lahr) = cheek tooth found between the canine teeth and molars; also called **bicuspids** because they have two points; abbreviated P.
- **molar** (mō-lahr) = most caudally located permanent cheek tooth used for grinding; *molar* comes from the Latin term *to grind*; abbreviated M.

When the number and type of teeth found in an animal are written, a shorthand method is used. This method is called the dental formula. The **dental formula** of an animal represents the type of tooth and the number of each tooth type found in that species. In the dental formula, only one side of the mouth is represented and is preceded by a 2 to indicate that this arrangement is the same on both sides. For example, the dental formula for the adult dog is 2(I 3/3, C 1/1, P 4/4, M 2/3). This means that an adult dog has 3 upper incisors, 1 upper canine, 4 upper premolars, and 2 upper molars on the left and right sides of its mouth. Similarly, an adult dog has 3 lower incisors, 1 lower canine, 4 lower premolars, and 3 lower molars

on the left and right sides of its mouth. Adding all of the numbers together gives the total dentition, which in the adult dog is 42 (Figure 6–4 and Table 6–1).

(a) Carnivore

(b) Herbivore

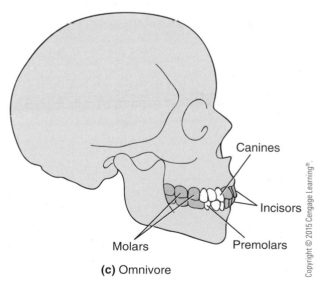

(c) Omnivore

FIGURE 6–3 Comparison of skull and teeth. (a) Carnivores (e.g., cats) have pointed canines and incisors for obtaining and tearing flesh. Sharp molars and premolars are essential for shearing flesh. (b) Herbivores (e.g., horses) have teeth adapted to biting off plant material and grinding the food into smaller pieces. (c) Omnivores (e.g., humans) have teeth adapted for eating a variety of foods.

Occasionally, teeth are referred to by terms other than those seen in dental formulas. Examples of lay terms for teeth include the following:

- **cheek teeth**: premolars and molars
- **needle teeth**: deciduous canines and third incisor of pigs
- **wolf teeth**: rudimentary premolar 1 in horses
- **milk teeth**: first set of teeth
- **tusks**: permanent canine teeth of pigs
- **carnassial** (kahr-nā-zē-ahl) **tooth**: large, shearing cheek tooth; upper P4 and lower M1 in dogs; upper P3 and lower M1 in cats
- **fighting teeth**: set of six teeth in llamas that include upper vestigial incisors and upper and lower canines on each side

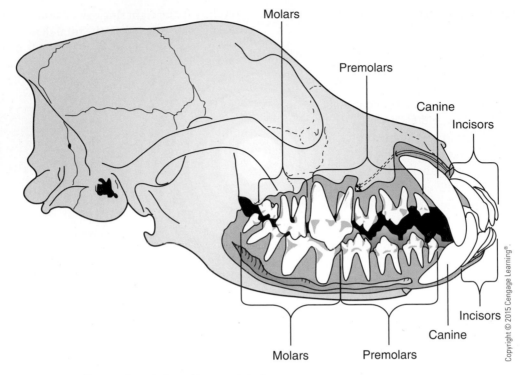

FIGURE 6–4 Types of teeth found in an adult dog.

TABLE 6-1 Dental Formulas of Select Adult Animals

Animal	Formula	Total Number of Adult Teeth
Dog	2(I 3/3, C 1/1, P 4/4, M 2/3)	42
Cat	2(I 3/3, C 1/1, P 3/2, M 1/1)	30
Bovine, sheep, goat	2(I 0/4, C 0/0, P 3/3, M 3/3) or 2(I 0/3, C 0/1, P 3/3, M 3/3)	32
Horse	2(I 3/3, C 1/1, P 3-4/3, M 3/3)	40 or 42
Pig	2(I 3/3, C 1/1, P 4/4, M 3/3)	44
Llama (female) (male)	2(I 1/3, C 1/1, P 2/1, M 3/2) 2(I 1/3, C 1/1, P 2/1, M 3/3)	28 30
Rabbit	2(I 2/1, C 0/0, P 3/2, M 3/3)	28
Ferret	2(I 3/3, C 1/1, P 3/3, M 1/2)	34
Chinchilla	2(I 1/1, C 0/0, P 1/1, M 3/3)	20
Gerbil	2(I 1/1, C 0/0, P 0/0, M 3/3)	16
Hamster	2(I 1/1, C 0/0, P 0/0, M 3/3)	16
Rat and mouse	2(I 1/1, C 0/0, P 0/0, M 3/3)	16
Guinea pig	2(I 1/1, C 0/0, P 1/1, M 3/3)	20

Teeth also can be identified using the triadan (trī-ə-dahn) system. In this numbering system, each tooth has a three-digit number. The first digit (the quadrant number) represents the quadrant in which the tooth is located. The quadrant numbers start on the animal's upper right side (right side of the maxilla), continues to the animal's upper left side (left side of the maxilla), then the animal's lower left side (left side of the mandible), and ends on the lower right side (right side of the mandible). The second and third digits represent individual teeth. The numbering of individual teeth starts at rostral midline and increases in number from rostral to caudal. For example, all central incisor teeth in dogs and cats are numbered 01; therefore, the central incisor on the right side of the maxilla is 101, the central incisor on the left side of the maxilla is 201, the central incisor on the left side of the mandible is 301, and the central incisor on the right side of the mandible is 401. Likewise, all canine teeth in dogs and cats are numbered 04; therefore, the canine tooth on the right side of the maxilla is 104, the canine tooth on the left side of the maxilla is 204, the canine tooth on the left side of the mandible is 304, and the canine tooth on the right side of the mandible is 404 (Figure 6–5).

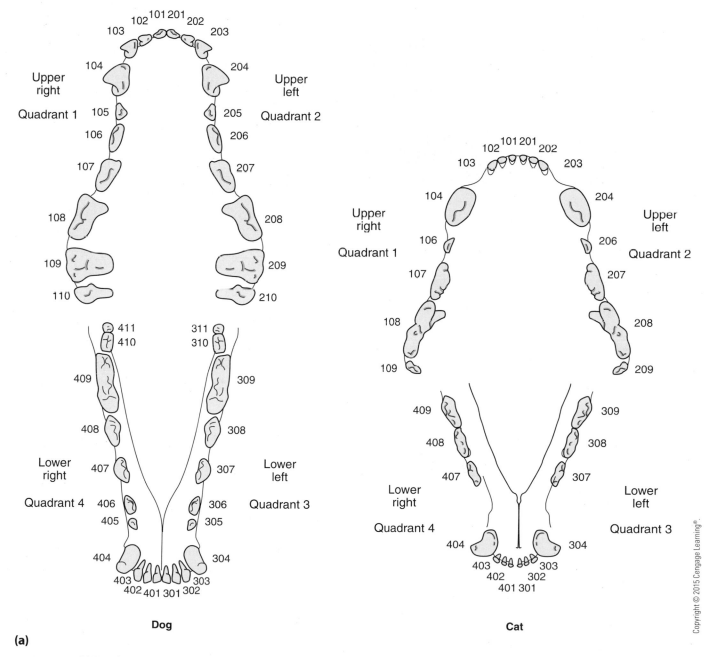

FIGURE 6–5 (a) Triadan system of tooth identification of dogs and cats. (b) Canine dental model A = mandible and B = maxilla. (c) Feline dental model A = mandible and B = maxilla.

(b)

(c)

FIGURE 6.5 (continued)

Animals may be grouped based on the types of teeth they have. Examples include the following:

- **selenodont** (sē-lēn-ō-dohnt) = animals with teeth that have crescents on their grinding surfaces (e.g., ruminants)
- **lophodont** (lō-fō-dohnt) = animals with teeth that have ridged occlusal surfaces (e.g., equine)
- **bunodont** (boon-ō-dohnt) = animals with teeth that have worn, rounded surfaces (e.g., swine)
- **hypsodont** (hihps-ō-dohnt) = animals with continuously erupting teeth (e.g., cheek teeth of ruminants)
- **pleurodont** (pluhr-ō-dohnt) = animals with teeth attached by one side on the inner jaw surface (e.g., lizards)
- **brachydont** (brā-kē-dohnt) = animals with permanently rooted teeth (e.g., carnivores)

The anatomy of the tooth consists of the enamel (located in the crown) or cementum (located in the root), dentin, and pulp (Figure 6–6). The **enamel** (ē-nahm-ahl) is the hard white substance covering the dentin of the crown of the tooth. **Cementum** (sē-mehn-tuhm) is the bonelike connective tissue that covers the root of the tooth. The **dentin** (dehn-tihn) is the connective tissue surrounding the tooth pulp. The tooth **pulp** consists of nerves, blood vessels, and loose connective tissue. The hole at the tip of the root where nerves and blood vessels enter the tooth is the **apical foramen** (ā-pih-kahl fō-rā-mehn). The **periodontia** (pehr-ē-ō-dohn-shē-ah) are the structures that support the teeth. Teeth are situated in sockets or saclike dilations called **alveoli** (ahl-vē-ō-lī). A single socket is called an alveolus (ahl-vē-ō-luhs). The thin layer of compact bone that forms the tooth socket is **alveolar bone** (ahl-vē-ō-lahr bōn). Alveolar bone surrounds the roots of the teeth. The fibrous structure that holds the tooth in the alveolus is the **periodontal** (pehr-ē-ō-dohn-tahl) **ligament.** The periodontal ligament contains collagen fibers that are anchored to the cementum of the tooth and to the alveolar bone.

The **gingiva** (jihn-jih-vah) is the mucous membrane that surrounds the teeth and forms the mouth lining. The gingiva is sometimes called the gums. The combining form **gingiv/o** means gums. The space that surrounds the tooth is the **gingival sulcus** (jihn-jih-vahl suhl-kuhs). The gingival sulcus is located between the tooth and gingival margin.

Salivary glands (sahl-ih-vahr-ē glahndz) are a group of cells located in the oral cavity that secrete a clear substance containing digestive enzymes (saliva). The **saliva** (sah-lī-vah) moistens food, begins the digestive process by aiding in bolus formation and some digestive enzyme activity (amylase in some animal species), and cleanses the mouth. Different salivary glands are named for the location in which they are found: the **mandibular** (mahn-dihb-yoo-lahr) **salivary glands** are found near the mandible the **sublingual** (suhb-lihn-gwahl) **salivary glands** are found under the tongue, the **zygomatic** (zī-gō-mah-tihck) **salivary glands** are found medial to the zygomatic arch, and the **parotid** (pah-roht-ihd) **salivary glands** are found near the ear. **Para-** is the prefix for near, and the combining form **ot/o** means ear. The combining forms for salivary glands are **sialaden/o** and **sial/o** (which also means saliva) (Figure 6–7).

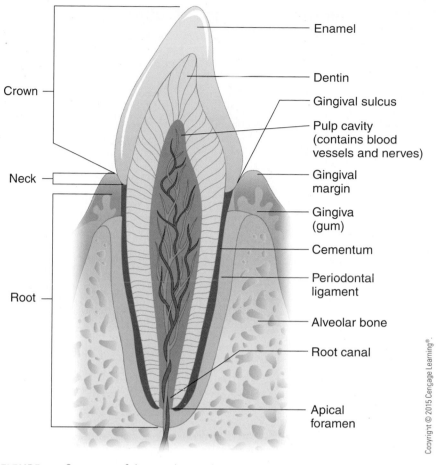

FIGURE 6–6 Structure of the tooth in a dog or cat.

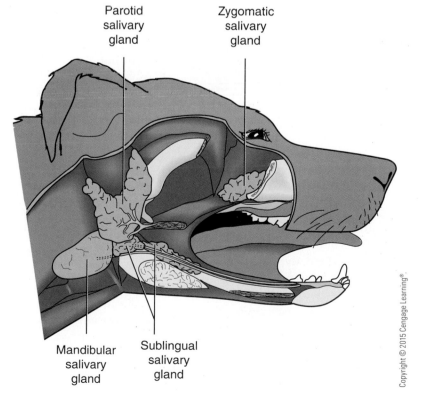

FIGURE 6–7 Salivary glands of the dog.

Branches of Dentistry

Veterinary dental specialties consist of the following:

- **endodontics** (ehn-dō-dohn-tihks) = branch of dentistry that involves treatment of diseases that affect the tooth pulp
- **exodontics** (ehcks-ō-dohn-tihks) = branch of dentistry that involves extraction of teeth and related procedures
- **oral surgery** = branch of surgery that involves surgical correction of the jaw, gums, and inside of the mouth
- **orthodontics** (ohrth-ō-dohn-tihks) = branch of dentistry that involves the guidance and correction of malocclusion
- **periodontics** (pehr-ē-ō-dohn-tihks) = branch of dentistry that studies and treats the diseases of tooth-supporting structures

Pharynx

The **pharynx** (fār-ihncks) is the cavity in the caudal oral cavity that joins the respiratory and gastrointestinal systems. The pharynx also is called the **throat**. The combining form for pharynx is **pharyng/o**. The pharynx is covered in Chapter 9.

Hard to Swallow

Chewing, also called **mastication** (mahs-tih-kā-shuhn), makes food easier to swallow by breaking it into smaller pieces and by increasing the surface area of food particles or ingesta. **Ingesta** (ihn-jehst-ah) is the material taken in orally. This increased surface area increases the contact between digestive enzymes and the food and may speed up the breakdown of food. The first digestive enzymes with which food comes in contact are found in saliva. Sometimes animals salivate or drool when they smell food. **Hypersalivation** (hī-pər-sahl-ih-vā-shuhn) is excessive production of saliva. Hypersalivation also is called **ptyalism** (tī-uh-lihz-uhm) and **hypersialosis** (hī-pər-sī-ahl-ō-sihs). The combining forms for saliva are **sial/o** and **ptyal/o**.

The process of swallowing is called **deglutition** (dē-gloo-tih-shuhn). The combining form **phag/o** means eating or ingestion. Swallowed food passes from the mouth to the pharynx and then to the esophagus.

Gullet

The **esophagus** (ē-sohf-ah-guhs) is a collapsible, muscular tube that leads from the oral cavity to the stomach. The esophagus is located dorsal to the trachea. The combining form **esophag/o** means esophagus. The esophagus enters the stomach through an opening that is surrounded by a **sphincter**. A sphincter (sfihngk-tər) is a ringlike muscle that constricts an opening.

Stomach

After the esophagus, the remaining organs of digestion are located in the abdominal cavity (Figure 6–8). The **abdomen** is the cavity located between the diaphragm and pelvis. The combining forms for abdomen are **abdomin/o** and **celi/o**. **Lapar/o** is the combining form for abdomen and flank. The **peritoneum** (pehr-ih-tō-nē-uhm) is the membrane lining that covers the abdominal and pelvic cavities and some of the organs in that area. The layer of the peritoneum that lines the abdominal and pelvic cavities is called the **parietal peritoneum**, and the layer of the peritoneum that covers the abdominal organs is called the **visceral peritoneum**.

Food enters the stomach from the esophagus, where it is stored, and the act of digestion begins. The stomach is connected to other visceral organs by a fold of peritoneum called the **lesser omentum** (ō-mehn-tuhm) and to the dorsal abdominal wall by another fold of peritoneum called the **greater omentum** (Figure 6–9). *Omentum* is Latin for apron.

The combining form for stomach is **gastr/o**. Animals can be classified as **monogastric** (mohn-ō-gahs-trihck) or **ruminant** (roo-mihn-ehnt). Monogastric animals have one true, or glandular, stomach. The **glandular stomach** is the one that produces secretions for digestion. Ruminants also have one true, or glandular, stomach (the abomasum), but they also have three forestomachs (the rumen, reticulum, and omasum). The forestomachs of ruminants are actually outpouchings of the esophagus. Because all animals have only one true or glandular stomach, monogastric animals often are called simple nonruminant animals. The parts of the true stomach are as follows (Figure 6–10):

- **cardia** (kahr-dē-ah) = entrance area located nearest the esophagus.
- **fundus** (fuhn-duhs) = base of an organ, which is the cranial, rounded part.
- **body** (boh-dē) = main portion of an organ, which is the rounded base or bottom; also called the **corpus** (kōr-puhs).
- **antrum** (ahn-truhm) = caudal part, which is the constricted part of the stomach that joins the pylorus.

- **pylorus** (pī-lōr-uhs) = narrow passage between the stomach and the duodenum. The combining form **pylor/o** means gatekeeper and refers to the narrow passage between the stomach and duodenum. The **pyloric sphincter** is the muscle ring that controls the flow of material

from the stomach to the duodenum of the small intestine.

- **rugae** (roo-gā) = folds present in the mucosa of the stomach. Rugae contain glands that produce gastric juices that aid in digestion and mucus that forms a protective coating for the stomach lining.

(a) Dog

(b) Horse

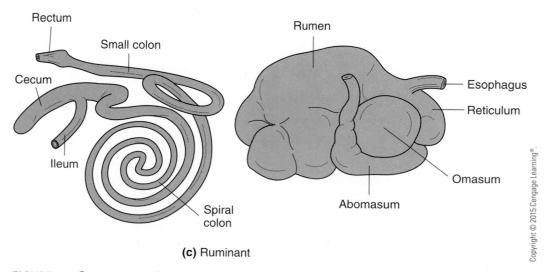

(c) Ruminant

FIGURE 6–8 Gastrointestinal tracts. (Small intestinal segments were omitted for clarity.) (a) Dog. (b) Horse. (c) Ruminant.

(a)

(b)

Copyright © 2015 Cengage Learning®.

Copyright © 2015 Cengage Learning®.

FIGURE 6–9 (a) Omentum in a preserved cat. (b) Intraoperative photograph demonstrating numerous dark red tumors in the omentum of a dog with a splenic tumor.

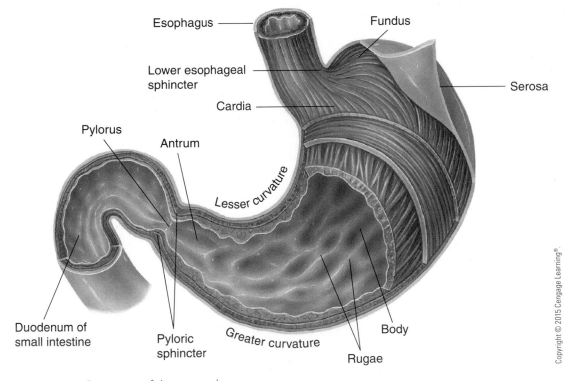

Esophagus — Fundus

Lower esophageal sphincter — Serosa

Cardia

Pylorus

Antrum

Lesser curvature

Duodenum of small intestine

Pyloric sphincter

Greater curvature

Body

Rugae

Copyright © 2015 Cengage Learning®.

FIGURE 6–10 Structures of the stomach.

Ruminants

Ruminants are animals that can **regurgitate** (rē-guhr-jih-tāt) and **remasticate** (rē-mahs-tih-kāt) their food. The ruminant stomach is adapted for fermentation of ingested food by bacterial and protozoan microorganisms. Normal microorganisms residing in the gastrointestinal tract are called **intestinal flora** (ihn-tehs-tih-nahl flō-rah). These microbes produce enzymes that can digest plant cells through fermentation. Fermentation is aided by regurgitation (return of undigested material from the rumen to the mouth) and remastication (rechewing). Regurgitation and remastication provide finely chopped material with a greater surface area to the stomach. Regurgitated food particles, fiber, rumen fluid, and rumen microorganisms are called **cud** (kuhd).

BIOLOGY BONUS

The wall of the digestive tract has four layers: mucosa, submucosa, muscular layer, and serosa (Figure 6–11). The mucosa is the lining of the digestive system and is the innermost layer. The mucosa of the mouth and anus is thick for protection while in the stomach and intestines it is thin for secretion and absorption. The submucosa is a thick layer surrounding the mucosa. The submucosa contains blood vessels, lymphatic vessels, and nerves. The muscular layers, called the muscularis, contain smooth muscle responsible for movements of the digestive tract and is arranged in two layers. The serosa is the connective tissue layer covering the digestive tract and secretes serous fluid for lubrication so that the abdominal organs move smoothly against each other without friction.

Serosa

Muscularis

Submucosa

Mucosa

Copyright © 2015 Cengage Learning®.

FIGURE 6–11 Basic structure of the digestive tract wall.

The ruminant stomach is divided into the following four parts (Figures 6–12 and 6–13):

- **rumen** (roo-mehn) = largest compartment of the ruminant stomach that serves as a fermentation vat; also called the **paunch**. The rumen is divided into a ventral sac and a dorsal sac.
- **reticulum** (rē-tihck-yoo-luhm) = most cranial compartment of the ruminant stomach; also called the honeycomb because it is lined with a mucous membrane that contains numerous intersecting ridges.
- **omasum** (ō-mā-suhm) = third compartment of the ruminant stomach. The omasum has short, blunt papillae that grind food before it enters the abomasum. Omasal contractions also squeeze fluid out of the food bolus.
- **abomasum** (ahb-ō-mā-suhm) = fourth compartment of the ruminant stomach; also called the true stomach. The abomasum is the glandular portion that secretes digestive enzymes.

The layout of the ruminant stomach depends on the age of the animal. In adult ruminants, the rumen is the largest compartment and occupies a prominent portion of the left side of the animal. For the most part, the abomasum is on the right side of the animal. In young ruminants, the abomasum is the largest compartment. Forestomach development is associated with roughage intake, and calves are fed only milk for a period of time after birth.

In adult ruminants, food enters the rumen. A contraction transfers the rumen contents into the reticulum. The foodstuff is then regurgitated or directed toward the caudal part of the rumen or the omasum.

(a)

(b)

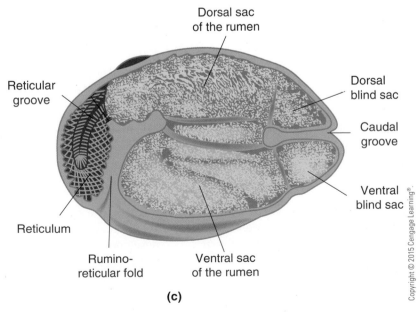

(c)

FIGURE 6–12 (a) Parts of the ruminant stomach. (b) Left side of ruminant stomach. (c) Interior view of the rumen and reticulum.

FIGURE 6–13 Linings of the four stomachs of cattle. (a) Rumen. Note the small fingerlike projections, the rumen papillae, that increase the surface area for absorption. (b) Reticulum. The distinct lining of the reticulum looks like a honeycomb. (c) Omasum. The lining has large sheets with ingesta between layers. (d) Abomasum. This smooth glandular lining appears very similar to the monogastric stomach.

The plies of the omasum grind the food, and water is removed. Food enters the abomasum, or true stomach, which is similar to the glandular stomach of the monogastric. In young ruminants, a reticular groove shuttles milk from the esophagus to the abomasum.

Small Intestine

The small intestine, or small bowel, extends from the pylorus to the proximal part of the large intestine. The small intestine is attached to the dorsal abdominal wall by a fold of the peritoneum called the **mesentery** (mehs-ehn-tehr-ē) (Figure 6–14). Digestion and absorption of food takes place in the small intestine for animals not needing extensive fermentation of their ingested food. The combining form **enter/o** means small intestine. **Gastroenterology**

FIGURE 6–14 The mesentery is a fold of peritoneum that attaches the small intestine to the dorsal abdominal wall.

Rumination (roo-mehn-ā-shuhn) is the process of bringing up food material from the stomach to the mouth for further chewing. Rumination is a cycle of four phases: regurgitation, remastication, resalivation, and redeglutition.

Step 1: Regurgitation. The animal takes a deep breath (glottis is closed); the thoracic cavity enlarges; intrapleural pressure decreases; the cardia opens; and because of the low pressure in the esophagus, the rumen content is aspirated into the esophagus. Reverse peristalsis occurs, and the food bolus quickly enters the mouth.

Steps 2 and 3: Remastication and resalivation. Liquid is squeezed out of bolus, and the liquid is reswallowed. Remastication and resalivation occur together, and the animal may chew its cud 100 or more times before swallowing.

Step 4: Redeglutition. The bolus is reswallowed, and the next rumination cycle begins. Rumination usually occurs when the animal is quiet. The time spent ruminating each day varies with species and diet. Smaller particles (such as ground feed) take less rumination time.

A **herbivore** (hərb-ih-vōr) is an animal that is able to sustain life by eating only plants.

An **omnivore** (ohm-nih-vōr) is an animal that sustains life by eating plant and animal tissue.

A **carnivore** (kahr-nih-vōr) is an animal that is able to sustain life by eating only animal tissue. A carnivore may eat a plant, but that does not make it an omnivore. A cat is a carnivore, but cats may occasionally eat grass. This does not make them omnivores.

(gahs-trō-ehn-tər-ohl-ō-jē) is the study of the stomach and small intestine.

The small intestine has the following three segments (Figure 6–15):

- **duodenum** (doo-ō-də-nuhm or doo-wahd-nuhm) = proximal or first portion of the small intestine. The proximal portion also is known as the most oral portion. The duodenum is the segment of the small intestine located nearest the mouth. The combining forms **duoden/i** and **duoden/o** mean duodenum.
- **jejunum** (jeh-joo-nuhm) = middle portion of the small intestine. The combining form **jejun/o** means jejunum.

- **ileum** (ihl-ē-uhm) = distal or last portion of the small intestine. The distal portion also is known as the most **aboral** (ahb-ōr-ahl) portion. The ileum is the segment of the small intestine located furthest from the mouth. The combining form **ile/o** means ileum. (Remember that *ileum* is spelled with an *e*, as in *eating* or *entero*. The ilium is part of the hip bone.)

Once food is digested in the small intestine, it forms a milky fluid. This milky fluid is called **chyle** (kī-uhl). Chyle is absorbed through the intestinal wall and travels via the thoracic duct, where it is passed into veins.

Large Intestine

The large intestine, or large bowel, extends from the ileum to the anus. The large intestine consists of the **cecum** (sē-kuhm), **colon** (kō-lihn), **rectum** (rehck-tuhm), and **anus** (ā-nuhs). Development of the large intestine varies among species. Fermentation occurs to some extent in the large intestine of all animals but is a more consuming process in herbivorous animals. In ruminants, the forestomach constitutes a fermentation vat (hence, they are called foregut fermenters) and in nonruminant herbivores (e.g., rabbits and horses), the cecum and colon provide fermentation (hence, they are called hindgut fermenters).

Food enters a pouch, called the cecum, from the ileum. The cecum may be poorly developed, as in the dog, or large, as in the horse, or double-pouched, as in birds. **Cec/o** is the combining form for cecum. The ileocecal junction is the area of the gastrointestinal tract where the small intestine (ileum) opens into the large intestine (cecum).

The colon continues from the cecum to its termination at the rectum. The colon consists of ascending, transverse, and descending portions. All animals have a transverse and descending portion. The arrangement from the cecum to the transverse colon varies among species. Dogs and cats have an ascending colon, pigs and ruminants have spiral colons, and horses have large colons. **Col/o** is the combining form for colon. The cecum and colon of pigs and horses are sacculated; these sacculations are called **haustra** (haw-strah) (Figure 6–16). Haustra act as buckets and prolong retention of material so that the microbes have more time for digestion. Haustra are formed because of the longitudinal smooth muscle bands, or **teniae** (tehn-ē-ā), in the cecal wall.

The rectum is the caudal portion of the large intestine. The combining form **rect/o** means rectum. The anus is the caudal opening of the gastrointestinal

(a) Dog

(b) Horse

Dorsal colon

Left ventral colon

Right ventral colon

Spiral loops

(c) Ruminant

A = Duodenum
B = Jejunum
C = Ileum
D = Cecum
E = Colon
F = Rectum
G = Anus

FIGURE 6–15 Gastrointestinal tracts of various animals. (a) Dog. (b) Horse. (c) Ruminant.

Courtesy of Laura Lien, CVT, BS, MS

FIGURE 6–16 Haustra and teniae of the equine large intestine.

The Colon

The colon is divided into three parts: ascending, transverse, and descending. **Ascending** is the part that progresses upward, or cranially. *Ascend* means to move up. **Transverse** is the part that travels across. **Trans-** is the prefix for across. **Descending** is the part that progresses downward, or caudally. To descend a flight of stairs is to move down. Bends or curves are called **flexures** (flehck-shərz); therefore, the **pelvic flexure** is a bend in the colon near the pelvis, and the **diaphragmatic flexure** is a bend in the colon near the diaphragm.

tract. The anus is controlled by two anal sphincter muscles that tighten or relax to allow or control defecation. The combining form for anus is **an/o. Anorectal** (ā-nō-rehck-tahl) is a term that means pertaining to the anus and rectum. The combining form **proct/o** refers to the anus and rectum collectively.

The anal canal is a short terminal portion of the digestive tract. Dogs and cats have a pair of pouches in the skin between the internal and external anal sphincters. These pouches are called **anal sacs**. Anal sacs are lined with microscopic anal glands that secrete a foul-smelling fluid. Normally, the anal sacs are compressed during defecation. The fluid may be related to social recognition in dogs and cats.

Accessories

Accessory organs aid the digestive tract in different ways. The accessory organs of the digestive tract include the salivary glands (covered previously), the liver, the gallbladder, and the pancreas (Figure 6–17).

Liver

The liver is located caudal to the diaphragm and has several important functions. The combining form for liver is **hepat/o**. The liver removes excess **glucose** (gloo-kohs) from the bloodstream and stores it as **glycogen** (glī-koh-jehn). When blood sugar is low, a condition called **hypoglycemia**

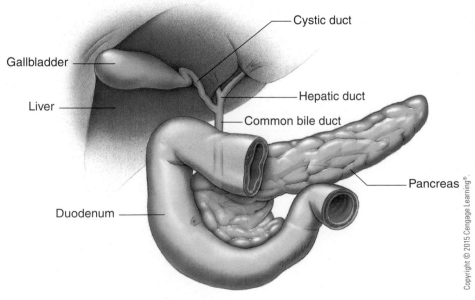

FIGURE 6–17 Accessory organs of digestion.

(hī-pō-glī-sē-mē-ah) results, and the liver converts glycogen back into glucose and releases it. The liver also destroys old erythrocytes; removes toxins from the blood; produces some blood proteins; and stores iron and vitamins A, B$_{12}$, and D. Liver cells are called **hepatocytes** (heh-paht-ō-sītz). The liver also contains **sinusoids** (sīn-yoo-soydz), or channels. The hepatocytes make up the liver **parenchyma** (pahr-ehnk-ih-mah), or the functional elements of a tissue or organ.

The digestive function of the liver involves the production of **bile** (bīl). The term **biliary** (bihl-ē-ār-ē) means pertaining to bile. Bile travels down the hepatic duct to the cystic duct, which leads to the gallbladder (in species that have a gallbladder). Bile alkalinizes the small intestine, and bile salts play a part in fat digestion. Fat digestion is called **emulsification** (ē-muhl-sih-fih-kā-shuhn). **Bilirubin** (bihl-ē-roo-bihn) is a pigment produced from the destruction of hemoglobin that is released by the liver in bile.

Gallbladder

The gallbladder is a sac embedded in the liver that stores bile for later use. When bile is needed, the gallbladder contracts, which forces bile out of the cystic duct into the common bile duct. The common bile duct carries bile into the duodenum. The rat and horse do not have gallbladders but have a continuous flow of bile into the duodenum. The combining form **cyst/o** means cyst, sac of fluid, or urinary bladder. The combining form **chol/e** means bile or gall. The combining form **doch/o** means receptacle. Together these combining forms are used to refer to the bile and its associated structures. **Cholecystic** (kō-lē-sihst-ihck) means pertaining to the gallbladder. **Choledochus** (kō-lehd-uh-kuhs) means common bile duct.

Pancreas

The **pancreas** (pahn-krē-ahs) is an elongated gland located near the cranial portion of the duodenum. The main pancreatic duct enters the duodenum close to the common bile duct. (In species such as sheep and goats, it empties directly into the common bile duct.)

Pancreat/o is the combining form for pancreas. The pancreas is an organ that has both exocrine and endocrine functions. The endocrine functions are covered in Chapter 11. The exocrine function of the pancreas involves the production of pancreatic juices, which are filled with digestive enzymes. **Trypsin** (trihp-sihn) is an enzyme that digests protein, **lipase** (lī-pās) is an enzyme that digests fat, and **amylase** (ahm-ih-lās) is an enzyme that digests carbohydrates. All are produced in the pancreas.

DIGESTION

Digestion (dī-jehst-shuhn) is the process of breaking down foods into nutrients that the body can use. **Enzymes** (ehn-zīmz) are substances that chemically change another substance. Digestive enzymes are responsible for the chemical changes that break foods into smaller forms for the body to use. Enzymes typically end with **-ase**. For example, proteases are enzymes that work on proteins, and lipases are enzymes that work on fats (lipids). One enzyme that does not end in *-ase* is pepsin, which is an enzyme that digests protein.

Metabolism (meh-tahb-ō-lihzm) is the processes involved in the body's use of nutrients. The prefix **meta-** means change or beyond. **Anabolism** (ah-nahb-ō-lihzm) is the building of body cells and substances. **Catabolism** (kah-tahb-ō-lihzm) is the breaking down of body cells and substances.

Absorption (ahb-sōrp-shuhn) is the process of taking digested nutrients into the circulatory system. A **nutrient** (nū-trē-ehnt) is a substance that is necessary for normal functioning of the body. Absorption occurs mainly in the small intestine. The small intestine contains tiny hairlike projections called **villi** (vihl-ī). A single projection is called a **villus** (vihl-uhs). The combining form **vill/i** means tuft of hair. Because the small intestine has villi, or projections, it also has blind sacs, or valleys. These blind sacs are called **crypts** (krihptz). An intestinal crypt is a valley of the intestinal mucous membrane lining the small intestine.

Path of Digestion

- **Prehension** (prē-hehn-shuhn), or grasping of food, involves collecting food in the oral cavity. The material taken orally is known as **ingesta** (ihn-jehst-ah).
- **Mastication** breaks food into smaller pieces and mixes the ingesta with saliva.
- **Deglutition** moves chewed ingesta into the pharynx and into the esophagus. (The epiglottis closes off the entrance to the trachea and allows food to move into the esophagus.)
- **Peristalsis** (pehr-ih-stahl-sihs) is the series of wavelike contractions of smooth muscles (Figure 6–18). (The suffix **-stalsis** means contraction.) Food moves down the esophagus by gravity and peristalsis.

FIGURE 6–18 Peristalsis actively propels ingesta through the intestinal tract.

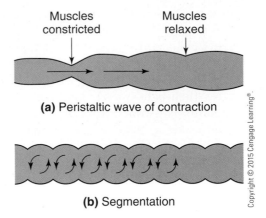

(a) Peristaltic wave of contraction

(b) Segmentation

FIGURE 6–19 Peristalsis versus segmentation. (a) Peristalsis propels food through the digestive system. (b) Segmentation helps break down and mix food through cement-mixer-type action.

The Scoop on Poop

The following terms are used for gastrointestinal waste:

- feces
- dung
- manure
- stool
- bowel movement
- excrement
- poop

Terms used to describe gastrointestinal waste include the following:

- solid
- loose
- soft
- watery
- mucoid
- bloody

Ingesta moves into the stomach. In ruminants, it enters the rumen, reticulum, and omasum before entering the true stomach, or abomasum. The true glandular stomach contains **hydrochloric** (hī-drō-klōr-ihck) **acid** and the enzymes **protease** (prō-tē-ās), **pepsin** (pehp-sihn), and **lipase** (lī-pās). The muscular action of the stomach mixes the ingesta with the gastric juices to convert the food to chyme. **Chyme** (kīm) is the semifluid mass of partly digested food that passes from the stomach.

Chyme passes from the stomach into the duodenum. Food moves through the small intestine by peristaltic action and **segmentation** (sehg-mehn-tā-shuhn). Peristalsis consists of contractile waves that propel ingesta caudally; segmentation mixes and thus delays movement of ingesta (Figure 6–19). Digestion is completed in the duodenum after chyme has mixed with bile and pancreatic secretions.

Digested food is absorbed in the small intestine. Another term for absorption is **assimilation** (ah-sih-mih-lā-shuhn).

The large intestine receives waste products of digestion and in some species is responsible for fermentation of plant material and vitamin synthesis. Excess water is absorbed from the waste, and solid feces are formed. **Defecation** (dehf-eh-kā-shuhn) is the emptying of the bowels.

TEST ME: DIGESTIVE SYSTEM

Diagnostic tests performed on the digestive system include the following:

- **ballottement** (bahl-oht-mehnt) = diagnostic technique of hitting or tapping the wall of a fluid-filled structure to bounce a solid structure against a wall; used for pregnancy diagnosis and determination of abdominal contents.
- **barium** (bār-ē-uhm) = contrast material used for radiographic studies. To evaluate the gastrointestinal tract, barium sulfate may be given orally (and the resulting test is called a **barium swallow** or

upper GI) or barium sulfate may be given rectally (and the resulting test is called a **barium enema** (BE) or **lower GI**). An **enema** (ehn-ah-mah) is introduction of fluid into the rectum (Figure 6–20).

- **biopsy** (bī-ohp-sē) = removal of tissue to examine. Biopsies can be **incisional** (ihn-sih-shuhn-ahl), in which part of the tissue is removed and examined, or they can be **excisional** (ehcks-sih-shuhn-ahl), in which the entire tissue is removed.

- **blood tests** = determination of blood parameters used to detect some diseases of the gastrointestinal tract. For example, bile acids are used to assess liver disease, and elevated amylase levels may indicate pancreatitis. Sometimes it is important to know whether the blood sample was taken as a preprandial or postprandial sample. **Preprandial** (prē-prahn-dē-ahl) is before a meal, and **postprandial** (pōst-prahn-dē-ahl) is after a meal (-**prandial** means meal).

- **colonoscopy** (kō-luhn-ohs-kō-pē) = endoscopic visual examination of the inner surface of the colon; the scope is passed from the rectum through the colon. An **endoscope** (ehn-dō-skōp) is a tubelike instrument with lights and refracting mirrors that is used to examine the body or organs internally. Endoscope is a generic term and obtains its specific name from the organ it is used to examine. For example, an instrument used to view the stomach is a gastroscope and an instrument used to view the colon is a colonoscope.

- **esophagoscopy** (ē-sohf-ah-gohs-kō-pē) = endoscopic visual examination of the esophagus; the scope is passed from the oral cavity through the esophagus (Figure 6–21).

(a)

(b)

(a)

(b)

Copyright © 2015 Cengage Learning®.

FIGURE 6–20 Radiograph of a cat during a barium series. The bright white areas are barium in the gastrointestinal tract. (a) Ventrodorsal view. (b) Lateral view.

Courtesy of Lodi Veterinary Hospital S.C.

Courtesy of Dr. Mark Jackson University of Glasgow

(c)

Courtesy of Eli Larson, DVM

FIGURE 6–21 Endoscopy. (a) An endoscope is passed through the esophagus (esophagoscopy) to view parts of the gastrointestinal tract. (b) View of the intestine from an endoscope. A string foreign body is identified. (c) View of the intestine during surgery to remove the string foreign body.

FIGURE 6–22 A fecal flotation is a basic test that checks for intestinal parasites (usually by identification of the parasite eggs, or oocysts). Flotation methods are based on the use of flotation fluid with a specific gravity suitable to float the eggs, or oocysts.

FIGURE 6–23 Parasite eggs commonly found on fecal flotation. (a) Roundworm eggs (*Toxocara canis*, left; *Toxocaris leonina*, right). (b) Whipworm, left (*Trichuris vulpis*). (c) Lungworm, right (*Capillara aerophila*). (d) Hookworm (*Ancylostoma caninum*). (e) Tapeworm (*Dipylidium caninum*).

- **fecal examinations** = various procedures used to detect parasitic diseases of animals (Figures 6–22 and 6–23). Specialized fecal tests also can identify bacteria, isolate viruses, or demonstrate abnormal substances present in the stool.
- **gastroscopy** (gahs-trohs-kō-pē) = endoscopic visual examination of the inner surface of the stomach; the scope is passed from the oral cavity through the stomach.
- **hemoccult** (hēm-ō-kuhlt) = test for hidden blood in the stool. **Occult** means hidden.
- **radiography** (rā-dē-ohg-rah-fē) = imaging of internal structures is created by the exposure of sensitized film to X-rays. Radiographs of the gastrointestinal system demonstrate foreign bodies, torsions, organ distention or enlargement, and some masses (Figure 6–24).
- **ultrasound** (uhl-trah-sownd) = imaging of internal body structures by recording echoes of sound waves.

PATHOLOGY: DIGESTIVE SYSTEM

Pathologic conditions of the digestive system include the following:

- **achalasia** (ahk-ah-lahz-ah) = inability to relax the smooth muscle of the gastrointestinal tract.
- **adontia** (ā-dohnt-shah) = absence of teeth.
- **aerophagia** (ār-ō-fā-jē-ah) = swallowing of air.

- **anal sacculitis** (ā-nahl sahck-yoo-lī-tihs) = inflammation of the pouches located around the anus. The term **inspissation** (ihn-spihs-sā-shuhn) is the process of rendering dry or thick by evaporation and is used to describe the anal sac fluid in animals with anal sacculitis.
- **anorexia** (ahn-ō-rehck-sē-ah) = lack or loss of appetite.
- **ascariasis** (ahs-kah-rī-ah-sihs) = parasitic infestation with roundworms of the genus *Ascaris* (Figure 6–25).
- **ascites** (ah-sī-tēz) = abnormal accumulation of fluid in the abdomen.
- **atresia** (ah-trēz-ah) = occlusion or absence of normal body opening or tubular organ.
- **bloat** (blōt) = accumulation of gas in the digestive tract. In monogastric animals, bloat is accumulation of gas in the stomach. In ruminants, bloat is accumulation of gas in the rumen, abomasum, or cecum. In ruminants, gas accumulation in the rumen is also called **ruminal tympany** (tihm-pahn-ē) (Figure 6–26).

FIGURE 6–24 Radiograph of a dog that ingested rocks.

FIGURE 6–25 Ascariasis is a parasitic infestation with roundworms.

FIGURE 6–26 Bloated thin steer in a chute. The left side of the abdomen bulges past the bars on the chute.

- **borborygmus** (bohr-bō-rihg-muhs) = gas movement in the gastrointestinal tract that produces a rumbling noise.
- **bruxism** (bruhck-sihzm) = involuntary grinding of the teeth.
- **cachexia** (kah-kehcks-ē-ah) = general ill health and malnutrition; used in describing the condition of cancer patients.
- **cholecystitis** (kō-lē-sihs-tī-tihs) = inflammation of the gallbladder.
- **cirrhosis** (sihr-rō-sihs) = degenerative disease that disturbs the structure and function of the liver. **Cirrh/o** means tawny, orange yellow, which is the color of cirrhotic livers.

- **colic** (kohl-ihck) = severe abdominal pain. Colic is a common condition in horses and may be caused by several factors including ingesting large amounts of grain, pasture, fresh grass, or sand; excessive gas; stress; internal parasites; dehydration; sudden dietary changes; and constipation or impaction. Signs of colic include flanking or looking at the abdomen, bloated or distended abdomen, frequent lying down and standing, rolling, kicking or biting at abdomen, sweating, restlessness, constipation or no bowel movements, and anorexia.
- **colitis** (kō-lī-tihs) = inflammation of the colon.

- **constipation** (kohn-stah-pā-shuhn) = condition of prolonged gastrointestinal transit time, making the stool hard, dry, and difficult to pass.
- **coprophagia** (kō-prō-fā-jē-ah or kohp-rohf-ah-jē-ah) = ingestion of fecal material. The combining form **copr/o** means feces. **Coprophagic therapy**, feeding a suspension of fecal pellets from a healthy animal to an ill animal, is used in some animal species such as guinea pigs and rabbits to treat gastrointestinal disease.
- **cribbing** (krihb-ihng) = vice of equine in which an object is grasped between the teeth, pressure is applied, and air is inhaled. See Figure 19–12.
- **dehydration** (dē-hī-drā-shuhn) = condition of excessive loss of body water or fluid. Dehydration status can be determined by how quickly the skin over the base of the neck or over the shoulder "snaps" back into place after this normally elastic skin is lifted. When an animal is dehydrated, the skin will stay "tented" or appear not to go back in its normal position as quickly as normal. Dehydration can also be observed when the eyes appear sunken into the head.
- **dental calculus** (dehn-tahl kahl-kyoo-luhs) = abnormal mineralized deposit that forms on teeth (Figure 6–27). Calculus is mineral deposit. Dental calculus also is called **tartar** (tahr-tahr).
- **dental caries** (dehn-tahl kār-ēz) = decay and decalcification of teeth, producing a hole in the tooth.
- **diarrhea** (dī-ah-rē-ah) = abnormal frequency and liquidity of fecal material (Figure 6–28). Diarrhea is caused by irritation of the intestinal lining that causes feces to pass through the intestine too quickly for adequate amounts of water to be reabsorbed.
- **displaced abomasum** (dihs-plāsd ahb-ō-mā-suhm) = disease of ruminants in which the fourth stomach compartment becomes trapped under the rumen; also called DA. It is denoted LDA (left displaced abomasum) or RDA (right displaced abomasum) depending on its location. LDAs are more common.
- **diverticulitis** (dī-vər-tihck-yoo-lī-tihs) = inflammation of a pouch or pouches occurring in the wall of a tubular organ. A **diverticulum** (dī-vər-tihck-yoo-luhm) is a pouch occurring on the wall of a tubular organ; diverticula are pouches occurring on the wall of a tubular organ.

FIGURE 6–27 Dental calculus and gingivitis in a dog.

Courtesy of Eli Larson, DVM

FIGURE 6–28 Bovine with diarrhea caused by *Salmonella* bacteria.

Courtesy of Dr. Ramos-Vara, College of Veterinary Medicine, Michigan State University

- **dyschezia** (dihs-kē-zē-ah) = difficulty defecating. *Chezein* is Greek for stool. The prefix **dys-** means difficult.
- **dysentery** (dihs-ehn-tər-ē) = number of disorders marked by inflammation of the intestine, abdominal pain, and diarrhea.
- **dysphagia** (dihs-fā-jē-ah) = difficulty swallowing or eating. The suffix **-phagia** means eating or swallowing.
- **emaciation** (ē-mā-sē-ā-shuhn) = marked wasting or excessive leanness (Figure 6–29).
- **emesis** (ehm-eh-sihs) = forcible expulsion of stomach contents through the mouth; also known as vomiting. The material vomited is known as

vomitus (voh-mih-tuhs). When an animal is vomiting, the recommendation is not to give it anything orally. The term for orally is **per os**, which is abbreviated **PO**. If the desire is to give nothing orally, the abbreviation is **NPO.**

- **enteritis** (ehn-tər-ī-tihs) = inflammation of the small intestine.
- **enterocolitis** (ehn-tehr-ō-kō-lī-tihs) = inflammation of the small intestine and large intestine.
- **epulis** (ehp-uhl-uhs) = benign tumor arising from periodontal mucous membranes.
- **eructation** (ē-ruhck-tā-shuhn) = belching or raising gas orally from the stomach.
- **esophageal reflux** (ē-sohf-ah-jē-ahl rē-fluhcks) = return of stomach contents into the esophagus; also called gastroesophageal reflux disease, or GERD. **Reflux** means a backward or return flow. When esophageal reflux occurs, the stomach acid irritates and damages the lining of the esophagus.
- **eviscerate** (ē-vihs-ər-āt) = remove or expose internal organs. Evisceration is used to describe the exposure of internal organs after unsuccessful surgical closure of the abdomen (or another area containing organs).
- **exocrine pancreatic insufficiency** (ehck-sō-krihn pahn-krē-ah-tihck ihn-suh-fih-shehn-sē) = metabolic disease in which the pancreas does not secrete adequate amounts of digestive enzymes and is associated with weight loss, fatty stools, and borborygmus; abbreviated EPI.
- **fecalith** (fēck-ah-lihth) = stonelike fecal mass. The suffix **-lithiasis** (lih-thī-ah-sihs) means presence of stones. **Coprolith** (kō-prō-lihth) is another name for a fecalith.

FIGURE 6–29 Emaciation in a Jersey cow with Johne's disease.

- **flatulence** (flaht-yoo-lehns) = excessive gas formation in the gastrointestinal tract.
- **gastric dilatation** (gahs-trihck dihl-ah-tā-shuhn) = condition usually seen in deep-chested canines in which the stomach fills with air and expands. Dilatation is stretching beyond normal.
- **gastric dilatation volvulus** (gahs-trihck dihl-ah-tā-shuhn vohl-vū-luhs) = condition usually seen in deep-chested canines in which the stomach fills with air, expands, and twists on itself; abbreviated GDV (Figures 6–30 and 6–31).
- **gastritis** (gahs-trī-tihs) = inflammation of the stomach.
- **gastroenteritis** (gahs-trō-ehn-tehr-ī-tihs) = inflammation of the stomach and small intestine. Note that anatomically the stomach occurs first, followed by the small intestine. This order is reflected in the order of the medical term as well.
- **gingival hyperplasia** (jihn-jih-vahl hī-pər-plā-zē-ah) = overgrowth of the gingiva characterized by firm, nonpainful swellings associated with the gingiva.
- **gingivitis** (jihn-jih-vī-tihs) = inflammation of the gums (Figure 6–27).
- **glossitis** (glohs-ī-tihs) = inflammation of the tongue.
- **hematemesis** (hēm-ah-tehm-eh-sihs) = vomiting blood.
- **hematochezia** (hēm-aht-ō-kē-zē-ah) = passage of bloody stool.
- **hemoperitoneum** (hēm-ō-pehr-ih-tō-nē-uhm) = blood in the peritoneum.
- **hepatitis** (hehp-ah-tī-tihs) = inflammation of the liver.
- **hepatoma** (heh-pah-tō-mah) = tumor of the liver.
- **hepatomegaly** (hehp-ah-tō-mehg-ah-lē) = abnormal enlargement of the liver.
- **hiatal hernia** (hī-ā-tahl hər-nē-ah) = protrusion of part of the stomach through the esophageal opening in the diaphragm.
- **hydrops** (hī-drohps) = abnormal accumulation of fluid in tissues or a body cavity; also called **dropsy** (drohp-sē).
- **hyperglycemia** (hī-pər-glī-sē-mē-ah) = elevated blood sugar levels.
- **hypoglycemia** (hī-pō-glī-sē-mē-ah) = lower-than-normal blood sugar levels.
- **ileitis** (ihl-ē-ī-tihs) = inflammation of the ileum.

Gastric Dilatation Volvulus

Pyloric antrum
is displaced
aborally

Pylorus crosses
midline, passes
underneath
distended oral
part of stomach

Fundus
moves ventrally
and becomes
located in
ventral abdomen

Gastric dilatation
displaces greater
curvature ventrally

Necrotic
tissue

Copyright © 2015 Cengage Learning®

FIGURE 6–30 Gastric dilatation volvulus formation. *Gastric dilatation volvulus* is a term commonly used to describe a condition in deep-chested dogs in which the stomach fills with air, expands, and twists on itself. Because the stomach is "attached" at only two points, it should be more accurately described as a torsion.

Copyright © 2015 Cengage Learning®

FIGURE 6–31 Radiograph of a dog with GDV. Note the large gas-distended stomach.

- **ileus** (ihl-ē-uhs) = stoppage of intestinal peristalsis.
- **impaction** (ihm-pahck-shuhn) = obstruction of an area, usually when feed is too dry.
- **inappetence** (ihn-ahp-eh-tehns) = lack of desire to eat.
- **incontinence** (ihn-kohn-tihn-ehns) = inability to control. A descriptive term usually is applied in front of it; for example, fecal incontinence is the inability to control bowel movements.
- **inflammatory bowel disease** = a group of chronic gastrointestinal disorders that are characterized microscopically by the infiltration of inflammatory cells into the lining of the digestive tract; abbreviated IBD.
- **inguinal hernia** (ihng-gwih-nahl hər-nē-ah) = protrusion of bowel through the inguinal canal; protrusion is seen in the groin.
- **intussusception** (ihn-tuhs-suhs-sehp-shuhn) = telescoping of one part of the intestine into an adjacent part (Figure 6–32).
- **jaundice** (jawn-dihs) = yellow discoloration of the skin and mucous membranes caused by elevated bilirubin levels, also called **icterus** (ihck-tər-uhs) (Figure 6–33).
- **lethargy** (lehth-ahr-jē) = condition of drowsiness or indifference.
- **malabsorption** (mahl-ahb-sōrp-shuhn) = impaired uptake of nutrients from the intestine.
- **malnutrition** (mahl-noo-trih-shuhn) = lack of proper food or nutrients in an animal's body due to food shortage, poor eating habits, or the inability to digest, absorb, and distribute these nutrients due to disease conditions. Animals with malnutrition break down their own tissues to meet their nutritional and metabolic needs. Malnourished animals may be given **total parenteral** (pah-rehn-tər-ahl) **nutrition** or TPN in which all of their nutritional requirements are met through a nutritional liquid administered intravenously. **Parenteral** means not in the digestive tract.
- **malocclusion** (mahl-ō-kloo-shuhn) = abnormal contact between the teeth. The prefix **mal-** means bad, and **occlusion** means any contact between the chewing surfaces of the teeth.

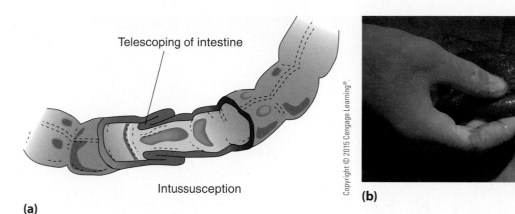

Telescoping of intestine

Intussusception

(a)

(b)

FIGURE 6–32 (a) Intussusception is the telescoping of the bowel onto itself. (b) Intraoperative view of an intussusception.

FIGURE 6–33 Jaundice is the yellow discoloration of the skin and mucous membranes.

- **megacolon** (mehg-ah-kō-lihn) = abnormally large colon (Figure 6–34a).
- **megaesophagus** (mehg-ah-ē-sohf-ah-guhs) = abnormally large esophagus (Figure 6–34b).
- **melena** (meh-lē-nah) = black, tarry stools containing digested blood. Melena suggests a bleeding problem in the upper gastrointestinal tract.
- **nausea** (naw-sē-ah) = stomach upset or sensation of urge to vomit; difficult to use descriptively in animals.
- **obstruction** (ohb-struhck-shuhn) = complete stoppage or impairment to passage. Obstructions usually are preceded by a term that describes its location, as in **intestinal obstruction** (Figure 6–35). When the obstruction is not complete, it is called a **partial obstruction.**

- **oronasal fistula** (ohr-ō-nā-zahl fihs-tyoo-lah) = abnormal opening between the nasal cavity and the oral cavity. Oronasal fistulas may be congenital, traumatic, or associated with dental disease.
- **palatoschisis** (pahl-ah-tohs-kih-sihs) = congenital fissure of the roof of the mouth that may involve the upper lip, hard palate, and soft palate; more commonly called a **cleft palate.**
- **perforating ulcer** (pər-fohr-āt-ihng uhl-sihr) = erosion through the entire thickness of a surface.
- **periapical abscess** (pehr-ē-ā-pih-kahl ahb-sehsz) = inflammation of tissues and collection of pus surrounding the apical portion of a tooth root due to pulpal disease.
- **periodontitis** (pehr-ē-ō-dohn-tī-tihs) = inflammation of the tissue surrounding and supporting the teeth; also called **periodontal disease. Inflammation** is a localized protective response elicited by injury or destruction of tissue. The signs of inflammation are heat, redness, pain, swelling, and loss of function.
- **pica** (pī-kah) = eating and licking abnormal substances or a depraved appetite.
- **plaque** (plahck) = small, differentiated area on a body surface. In the gastrointestinal system, it is used to refer to the mixed colony of bacteria, leukocytes, and salivary products that adhere to the tooth enamel; also called **dental plaque.**
- **polydipsia** (pohl-ē-dihp-sē-ah) = excessive thirst or drinking; abbreviated PD. **Dips/o** means thirst.
- **polyp** (pohl-uhp) = small growth on a mucous membrane.
- **polyphagia** (pohl-ē-fā-jē-ah) = excessive eating or swallowing. The prefix **poly-** means many or much.

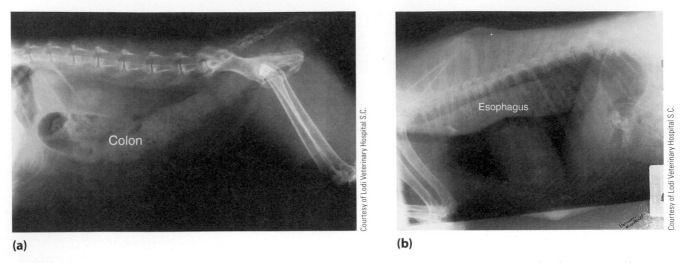

(a)

(b)

FIGURE 6–34 Contrast radiography. (a) Megacolon is diagnosed in the dog using contrast radiography. (b) Megaesophagus is diagnosed in the dog using contrast radiography.

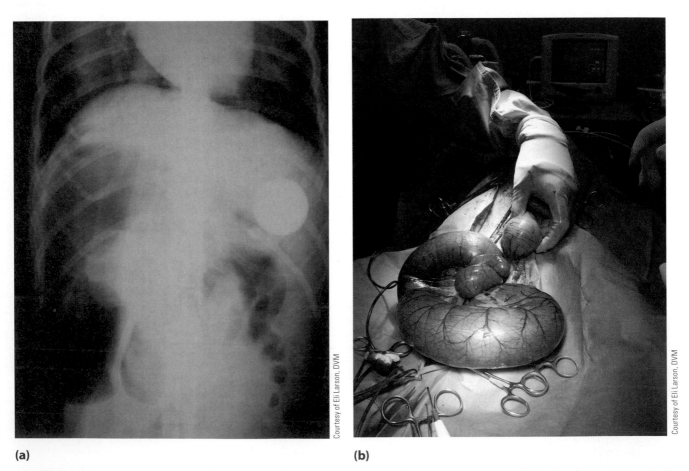

(a)

(b)

FIGURE 6–35 (a) Radiograph of a dog with an intestinal obstruction (golf ball). (b) Surgical view of the dog with an intestinal obstruction.

FIGURE 6–36 Salivary mucocele in a dog.

Volvulus

Torsion

FIGURE 6–37 Different types of twists. Volvulus is a twist around the long axis of the mesentery; torsion is a twist around the long axis of the gut.

- **prolapse** (prō-lahpz) = protrusion of viscera. A descriptive term usually precedes the term *prolapse*. For example, a rectal prolapse is protrusion of the rectum through the anus.
- **quidding** (kwihd-ihng) = condition in which food is taken into the mouth and chewed but falls from the mouth.
- **regurgitation** (rē-gərj-ih-tā-shuhn) = return of swallowed food into the oral cavity; a passive event compared with the force involved with vomiting.
- **salivary mucocele** (sahl-ih-vahr-ē myoo-kō-sēl) = collection of saliva that has leaked from a damaged salivary gland or duct and is surrounded by granulation tissue (Figure 6–36).
- **scours** (skowrz) = diarrhea in livestock.
- **shunt** (shuhnt) = to bypass or divert. In a **portosystemic** (poor-tō-sihs-tehm-ihck) **shunt**, blood vessels bypass the liver and the blood is not detoxified properly.
- **stenosis** (steh-nō-sihs) = narrowing of an opening. The term *stenosis* usually is used with a descriptive term in front of it. For example, a pyloric stenosis is narrowing of the pylorus as it leads into the duodenum.
- **stomatitis** (stō-mah-tī-tihs) = inflammation of the mouth.
- **tenesmus** (teh-nehz-muhs) = painful, ineffective defecation. Tenesmus also means painful, ineffective urination but is rarely used in this context.
- **torsion** (tōr-shuhn) = axial twist; twist around the long axis of gut.

- **trichobezoar** (trī-kō-bē-zōr) = hairball. The combining form **trich/o** means hair.
- **ulcer** (uhl-sihr) = erosion of tissue.
- **volvulus** (vohl-vū-luhs) = twisting on itself (end-to-end twist); twist around long axis of mesentery (Figure 6–37).

PROCEDURES: DIGESTIVE SYSTEM

Procedures performed on the digestive system include the following:

- **abdominocentesis** (ahb-dohm-ihn-ō-sehn-tē-sihs) = surgical puncture to remove fluid from the abdomen.
- **abomasopexy** (ahb-ō-mahs-ō-pehcks-ē) = surgical fixation of the abomasum of ruminants to the abdominal wall (Figure 6–38). The suffix **-pexy** means to surgically fix something to a body surface.

FIGURE 6–38 A displaced abomasum is repaired surgically with a procedure called an abomasopexy.

- **anastomosis** (ah-nahs-tō-mō-sihs) = surgical connection between two tubular or hollow structures.
- **anoplasty** (ah-nō-plahs-tē) = surgical repair of the anus.
- **antidiarrheal** (ahn-tih-dī-ə-rē-ahl) = substance that prevents frequent and extremely liquid stool.
- **bolus** (bō-luhs) = rounded mass of food or large pharmaceutical preparation or to give something rapidly.
- **cholecystectomy** (kō-lē-sihs-ehck-tō-mē) = surgical removal of the gallbladder.
- **colectomy** (kō-lehck-tō-mē) = surgical removal of the colon.
- **colostomy** (kō-lahs-tō-mē) = surgical production of an artificial opening between the colon and the body surface.
- **colotomy** (kō-loht-ō-mē) = surgical incision into the colon.
- **crown** (krown) = restoration of teeth using materials that are cemented into place; used to cap or completely cover a tooth; also called a **cap.**
- **drench** (drehnch) = to give medication in liquid form by mouth and forcing the animal to drink.
- **emetic** (ē-meh-tihck) = producing vomiting. An **antiemetic** (ahn-tih-ē-meh-tihck) prevents vomiting.
- **enterostomy** (ehn-tər-ohs-tō-mē) = surgical production of an artificial opening between the small intestine and the abdominal wall.
- **esophagoplasty** (ē-sohf-ah-gō-plahs-tē) = surgical repair of the esophagus.
- **extraction** (ehcks-trahck-shuhn) = removal; used to describe surgical removal of a tooth.

- **fistula** (fihs-tyoo-lah) = abnormal passage from an internal organ to the body surface or between two internal organs. A ruminant that has an artificial opening created between the rumen and the body surface has a rumen fistula (Figure 6–39). This also is called a **rumenostomy** (roo-mehn-ah-stō-mē). A **perianal fistula** (pehr-ih-ā-nahl fihsh-too-lah) is an abnormal passage around the caudal opening of the gastrointestinal tract. **Perianal** means around the anus.
- **float** (flōt) = instrument used to file or rasp an equine's premolar or molar teeth (Figure 6–40); also used to describe the procedure of filing equine teeth.
- **gastrectomy** (gahs-trehck-tō-mē) = surgical removal of all or part of the stomach. To clarify the extent of the excision, the term **partial gastrectomy** is used to denote surgical removal of part of the stomach.
- **gastroduodenostomy** (gahs-trō-doo-ō-deh-nohs-tōm-ē) = removal of part of the stomach and duodenum and making a connection between them.
- **gastropexy** (gahs-trō-pehcks-ē) = surgical fixation of the stomach to the abdominal wall.
- **gastrostomy** (gahs-trohs-tō-mē) = surgical production of an artificial opening between the stomach and abdominal wall. The suffix **-stomy** means surgical production of an opening between an organ and a body surface. The opening created during this procedure is a **stoma** (stō-mah). **Effluent** (ehf-floo-ehnt) means discharge and an effluent flows from the stoma created by a -stomy surgery (e.g., gastrostomy).

FIGURE 6–39 A rumen fistula is an artificial opening created between the rumen and the body surface. Fluid from the healthy rumen can be transferred to a sick bovine to reestablish the normal microbes in the rumen.

FIGURE 6–40 A tooth float used to file the teeth of horses. A gag is used to hold the mouth open for the procedure.

- **gastrotomy** (gahs-troht-ō-mē) = surgical incision into the stomach (Figure 6–41).
- **gavage** (gah-vahzh) = forced feeding or irrigation through a tube passed into the stomach.
- **gingivectomy** (jihn-jih-vehck-tō-mē) = surgical removal of the gum tissue.
- **hepatotomy** (hehp-ah-toht-ō-mē) = surgical incision into the liver.
- **ileectomy** (ihl-ē-ehck-tō-mē) = surgical removal of the ileum.
- **ileostomy** (ihl-ē-ohs-tō-mē) = surgical production of an artificial opening between the ileum and abdominal wall.
- **laparotomy** (lahp-ah-roht-ō-mē) = surgical incision into the abdomen; **lapar/o** is the combining form for abdomen or flank.

- **nasogastric intubation** (nā-zō-gahs-trihck ihn-toob-ā-shuhn) = placement of a tube through the nose into the stomach (Figure 6–42). Nasogastric is abbreviated NG.
- **orogastric intubation** (ōr-ō-gahs-trihck ihn-too-bā-shuhn) = passage of a tube from the mouth to the stomach; also called a **stomach tube**. *Orogastric* means pertaining to the mouth and stomach.
- **palatoplasty** (pahl-ah-tō-plahs-tē) = surgical repair of a cleft palate.
- **transfaunation** (trahns-fah-nā–shuhn) = transferring microbes from one animal to another to reinoculate the ill animal with a healthy microbial population. Transfaunation is used to treat cattle that are severely off-feed, particularly in cases of acidosis. The healthy microbes are found in rumen content and are typically transferred from a bovine that has a rumen fistula (see Figure 6–39).
- **trocarization** (trō-kahr-ih-zā-shuhn) = insertion of a pointed instrument (trocar) into a body cavity or an organ. The trocar usually is inside a cannula so that once the trocar penetrates the membrane, it can be withdrawn and the cannula remains in place. Trocarization usually is performed for acute cases of bloat to relieve pressure. When trocarization is performed for treatment of ruminal bloat, it may be called ruminal **paracentesis** (pahr-ah-sehn-tē-sihs).

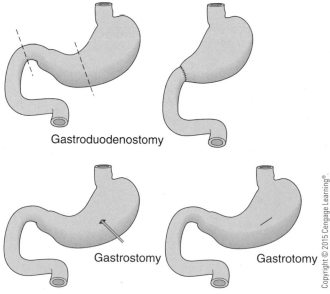

Gastroduodenostomy

Gastrostomy

Gastrotomy

FIGURE 6–41 (a) Gastroduodenostomy, (b) Gastrostomy, and (c) Gastrotomy.

FIGURE 6–42 Nasogastric tube placement in a cat.

ABBREVIATIONS RELATED TO THE DIGESTIVE SYSTEM

- BE = barium enema
- BM = bowel movement
- C = canine (tooth)
- DA = displaced abomasum
- EPI = exocrine pancreatic insufficiency
- GDV = gastric dilatation volvulus
- GERD = gastroesophageal reflux disease
- GI = gastrointestinal

- I = incisor (tooth)
- IBD = inflammatory bowel disease
- LDA = left displaced abomasum
- M = molar (tooth)
- NG = nasogastric
- NPO = nothing by mouth (non per os)
- P = premolar (tooth)
- PD = polydipsia
- PO = orally (per os)
- RDA = right displaced abomasum
- TPN = total parenteral nutrition

REVIEW EXERCISES

Multiple Choice

Choose the correct answer.

1. Mixing of ingesta in the intestine is called
   a. propulsion
   b. peristalsis
   c. segmentation
   d. separation

2. Abnormal accumulation of fluid in the abdomen is called
   a. ascites
   b. effusion
   c. icterus
   d. bloat

3. Telescoping of one part of the intestine into an adjacent part is called
   a. volvulus
   b. diverticulum
   c. parenchyma
   d. intussusception

4. The small intestine is attached to the dorsal abdominal wall by the
   a. peritoneum
   b. emesis
   c. mesentery
   d. omentum

5. Eating and licking of abnormal substances is called
   a. coprophagy
   b. pica
   c. dysphagia
   d. polyphagia

6. Inflammation of the mouth is
   a. stomatitis
   b. orititis
   c. dentitis
   d. osititis

7. Straining, painful defecation is called
   a. strangstolia
   b. colostrangia
   c. tenesmus
   d. epulis

8. A tumor of the liver is a
   a. hematoma
   b. hemoma
   c. hepatoma
   d. hemotoma

9. Marked wasting or excessive leanness is
   a. evaluation
   b. elimination
   c. emesis
   d. emaciation

10. Forced feeding or irrigation through a tube passed into the stomach is called
    a. gavage
    b. drench
    c. bolus
    d. cachexia

11. The combining form for the first part of the large intestine is
    a. ile/o
    b. cec/o
    c. duoden/o
    d. jejun/o

12. The muscular, wavelike movement used to transport food through the digestive system is
    a. mastication
    b. peristalsis
    c. anastomosis
    d. regurgitation

13. The part of the tooth that contains a rich supply of nerves and blood vessels is the
    a. enamel
    b. dentin
    c. pulp
    d. cementum

14. *Buccal* means
    a. pertaining to the cheek
    b. pertaining to the tongue
    c. pertaining to the throat
    d. pertaining to the palate

15. *Stomat/o* means
    a. mouth
    b. cheek
    c. stomach
    d. intestine

16. The term for erosion of tissue is
    a. melena
    b. plaque
    c. shunt
    d. ulcer

17. The narrow passage between the stomach and the duodenum is the
    a. ileum
    b. pylorus
    c. peritoneum
    d. mesentery

18. Incontinence means
    a. prolonged transit time
    b. dribbling urine
    c. loose stool
    d. inability to control

19. A drug used to prevent vomiting is known as an
    a. emetic
    b. atresic
    c. antiemetic
    d. antiemaciation

20. The formation of a new opening from the large intestine to the surface of the body is known as a(n)
    a. enterostomy
    b. colostomy
    c. colectomy
    d. enterotomy

21. Proofread the following statements. Which is the only correct statement?
    a. The lower jaw is the maxilla.
    b. Difficulty in defecating is dysphagia.
    c. Melena is the passage of black, tarry stools.
    d. A perforating ulcer is a raised lesion on a mucous membrane.

22. Cholecystectomy and cholecystitis are terms used to describe a procedure and condition of the
    a. pancreas
    b. gallbladder
    c. liver
    d. colon

23. The one term that is *not* a body process is
    a. peristalsis
    b. deglutation
    c. mastication
    d. fecalith

24. *Postprandial* vomiting occurs
    a. in the morning
    b. in the evening
    c. before a meal
    d. after a meal

25. The set of terms that are all diagnoses is
    a. endoscopy, icterus, hyperglycemia
    b. hepatitis, bloat, gastric dilatation
    c. pancreatitis, jaundice, gastropexy
    d. colic, hiatal hernia, esophagoscopy

Matching

Match the term in Column I with the definition in Column II.

Column I	Column II
1. _____ chyme	a. absorption
2. _____ bile	b. tiny hairlike projection on the small intestine
3. _____ glycogen	c. semifluid mass of partly digested food
4. _____ rugae	d. substance made by the liver that helps with fat digestion
5. _____ mastication	e. folds present in the mucosa of the stomach
6. _____ aerophagia	f. stored energy present in the liver
7. _____ trypsin	g. swallowing of air
8. _____ amylase	h. breaking food into smaller pieces and mixing with saliva
9. _____ villus	i. enzyme that digests carbohydrates
10. _____ assimilation	j. enzyme that digests protein
11. _____ bilirubin	k. sacculations of cecum and colon in horses and pigs
12. _____ haustra	l. fold of peritoneum
13. _____ teniae	m. hard white covering the dentin of the crown of the tooth
14. _____ omentum	n. pigment produced by destruction of hemoglobin
15. _____ pyloric sphincter	o. longitudinal smooth muscle bands in cecal wall of horses and pigs
16. _____ jejunum	p. controls flow from stomach to duodenum
17. _____ enamel	q. connective tissue surrounding tooth pulp
18. _____ dentin	r. fermentation vat of ruminants
19. _____ abomasum	s. middle part of small intestine
20. _____ rumen	t. true stomach of ruminants

Fill in the Blanks

Write the correct term in the space provided.

1. Gloss/o and lingu/o mean _____.

2. Or/o and stomat/o mean _____.

3. Cheil/o and labi/o mean _____.

4. Abdomin/o, celi/o, and lapar/o mean _____.

5. Dent/o, dent/i, and odont/o mean _____.

6. The dog's gingival tissue was severely infected and he was diagnosed with _____.

7. The return of swallowed food into the mouth is known as _____.

8. Twisting around the long axis of mesentery is _____.

9. Twisting around the long axis of the gut is _____.

10. Erosion of tissue is a/an _____.

True or False

If the statement is true, write T on the line. If the statement is false, write F on the line.

1. _____The medical term meaning inflammation of the small intestine is colitis.

2. _____Vomiting blood is hematemesis.

3. _____Borborygmus is telescoping of one part of the intestine into another.

4. _____Alveolar bone is the thin layer of compact bone that forms the tooth socket.

5. _____The tongue is connected to the dorsal surface of the oral cavity by a band of connective tissue called the frenulum.

Spelling

Circle the term that is spelled correctly.

1. twisting of the intestine:	vulvulus	volvulus	volvolus
2. inability to relax the smooth muscles of the digestive tract:	achalsia	achalasia	achaelasia
3. yellow discoloration of the skin and mucous membranes:	jaundise	jawndise	jaundicc
4. remove or expose internal organs:	eviserate	eviscerate	evicerate
5. abnormal accumulation of fluid in the abdomen:	ascites	asites	ascetes
6. hidden:	ocult	occult	okult
7. endoscopic visual examination of the esophagus:	esophagoscopy	esofagoscopy	esphagoscopy
8. lack of appetite:	anorexhia	anorrhexia	anorexia
9. building up of body cells:	annabolizm	anabolizm	anabolism
10. abnormal frequency and liquidity of feces:	diarrhea	diarhea	diarhhea

Crossword Puzzles
Disease Terms

Supply the correct term in the appropriate space for the definition listed.

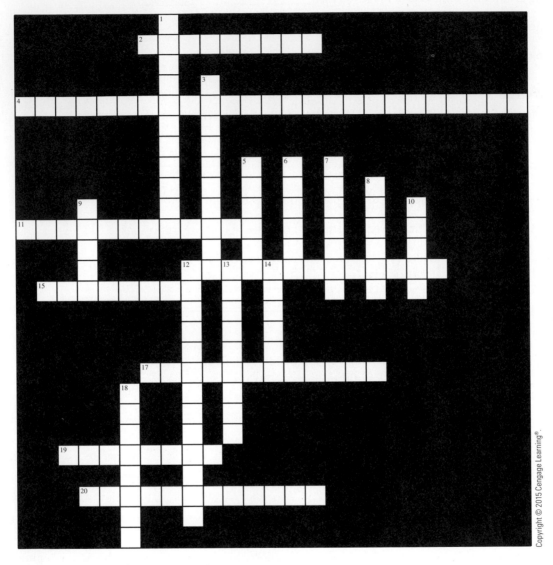

Across

2 abnormally large portion of the large intestine located between the cecum and rectum

4 condition usually seen in deep-chested canines in which the stomach fills with air, expands, and twists on itself

11 inability to control

12 abnormally large tube that connects the oral cavity to the stomach

15 general ill health and malnutrition

17 process of rendering dry or thick by evaporation

19 stonelike fecal mass

20 hairball

Down

1 condition of excessive loss of body water or fluid

3 excessive gas formation in the GI tract

5 black stools containing digested blood

6 diarrhea in livestock

7 abnormal accumulation of fluid in tissues or a body cavity

8 another term for dental calculus

9 severe abdominal pain

10 rounded mass of food, large pharmaceutical preparation, or to give something rapidly

12 impaired uptake of nutrients from the intestine

13 inflammation of the tongue

14 benign tumor arising from periodontal mucous membranes

18 excessive salivation

Digestive Organs

Supply the correct term in the appropriate space for the definition listed.

Across

1 throat

4 caudal portion of the small intestine

6 caudal portion of the large intestine

8 organ that produces bile

11 one of the combining forms for mouth

14 cheek teeth

15 cranial portion of the small intestine

Down

1 gland that secretes digestive juices as well as hormones

2 combining form for nourishment

3 enter/o is the combining form for this GI organ

5 collapsible muscular tube that leads from the oral cavity to the stomach

6 most cranial compartment of the ruminant stomach

7 one of the combining forms for tongue

9 organ that stores bile

10 part of large intestine located between the cecum and rectum

12 middle portion of the small intestine

13 combining form for stomach

14 roof of the mouth

Word Search

Define the following terms; then find each term in the puzzle.

```
R T Y T N N D L T Y U B T Y I M A S N N O O
Y A M A O A C A T E E H Y M Y C I I T T E N
O E O N A N A P O Y S I S O M O T S A N A A
A Y N E M Y O A A T E H M T T E T E T E B S
D T G O I T T R O T E S L S T N A T N N D O
O T N T E N T O Y O A P Y O E O I N O O A G
A O A T O O O T T B T T A L O I O E O D M A
T N N H O N M O O N Y N T O Y S S C G T Y S
E D T O C O O M N N T B I C E B I O P S Y T
O N O E T O A Y Y I O Y O N O C D N R E S R
S O I S O S M O D N M E M E T I C I Y T S I
O T M E O O D I S M T A A O A G N M H N C C
L C S P B R A O O D N A O A N Y H O T T N I
E M E O M R M I N T T A M G M T E D S N T N
M X Y C R O O N T Y S A O R C O O B T T E T
Y S C H T T T Y E E O C A G A P O A O O B U
O Y E I A Y I N E T O S A C A N E M E N T B
T A S O O E E N T H S C Y G I O Y T O A A A
L X N T T M S A L E M N T E T A O N T N G T
H T D M A A N O M C N E X T R A C T I O N I
A M O A N C Y N S G A S T R E C T O M Y N O
A N O P L A S T Y O T N O N A B S I I T M N
```

surgical connection between two tubular or hollow organs _____

surgical incision into the abdomen _____

surgical removal of all or part of the stomach _____

produces vomiting _____

to remove (teeth) _____

surgical repair of the anus _____

surgical production of an artificial opening between the colon and the body surface _____

surgical fixation in ruminants of the fourth stomach compartment to the abdominal wall _____

removing tissue to examine _____

substance that prevents frequent and extremely liquid stool _____

placement of a tube through the nose into the stomach _____

surgical puncture to remove fluid from the abdomen_____

Label the Diagram

Label the diagram in Figure 6–43.

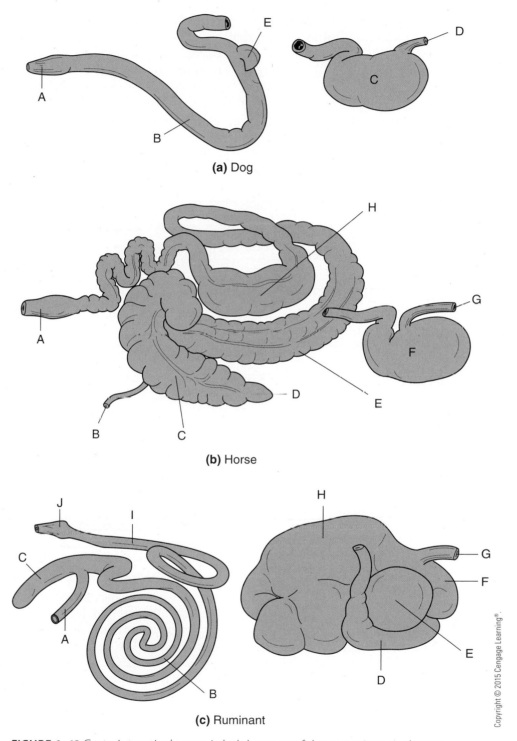

(a) Dog

(b) Horse

(c) Ruminant

FIGURE 6–43 Gastrointesetinal tracts. Label the parts of the gastrointestinal tracts.

Case Studies

Using terms presented in this chapter, the appendices, and a dictionary, define the underlined terms in each case study.

A 3-<u>yr</u>-old <u>F/S</u> New Zealand white rabbit is presented for a few months <u>hx</u> of <u>quidding</u> and <u>ptyalism</u>. The problem has become progressively worse, and although the rabbit has a good appetite, it has difficulty <u>prehending</u> food. On PE, it is noted that there is severe overgrowth of the upper and lower <u>incisors</u>. This problem in rabbits is most often caused by <u>malocclusion</u> of the <u>dental arcades</u>. The <u>premolars</u> and <u>molars</u> are found to be normal. Conservative treatment includes periodic trimming of the overgrown teeth. A more permanent solution involves the <u>extraction</u> of the incisors. Under general <u>anesthesia</u>, all incisors are removed. <u>Analgesics</u> are provided for the first few days <u>postoperatively</u> because rabbits are easily stressed by pain. Once healed, the rabbit returned to its normal eating habits.

1. yr _____

2. F/S _____

3. hx _____

4. quidding _____

5. ptyalism _____

6. prehending _____

7. incisors _____

8. malocclusion _____

9. dental arcades _____

10. premolars _____

11. molars _____

12. extraction _____

13. anesthesia _____

14. analgesics _____

15. postoperatively _____

An 8-<u>wk</u>-old M coonhound is presented to the clinic with an <u>acute</u> history of <u>emesis, hemorrhagic diarrhea,</u> <u>lethargy</u>, and <u>anorexia</u>. The pup was not vaccinated and was healthy until yesterday. Upon PE, it was noted that the pup was <u>pyrexic, dehydrated</u>, and lethargic. Heart and lungs <u>ausculted</u> normally. <u>Stool</u> was collected for parasitic examination, and blood was collected for a CBC and chemistry panel. The stool was negative for parasites, the blood count revealed <u>lymphopenia</u>, and the chemistry panel was normal except for indications of dehydration. A <u>dx</u> of <u>canine</u> parvoviral <u>enteritis</u> was suspected because of the lymphopenia and clinical signs, so virus isolation was performed on a stool sample. Pending virus isolation results, the pup was hospitalized and isolated, <u>IV</u> fluids were administered, and antibiotics were given to prevent a <u>secondary septicemia</u>. Twelve hours after hospitalization the pup <u>expired</u>. A necropsy was done, and the intestines demonstrated loss of <u>intestinal villi</u> and <u>crypt</u> necrosis. The virus isolation test was positive for canine parvovirus infection. The facility was thoroughly <u>disinfected</u>, and the owners were advised to disinfect their facility and vaccinate any future pups.

16. wk _____

17. acute _____

18. emesis _____

19. hemorrhagic _____

20. diarrhea _____

21. lethargy _____

22. anorexia _____

23. pyrexic _____

24. dehydrated _____

25. ausculted _____

26. stool _____

27. lymphopenia _____

28. dx _____

29. canine _____

30. enteritis _____

31. IV _____

32. secondary _____

33. septicemia _____

34. expired _____

35. necropsy _____

36. intestinal villi _____

37. crypt _____

38. disinfected _____

Critical Thinking Exercise

The following story and questions contain words and concepts from this chapter and are designed to stimulate critical thinking through class discussion, Internet searches, or brief writing responses.

Frieda, an 8-yr-old F/S black DSH, weighs 14# and was diagnosed with diabetes mellitus 1 year ago when she weighed 18#. She still has episodes of hyperglycemia (during which time she is PU/PD) because her owner has difficulty administering her medication and feeding her at the same time every day. Her veterinarian recommended that her owner feed Frieda a special diet that is high in fiber to help her lose weight.

1. What is diabetes mellitus? (You may want to refer to Chapter 11.)

2. What part of her history correlates to the diagnosis of diabetes mellitus?

In addition to her uncontrolled diabetes, Frieda has had spasmodic episodes of diarrhea and flatulence for the past year. During these episodes her stool is black and tarry. Based on the duration of illness, her veterinarian recommended passing an endoscope to look at her intestinal mucosa.

3. List all of Frieda's signs of disease.

4. Define diarrhea and flatulence.

5. Because her stool is black and tarry, what word is used to describe it?

6. What part of the gastrointestinal tract is affected when stool is black and tarry?

7. Based on the black and tarry stool, what type of lesion would the veterinarian look for with the endoscope?

8. Do you think the endoscope would be passed rectally or orally?

9. Define spasmodic.

10. What additional information would you like to ask the client about her cat?

Chapter 7
NULL AND VOID

Objectives

Upon completion of this chapter, the reader should be able to:

- Identify the major organs and tissues of the urinary system

- Describe the major functions of the urinary system

- Identify properties of urine from different animal species

- Recognize, define, spell, and pronounce terms relating to diagnosis, pathology, and treatment of the urinary system

- Interpret the meaning of abbreviations of the urinary system

FUNCTIONS OF THE URINARY SYSTEM

The urinary system's main responsibility is the removal of wastes from the body. The **urinary** (yoo-rih-nār-ē) **system** removes wastes from the body by constantly filtering blood. The major waste product of protein metabolism is **urea** (yoo-rē-ah), which is filtered by the kidney and used in some diagnostic tests to determine the health status of the kidney.

In addition to filtering wastes, the urinary system also maintains proper balance of water, electrolytes, and acids in body fluids and removes excess fluids from the body. Maintaining a proper balance of water, electrolytes, and acids allows the body to have a stable internal environment. This stable internal environment is called **homeostasis** (hō-mē-ō-stā-sihs), and it involves continually adjusting to conditions to maintain a relatively constant internal environment. The combining form **home/o** means unchanging.

One part of the urinary system, the kidney, also produces hormones and affects the secretory rate of other hormones.

Urin/o and **ur/o** are combining forms meaning urine or pertaining to the urinary organs.

STRUCTURES OF THE URINARY SYSTEM

The structures of the normal urinary system include a pair of kidneys, a pair of ureters, a single urinary bladder, and a single urethra (Figure 7–1).

> Urine is formed in the kidneys, flows through the ureters to the urinary bladder, is stored in the urinary bladder, and flows through the urethra and outside the body.

Kidneys

Kidneys are located **retroperitoneally** (reh-trō-peh-rih-tō-nē-ah-lē), which means they are located behind the lining of the abdominal cavity or outside the peritoneal cavity. One kidney sits on each side of the vertebral column below the diaphragm. **Ren/o** (Latin for kidney) and **nephr/o** (Greek for kidney) are combining forms for kidney. Ren/o (the Latin form) is used

as an adjective, as in *renal pelvis* and *renal disease*. Nephr/o (the Greek term) tends to be used to describe pathologic conditions and surgical procedures, as in *nephritis* and *nephrectomy*.

Blood flows into each kidney through the renal artery and leaves the kidney via the renal vein. Filtration of waste products by the kidney depends on this blood flow; therefore, blood pressure can affect the rate at which filtration takes place.

Each kidney has two layers that surround the renal pelvis. The outer layer of the kidney is known as the **cortex** (kōr-tehckz). **Cortic/o** means outer region and is used to describe the outer region of many organs. The **medulla** (meh-doo-lah) of the kidney is the inner layer. **Medull/o** means middle or inner portion and is used to describe the middle or inner region of many organs. The cortex contains the majority of the nephron, and the medulla contains most of the collecting tubules.

The **nephron** (nehf-rohn) is the functional unit of the kidney. The nephron consists of the glomerulus, Bowman's capsule (also called the glomerular capsule), a proximal convoluted tubule, a loop of Henle, a distal convoluted tubule, and a collecting duct. The nephrons form urine by the processes of filtration,

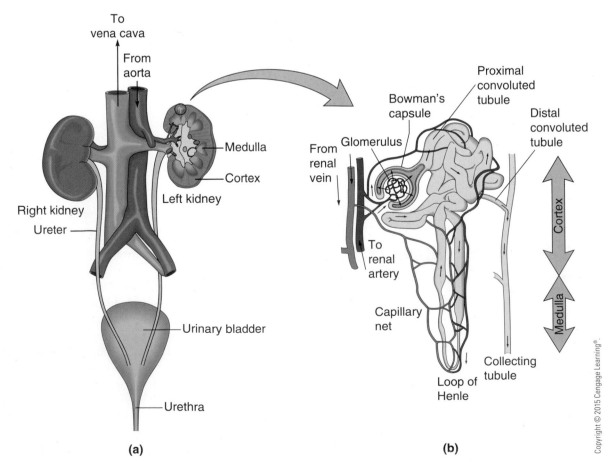

(a) (b)

Copyright © 2015 Cengage Learning®.

FIGURE 7–1 (a) Structures of the urinary tract. (b) Structures of a nephron.

TABLE 7–1 Parts of the Nephron

glomerulus	cluster of capillaries that filter blood
Bowman's capsule	cup-shaped structure that contains the glomerulus (the glomerulus and Bowman's capsule together are called the **renal corpuscle**)
proximal convoluted tubules	hollow tubes located between Bowman's capsule and loops of Henle that are involved in reabsorption
loop of Henle	U-shaped turn in the convoluted tubule of the kidney located between the proximal and distal convoluted tubules that is involved in reabsorption; has ascending and descending loop
distal convoluted tubules	hollow tubes located between the loops of Henle and the collecting tubules that are involved in secretion
collecting tubules	hollow tubes that carry urine from the cortex to the renal pelvis

reabsorption, and secretion. Filtration occurs in the glomerulus; reabsorption occurs in the proximal convoluted tubule, loop of Henle, and collecting tubule; and secretion occurs in the distal convoluted tubule (Table 7–1).

The **glomerulus** (glō-mər-yoo-luhs) is a cluster of capillaries surrounded by Bowman's capsule. The combining form for glomerulus is **glomerul/o**, which means to wind into a ball, which is what the glomerulus looks like microscopically. The plural form of glomerulus is **glomeruli** (glō-mər-yoo-lī).

The **renal pelvis** is the area of the kidney where the nephrons collect urine before entering the ureters. The combining form for renal pelvis is **pyel/o**. The renal pelvis collects urine from the **calyces** (kā-lah-sēz), which are irregular cuplike spaces that collect urine from the kidney. **Calyx** (kā-lihcks) is the singular form of *calyces.*

Each kidney has a concave depression called the **hilus** (hī-luhs) that serves as the point of attachment of the renal blood vessels, nerves, and ureter (Figure 7–2a and b). The hilus is located on the medial surface of the

kidney and gives some kidneys their bean shape. Not all species have kidneys that are bean-shaped. Cattle have lobulated kidneys (and no renal pelvis), and the right kidney of horses is heart-shaped (Figure 7–3).

Descriptive Structural Terms of the Kidney

- **distal** = farthest from midline
- **proximal** = closest to midline
- **convoluted** (kohn-vō-lūt-ehd) = rolled or coiled
- **cortex** = outer
- **medulla** = middle or inner
- **hilus** = point of attachment or depression
- **calyx** or **calix** = cuplike organ
- **ascending** (ā-sehnd-ihng) = moving upward or cranially
- **descending** (dē-sehnd-ihng) = moving downward or caudally

BIOLOGY BONUS

The two functions of the kidney are to form urine for excretion and to retain essential substances the body needs through reabsorption. Blood enters each kidney through the renal artery and leaves through the renal vein. Once inside the kidney, the renal artery branches into arterioles that lead to the nephron. The glomerulus *filters* substances such as water, salts, sugar, urea, and other nitrogenous waste from the blood and is the first place that urine is formed. The filtered substances from blood collect in Bowman's capsule and then pass into the renal tubules. Substances in the renal tubule that can be used by the body are *reabsorbed* back into the bloodstream, while any substance not reabsorbed becomes urine. Additional substances, mainly hydrogen ion, are *secreted* from the capillaries into the tubules and collecting ducts. Formed urine then travels to and is collected in the renal pelvis prior to entering the ureters.

(a)

FIGURE 7–3 External appearance of a bovine kidney.

(b)

FIGURE 7–2 (a) Structures of the kidney. (b) Sectioned sheep kidney.

Ureters

The **ureters** (yoo-rē-tərz) are a pair of narrow tubes that carry urine from the kidneys to the urinary bladder. The combining form for ureter is **ureter/o**. The ureters enter the urinary bladder at the **trigone** (trī-gōn). *Trigone* comes from the Greek term meaning triangle. The trigone of the urinary bladder is a triangular portion at the base of that organ where the three angles are marked by the two ureteral openings and one urethral opening.

Urinary Bladder

The **urinary bladder** (yoo-rihn-ār-ē blah-dər) is a singular hollow muscular organ that holds urine (Figure 7–4). **Cyst/o** is the combining form for urinary bladder. The urinary bladder is very elastic, and its shape and size depend on the amount of urine it is holding. The flow of urine into the urinary bladder enters from the ureters at such an angle that it serves as a natural valve to control backflow. The flow of urine out of the urinary bladder to the urethra is controlled

Five Words, One Meaning?

The terms **urination** (yoo-rih-nā-shuhn), **excretion** (ehcks-krē-shuhn), **voiding** (voy-dihng), **elimination** (eh-lih-mə-nā-shuhn), and **micturition** (mihck-too-rih-shuhn) have been used to describe the act of excreting urine. Actually, the terms *excretion*, *elimination*, and *voiding* mean elimination of a substance and can be used in other body systems. The term *urination* means the elimination of urine from the body. The term *micturition* means the elimination of urine from the body; however, it implies voluntary control of the sphincter muscles of the urinary tract. This voluntary control of voiding urine is learned and implies a more intelligent form of animal life.

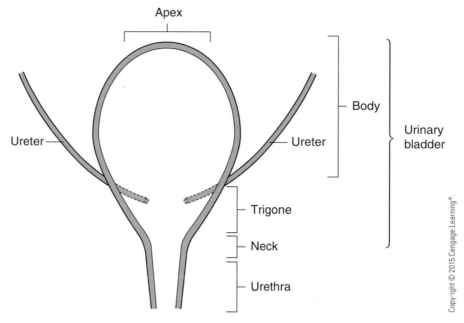

FIGURE 7–4 Divisions of the urinary bladder. The divisions of the urinary bladder include the apex (cranial free end), the body (central main part), the trigone (triangular portion where ureters enter the urinary bladder), and the neck (constricted portion that joins the urethra).

by **sphincters** (sfihngk-tərz). Sphincters are ringlike muscles that close a passageway.

Urethra

The **urethra** (yoo-rē-thrah) is a tube extending from the urinary bladder to the outside of the body. **Urethr/o** is the combining form for urethra. The external opening of the urethra is the **urethral meatus** (yoo-rē-thrahl mē-ā-tuhs), or urinary meatus. The combining form **meat/o** means opening.

In females, the only function of the urethra is to transport urine from the urinary bladder to the outside of the body. In males, the urethra transports urine from the urinary bladder and reproductive fluids from the reproductive organs out of the body.

URINE

The end product of renal filtration of wastes is urine. The process of urine production is **uropoiesis** (yoo-rō-pō-ē-sihs). The suffix **-poiesis** means formation. Normal urine of most species is clear and pale yellow in color. The suffix for color is **-chrome.** However, some species may have normal urine that is **turbid** (tər-bihd), or cloudy, and may be brown, white, or another color. Sometimes the color of urine depends on the diet (as in rabbits) or the reproductive cycle. Urine color also can

Urinary System Hormones

Hormones that affect or are produced by the urinary system include the following:

- **erythropoietin** (ē-rihth-rō-poy-ē-tihn) = hormone produced by the kidney that stimulates red blood cell production in the bone marrow. Also pronounced ē-rihth-rō-pō-ih-tihn.
- **antidiuretic** (ahn-tih-dī-yoo-reht-ihck) hormone = hormone released by the posterior pituitary gland that suppresses urine formation by reabsorbing more water; abbreviated as **ADH.**
- **aldosterone** (ahl-dah-stər-own) = hormone secreted by the adrenal cortex that regulates electrolyte balance via the reabsorption of sodium (Figure 7–5).

reflect hydration status. In dehydrated animals, urine is more concentrated and therefore a deeper shade of yellow. Urine also has a pH, which depends on species and diet. Herbivores tend to have basic urine (a higher pH), whereas carnivores tend to have acidic urine (a lower pH). Through dietary management, urine pH can be manipulated to treat or prevent disease.

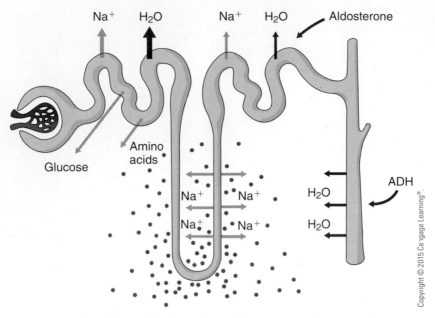

Na⁺   H₂O   Na⁺   H₂O   Aldosterone

Glucose

Amino acids

Na⁺   Na⁺

Na⁺   Na⁺

H₂O   ADH

H₂O

FIGURE 7–5 The effects of the hormones aldosterone and ADH on the nephron.

BIOLOGY BONUS

The kidneys maintain blood pressure through the filtration rate of the glomeruli. The constant flow of water and its substances back into the bloodstream and the flow of water and waste substances into the renal tubule maintain the body's balance of water, salts like sodium chloride, sugars like glucose, and other substances. The kidneys do this in two ways: first by the automatic dilating (widening) and constricting (narrowing) of the blood vessels to increase or decrease blood flow to the glomeruli and the second is by releasing the enzyme renin to increase the blood pressure and thus the filtration rate of blood to maintain a constant environment. Maintaining homeostasis affects blood pressure either by lowering it when blood is flowing too quickly or by increasing it when blood is flowing too slowly. Keep in mind that high blood pressure could be caused by poorly functioning kidneys to maintain homeostasis.

TEST ME: URINARY SYSTEM

Diagnostic procedures performed on the urinary system include the following:

- **cystocentesis** (sihs-tō-sehn-tē-sihs) = surgical puncture of the urinary bladder, usually to collect urine. A cystocentesis usually is performed with a needle and syringe (Figure 7–6).
- **cystography** (sihs-tohg-rah-fē) = radiographic study of the urinary bladder after contrast material has been placed in the urinary bladder via a urethral catheter. The contrast material used in the urinary bladder is water-soluble. Cystography can be single-contrast, when one contrast material is used, or double-contrast, when more than one contrast material is used. Double-contrast cystography is a radiographic study of the urinary bladder after air and contrast material have been placed in the urinary bladder via a urethral catheter. A **cystogram** (sihs-tō-grahm) is the radiographic film of the urinary bladder after contrast material has been placed in the urinary bladder via a urethral catheter. **Retrograde** (reh-trō-grād) means going backward and can be used to describe the path that contrast material takes. If the contrast material goes in reverse order of how urine normally flows in the body, it is referred to as retrograde.

FIGURE 7–6 Cystocentesis in a dog.

- **cystoscopy** (sihs-tohs-kō-pē) = visual examination of the urinary bladder using a fiberoptic instrument. A **cystoscope** (sihs-toh-skōp) is the fiberoptic instrument used to access the interior of the urinary bladder.
- **intravenous pyelogram** (ihn-trah-vē-nuhs pī-eh-lō-grahm) = radiographic study of the kidney (especially the renal pelvis) and ureters in which a dye is injected into a vein to define structures more clearly. The urinary bladder also may be visualized better with an intravenous pyelogram. It is abbreviated IVP (Figure 7–7).
- **pneumocystography** (nū-mō-sihs-tohg-rah-fē) = radiographic study of the urinary bladder after air has been placed in the bladder via a urethral catheter.
- **radiography** (rā-dē-ohg-rah-fē) = imaging of internal structures that is created by exposing specialized film to X-rays. A **scout film** is a plain X-ray made without the use of contrast material (Figure 7–8).
- **retrograde pyelogram** (reh-trō-grād pī-eh-lō-grahm) = radiographic study of the kidney and ureters in which contrast material is placed directly in the urinary bladder.
- **urinalysis** (yoo-rih-nahl-ih-sihs) = examination of urine components. It is abbreviated UA. There are three parts of a UA: macroscopic, chemical, and microscopic. The macroscopic analysis (also called the gross exam) is visual observation of the urine in which things like color, transparency (turbidity (cloudiness) is one way to describe transparency), odor, and foam are recorded. The chemical analysis consists of applying urine to a reagent strip impregnated with chemicals that react with substances in the urine and change colors. The colors are compared to a set of standards to determine whether or not there are abnormal levels

FIGURE 7–7 IVP in a dog. Contrast material outlines the renal pelvis in this radiograph. Some contrast material also is present in the urinary bladder.

FIGURE 7–8 Radiograph of the urinary bladder. Urinary calculi in the canine urinary bladder are seen on this lateral scout radiograph.

of some substances in the sample. The chemical analysis can tell us about pH (hydrogen ion concentration that indicates acidity or alkalinity), leukocytes, erythrocytes, protein, glucose, specific gravity (measurement that reflects the amount of wastes, minerals, and solids in urine), and other factors (Tables 7–2 and 7–3; Figure 7–9).

TABLE 7–2 Descriptive Terms for Urine and Urination

albuminuria (ahl-bū-mihn-yoo-rē-ah)	presence of the major blood protein in urine
anuria (ah-nū-rē-ah)	complete suppression of urine production
bacteriuria (bahck-tē-rē-yoo-rē-ah)	presence of bacteria in urine
crystalluria (krihs-tahl-yoo-rē-ah)	urine with naturally produced angular solid of definitive form (crystals)
dysuria (dihs-yoo-rē-ah)	difficult or painful urination
glucosuria (gloo-kohs-yoo-rē-ah)	glucose (sugar) in urine
glycosuria (glī-kohs-yoo-rē-ah)	glucose (sugar) in urine
hematuria (hēm-ah-toor-ē-ah)	blood in urine
ketonuria (kē-tō-nū-rē-ah)	presence of ketones in urine (ketones are produced during increased fat metabolism)
nocturia (nohck-too-rē-ah)	excessive urination at night (*nocti* means night)
oliguria (ohl-ih-goo-rē-ah)	scanty or little urine
pollakiuria (pōl-lahck-ē-yoo-rē-ah)	frequent urination (*pollaki* means many times)
polyuria (pohl-ē-yoo-rē-ah)	excessive urination
proteinuria (prō-tēn-yoo-rē-ah)	presence of proteins in urine
pyuria (pī-yoo-rē-ah)	pus in urine
stranguria (strahng-yoo-rē-ah)	slow or painful urination

The microscopic analysis (also called the urine sediment exam) is used to detect the presence of cellular material in urine that has been centrifuged and is viewed under the microscope.

- **urinary catheterization** (kahth-eh-tər-ih-zā-shuhn) = insertion of a tube through the urethra into the urinary bladder (usually to collect urine); abbreviated cath. A **catheter** (kahth-eh-tər) is the hollow tube that is inserted into a body cavity to inject or remove fluid (Figure 7–10a and b).

FIGURE 7–9 Manual testing of urine by reagent strip: (a) Hold the reagent strip by the top end and dip the strip into the urine specimen; (b) remove the strip from the urine; blot edge of strip to remove excess urine, and begin timer; and (c) read reactions at appropriate time intervals by comparing reagent pad colors to color chart. Chemical properties of urine, such as pH, glucose, ketones, and bilirubin, are tested with a dipstick.

TABLE 7–3 Urine Test Strips: Common Tests and Significance

Test	Significance
Urobilinogen	This test is used more commonly in human medicine to evaluate liver disease or the breakdown of red blood cells. This test is less useful in veterinary medicine. (The test strips used are designed for human urinalysis.)
Glucose	This test is used to screen for diabetes mellitus. Diabetics have elevated blood sugar. The kidneys are unable to conserve all the sugar once the level becomes too high. The test is also used to monitor diabetics once treatment has been started.
Ketones	In dogs and cats, the presence of ketones is typical of an animal with uncontrolled diabetes mellitus. Ketones result when metabolism is shifted from carbohydrates to lipids.
Bilirubin	Aged red blood cells are removed from the circulation in organs such as the spleen. Bilirubin is formed in the breakdown process of hemoglobin. (Normally bilirubin is cleared from the blood by the liver and excreted in bile.) Bilirubin is found in the urine in liver disease or excessive blood cell breakdown.
Protein	Proteins are large molecules that are not normally filtered into the urine. Protein in the urine can be present with a disease of the glomerulus (making it leaky) or with inflammation of the urinary tract (such as a bladder infection). A small amount of protein may normally be detected in very concentrated urine.
Blood	The test strip detects occult blood (i.e., blood that cannot be visibly seen in the urine). Blood can be present in diseases that cause inflammation of the urinary tract, much like protein. Bladder infections, stones, tumors, and trauma (e.g., hit by car) can all cause blood to be present in the urine. Bleeding disorders may also cause the test to be positive.
pH	Urine pH is influenced by diet and disease states in the body. Acidic pH is typical in animals with a meat diet or with acidosis (the kidney attempting to rid the body of excess acid). Basic or alkaline pH is typical in animals with a cereal grain diet, some urinary tract infections, and alkalosis in the body.

Courtesy of Teri Raffel, CVT

Copyright © 2015 Cengage Learning®

(a)   **(b)**

FIGURE 7–10 (a) Urinary catheterization in a male dog. (b) Examples of urinary catheters. The larger red rubber catheter would be used on a larger dog and the small clear catheter is typically used to catheterize male cats.

PATHOLOGY: URINARY SYSTEM

Pathologic conditions of the urinary system include the following:

- **azotemia** (ā-zō-tē-mē-ah) = presence of urea or other nitrogenous elements in the blood. **Azot/o** means urea or nitrogen.
- **calculus** (kahl-kyoo-luhs) = abnormal mineral deposit. Urine may contain crystals that remain in solution in urine. Stones, calculi, and liths are formed when crystals precipitate and form solids. **Lith/o** is the combining form for stone or calculus, and -**lith** is the suffix for stone or calculus. When used in relationship to the urinary system, *calculus* is modified to *urinary calculus*, *lith* is modified to *urolith*, and *stone* is modified to *urinary stone* or *kidney stone* to clarify which system is involved.
- **casts** (kahstz) = fibrous or protein materials found in the urine with renal disease or another abnormality (Figure 7–11).
- **crystals** (krihs-tahlz) = naturally produced angular solid of definitive form (Figure 7–11).
- **cystalgia** (sihs-tahl-jē-ah) = urinary bladder pain; also called **cystodynia** (sihs-tō-dihn-ē-ah).
- **cystitis** (sihs-tī-tihs) = inflammation of the urinary bladder.
- **cystocele** (sihs-tō-sēl) = displacement of the urinary bladder through the vaginal wall.
- **epispadias** (ehp-ih-spā-dē-uhs) = abnormal condition in which the urethra opens on the dorsum of the penis.
- **feline lower urinary tract disease** = common disease of cats in which cystitis, urethritis, and crystalluria are found; formerly called feline urologic syndrome (FUS). In male cats, urethral obstruction is commonly associated with this disease, abbreviated FLUTD.
- **glomerulonephritis** (glō-mər-yoo-lō-nehf-rī-tihs) = inflammation of the kidney involving the glomeruli.
- **hydronephrosis** (hī-drō-nehf-rō-sihs) = dilation of the renal pelvis as a result of an obstruction to urine flow (Figure 7–12).
- **hydroureter** (hī-drō-yoo-rē-tər) = distention of the ureter with urine caused by any blockage.
- **hypospadias** (hī-pō-spā-dē-uhs) = abnormal condition in which the urethra opens on the ventral surface of the penis.

- **inappropriate urination** (ihn-ah-prō-prē-ət yoo-rihn-ā-shuhn) = eliminating urine at the wrong time or in the wrong place.
- **incontinence** (ihn-kohn-tih-nehns) = inability to control excretory functions. The term *urinary* is applied in front of this term to refer to the inability to control urine.
- **interstitial cystitis** (ihn-tər-stihsh-ahl sihs-tī-tihs) = inflammation within the wall of the urinary bladder.
- -**lithiasis** (lih-thī-ah-sihs) = suffix meaning the presence of stones or calculi, as in nephrolithiasis (disorder characterized by the presence of kidney stones), urolithiasis (disorder characterized by the presence of urinary bladder stones), and ureterolithiasis (disorder characterized by the presence of stones in the ureters).
- **nephrectasis** (neh-frehck-tah-sihs) = distention of the kidneys. **Distention** means enlargement, and the suffix -**ectasis** means distention or stretching.
- **nephritis** (neh-frī-tihs) = inflammation of the kidneys.
- **nephrolith** (nehf-rō-lihth) = kidney stone or renal calculus.
- **nephromalacia** (nehf-rō-mah-lā-shē-ah) = abnormal softening of the kidney.
- **nephropathy** (neh-frohp-ah-thē) = disease of the kidneys.
- **nephrosclerosis** (nehf-rō-skleh-rō-sihs) = abnormal hardening of the kidney.
- **nephrosis** (neh-frō-sihs) = abnormal condition of the kidney(s).
- **prerenal** (prē-rē-nahl) and **postrenal** (pōst-rē-nahl) = before and after the kidney, respectively. These terms are used to describe other pathologic conditions, such as prerenal azotemia and postrenal obstruction.
- **prolapse** (prō-lahps) = downward displacement of a body organ. The suffix -**ptosis** means drooping or dropping down, and **nephroptosis** (neh-frohp-tō-sihs) may be used to describe a prolapsed kidney.
- **pyelitis** (pī-eh-lī-tihs) = inflammation of the renal pelvis.
- **pyelonephritis** (pī-eh-lō-nehf-rī-tihs) = inflammation of the renal pelvis and kidney.

Courtesy of Hycor Biomedical Inc., Garden Grove, CA

A. Cells found in urine sediment (a) two squamous epithelial cells (upper right), red blood cells, and white blood cells; (b) white blood cells; (c) renal tubular cell; (d) budding yeast cells (arrow) and squamous epithelial cells; (e) bacteria; (f) Trichomas (arrow); (g) spermatozoa (arrows).

Courtesy of Bayer Healthcare

B. Casts found in urine sediment (a) hyaline (arrow); (b) granular; (c) cellular; (d) waxy.

Courtesy of Bayer Healthcare and Centers for Disease Control (a–d)

Courtesy of Hycor Biomedical Inc., Garden Grove, CA and Bayer Healthcare (e–g)

C. Crystals found in urine sediment (a) uric acid; (b) triple phosphate; (c) calcium oxalate; (d) ammonium biurate; (e) cystine; (f) leucine; (g) tyrosine.

Courtesy of Hycor Biomedical Inc., Garden Grove, CA and Bayer Healthcare

D. Other substances found in urine sediment (a) starch granules (arrows); (b) cotton fibers

FIGURE 7–11 Substances found in urine.

FIGURE 7–12 Hydroureter and hydronephrosis caused by urethral stricture.

- **renal failure** (rē-nahl fāl-yər) = inability of the kidneys to function. Renal failure may be **acute** (ā-kūt) or **chronic** (krohn-ihck). Acute means occurring suddenly or over a short period. Acute renal failure (ARF) is the sudden onset of the inability of the kidneys to function. ARF may be caused by a **nephrotoxin** (nehf-rō-tohcks-ihn), which is a poison having destructive effects on the kidneys. Chronic means having a longer onset. Chronic renal failure (CRF) is the progressive onset of the inability of the kidneys to function. Signs of renal failure may include **polyuria** and **polydipsia** (pohl-ē-dihp-sē-ah); polyuria/polydipsia is abbreviated PU/PD. Polyuria is elevated urine production, and polydipsia is excessive thirst or drinking. Animals with renal failure may undergo **diuresis** (dī-yoo-rē-sihs). Diuresis is the elevated excretion of urine. Diuresis may be produced by fluid therapy or drug therapy. Drugs that increase urine production are **diuretics**.
- **renal infarction** (rē-nahl ihn-fahrck-shuhn) = obstruction of blood flow to the kidney(s).
- **uremia** (yoo-rē-mē-ah) = waste products in the blood. Uremia is seen with many types of kidney disease.
- **ureterectasis** (yoo-rē-tər-ehck-tah-sihs) = distention of the ureter.
- **ureterolith** (yoo-rē-tər-ō-lihth) = stone in the ureter.

- **urethritis** (yoo-rē-thrī-tihs) = inflammation of the urethra.
- **urethrolithiasis** (yoo-rē-thrō-lih-thī-ah-sihs) = disorder characterized by the presence of urethral stones. An **urethrolith** is a urethral stone.
- **urethrostenosis** (yoo-rē-thrō-steh-nō-sihs) = stricture of the urethra. A **stricture** (strihck-shər) is an abnormal band of tissue narrowing a passage.
- **urinary retention** (yoo-rih-nār-ē rē-tehn-shuhn) = inability to completely empty the urinary bladder.
- **urinary tract infection** (yoo-rihn-ār-ē trahckt ihn-fehck-shuhn) = invasion of microorganisms in the urinary system, which results in local cellular injury; abbreviated UTI. UTIs are diagnosed via a urinalysis that may be performed on voided or midstream urine samples (Figure 7–13).
- **urolith** (yoo-rō-lihth) = urinary bladder stone; also called **cystolith** (sihs-tō-lihth) (Figure 7–14).

FIGURE 7–13 Midstream urine samples are collected shortly after urination begins and just prior to the process ending.

P015662

FIGURE 7–14 Uroliths.

PROCEDURES: URINARY SYSTEM

Procedures performed on the urinary system include the following:

- **cystectomy** (sihs-tehck-tō-mē) = surgical removal of all or part of the urinary bladder.
- **cystopexy** (sihs-tō-pehck-sē) = surgical fixation of the urinary bladder to the abdominal wall.
- **cystoplasty** (sihs-tō-plahs-tē) = surgical repair of the urinary bladder.
- **cystostomy** (sihs-tohs-tō-mē) = surgical creation of a new opening between the skin and urinary bladder (Figure 7–15).
- **cystotomy** (sihs-tah-tō-mē) = surgical incision into the urinary bladder.
- **dialysis** (dī-ahl-ih-sihs) = procedure to remove blood waste products when the kidneys are no longer functioning. **Peritoneal dialysis** is the removal of blood waste products by fluid exchange through the peritoneal cavity; **hemodialysis** (hēm-ō-dī-ahl-ih-sihs) is the removal of blood waste products by filtering blood through a machine; abbreviated HD.
- **lithotripsy** (lihth-ō-trihp-sē) = destruction of stone using ultrasonic waves traveling through water (the suffix **-tripsy** means to crush).
- **nephrectomy** (neh-frehck-tō-mē) = surgical removal of a kidney.
- **nephropexy** (nehf-rō-pehcks-sē) = surgical fixation of a kidney to the abdominal wall.

- **nephroplasty** (nehf-rō-plahs-tē) = surgical repair of the kidney.
- **nephrostomy** (nehf-rōs-tō-mē) = surgical creation of a new opening between the skin and kidney.
- **nephrotomy** (nehf-rō-tō-mē) = surgical incision into the kidney.
- **ureterectomy** (yoo-rē-tər-ehck-tō-mē) = surgical removal of the ureter.
- **ureteroplasty** (yoo-rē-tər-ō-plahs-tē) = surgical repair of the ureter.
- **ureterotomy** (yoo-rē-təroh-tō-mē) = surgical incision into the ureter.
- **urethroplasty** (yoo-rē-throh-plahs-tē) = surgical repair of the urethra.
- **urethrostomy** (yoo-rē-throhs-tō-mē) = surgical creation of a permanent opening between the urethra and the skin. **Perineal urethrostomy** (pər-ih-nē-ahl yoo-rē-throhs-tō-mē) is the surgical creation of a permanent opening between the urethra and the skin between the anus and scrotum.
- **urethrotomy** (yoo-rē-throh-tō-mē) = surgical incision into the urethra.

ABBREVIATIONS RELATED TO THE URINARY SYSTEM

- ADH = antidiuretic hormone
- AKI = acute kidney injury
- ARF = acute renal failure
- BUN = blood urea nitrogen
- cath = catheterization
- CRF = chronic renal failure
- CRRT = continuous renal replacement therapy
- FLUTD = feline lower urinary tract disease
- FUS= feline urologic syndrome
- GFR = glomerular filtration rate
- HD = hemodialysis
- IHD = intermittent hemodialysis
- IVP = intravenous pyelogram
- K^+ = potassium
- Na^+ = sodium
- PU = polyuria
- SG = specific gravity
- UA = urinalysis
- UTI = urinary tract infection

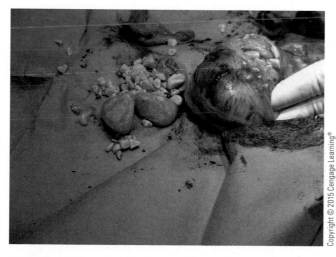

Copyright © 2015 Cengage Learning®

FIGURE 7–15 A cystotomy was performed to remove uroliths from the urinary bladder.

REVIEW EXERCISES

Multiple Choice

Choose the correct answer.

1. The combining forms for kidney are
   a. ren/o and ureter/o
   b. ren/o and nephr/o
   c. ren/o and cyst/o
   d. ren/o and periren/o

2. Inflammation of the kidney is
   a. nephrosis
   b. nephroptosis
   c. nephritis
   d. nephropathy

3. Insertion of a hollow tube through the urethra into the urinary bladder is called
   a. cystocentesis
   b. cystogram
   c. urinary injection
   d. urinary catheterization

4. The hormone produced by the kidney that stimulates red blood cell production in the bone marrow is
   a. ADH
   b. erythropoietin
   c. aldosterone
   d. renerythogenin

5. Retrograde means
   a. going backward
   b. going forward
   c. going sideways
   d. repeating

6. Examination of the components of urine is a
   a. urinoscopy
   b. cystoscopy
   c. urinalysis
   d. cystolysis

7. Inflammation of the urinary bladder is
   a. cystitis
   b. urolithiasis
   c. urology
   d. uritis

8. UTI is the abbreviation for
   a. urinary treatment for infection
   b. urinary tract infection
   c. urinary tract inflammation
   d. urinary trigone infarct

9. Stable internal environment is
   a. stricture
   b. status
   c. homeostasis
   d. isostatic

10. Diuretics are chemical substances that
    a. cause painful urination
    b. cause complete cessation of urine
    c. cause nighttime urination
    d. cause an increase in urine production

11. The triangular part of the urinary bladder is the
    a. calyx
    b. hilus
    c. medulla
    d. trigone

12. The term for production of urine is
    a. urinogenesis
    b. uropoiesis
    c. turbidity
    d. renogenesis

13. Inability to control excretory functions is
    a. inappropriate urination
    b. urinary retention
    c. incontinence
    d. urethritis

14. The medical term for excessive urination is
    a. pyuria
    b. polyuria
    c. polydipsia
    d. pollakiuria

15. Oliguria means
    a. scanty or little urine
    b. blood in urine
    c. frequent urination
    d. excessive urination

16. The term for frequent urination is
    a. scanty or little urine
    b. dysuria
    c. stranguria
    d. pollakiuria

17. A surgical incision into the urinary bladder is known as
    a. cystectomy
    b. cystotomy
    c. cystopexy
    d. cystostomy

18. The presence of urea or other nitrogenous elements in the blood is called
    a. uremia
    b. diuresis
    c. azotemia
    d. proteinuria

19. Obstruction of blood flow to the kidney(s) is
    a. dialysis
    b. azotemia
    c. prerenal infarction
    d. renal infarction

20. The term that means no urine production is
    a. diuresis
    b. anuria
    c. nocturia
    d. hematuria

21. What does the N in BUN stand for?
    a. nocturia
    b. nitrogen
    c. nephron
    d. neuron

22. Renal calculus is another term for
    a. kidney stone
    b. cystitis
    c. azotemia
    d. bladder crystals

23. An animal with urinary retention has a urinary bladder that
    a. has urine flowing backward into the ureter
    b. spasms frequently causing abdominal pain
    c. is unable to completely empty
    d. is enlarged and contains uroliths

24. The term for rolled or coiled is
    a. calyx
    b. ascending
    c. descending
    d. convoluted

25. The point of attachment of the renal blood vessels, nerves, and ureter to
    a. the kidney is the calyx
    b. convoluted
    c. hilus
    d. cortex

Matching

Match the term in Column I with the definition in Column II.

Column I

1. _____ bacteriuria
2. _____ glycosuria or glucosuria
3. _____ nocturia
4. _____ proteinuria
5. _____ anuria
6. _____ oliguria
7. _____ albuminuria
8. _____ stranguria
9. _____ polyuria
10. _____ pyuria
11. _____ pollakiuria
12. _____ ketonuria
13. _____ crystalluria
14. _____ dysuria
15. _____ hematuria
16. _____ diuresis
17. _____ calculus
18. _____ cortical
19. _____ erythropoietin
20. _____ cystitis

Column II

a. complete suppression of urine production

b. difficult or painful urination

c. scanty or little urine

d. excessive urination at night

e. slow or painful urination

f. blood in urine

g. frequent urination

h. presence of bacteria in urine

i. glucose (sugar) in urine

j. presence of ketones in urine

k. presence of proteins in urine

l. presence of the major blood protein in urine

m. pus in urine

n. increased urination

o. crystals in urine

p. pertaining to the outer portion of an organ

q. stone

r. inflammation of the urinary bladder

s. elevated urine excretion

t. hormone produced by the kidney that stimulates red blood cell production

Fill in the Blanks

Write the correct term in the space provided.

1. Urin/o and ur/o mean _____.

2. Ren/o and nephr/o mean _____.

3. Glycosuria and glucosuria mean _____.

4. Cystolith and urolith mean _____.

5. Excretion, elimination, and voiding mean _____.

Ordering: Go with the Flow

Put the following steps of urine flow in order by placing the numbers 1–7 in the space provided.

1. Urine exits the body _____

2. Fluid flows from Bowman's capsule to the renal tubule _____

3. Urine flows from the kidneys into the ureter _____

4. Urine flows from the ureters into the urinary bladder _____

5. Urine flows through the renal tubules to the collecting duct in the kidney _____

6. Fluid collects in Bowman's capsule _____

7. Urine flows from the urinary bladder to the urethra _____

Spell Check

Circle the term that is spelled correctly.

1. examination of urine components:	urinanalysis	urinalysis	urinealisis
2. functional unit of the kidney:	nephron	nefron	nephrone
3. constant internal environment:	homiostasis	homeostasis	homeostatis
4. excessive thirst:	polydipsia	polydypsia	polydyspsia
5. abnormal mineral deposit:	calculis	kalkulus	calculus
6. inflammation of the urethra:	urethritis	ureteritis	urethratis
7. difficult or painful urination:	disuria	dysuria	dyshuria
8. blood in urine:	hemoturia	hematouria	hematuria
9. inability to control excretory functions:	incontinence	inkontinence	incontinance
10. inflammation of the renal pelvis and kidney:	pylonephritis	pyelonephritis	pilonephritis

Crossword Puzzles

Urinary System Combining Forms

Supply the correct term in the appropriate space for the definition listed.

Across

2 combining form for inner or middle

5 combining form for urinary bladder

7 combining form for opening

8 combining forms for kidney

9 combining form for renal pelvis

10 combining forms for urine/urinary organs

Down

1 combining form for cluster of capillaries

3 combining form for stone/calculus

4 combining form for the tube that takes urine from the urinary bladder to the outside

5 combining form for outer

6 combining form for paired narrow tubes that are located between the kidneys and urinary bladder

Disease and Procedure Terms

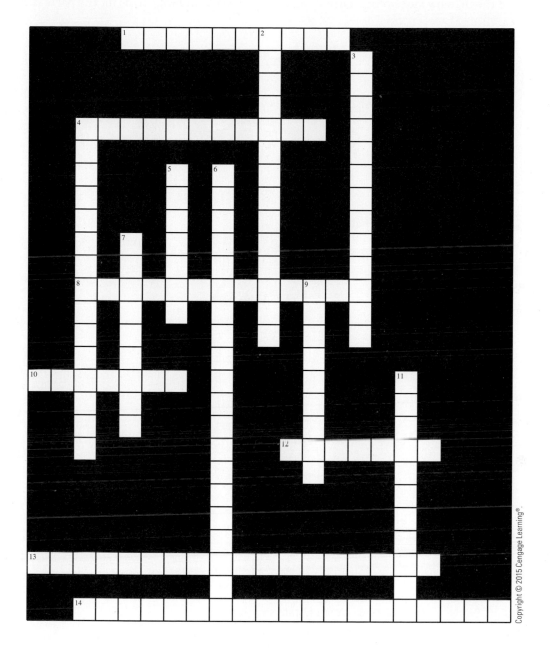

Across

1 surgical fixation of the kidney to the abdominal wall

4 surgical removal of a kidney

8 surgical puncture of the urinary bladder (to remove fluid)

10 study of the urinary system

12 urinary bladder stone or calculus

13 inflammation of the kidney involving the cluster of capillaries

14 inability to control the urine

Down

2 inflammation of the renal pelvis and kidney

3 surgical repair of the hollow tube that connects the urinary bladder to the outside of the body

4 abnormal hardening of the kidney

5 functional unit of the kidney

6 surgical creation of a permanent opening between the hollow tube that connects the urinary bladder to the outside of the body and the skin between the anus and scrotum

7 surgical incision into the urinary bladder

9 abnormal band of tissue narrowing a passage

11 voluntary control of urination

Label the Diagrams

Label the parts of the urinary system in Figure 7–16. Provide the combining forms for the parts labeled.

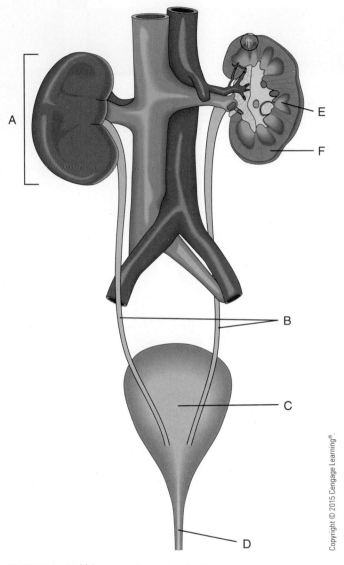

FIGURE 7–16 Urinary system structures.

Case Studies

Supply the proper abbreviation or medical term for the lay terms and definitions underlined.

A 5-year-old male neutered springer spaniel was presented with difficulty urinating and blood in the urine. The veterinarian performed examination of the urine by breaking it into its components after obtaining a urine sample by inserting a needle in the urinary bladder and withdrawing urine.

The test revealed large numbers of erythrocytes and struvite crystals. Suspecting more than inflammation of the urinary bladder, the veterinarian took an X-ray of the dog's urinary bladder, and urinary bladder stones were detected (Figure 7–17). An incision into the urinary bladder was performed to remove the urinary stones, and the dog recovered uneventfully.

1. year   _____

2. male neutered   _____

3. difficulty urinating   _____

4. blood in the urine   _____

5. examination of the urine by breaking it into its components   _____

6. inserting a needle in the urinary bladder and withdrawing urine   _____

7. erythrocytes   _____

8. inflammation of the urinary bladder   _____

9. X-ray   _____

10. urinary bladder stones   _____

11. incision into the urinary bladder   _____

Urinary bladder stones

FIGURE 7–17 Radiograph showing urinary bladder stones.

Define the underlined terms in each case study.

A 2-yr-old <u>M/N DLH</u> cat was presented to the clinic for crying when urinating in the litter box. The owner stated that the cat had been <u>inappropriately urinating</u> on the carpet for the past few days but otherwise seemed normal. Upon examination, the cat was about 7% <u>dehydrated,</u> T = 102° F, HR = 180 bpm, and RR = panting. Heart and lungs <u>auscultated</u> normally. Abdominal palpation revealed <u>cystomegaly</u>. The <u>urethra</u> was red and swollen. When the cat was put in a litter box, he was straining but was unable to <u>urinate</u>. The cat was <u>dx'd</u> with <u>urethral obstruction</u>. The owner was informed of the emergency status of relieving the obstruction, and the cat was hospitalized. The cat was <u>anesthetized</u> using <u>inhalant anesthesia</u> via a mask, and a <u>urinary catheter</u> was passed and sutured to the <u>perineum. IV</u> fluids were administered to reverse the cat's dehydration. Urine was collected via the urinary catheter, and a UA was performed. The <u>UA</u> revealed a <u>pH</u> = 7.0, large amounts of blood in the urine, a large number of <u>leukocytes</u> in the urine, a <u>specific gravity</u> of 1.040, and large amounts of <u>struvite crystals</u> and bacteria in the urine. In addition to the urethral obstruction, the cat also had <u>cystitis</u> and was treated with antibiotics and a urinary acidifying diet. The cat was hospitalized until the urinary catheter was removed. The cat was discharged with its medication, and the owner was advised to have a recheck UA when the medication was completed.

12. M/N _____

13. DLH _____

14. inappropriately urinating _____

15. dehydrated _____

16. auscultated _____

17. cystomegaly _____

18. urethra _____

19. urinate _____

20. dx'd _____

21. urethral obstruction _____

22. anesthetized _____

23. inhalant anesthesia _____

24. urinary catheter _____

25. perineum _____

26. IV _____

27. UA _____

28. pH _____

29. leukocytes _____

30. specific gravity _____

31. struvite crystals _____

32. cystitis _____

A 2-yr-old milking <u>cow</u> was showing a decrease in milk production and weight loss over the past few weeks. She was <u>off feed</u> and restless according to the farmer. Upon arrival at the farm, the veterinarian noted that the cow was switching its tail, vital signs were normal, and the cow was <u>polyuric. Hematuria</u> was noted, and urine was collected for analysis. <u>Rectal palpation</u> revealed <u>nephromegaly</u>. A tentative <u>diagnosis</u> of <u>pyelonephritis</u> was made. Antibiotics were given to this cow, the cow's milk was withheld for antibiotic withdrawal times, and the farmer was advised to monitor the other cows for similar abnormalities. The veterinarian returned to the clinic and performed a <u>UA</u>, which showed a large amount of blood in the urine and rod-shaped bacteria on microscopic examination. The sample was sent in for bacterial <u>culture and sensitivity</u> to determine whether the cow had a *Corynebacterium renale* infection. *C. renale* infections may indicate a herd with subclinical infection, which would affect the treatment and hygiene practices on this farm.

33. cow   _____

34. off feed   _____

35. polyuric   _____

36. hematuria   _____

37. rectal palpation   _____

38. nephromegaly   _____

39. diagnosis   _____

40. pyelonephritis   _____

41. UA   _____

42. culture and sensitivity   _____

Critical Thinking Exercise

The following story and questions contain words and concepts from this chapter and are designed to stimulate critical thinking through class discussion, Internet searches, or brief writing responses.

The owner of a 4-yr-old Rambouillet wether called the veterinary clinic because his sheep had been dribbling blood-tinged urine and had stranguria for 24 hours. The wether was now anorexic and was exhibiting tenesmus, tail twitching, and weight shifting. The veterinarian drove to the farm and did an exam on the wether, noting that urine had dried on the hair around the penis and that mineral deposits were present. The veterinarian palpated abnormal pulsations and swelling of the urethra. On rectal palpation the urinary bladder felt enlarged. The veterinarian diagnoses urethrolithiasis in this sheep.

1. What is stranguria?

2. Why may the urine be blood-tinged?

3. What does the presence of mineral deposits near the penis suggest?

4. Why might the urinary bladder be enlarged in this animal?

5. A sequela (condition following as a consequence of a disease) of urethrolithiasis is urinary retention or urinary bladder rupture. Explain how these conditions may develop.

The veterinarian explains to the owner that rams and wethers on high concentrate rations are susceptible to urinary calculi (also known as water belly). Rations high in phosphorus increase the chances of calculi being formed in the urinary tract. Calculi are formed from salts that are normally excreted but, because of their abnormally high concentration in the urine, they precipitate out and become stones. These stones frequently lodge in the urethra, blocking the flow of urine. Reduced water intake tends to aggravate the condition.

6. What management changes will the veterinarian recommend to this owner?

7. Urinary bladder rupture can result in death from uremia. What is uremia?

Chapter 8
HAVE A HEART

Objectives

Upon completion of this chapter, the reader should be able to:

- Identify the structures of the cardiovascular system

- Differentiate between the types of blood vessels and describe their functions

- Describe the functions of the cardiovascular system

- Describe the flow of blood throughout the body

- Recognize, define, spell, and pronounce the terms related to the diagnosis, pathology, and treatment of the cardiovascular system

- Interpret the meaning of abbreviations of the cardiovascular system

FUNCTIONS OF THE CARDIOVASCULAR SYSTEM

The cardiovascular system delivers oxygen, nutrients, and hormones to the various body tissues and transports waste products to the appropriate waste removal systems. Occasionally, the cardiovascular system is called the circulatory system; however, the circulatory system is divided into systemic circulation (blood flow to all parts of the body except the lungs) and pulmonary circulation (blood flow out of the heart through the lungs and back to the heart).

STRUCTURES OF THE CARDIOVASCULAR SYSTEM

Cardiovascular (kahr-dē-ō-vahs-kyoo-lər) means pertaining to the heart and vessels (blood vessels in this context). There are three major parts of this system: the heart, the blood vessels, and the blood. Blood is discussed in Chapter 15.

The Heart

The heart is a hollow muscular organ that provides the power to move blood through the body. The combining form for heart is **cardi/o**. The heart is located inside the **thoracic** (thō-rahs-ihck) **cavity**, or **chest cavity**. The heart lies between the lungs in a cavity called the **mediastinum** (mē-dē-ahs-tī-nuhm). The mediastinum also contains the large blood vessels, trachea, esophagus, lymph nodes, and other structures.

The Pericardium

Surrounding the heart is a double-walled membrane called the **pericardium** (pehr-ih-kahr-dē-uhm) (Figure 8–1). The two layers of the pericardium are the fibrous and serous. The **fibrous pericardium** is the tough external layer. The **serous layer** is the inner layer and is divided into two parts: the **parietal** (pahr-ī-ih-tahl) **layer** and the **visceral** (vihs-ər-ahl) **layer**. The

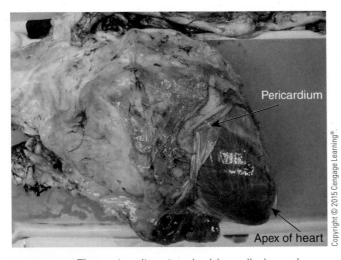

FIGURE 8–1 The pericardium is a double-walled membrane that surrounds the heart.

parietal layer (in Latin, *parietal* means belonging to the wall) is the serous layer that lines the fibrous pericardium. The visceral layer (*viscus* is Latin for an organ) is the serous layer that lines the heart. The visceral layer also is called the **epicardium** (ehp-ih-kahr-dē-uhm). Between the two serous layers of the pericardium is a space called the **pericardial** (pehr-ih-kahr-dē-ahl) **space**. **Pericardial fluid** is a liquid in the pericardial space. Pericardial fluid prevents friction between the heart and the pericardium when the heart beats.

Layers of the Heart Walls

The heart wall is made up of the following three layers (Figure 8–2):

- **epicardium** (ehp-ih-kahr-dē-uhm) = external layer of the heart; also known as the visceral layer of pericardium. The prefix **epi-** means upper.
- **myocardium** (mī-ō-kahr-dē-uhm) = middle and thickest layer of the heart; the actual heart muscle. The combining form **my/o** means muscle.
- **endocardium** (ehn-dō-kahr-dē-uhm) = inner layer of the heart; lines the heart chambers and valves. The prefix **endo-** means within.

Blood Supply

Heart tissue beats constantly and must have a continuous supply of oxygen and nutrients and the prompt removal of waste. The problem is that it cannot fulfill these needs from the blood it is pumping. These needs are supplied by its own arteries, and its wastes are removed by its own veins. The arteries that serve the heart are known as the **coronary arteries** (kōr-oh-nār-ē ahr-tər-ēz) because they resemble a crown. **Coron/o** is the combining form meaning crown. The **coronary veins** remove waste products from the myocardium.

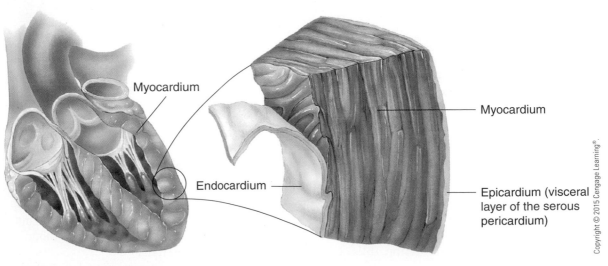

FIGURE 8–2 Layers of the heart.

If the blood supply to the heart is disrupted, the myocardium cannot function. Disruption of blood to the myocardium may be caused by **coronary occlusion** (ō-kloo-shuhn). *Occlusion* means blockage. Coronary occlusion may lead to **ischemia** (ihs-kē-mē-ah), which is deficiency in the blood supply to an area. Ischemia can lead to **necrosis** (neh-krō-sihs), which is tissue death. The area of necrosis due to ischemia is called an **infarct** (ihn-fahrckt) or **infarction** (ihn-fahrck-shuhn). An infarct is a localized area of necrosis caused by an interrupted blood supply.

Chambers

The heart is divided into right and left sides. The right and left sides are further subdivided into chambers. Mammalian and avian hearts are four-chambered; most reptile hearts have three chambers. The craniodorsal chambers of the heart are called the **atria** (ā-trē-ah); the singular form of *atria* is **atrium** (ā-trē-uhm). All vessels coming into the heart enter here. **Atri/o** is the combining form for atria. The left and right atria are separated by the **interatrial septum** (ihn-tər-ā-trē-ahl sehp-tuhm). A septum is a separating wall or partition.

The caudoventral chambers of the heart are **ventricles** (vehn-trih-kuhlz). **Ventricul/o** is the combining form for ventricle. The ventricles are separated from each other by the **interventricular septum.** The interventricular septum in reptiles is not complete, so the ventricles are open to each other and count as only one heart chamber. The ventricles are the pumping chambers of the heart, and all vessels leave the heart via the ventricles.

The narrow tip of the heart is called the **apex** (ā-pehcks) or cardiac apex (Figures 8–1 and 8–3).

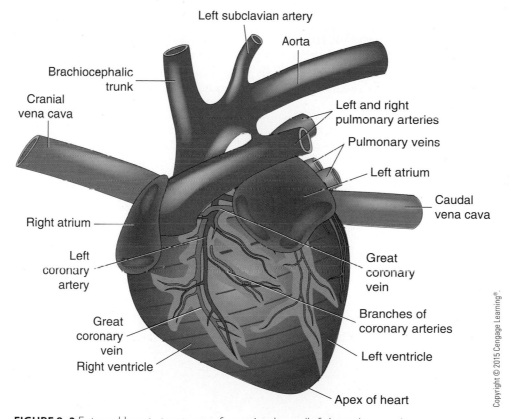

FIGURE 8–3 External heart structures of a canine heart (left lateral aspect).

BIOLOGY BONUS

The ventricles are pumping chambers of the heart, which is evident by the thickness of their myocardium. Because the left ventricle pumps blood to the entire body, the left ventricle has to generate much more pumping force than the right ventricle, which only pumps blood to the lungs. This is reflected in the fact that the left ventricular wall has a thicker myocardium than the right even though both ventricles hold about the same volume of blood.

Valves

Blood flow through the heart is controlled by valves (Figure 8–4 and Table 8–1). A **valve** (vahlv) is a membranous fold. The combining form for valve is **valv/o** or **valvul/o**. The four heart valves are as follows:

- **right atrioventricular valve** (ā-trē-ō-vehn-trihck-yoo-lahr vahlv), or right AV valve. This valve controls the opening between the right atrium and right ventricle (Figure 8–5). It also is called the **tricuspid valve** (trī-kuhs-pihd vahlv) because it has three points, or cusps (tri- = three; cusps = points).
- **pulmonary semilunar valve** (puhl-mah-nār-ē sehm-ē-loo-nahr vahlv), or pulmonary valve. This valve is located between the right ventricle and the pulmonary artery and controls blood entering the lungs. *Semilunar* means half-moon, and this valve is shaped like a half-moon.
- **left atrioventricular valve** (ā-trē-ō-vehn-trihck-yoo-lahr vahlv), or left AV valve. This valve controls the opening between the left atrium and left ventricle. It also is called the **mitral valve** (mī-trahl vahlv) or **bicuspid** (bī-kuhs-pihd) because it has two points (bi- = two).

- **aortic semilunar valve** (ā-ōr-tihck sehm-ē-loo-nahr vahlv), or aortic valve. The aortic valve is located between the left ventricle and the aorta and controls blood entering the arterial system. It also is half-moon shaped.

Rhythm

A rhythm is the recurrence of an action or a function at regular intervals. The heart's contractions are supposed to be rhythmic. The rate and regularity of the heart rhythm, called the **heartbeat**, are modified by electrical impulses from nerves that stimulate the myocardium. The heartbeat or cardiac cycle is an alternating sequence of relaxation and contraction of the heart chambers. **Cardiac output** is the volume of blood pumped by the heart per unit time. The volume of blood ejected from the ventricles during each heartbeat is **stroke volume**. To effectively pump blood throughout the body, contraction and relaxation of the heart must be synchronized accurately. These electrical impulses, also called the conduction system, are controlled by the sinoatrial node, atrioventricular node, bundle of His, and Purkinje fibers (Figure 8–6).

FIGURE 8–4 Internal heart structures of a canine heart (left lateral aspect).

TABLE 8–1 Blood Flow through the Heart

1. The right atrium receives blood from all tissues, except the lungs, through the cranial and caudal venae cavae. Blood flows from the right atrium through the tricuspid valve into the right ventricle. (This is systemic circulation.)

2. The right ventricle pumps blood through the pulmonary semilunar valve and into the pulmonary artery, which carries it to the lungs. (This is pulmonary circulation.)

3. The left atrium receives oxygenated blood from the lungs through the four pulmonary veins. The blood flows through the mitral valve into the left ventricle. (This is pulmonary circulation.)

4. The left ventricle receives blood from the left atrium. From the left ventricle, blood goes out through the aortic semilunar valve into the aorta and is pumped to all parts of the body except the lungs. (This is systemic circulation.)

5. Blood is returned by the venae cavae to the right atrium, and the cycle continues.

Chordae tendineae

FIGURE 8–5 Atrioventricular valve. The hemostat is under the leaflet of the valve. AV valves attach to papillary muscles in the ventricles by connective tissue strings called chordae tendineae (kohr-dā tehn-dih-nā).

The **sinoatrial node** (sī-nō-ā-trē-ahl nōd), or SA node, is located in the wall of the right atrium near the entrance of the superior vena cava. The SA node, along with atypical cardiac muscle cells called **Purkinje** (pər-kihn-jē) **fibers**, establishes the basic rhythm of the heart and is called the pacemaker of the heart. Purkinje fibers are less developed in the atria and are usually associated with the ventricles.

Electrical impulses from the SA node start waves of muscle contractions in the heart. The impulse in

Aorta

Cranial vena cava

Pulmonary artery

Caudal vena cava

Sinoatrial (SA) node

Right atrium

Atrioventricular (AV) node

Bundle of His

Right ventricle

Purkinje fibers

Right and left bundle branches

R

P

Q S

T

.... Atrial stimulation

– – – Stimulus reaches the AV node

------- Stimulus spreads through bundle of His

FIGURE 8–6 Conduction systems of the heart.

the right atrium spreads over the muscles of both atria, causing them to contract simultaneously. This contraction forces blood into the ventricles. Atrial contraction is called **atrial systole** (sihs-stohl-ē). **Inotropy** (ihn-ō-trōp-ē) is the term meaning force of contraction.

The electrical impulses from the SA node continue to travel to the **atrioventricular node** (ā-trē-ō-vehn-trihck-yoo-lahr nōd), or AV node. The AV node is located in the interatrial septum.

The AV node conducts impulses more slowly than the SA node does. This slower conduction of the AV node causes a pause after atrial contraction to allow the ventricles to fill with blood. The AV node transmits the electrical impulses to the bundle of His (also called the AV bundle).

The **bundle of His** (hihs) is located in the interventricular septum. The bundle of His continues on through the ventricle as ventricular Purkinje fibers, which carry the impulse through the ventricular muscle, causing the ventricles to contract. Ventricular contraction is called **ventricular systole**. Ventricular contraction forces blood into the aorta and pulmonary arteries.

The normal heart rhythm is called the **sinus rhythm** because it starts in the sinoatrial node. If the SA node does not function properly and is unable to send the impulse to the rest of the heart, other areas of the conduction system can take over and initiate a heartbeat. The resulting abnormal rhythm is called an **arrhythmia** (ā-rihth-mē-ah) or **dysrhythmia** (dihs-rihth-mē-ah) (Table 8–2). Antiarrhythmic drugs are substances that control heartbeat irregularities.

What Is Systole?

Systole (sihs-tō-lē) is a term that generally means contraction, derived from the Greek word for drawing together (think systole = squeeze). However, the term *systole* is used to denote ventricular contraction. Remember that the atria and ventricles cannot contract together, although they follow in rapid succession. **Diastole** (dī-ah-stō-lē) means expansion, derived from the Greek word for drawing apart (think diastole = dilate).

The term *diastole* is used to denote relaxation, or the time when the chambers are expanded. During diastole, the atria fill with blood. Then the atria contract, forcing blood into the ventricles, and the ventricles contract. If someone refers to systole and diastole without the modifiers *atrial* or *ventricular*, it is assumed that the person means ventricular contraction (systole) and ventricular relaxation (diastole).

Electrical Activity

The electrical events in the conduction system can be visualized by wave movement on an **electrocardiogram** (ē-lehck-trō-kahr-dē-ō-grahm). An electrocardiogram, abbreviated ECG or EKG, is the record of the electrical activity of the myocardium. The ECG or EKG is a tracing that shows the changes in voltage and polarity (positive and negative) over time

TABLE 8–2 Terms Relating to Rhythm

palpitation (pahl-pih-tā-shuhn) = heartbeat sensations that feel like pounding with or without irregularity in rhythm.
fibrillation (fih-brih-lā-shuhn) = rapid, random, and ineffective heart contractions.
flutter (fluht-tər) = cardiac arrhythmia in which atrial contractions are rapid but regular.
bradycardia (brā-dē-kahr-dē-ah) = abnormally slow heartbeat.
tachycardia (tahck-ē-kahr-dē-ah) = abnormally rapid heartbeat.
paroxysm (pahr-ohck-sihzm) = sudden convulsion or spasm.
normal sinus arrhythmia = irregular heart rhythm resulting from variation in vagal nerve tone as a result of respiration (a nonpathologic arrhythmia).
asystole (ā-sihs-tō-lē) = without contraction or lack of heart activity; flat line on an ECG.
syncope (sihn-kō-pē) = temporary suspension of respiration and circulation.
gallop (gahl-ohp) = low-frequency vibrations occurring during early diastole and late diastole.

Loading

The workload of the heart is divided into preload and afterload. **Preload** (prē-lōd) is the ventricular end-diastolic volume, or the volume of blood entering the right side of the heart. **Afterload** (ahf-tər-lōd) is the impedance to ventricular emptying presented by aortic pressure.

Preload problems usually are associated with right-sided heart disease, whereas afterload usually is associated with left-sided heart disease. Drugs used to treat heart disease may alter preload or afterload.

(Figure 8–7). The process of recording the electrical activity of the myocardium is **electrocardiography** (ē-lehck-trō-kahr-dē-ohg-rah-fē).

Electrocardiography produces a tracing that represents the variations in electric potential caused by excitation of heart muscle and is detected at the body surface. These variations in electric potential are detected by conductors called **leads** (lēdz) (Figure 8–8).

Heart Sounds

Auscultation (aws-kuhl-tā-shuhn) is the act of listening to body sounds and usually involves the use of a stethoscope. A **stethoscope** (stehth-ō-skōp) is an instrument used to listen; however, when the term is broken down into its basic components, **stetho-** means chest and **-scope** means instrument to visually examine or monitor. Traditionally, the stethoscope is considered to be a monitoring device of the chest area, but it also is used to auscultate other body parts.

When auscultating the heart, one hears a lub-dub sound. The lub is the first sound heard (called the first heart sound) and is caused by closure of the AV valves. The dub is the second sound heard (called the second heart sound) and is caused by closure of the semilunar valves. Systole, or ventricular contraction, occurs

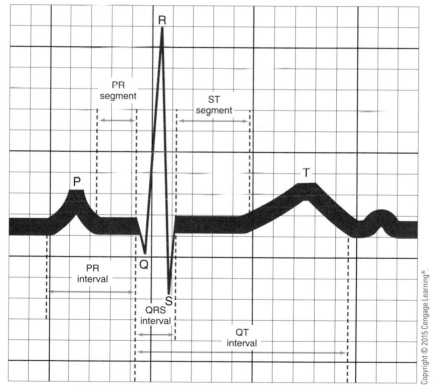

Copyright © 2015 Cengage Learning®.

FIGURE 8–7 Anatomy of an electrocardiogram. The first deflection, the P wave, represents excitation (depolarization) of the atria. The PR interval represents conduction through the atrioventricular valve. The QRS complex results from excitation of the ventricles. The QT interval represents ventricular depolarization and repolarization. The ST segment represents the end of ventricular depolarization to the onset of ventricular repolarization. The T wave results from recovery (repolarization) of the ventricles.

Courtesy of Lodi Veterinary Hospital S.C.

(a)

Courtesy of Kimberly Kruse Sprecher, CVT

(b)

FIGURE 8–8 (a) Electrocardiography in a dog. (b) ECG monitor displaying the electrical activity of the heart.

between the first and second heart sounds, whereas diastole, or ventricular relaxation, occurs between the second and first heart sounds.

A **heart murmur** (mər-mər) is an abnormal sound associated with the turbulent flow of blood. A murmur may be caused by a leak in a valve. A leak results in the inability of the valve to perform at the proper level, and this inability to perform at the proper level is called **insufficiency** (ihn-sah-fihsh-ehn-sē). Narrowing of a valve also may result in turbulent blood flow, causing a murmur.

Murmurs are described as systolic (the swooshing noise occurring between the first and second heart sounds) or diastolic (the swooshing noise occurring between the second and first heart sounds). Murmurs may be further described as **holosystolic** (hō-lō-sihs-stohl-ihck) or **pansystolic** (pahn-sihs-stohl-ihck), meaning that they occur during the entire ventricular contraction phase. The prefixes **holo-** and **pan-** mean all. These different phases in which a murmur occurs aid in identification of the cause of the murmur.

For example, murmurs heard during systole may be atrioventricular insufficiency or aortic or pulmonic valve **stenosis** (stehn-ō-sihs), or narrowing. Murmurs heard during diastole may be atrioventricular stenosis or aortic or pulmonic valve insufficiency.

In addition, murmurs can be described as crescendo and decrescendo. **Crescendo** (kreh-shehn-dō) **murmurs** are abnormal swooshing cardiac sounds that progressively increase in loudness, and **decrescendo** (deh-kreh-shehn-dō) **murmurs** progressively decrease in loudness.

The location of the murmur also is helpful in determining its cause. Where the murmur is heard the loudest is the point of maximal intensity (PMI). The PMI usually is located at the auscultation site of the defective valve. Occasionally, murmurs may result in vibrations felt on palpation of the chest. This vibration felt on palpation is called **thrill** (thrihl).

Clicks also may be heard during an examination. Clicks may be a sign of mitral insufficiency or may be of unknown origin. Other sounds that may be heard are split heart sounds (heartbeat sounds that are divided), crackles (which may be associated with movement or respiratory sounds), and rumbles (usually caused by shivering).

Blood Vessels

There are three major types of blood vessels: arteries, veins, and capillaries. The combining forms for vessel are **angi/o** and **vas/o**. These combining forms generally are used in reference to blood vessels but may be used to describe other types of vessels as well.

What Are We Hearing?

When auscultating the heart, there are various things on which to focus. One is heart rate (abbreviated HR), or the speed at which the heart is beating. Tachycardia and bradycardia can affect how blood is transported through the body. Abnormal sounds such as murmurs, clicks, and splits also should be assessed. Rhythm must be checked as well. Auscultation should be done in conjunction with pulse detection. The pulse is determined by feel and can be categorized as strong, weak, or thready. A pulse deficit occurs when the ventricles contract without enough force to propel blood to the periphery.

The **lumen** (loo-mehn) is the opening in a vessel through which fluid flows. The diameter of the lumen is affected by **constriction** (kohn-strihckt-shuhn), or narrowing of the vessel diameter, and **dilation** (dī-lā-shuhn), or widening of the vessel diameter. **Vasoconstrictors** (vahs-ō-kohn-strihck-tərz or vās-ō-kohn-strihck-tərz) are things that narrow a vessel's diameter; **vasodilators** (vahs-ō-dī-lāt-ərz or vās-ō-dī-lāt-ərz) are things that widen a vessel's diameter.

The **hilus** (hī-luhs) is the depression where vessels and nerves enter an organ.

The pumping action of the heart drives blood into the **arteries** (ahr-tər-ēz). An artery is a blood vessel that carries blood away from the heart. Blood in the arteries usually is oxygenated (the main exception is the pulmonary artery) and is bright red. The combining form for artery is **arteri/o**.

The **aorta** (ā-ōr-tah) is the main trunk of the arterial system that begins from the left ventricle of the heart.

The combining form for aorta is **aort/o**. After leaving the left ventricle, the aorta arches dorsally and then progresses caudally. The aorta is located ventral to the vertebrae. The aorta branches into other arteries that supply many muscles and organs of the body. The branches from the aorta usually are named for the area in which they supply blood. For example, the **celiac** (sē-lē-ahck) **artery** supplies the liver, stomach, and spleen (**celi/o** is derived from the Greek term *koilia*, meaning belly); the **renal arteries** supply the kidneys; and the **ovarian** (or **testicular**) **arteries** supply the ovaries (or testicles). Occasionally, arteries are named for their location, as in the **subclavian** (suhb-klā-vē-ahn) **artery**, which is located under the collarbone (Figure 8–9).

The **arterioles** (ahr-tē-rē-ōlz) are smaller branches of arteries. The combining form **arter/i** means vessel that carries blood away from the heart, and the suffix **-ole** means small. Arterioles are smaller and thinner than arteries and carry blood to the capillaries.

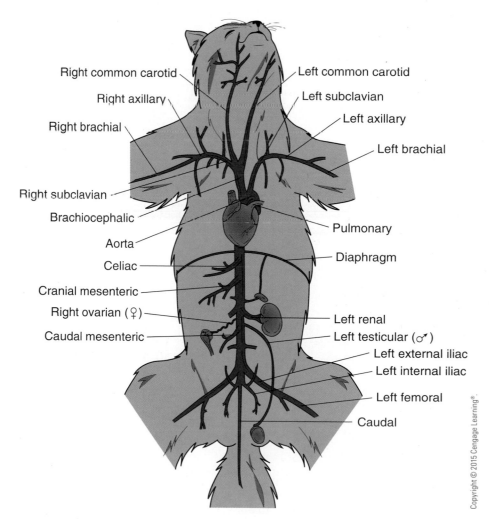

FIGURE 8–9 Major arteries in the cat.

The **capillaries** (kahp-ih-lār-ēz) are single-cell-thick vessels that connect the arterial and venous systems. Blood flows rapidly through arteries and veins; however, blood flow is slower through the capillaries due to their smaller diameter. This slower flow allows time for the diffusion of oxygen, nutrients, and waste products. Blood in the alveolar capillaries picks up oxygen and gives off carbon dioxide. In the rest of the body, oxygen diffuses (passes through) from the capillaries into tissue and carbon dioxide diffuses from tissue into the capillaries. Blood flow through tissues is called **perfusion** (pər-fū-shuhn). An indicator of perfusion is **capillary refill time**, or CRT. A CRT can be obtained by applying pressure to mucous membranes and timing how long it takes for the pink color to return (Figure 8–10).

Capillaries connect with **venules** (vehn-yoolz), which are tiny blood vessels that carry blood to the veins. **Veins** (vānz) form a low-pressure collecting system that returns blood to the heart. Veins have thinner walls and are less elastic than arteries, which have muscular walls to allow contraction and expansion to move blood throughout the body (Figure 8–11). Because the veins do not have muscular walls, contractions of the skeletal muscles cause the blood to flow through the veins toward the heart. Veins also have valves that permit blood flow toward the heart and prevent blood from flowing away from the heart (Figure 8–12). The combining forms for vein are **ven/o** and **phleb/o**.

Similar to arteries, veins usually are named for the area from which they take blood away. The **jugular**

FIGURE 8–11 Cross section of a large vein (left) and artery (right). Compare the differences between the thin-walled vena cava on the left and the thick-walled aorta on the right.

(a)   (b)   (c)

FIGURE 8–12 Veins contain valves to prevent the backward flow of blood. (a) External view of the vein shows wider area of the valve. (b) Internal view with the valve open as blood flows through. (c) Internal view with the valve closed.

FIGURE 8–10 The dog's mucous membranes are checked for color and capillary refill time. Mucous membranes can be described as normal (moist), tacky (slightly dry), or dry (no moisture).

Vessels are used to distribute substances throughout the body. Many drugs are injected into vessels and transported to different areas of the body. Examples include the following:

- **intravenous** (ihn-trah-vē-nuhs) = within a vein. **Perivascular** (pehr-ih-vahs-kyoo-lahr), or around the vessels, is an undesired route of administration and usually is an error of intravenous injection.
- **intra-arterial** (ihn-trah-ahr-tēr-ē-ahl) = within an artery.

(juhg-yoo-lahr) **vein** drains the head and neck area (*jugulum* is Latin for throat), **femoral veins** drain the legs, and **renal veins** drain the kidneys. An exception to this naming structure is the azygous vein. The **azygous** (ahz-ih-gihs) **vein** is a single vein that drains the chest wall and adjacent structures (the prefix **a-** means without, and the combining form **zygon** means yoke or pair) and is named based on the fact that it is not paired in the body (Figure 8–13).

Pressure

Blood pressure is the tension exerted by blood on the arterial walls. It is determined by the energy produced by the heart, the elasticity of the arterial walls, and the volume and **viscosity** (vihs-koh-siht-ē) (resistance to flow) of the blood. The **pulse** (puhlz) is the rhythmic expansion and contraction of an artery produced by pressure.

Blood pressure is measured by a **sphygmomanometer** (sfihg-mō-mah-nohm-eh-tər), which measures the amount of pressure exerted against the walls of the vessels.

Sphygm/o is the combining form for pulse, **man/o** is the combining form for pressure, and the suffix **-meter** means device. **Systolic** (sihs-stohl-ihck) **pressure** occurs when the ventricles contract and is highest toward the end of the stroke output of the left ventricle. **Diastolic** (dī-ah-stohl-ihck) **pressure** occurs when the ventricles relax and is lowest late in ventricular dilation.

The combining form **tensi/o** means pressure or tension and is used when describing blood pressure. **Hypertension** is high blood pressure (**hyper-** = excessive or above normal), and **hypotension** is low blood pressure (**hypo-** = deficient or less than normal). Drugs used to lower blood pressure are called **antihypertensives** (ahn-tih-hī-pər-tehns-ihvs).

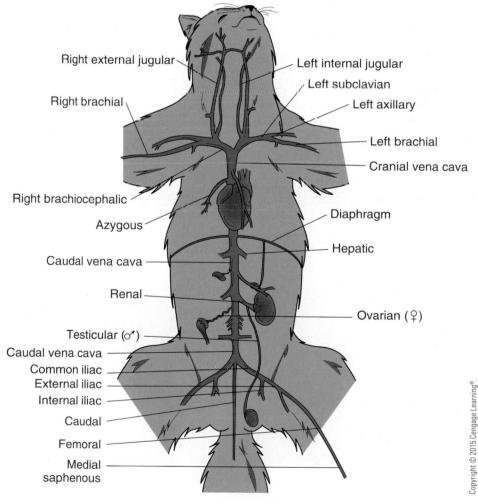

FIGURE 8–13 Major veins in the cat.

TEST ME: CARDIOVASCULAR SYSTEM

Diagnostic tests performed on the cardiovascular system include the following:

- **angiocardiography** (ahn-jē-ō-kahr-dē-ohg-rah-fē) = radiographic study of the blood vessels and heart using contrast material. The resulting film is an **angiocardiogram** (ahn-jē-ō-kahr-dē-ō-grahm).

FIGURE 8–14 Angiogram of a dog showing the femoral arteries. The use of a contrast medium makes the arteries visible.

- **angiography** (ahn-jē-ohg-rah-fē) = radiographic study of the blood vessels following injection of radiopaque material. An **angiogram** (ahn-jē-ō-grahm) is the film produced from this radiographic procedure (Figure 8–14).
- **cardiac catheterization** (kahr-dē-ahck kahth-eh-tər-ih-zā-shuhn) = radiographic study in which a catheter is passed into a blood vessel and is guided into the heart to detect pressures and patterns of blood flow.
- **echocardiography** (ehck-ō-kahr-dē-ohg-rah-fē) = process of evaluating the heart structures using sound waves. (**Ech/o** is a combing form for sound.) **Doppler** (dohp-lər) **echocardiography** uses the differences in frequency between sound waves and their echoes to measure the velocity of a moving object (Figure 8–15).
- **electrocardiography** (ē-lehck-trō-kahr-dē-ohg-rahf-ē) = process of recording the electrical activity of the heart. An **electrocardiogram** (ē-lehck-trō-kahr-dē-ō-grahm) is the record of the electrical activity of the heart and is abbreviated ECG or EKG (Figure 8–16). The machine that records the electrical activity of the heart is an **electrocardiograph** (ē-lehck-trō-kahr-dē-ō-grahf).
- **Holter monitor** (hōl-tər mohn-ih-tər) = 24-hour ECG that records the heart rates and rhythms onto a specialized tape recorder.
- **radiography** (rā-dē-ohg-rah-fē) = procedure of imaging objects by exposing sensitized film to X-rays. The resulting film is called a **radiograph** (rā-dē-ō-grahf) or X-ray (Figure 8–17). (This is an exception to the **-gram, -graph, -graphy** organization.)
- **tourniquet** (toor-nih-keht) = constricting band applied to a limb to control bleeding or to assist in drawing blood.

FIGURE 8–15 Echocardiogram of a canine heart.

II
Sinus tachycardia

V_1
Sinus bradycardia

V_1
Sinus arrhythmia

V_1
Atrial fibrillation

II
Ventricular fibrillation

V_1
Asystole

FIGURE 8–16 ECG tracings of common arrhythmias.

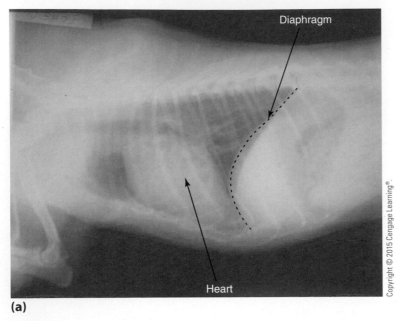

Diaphragm

Heart

(a)

Copyright © 2015 Cengage Learning®

(b)

Courtesy of Country Hills Pet Hospital

FIGURE 8–17 (a) Radiograph of a dog's chest showing the position of the heart in the thoracic cavity. (b) Radiograph of a dog's heart with cardiomegaly. Right-sided heart enlargement in a dog with heartworm disease gives the reverse D appearance of the heart in this radiograph.

Doppler echocardiography is based on the Doppler effect. An example of the Doppler effect (differences in frequency between sound waves and their echoes to measure the velocity of a moving object) is a police siren. A police siren seems to have a higher pitch when the police car is approaching than after it passes. Similarly, the pitch of a train whistle is higher when the train is approaching than after it passes. Medical use of the Doppler effect enables detection of blood flow based on various pitches after the blood passes by a transducer. Christian Doppler, an American mathematician, was the first person to describe this effect.

PATHOLOGY: CARDIOVASCULAR SYSTEM

Pathologic conditions of the cardiovascular system include the following:

- **aneurysm** (ahn-yoo-rihzm) = localized balloon-like enlargement of an artery.
- **angiopathy** (ahn-jē-ohp-ah-thē) = disease of vessels.
- **aortic insufficiency** (ā-ōr-tihck ihn-sah-fihsh-ehns-ē) = inability of the aortic valve to perform at the proper levels, which results in blood flowing back into the left ventricle from the aorta.

- **atherosclerosis** (ahth-ər-ō-skleh-rō-sihs) = hardening and narrowing of the arteries. This may be caused by **plaque** (plahck), which is a patch or raised area. **Ather/o** is the combining form for plaque or fatty substance.
- **atrial septal defect** (ā-trē-ahl sehp-tahl dē-fehckt) = opening in the wall dividing the right and left atria that may allow blood to move from the high-pressure right atrium to the low-pressure left atrium.
- **cardiac tamponade** (kahr-dē-ahck tahm-pō-nohd) = compression of the heart due to fluid or blood collection in the pericardial sac.
- **cardiomegaly** (kahr-dē-ō-mehg-ah-lē) = heart enlargement (Figure 8–17b).
- **cardiomyopathy** (kahr-dē-ō-mī-ohp-ah-thē) = disease of heart muscle. May be further classified as **hypertrophic** (hī-pər-trō-fihck), which is excessive growth of the left ventricle, or **dilated** (dī-lāt-ehd), which is characterized by a thin-walled left ventricle. Dilated cardiomyopathy also is known as **congestive** (kohn-jēhs-tihv) (Figure 8–18a and b).
- **carditis** (kahr-dī-tihs) = inflammation of the heart.
- **congestive** (kohn-jēhs-tihv) **heart failure** = syndrome that reflects insufficient cardiac output to meet the body's needs; abbreviated CHF. **Congestion** (kohn-jehs-chuhn), which is accumulation of fluid, and **edema** (eh-dē-mah),

Dilated ventricular
chambers

(a) Atrophicd papillary muscle   Abnormally thin ventricular walls

Abnormally increased muscle
mass due to a hypertrophied,
nondilated left ventricle

(b)

FIGURE 8–18 (a) Dilated cardiomyopathy. (b) Hypertrophic cardiomyopathy.

FIGURE 8-19 Heart with heartworms.

Copyright © 2015 Cengage Learning®

which is accumulation of fluid in the intercellular spaces, may be seen with CHF. **Ascites** (ah-sī-tēz) is fluid accumulation in the peritoneal cavity seen in dogs secondary to CHF and other diseases. **Pleural effusion** (ploor-ahl eh-fū-zhuhn) is abnormal fluid accumulation between the layers of the membrane encasing the lungs and is seen in cats secondary to CHF. Fluid accumulation can be relieved with the use of diuretics. **Diuretics** (dī-yoo-reht-ihcks) are substances that increase urine excretion.

- **cor pulmonale** (kōr puhl-mah-nahl-ē) = alterations in the structure or function of the right ventricle caused by pulmonary hypertension; also called pulmonary heart disease. **Cor** means heart, and **pulmon/o** is the combining form for lung.

- **dirofilariosis** (dī-rō-fihl-ahr-ē-ō-sihs) = heartworm infection; formerly called **dirofilariasis** (dī-rō-fihl-ahr-ē-ah-sihs). The scientific name of heartworm is *Dirofilaria immitis* (dī-rō-fihl-ahr-ē-ah ihm-ih-tihs), from which dirofilariosis is derived. Heartworm disease is found in dogs, cats, and ferrets (Figure 8–19). Heartworms mature and breed in the larger blood vessels. Mature heartworms produce tiny larvae called **microfilariae** (mī-krō-fihl-ahr-ē-ah). Mature heartworms may obstruct blood flow through the heart and blood vessels. A dead heartworm can cause pulmonary embolism. Obstruction of blood flow from the vena cava caused by heavy heartworm infestation is called **caval** (kā-vahl) **syndrome**. Heartworm disease can be prevented by the use of **prophylactic** (prō-fih-lahck-tihck) medication. **Prophylaxis**

(prō-fih-lahck-sihs) means prevention. If an animal has heartworm disease, treatment includes use of an **adulticide** (ah-duhlt-ih-sīd), or substance that kills mature or adult heartworms, and a **microfilaricide** (mī-krō-fihl-ahr-ih-sīd), or substance that kills larvae or juvenile heartworms.

- **embolus** (ehm-bō-luhs) = foreign object (e.g., a clot, air, or tissue) that is circulating in blood. An **embolism** (ehm-bō-lihzm) is blockage of a vessel by a foreign object.

- **endocarditis** (ehn-dō-kahr-dī-tihs) = inflammation of the endocardium and sometimes the heart valves (Figure 8–20). Endocarditis may be further modified, as in bacterial endocarditis.

- **heart block** = interference with the electrical conduction of the heart. Heart block may be partial or complete and is graded in degrees based on the characteristics of the block.

- **hemangioma** (hē-mahn-jē-ō-mah) = benign tumor comprised of newly formed blood vessels.

- **hematoma** (hē-mah-tō-mah) = collection of blood.

- **hypercapnia** (hī-pər-kahp-nē-ah) = above-normal levels of carbon dioxide. Hypercapnia results in reduced levels of oxygen (**hypoxia** = hī-pohck-sē-ah) and may cause a bluish tinge to the skin and mucous membranes. This bluish tinge is called **cyanosis** (sī-ah-nō-sihs). (The combining form **cyan/o** means blue.)

- **hypocapnia** (hī-pō-kahp-nē-ah) = below-normal levels of carbon dioxide.

- **hypoxia** (hī-pohck-sē-ah) = below-normal levels of oxygen.

Courtesy of Dr. Arthur Hattel, Pennsylvania State University

FIGURE 8-20 An AV valve with endocarditis. This valve would not create a tight seal, resulting in a murmur.

- **infarct** (ihn-fahrckt) = localized area of necrosis caused by an interrupted blood supply.
- **ischemia** (ihs-kē-mē-ah) = deficiency in blood supply (the combining form **isch/o** means hold back) (Figure 8–21).
- **mitral stenosis** (mī-trahl stehn-ō-sihs) = narrowing of the opening of the mitral valve (which affects the opening and closing of the mitral valve); abbreviated MS.
- **mitral valve insufficiency** (mī-trahl vahlv ihn-sah-fihsh-ehn-sē) = inability of the left atrioventricular valve to perform at the proper level; may be caused by fibrosis, endocarditis, or other conditions that occur in the mitral valve area.
- **mitral valve prolapse** (mī-trahl vahlv prō-lahps) = abnormal protrusion of the left atrioventricular valve that results in incomplete closure of the valve.
- **myocarditis** (mī-ō-kahr-dī-tihs) = inflammation of the myocardium.
- **occlusion** (ō-kloo-shuhn) = blockage in a vessel or passageway in the body.
- **patent ductus arteriosus** (pā-tehnt duhck-tuhs ahr-tē-rē-ō-sihs) = persistence of the fetal communication (ductus arteriosus) between the left pulmonary artery and aorta that should close shortly after birth; abbreviated PDA. (*Patent* means remaining open.) A PDA may cause overloading of the left ventricle, which may lead to left ventricular failure. A continuous heart murmur and enlarged heart are signs of a PDA.
- **pericarditis** (pehr-ih-kahr-dī-tihs) = inflammation of the pericardium (Figure 8–22).
- **pulmonic stenosis** (puhl-mah-nihck stehn-ō-sihs) = narrowing of the opening and valvular area between the pulmonary artery and right ventricle.

FIGURE 8–22 A bovine heart with pericarditis secondary to hardware disease.

- **regurgitation** (rē-gərj-ih-tā-shuhn) = backflow; used to describe backflow of blood caused by imperfect closure of heart valves.
- **shock** (shohck) = inadequate tissue perfusion. There are different types of shock, but one type occurs after cardiac arrest or cessation of heartbeat. Treatment of shock includes **resuscitation** (reh-suhs-ih-tā-shuhn), or the restoration of life. Resuscitative measures include fluid administration, cardiac massage, and artificial respiration. **Cardiopulmonary resuscitation** (kahr-dē-ō-puhl-mohn-ār-ē), or CPR, addresses only the cardiac and respiratory systems.
- **tetralogy of Fallot** (teht-rahl-ō-jē ohf fahl-ō) = congenital cyanotic cardiac condition that classically has four anatomical defects in the heart: pulmonary stenosis, ventricular septal defect, overriding aorta, and right ventricular hypertrophy.

FIGURE 8–21 Ischemia to a section of the small intestine of a horse (dark area on right).

Application of the Meaning of Regurgitation

Regurgitation is a term that can be applied to different body systems but in all cases it means essentially the same thing: something is flowing backward. In the gastrointestinal system, regurgitation means swallowed food is flowing backward into the oral cavity, while in the cardiovascular system blood is flowing backward into the previous chamber due to imperfect closure of heart valves.

- **thromboembolism** (throhm-bō-ehm-bō-lihzm) = the blocking of a blood vessel by a part that has broken away from a blood clot at its site of formation.
- **thrombus** (throhm-buhs) = blood clot attached to the interior wall of a vein or artery. A **thrombosis** (throhm-bō-sihs) is an abnormal condition in which a blood clot develops in a blood vessel. Substances that prevent blood clotting are called **anticoagulants** (ahn-tih-kō-āg-yoo-lahnts).
- **vasculitis** (vahs-kyoo-lī-tihs) = inflammation of a blood or lymph vessel.
- **ventricular septal defect** (vehn-trihck-yoo-lahr sehp-tahl dē-fehckt) = opening in the wall dividing the right and left ventricles that may allow blood to shunt from the right ventricle to the left ventricle without becoming oxygenated. To **shunt** (shuhnt) means to bypass or divert. A shunt resulting from a ventricular septal defect would bypass the lungs. Ventricular septal defect is abbreviated VSD. A harsh holosystolic murmur usually is a sign of a VSD.

PROCEDURES: CARDIOVASCULAR SYSTEM

Procedures performed on the cardiovascular system include the following:

- **angioplasty** (ahn-jē-ō-plahs-tē) = surgical repair of blood or lymph vessels. Angioplasties may be **transluminal** (trahnz-loo-mehn-ahl), which means the procedure is done through the opening of a vessel, or **percutaneous** (pehr-kyoo-tā-nē-uhs), which means the procedure is done through the skin.
- **angiorrhaphy** (ahn-jē-ōr-ah-fē) = suture of a vessel.
- **arteriectomy** (ahr-tē-rē-ehck-tō-mē) = surgical removal of part of a blood vessel that carries blood away from the heart.

FIGURE 8–23 A cat receiving a transfusion.

Courtesy of Kimberly Kruse Sprecher, CVT

- **arteriotomy** (ahr-tē-rē-oh-tō-mē) = incision of a blood vessel that carries blood away from the heart.
- **central venous pressure** (sehn-trahl vē-nuhs prehs-shər) = tension exerted by blood in the cranial vena cava; abbreviated CVP. CVP is monitored by catheterization of the cranial vena cava via the jugular vein. The catheter is connected to a fluid-filled column and a syringe or bag that serves as a fluid source.
- **defibrillation** (dē-fihb-rih-lā-shuhn) = use of electrical shock to restore the normal heart rhythm.
- **hemostasis** (hē-mō-stā-sihs) = control or stoppage of bleeding.
- **stent** (stehnt) = small expander implanted in a blood vessel to prevent it from collapsing. (A stent also is a device to hold tissue in place or to provide support for a graft.)
- **transfusion** (trahnz-fū-shuhn) = introduction of whole blood or blood components into the bloodstream of the recipient (Figure 8–23).
- **valvotomy** (vahl-vah-tō-mē) = surgical incision into a valve or membranous flap.

ABBREVIATIONS RELATED TO THE CARDIOVASCULAR SYSTEM

- AF = atrial fibrillation
- ASD = atrial septal defect
- AV = atrioventricular
- BP = blood pressure
- CHF = congestive heart failure
- CPR= cardiopulmonary resuscitation
- CRT = capillary refill time

Terms Used to Describe Disease

- **congenital** (kohn-jehn-ih-tahl) present at birth
- **hereditary** (hər-eh-dih-tahr-ē) genetically transmitted from parent to offspring
- **anomaly** (ah-nohm-ah-lē) deviation from normal
- **idiopathic** (ihd-ē-ō-pahth-ihck) of unknown cause
- **iatrogenic** (ī-aht-rō-jehn-ihck) produced by treatment

- CVP = central venous pressure
- ECG or EKG = electrocardiogram or electrocardiograph
- HR = heart rate
- LV = left ventricle
- MM = mucous membrane
- MS = mitral stenosis

- PDA = patent ductus arteriosis
- PMI = point of maximal intensity
- PVC = premature ventricular complex
- RV = right ventricle
- SA = sinoatrial
- VSD = ventricular septal defect
- VT = ventricular tachycardia

REVIEW EXERCISES

Multiple Choice

Choose the correct answer.

1. The right atrioventricular valve is also known as the
   a. mitral valve
   b. semilunar valve
   c. tricuspid valve
   d. bicuspid valve

2. The double-walled membranous sac enclosing the heart is the
   a. peritoneum
   b. pericardium
   c. perimyocardium
   d. pericardosis

3. A partition or wall separating something is called a
   a. septum
   b. valve
   c. lumen
   d. plaque

4. A localized area of necrosis caused by an interrupted blood supply is
   a. ischemia
   b. resuscitation
   c. pulse
   d. infarct

5. Introduction of whole blood or blood components into the bloodstream of the recipient is a(n)
   a. embolism
   b. thrombus
   c. transfusion
   d. stent

6. A bypass or diversion is called a
   a. preload
   b. shunt
   c. stent
   d. tourniquet

7. *Cor* means
   a. abnormality
   b. vessel
   c. heart
   d. valve

8. Disease of heart muscle is
   a. cardiopathy
   b. cor pulmonale
   c. cardiovalvopathy
   d. cardiomyopathy

9. Heart enlargement is
   a. cardiac swelling
   b. cardiac augmentation
   c. cardiac dilation
   d. cardiomegaly

10. Blood flow through tissue is
    a. ischemia
    b. infarct
    c. auscultation
    d. perfusion

11. The blood vessels that carry blood from the heart to the lungs are the
    a. pulmonary veins
    b. pulmonary arteries
    c. vena cava
    d. aorta

12. The contraction phase of the heartbeat is the
    a. septum
    b. diastole
    c. systole
    d. tachycardia

13. A disease produced by treatment is known as
    a. idiopathic
    b. iatrogenic
    c. congenital
    d. hereditary

14. The term for the external layer of the heart is the
    a. myocardium
    b. endocardium
    c. pericardium
    d. epicardium

15. The myocardium receives its blood supply from the
    a. aorta
    b. coronary arteries
    c. vena cava
    d. subclavian artery

16. Which heart valve is also known as the biscupid valve?
    a. aortic semilunar valve
    b. pulmonary semilunar valve
    c. mitral valve
    d. tricuspid valve

17. Which heart chamber pumps blood to the lungs?
    a. left atrium
    b. left ventricle
    c. right atrium
    d. right ventricle

18. Hemostasis means
    a. control of bleeding
    b. formation of new blood
    c. introduction of whole blood
    d. present at birth

19. A deficiency in blood supply is called
    a. infarct
    b. ischemia
    c. prophylaxis
    d. dilation

20. The opening in a vessel through which fluid flows is known as a
    a. hilum
    b. perfusion
    c. murmur
    d. lumen

21. Which two terms refer to chambers of the heart?
    a. tricuspid and pulmonic
    b. aortic and mitral
    c. atrium and ventricle
    d. apex and myocardium

22. Which of the following results in failure of the right ventricle to pump blood properly?
    a. ascites
    b. carditis
    c. emboli
    d. cor pulmonale

23. Pick the set of terms that are all blood vessels.
    a. vena cava, arteriole, apex
    b. venule, arteriole, capillary
    c. artery, edema, pulmonale
    d. embolism, hypertrophic, dilated

24. An instrument for measuring blood pressure is a/an
    a. stethoscope
    b. sphygmomanometer
    c. angiogram
    d. echocardiograph

25. The S in VSD stands for
    a. sclerosis
    b. stenosis
    c. septal
    d. stent

Matching

Match the term in Column I with the definition in Column II.

Column I

1. _____ perfusion

2. _____ heart murmur

3. _____ constriction

4. _____ vasculitis

5. _____ dilation

6. _____ hematoma

7. _____ capillaries

8. _____ veins

9. _____ arteries

10. _____ prophylaxis

11. _____ occlusion

12. _____ ischemia

13. _____ patent

14. _____ thrombus

15. _____ necrosis

16. _____ plaque

17. _____ insufficiency

18. _____ viscosity

19. _____ thrill

20. _____ stenosis

Column II

a. vessels that return blood to the heart

b. single-cell-thick vessels that connect the arterial and venous systems

c. vessels that carry blood away from the heart

d. blood flow through tissues

e. prevention

f. abnormal sound associated with the turbulent flow of blood

g. narrowing of a vessel diameter

h. widening of a vessel diameter

i. inflammation of a blood or lymph vessel

j. collection of blood

k. attached clot

l. resistance to flow

m. blocked

n. vibration felt on palpation

o. inability to perform at proper level

p. narrowing

q. patch or raised area

r. open

s. death of cells

t. lack of blood supply

Fill in the Blanks

Write the correct term in the space provided.

1. Angi/o and vas/o mean _____.

2. Ven/o and phleb/o mean _____.

3. Valv/o and valvul/o mean _____.

4. Cardi/o and cor mean _____.

5. Holo- and pan- mean _____.

Trace the flow of blood through the heart.

6. Blood enters the heart through two large veins called the _____ and _____, which empty blood into the _____.

7. Blood flows from the right atrium to the right ventricle through the _____, which also is called the _____.

8. After blood enters the right ventricle, it passes through the _____ and enters the lungs via the _____.

9. Blood becomes oxygenated in the lungs and returns to the heart via the _____.

10. Once blood enters the left atrium, it passes through the _____ or _____ on its way to the _____.

11. Blood passes through one last valve, the _____, before it enters the aorta and flows to various parts of the body.

True or False

If the statement is true, write T on the line. If the statement is false, write F on the line.

1. _____An animal with an abnormally large heart is diagnosed with carditis.

2. _____The term *atrioventricular* relates to the atria and ventricles of the heart.

3. _____A portable device that records a 24-hour electrocardiogram is a Holter monitor.

4. _____Substances that prevent blood clotting are anticoagulants.

5. _____A stent controls bleeding.

6. _____Edema is the accumulation of fluid in the intercellular spaces.

7. _____Microfilariae are mature heartworms that block blood flow through the heart in animals with dirofilariosis.

8. _____An animal with hypertrophic cardiomyopathy will have a thickened left ventricle.

9. _____An animal with dilated cardiomyopathy will have a thin-walled left ventricle.

10. _____A procedure that is done through the skin is percutaneous.

Spell Check

Circle the term that is spelled correctly.

1. bluish tinge to the skin and mucous membranes: cianosis   cyianosis   cyanosis
2. below-normal levels of carbon dioxide: hypokapnea   hypocapnia   hypocapnea
3. widening of a vessel: vasodialation   vasodilation   vasodilatation
4. relaxation of the heart: diastole   diastoly   diastooly
5. contraction of the ventricles: systole   sistolle   sistole

Crossword Puzzle

Cardiovascular Structure and Disease Terms

Supply the correct term in the appropriate space for the definition listed.

Across

2 to narrow

3 combining form for sound

6 vibration felt on palpation

11 abnormally rapid heart rate

12 deviation from normal

13 resistance to flow

14 craniodorsal chambers of the heart

16 to widen

17 above-normal levels of carbon dioxide

18 constricting band applied to limb to control bleeding or to assist in drawing blood

Down

1 caudoventral chamber of the heart

2 compression of the heart due to fluid or blood collection in the pericardial sac

4 blockage in a vessel

5 pertaining to the lungs

6 blood clot attached to the interior wall of a vessel

7 listening

8 external layer of the heart

9 blood vessel that returns blood to the heart

10 abnormally slow heart rate

12 without contraction

15 portion of the EKG that represents repolarization of the ventricules

Label the Diagrams

Label the diagrams in Figures 8–24 and 8–25.

FIGURE 8–24 External heart structures. Label the external heart structures to which the lines are pointing.

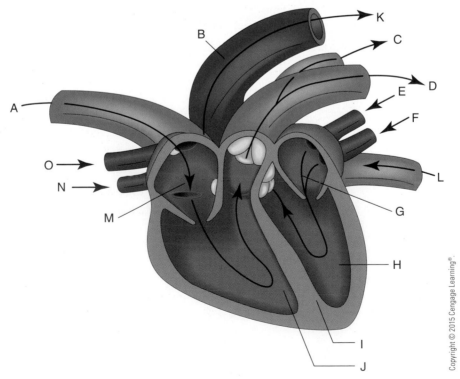

FIGURE 8–25 Internal heart structures. Label the internal heart structures to which the lines are pointing.

Case Studies

Define the underlined terms in each case study.

A 5-yr-old M/N Doberman was presented with signs of <u>lethargy</u>, <u>syncope</u>, and <u>cyanotic</u> mucous membranes. After auscultating the heart, the veterinarian detected a <u>cardiac arrhythmia</u> and <u>tachycardia</u>. An <u>ECG</u> and <u>radiograph</u> of the heart were ordered. The radiograph revealed <u>cardiomegaly</u>, which helped to support the veterinarian's <u>diagnosis</u> of <u>cardiomyopathy</u>.

1. lethargy _____

2. syncope _____

3. cyanotic _____

4. cardiac arrhythmia _____

5. tachycardia _____

6. ECG _____

7. radiograph _____

8. cardiomegaly _____

9. diagnosis _____

10. cardiomyopathy _____

A farmer called the clinic because one of his cows was suddenly <u>off feed</u> and had not been producing as much milk as before. Upon arrival at the farm, the veterinarian noted that the cow was reluctant to move, had an arched back, and appeared tachypnic. <u>PE</u> revealed <u>tachycardia</u>, <u>tachypnea</u>, <u>dyspnea</u>, <u>pyrexia</u>, and <u>abducted elbows</u>. Auscultation of the <u>thorax</u> revealed muffled lung and heart sounds. The farmer was questioned as to his use of magnets to prevent metallic objects from staying in the <u>rumen</u> or reticulum. The owner did not use magnets as a <u>prophylactic</u> measure, so the veterinarian suspected <u>acute traumatic reticuloperitonitis</u> in this cow. <u>Acute traumatic reticuloperitonitis</u> is commonly called hardware disease and is seen when swallowed metallic objects fall into the <u>reticulum</u> of the <u>ruminant</u> stomach, pierce the reticulum wall, and contaminate the <u>peritoneal cavity</u>. Occasionally, the object punctures the <u>diaphragm</u>, enters the thoracic cavity, and punctures the <u>pericardial sac</u>, causing <u>pericarditis</u>. A magnet was placed in this cow via a <u>balling gun</u> (Figure 8–26), and antibiotics were initiated. If the cow does not improve, she may be sent to slaughter because of the high cost of treating this disease.

11. off feed _____

12. PE _____

13. tachycardia _____

FIGURE 8–26 Balling gun used to administer oral medication to cattle. The rectangular gray magnet remains in the reticulum, where it can trap ingested metal.

14. tachypnea _____

15. dyspnea _____

16. pyrexia _____

17. thorax _____

18. rumen _____

19. prophylactic _____

20. acute _____

21. traumatic _____

22. reticuloperitonitis _____

23. acute traumatic reticuloperitonitis _____

24. reticulum _____

25. ruminant _____

26. peritoneal cavity _____

27. diaphragm _____

28. pericardial sac _____

29. pericarditis _____

30. balling gun _____

Critical Thinking Exercise

The following story and questions contain words and concepts from this chapter and are designed to stimulate critical thinking through class discussion, Internet searches, or brief writing responses.

Rosey, a 14#, 4-year-old F/S Siamese cat, presented to the clinic with weight loss and lethargy of 2 weeks' duration. The owner also stated that Rosey has had intermittent episodes of anorexia and weakness. On PE, she was tachycardic and had labored breathing. The veterinarian ausculted a systolic heart murmur and determined her CRT was 2 sec.

Define the following:

1. lethargy _____

2. anorexia _____

3. tachycardic _____

4. ausculted _____

5. systolic _____

6. murmur _____

7. CRT _____

8. Rosey's episodes of anorexia and weakness were intermittent. What does the term intermittent mean?

The veterinarian is concerned that Rosey may have cardiomyopathy. To determine whether she has this condition or not, the veterinarian requests that a thoracic radiograph and ECG be performed.

9. What are the different types of cardiomyopathy?

10. Cats with cardiomyopathy may develop congestive heart failure. Where does fluid typically accumulate in cats with cardiomyopathy? Where does fluid typically accumulate in dogs with cardiomyopathy?

Rosey is diagnosed with cardiomyopathy based on cardiomegaly on her radiographs. The veterinarian orders an echocardiogram and it is determined that Rosey has hypertrophic cardiomyopathy.

11. Because her heart muscle is severely thickened, why might this pose a problem for her?

12. One concern in cats with hypertrophic cardiomyopathy is that they develop an aortic thromboembolism. What is an aortic thromboembolism?

13. What clinical signs would Rosey have if she developed an aortic thromboembolism?

Chapter 9
A BREATH OF FRESH AIR

Objectives

Upon completion of this chapter, the reader should be able to:

- Identify the structures of the respiratory system

- Describe the functions of the respiratory system

- Differentiate between internal and external respiration

- Recognize, define, spell, and pronounce terms related to the diagnosis, pathology, and treatment of the respiratory system

- Interpret the meaning of abbreviations of the respiratory system

FUNCTIONS OF THE RESPIRATORY SYSTEM

The **respiratory** (rehs-pih-rah-tōr-ē) **system** is the body system that brings oxygen from the air into the body for delivery via the blood to the cells. Once the blood has delivered the oxygen to the cells, it picks up carbon dioxide and carries it back to the lungs, where this waste is expelled into the air. Carbon dioxide has acid properties and therefore also is involved in maintaining the body's acid-base status.

The term **respiration** means the diffusion of gases (oxygen and carbon dioxide) between the atmosphere and the cells of the body. The gas exchange between the blood and the cells is called internal or cellular respiration. External respiration is the absorption of atmospheric oxygen by the blood in the lungs and the diffusion of carbon dioxide from the blood in the lungs to atmospheric air.

Ventilation (vehn-tih-lā-shuhn) is a term that means the intake of fresh air. Ventilation is used to refer to breathing. Ventilation may be natural, as in normal breathing, or assisted, as in the use of a ventilator. Ventilators are devices that aid in breathing and therefore should not be called respirators.

STRUCTURES OF THE RESPIRATORY SYSTEM

The respiratory tract is routinely divided into the upper and lower respiratory tracts. The **upper respiratory tract** (uhp-pər rehs-pih-rah-tōr-ē trahckt) is the part of the respiratory system that consists of the nose, mouth, pharynx, epiglottis, and larynx. The **lower respiratory tract** (lō-ər rehs-pih-rah-tōr-ē trahckt) is the part of the respiratory system that consists of the trachea and within the lungs the bronchi, bronchioles, and alveoli. The trachea sometimes is considered part of the upper respiratory tract (Figure 9–1).

Upper Respiratory Tract

The conducting passages of the upper respiratory tract consist of the nose, mouth, pharynx, epiglottis, and larynx. These structures open to the outside and are lined with mucous membranes that may or may not have cilia to help filter the air.

Nose

Air enters and exits the body through the **nose**. **Nas/o** and **rhin/o** are combining forms for nose. In animals, not all noses look alike. The rigidity of the nose in swine has led to its being called the **snout** (snowt)

because it is so different from other species' noses. The nose consists of **nostrils**, or **nares** (nehr-ēz), which are the paired external openings of the respiratory tract.

Nares vary from species to species and may have the ability to open widely, as in equine, or may remain the same, as in the canine. In **endotherms** (ehn-dō-thərmz), or warm-blooded animals, the nasal passages contain **nasal turbinates** (tər-bih-nātz) (Figure 9–2). The nasal turbinates, sometimes called conchae (kohn-kā), are scroll-like cartilages covered with highly vascular mucous membranes. The nasal turbinates warm, humidify, and filter inspired air. Two nasal turbinates, the dorsal and ventral, separate the nasal cavity into passages. Each passage is called a **meatus** (mē-ā-tuhs) (**meat/o** is the combining form for opening or passageway), and they are named based on their location: dorsal meatus, middle meatus, ventral meatus, and common meatus. A **nasogastric** (nā-zō-gahs-trihck) **tube** is a tube that passes through the nose down to the stomach. A nasogastric tube is placed through the ventral nasal meatus.

Rhin/o is the Greek root for nose. Consider the term *rhinoceros*. Rhin/o means nose, and **cer/o** means horn; therefore, a rhinoceros is an animal with a horn on its nose. Where do you think a rhinovirus causes disease?

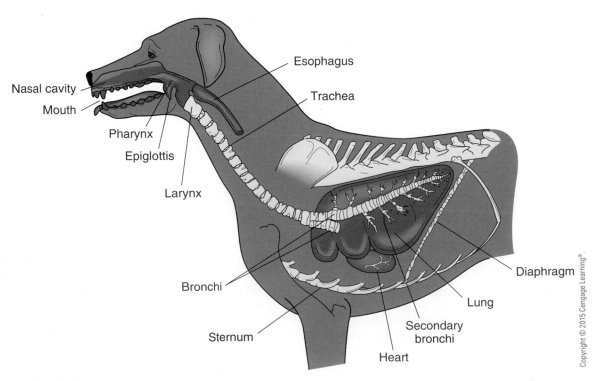

FIGURE 9–1 Structures of the respiratory system.

(a)

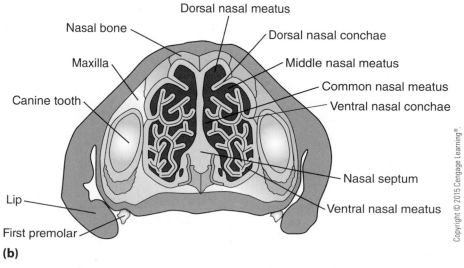

(b)

FIGURE 9–2 (a) Dorsal and sagittal views of the canine nasal cavity. (b) Cross section of the canine nasal cavity.

Cilia

Goblet cell

FIGURE 9–3 Respiratory epithelial tissue contains cilia and goblet cells.

Air passes from the nose through the nasal cavity. The rostral part of the nostrils and nasal cavity are called the **vestibule** (vehs-tih-buhl). The nose is divided by a wall of cartilage called the nasal septum (nā-zahl sehp-tuhm). The combining form **sept/o** means partition.

The respiratory system is lined with **mucous membrane** (myoo-kuhs mehm-brān), which is a specialized form of epithelial tissue. The mucous membranes secrete **mucus** (myoo-kuhs). Mucus is a slimelike substance that is composed of glandular secretions, salts, cells, and leukocytes. Mucus helps to moisten, warm, and filter the air as it enters the nose. **Cilia** (sihl-ē-ah) are thin hairs located inside the nostrils (Figure 9–3). Cilia filter the air to remove **debris** (deh-brē). Debris is the remains of something destroyed or damaged.

The **olfactory** (ohl-fahck-tōr-ē) **receptors** are responsible for the sense of smell. The combining form **olfact/o** means smell. Olfactory receptors are nerve endings located in the mucous membranes of the nasal cavity.

The **tonsils** (tohn-sihlz) are lymphatic tissue that protects the nasal cavity and proximal (upper) throat. The combining form for tonsil is **tonsill/o**. The tonsils are discussed in Chapter 15.

Sinuses

A **sinus** (sīn-uhs) is an air-filled or fluid-filled space. A sinus in the respiratory system is an air-filled or fluid-filled space in bone. Sinuses have a mucous membrane lining. The functions of sinuses are to provide mucus, to make bone lighter, and to help to produce sound. The combining form for sinus is **sinus/o** (Table 9–1).

Pharynx

Air passes through the nasal cavity to the **pharynx** (fār-ihnks). The pharynx is commonly called the throat. The pharynx is the common passageway for the upper respiratory and gastrointestinal tracts. The

TABLE 9–1 Sinuses and Their Locations

Sinus	Species Found	Location
frontal (frohn-tahl)	all domestic species	dorsal part of skull between nasal cavity and orbit
maxillary (mahx-ihl-ār-ē)	all domestic species	maxilla with nasal cavity on each side (maxillary recess in carnivores)
sphenoid (sfehn-oyd)	feline, bovine, equine, swine	sphenoid bone; opens to nasal cavity
palatine (pahl-eh-tīn)	ruminants, equine	palatine bone; communicates with maxillary sinus
lacrimal (lahck-rih-mahl)	swine, ruminants	lacrimal bone
conchal (kohn-kahl)	swine, ruminants, equine	formed by enclosure of conchae

pharynx extends from the caudal part of the nasal passages and mouth to the larynx and connects the nasal passages to the larynx and the mouth to the esophagus. The combining form for throat is **pharyng/o.**

The pharynx has three divisions:

* **nasopharynx** (nā-zō-fār-ihnckz) = portion of the throat posterior to the nasal cavity and dorsal to (above) the soft palate.
* **oropharynx** (ō-rō-fār-ihnckz) = portion of the throat between the soft palate and epiglottis.
* **laryngopharynx** (lah-rihng-gō-fār-ihnckz) = portion of the throat dorsal to the larynx that opens into the voice box and esophagus.

The nasopharynx is the passageway for air entering through the nose, whereas the oropharynx and laryngopharynx are passageways for air entering through the nose and food entering through the mouth, respectively. During swallowing, the **soft palate** (pahl-aht) moves dorsally and caudally to close off the nasopharynx to prevent food from going into the nasal cavity. *Palate* means roof of the mouth, and the combining form for palate is **palat/o.** The **epiglottis** (ehp-ih-gloht-ihs) acts like a lid and covers the larynx during swallowing. When the epiglottis covers the larynx, it also covers the opening to the trachea, preventing food from possibly entering the trachea and going into the lungs. The combining form for epiglottis is **epiglott/o** (Figure 9–4).

Larynx

The **larynx** (lār-ihnckz) is the part of the respiratory tract located between the pharynx and trachea (Figure 9–5). The larynx is commonly called the **voice box.** The larynx contains the **vocal cords** (vō-kahl kōrdz), which are paired membranous bands in the larynx that help produce sound. **Laryng/o** is the combining form for the voice box.

The vocal apparatus is found in the **glottis** (gloh-tihs). The glottis is the space between the vocal cords. Air passing through the glottis causes vibration of the vocal cords that produces sound. The combining form for glottis is **glott/o.** The vocal apparatus of avian species is the **syrinx** (sehr-ihnks), which is located between the trachea and bronchi.

Lower Respiratory Tract

The conducting passages of the lower respiratory tract consist of the trachea, bronchi, bronchioles, and alveoli. These structures enter or are found in the thoracic cavity.

Trachea

Air passes from the larynx to the **trachea** (trā-kē-ah). The trachea is commonly called the **windpipe** and extends from the neck to the chest. The

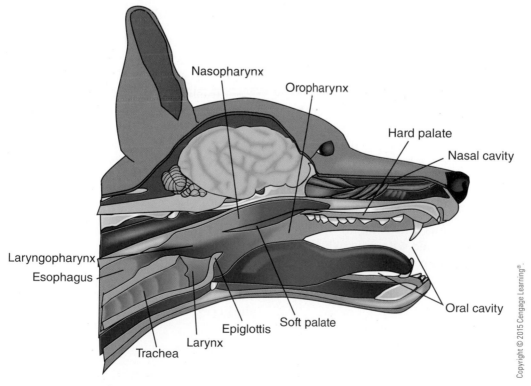

FIGURE 9–4 Structures of the nasal cavity and larynx.

Esophagus
Trachea
Laryngeal inlet
Vocal folds
Epiglottis

(a)

Vocal folds

Epiglottis

(b)

FIGURE 9–5 (a) Epiglottis and larynx. (b) Epiglottis and larynx of a bovine.

trachea attaches to the larynx in the neck and passes into the **thorax** (thōr-ahcks), or chest cavity, through the **thoracic** (thoh-rahs-ihck) **inlet**. The trachea is located ventral to the esophagus and is held open by a series of C-shaped cartilaginous rings. The open part of the *C*s typically are along the dorsal aspect, which is adjacent to the esophagus. This allows easier expansion of the esophagus when the animal swallows. The trachea also is lined with cilia, which help filter debris. The combining form for the windpipe is **trache/o**.

Bronchi

The distal end of the trachea (bottom of the trachea) divides into two branches at the **tracheal bifurcation** (trāk-ē-ahl bī-fər-kā-shuhn). The branches from the trachea are called **bronchi** (brohng-kī). The

combining form for bronchi is **bronch/o**. Each **bronchus** (brohng-kuhs) leads to a separate lung (right or left) and continues to divide. This continual division appears similar to a tree and its branches; therefore, the bronchi and its branches sometimes are called the **bronchial tree**. Each bronchus that leads to a separate lung is called a **principal** or **primary bronchus** (right principal or left principal bronchus). The principal bronchi divide into smaller branches called **secondary** (sehck-ohnd-ār-ē) or lobar (lō-bahr) **bronchi**. The secondary bronchi divide into **tertiary** (tər-shē-ār-ē) or **segmental** (sehg-mehn-tahl) **bronchi**. The tertiary bronchi are smaller units and are also called **bronchioles** (brohng-kē-ōlz) or **bronchiolus** (brohngk-ē-ō-luhs). The suffix **-ole** means small, indicating that bronchioles are smaller than bronchi. The combining form for bronchiole is **bronchiol/o**. Bronchioles contain no cartilage or glands. The bronchioles continue to divide. The terminal bronchioles are the last portion of a bronchiole that does not contain alveoli. The respiratory bronchioles are the final branches of the bronchioles. The respiratory bronchioles have alveolar outcroppings and branch into alveolar ducts (Figure 9–6).

Alveoli

Alveoli (ahl-vē-ō-lī) are air sacs in which most of the gas exchange occurs. An **alveolus** (ahl-vē-ō-luhs), which is Latin for small hollow thing, is a small grapelike cluster at the end of each bronchiole. The alveolus is connected to the bronchiole via an alveolar duct. The combining form **alveol/o** means small sac.

Alveoli have thin, flexible membrane walls that are surrounded by a network of microscopic capillaries. Gas exchange occurs across the alveolar membrane. Oxygen diffuses into the blood in alveolar capillaries and binds to the hemoglobin in erythrocytes. Carbon dioxide diffuses from the plasma across the alveolar membrane into the alveolus. When an animal exhales, much of this air is pushed out of the alveolus, back up through the respiratory tract, and out the nose or mouth.

Alveoli contain liquid that reduces alveolar surface tension. This liquid is called **surfactant** (sihr-fahck-tehnt). Surfactant prevents collapse of the alveoli during expiration.

Thorax

The thoracic cavity is contained within the ribs. The combining form **cost/o** means ribs. *Intercostal*

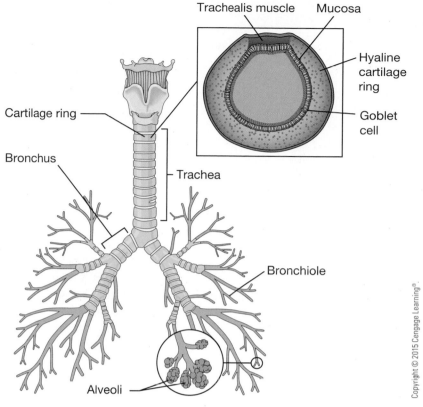

FIGURE 9–6 Lower respiratory tract.

means pertaining to between the ribs. The lungs are located in the thoracic cavity. The thoracic cavity also protects the lungs. The combining form **thorac/o** and the suffix **-thorax** mean chest cavity or chest.

The **lung** (luhng) is the main organ of respiration. There are two lungs (right and left) that are composed of divisions called **lobes** (lōbz). A lobe is a well-defined portion of an organ and is used in describing areas in the lung, liver, and other organs. The combining form **lob/o** means well-defined portion. The number and names of lung lobes vary between species. The combining forms **pneum/o**, **pneumon/o**, and **pneu** mean lung or air; the combining forms **pulm/o** and **pulmon/o** mean lung.

The term **parenchyma** (pahr-ehnk-ih-mah) refers to the functional elements of an organ, as opposed to its framework, or **stroma** (strō-mah). The functional elements of the lung are collectively called the lung parenchyma.

The region between the lungs is called the **mediastinum** (mē-dē-ahs-tī-nuhm). The mediastinum is the space between the lungs that houses the heart, aorta, lymph nodes, esophagus, trachea, part of the bronchial tubes, nerves, thoracic duct, and thymus (Figure 9–7).

Pleura

Each lung is encased in a membranous sac called the **pleura** (ploor-ah). The combining form **pleur/o** means membrane surrounding the lung; the plural form of pleura is **pleurae**.

- **Parietal pleura** (pah-rī-eh-tahl ploor-ah) is the outer layer of the membrane lining the inner wall of the thoracic cavity (Figure 9–7).
- **Visceral pleura** (vihs-ər-ahl ploor-ah) is the inner layer of the membrane lining the outside of the lung.
- **Pleural** (ploor-ahl) **space** is the potential space between the parietal pleura and visceral pleura. The pleural space or pleural cavity contains a small amount of lubricating fluid called pleural fluid. **Pleural fluid** prevents friction when the membranes rub together during respiration and provides adhesive force to keep the lungs in contact with the chest wall as it expands during inspiration.

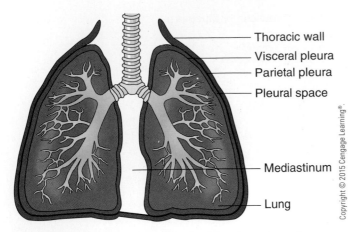

Thoracic wall
Visceral pleura
Parietal pleura
Pleural space

Mediastinum

Lung

FIGURE 9-7 Respiratory structures of the thoracic cavity.

Diaphragm

The thoracic and peritoneal cavities are separated from each other by the **diaphragm** (dī-ah-frahm) (Figure 9–1). The prefix **dia-** means across, and the combining form **phragm/o** means wall. The diaphragm is a muscle, and contraction of the diaphragm causes air pressure in the lungs to drop below atmospheric pressure. This produces a vacuum in the thoracic cavity to draw in air. When the diaphragm relaxes, this negative pressure is no longer generated and air is forced out of the lung. The combining forms **diaphragmat/o** and **phren/o** mean diaphragm. A **diaphragmatic hernia** (dī-ah-frahg-mah-tihck hər-nē-ah) is an abnormal displacement of organs through the muscle separating the chest and abdomen (Figure 9–8). The phrenic nerve innervates the diaphragm.

BREATHING

Breathing is the inhalation and exhalation of air. **Inhalation** (ihn-hah-lā-shuhn) is the drawing in of breath. **Inspiration** is another term for the drawing in of breath. **Exhalation** (ehcks-hah-lā-shuhn) is the release of breath. **Expiration** is another term for the release of breath. *Spirare* in Latin means to breathe. The combining form **spir/o** also means breath or breathing. Most medical terms use the Greek word **-pnea** to refer to breathing.

- **apnea** (ahp-nē-ah) = absence of breathing.
- **dyspnea** (dihsp-nē-ah) = difficult or labored breathing.
- **bradypnea** (brād-ihp-nē-ah) = abnormally slow respiratory rates.
- **tachypnea** (tahck-ihp-nē-ah) = abnormally rapid respiratory rates.
- **hyperpnea** (hī-pərp-nē-ah) = abnormal increase in the rate and depth of respirations.
- **hypopnea** (hī-pōp-nē-ah) = abnormally slow or shallow respirations.
- **hyperventilation** (hī-pər-vehn-tih-lā-shuhn) = abnormally rapid deep breathing, which results in decreased levels of cellular carbon dioxide.
- **agonal** (āg-uh-nuhl) **breathing** = respirations near death or during extreme suffering.

Respiration involves the diffusion of oxygen (O_2) and carbon dioxide (CO_2). The combining forms **ox/i**, **ox/o**, and **ox/y** refer to O_2, and **capn/o** refers to CO_2.

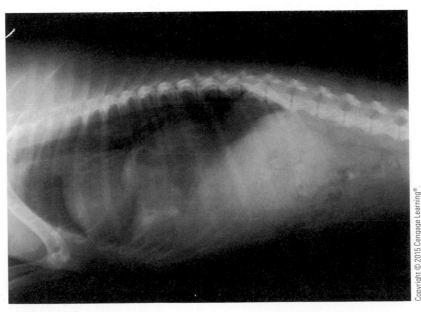

FIGURE 9-8 Radiograph of a diaphragmatic hernia. Intestinal organs are visible in the chest cavity.

The combining form **phren/o** also is used to refer to the mind. The ancient Greeks once thought that the spleen and kidneys were the organs responsible for emotion, and because the diaphragm sits across the cranial portion of these organs, phren/o was used for both the diaphragm and the mind.

Hypoxia (hī-pohck-sē-ah) refers to an inadequate supply of oxygen to tissue despite an adequate blood supply. **Hypercapnia** (hī-pər-kahp-nē-ah) refers to excessive amounts of carbon dioxide in the blood. **Hyperventilation** (hī-pər-vehn-tih-lā-shuhn), an abnormal increase in the rate or depth of breathing, may lead to **hypocapnia** (hī-pō-kahp-nē-ah), which is a decrease in the carbon dioxide levels in the blood.

When carbon dioxide dissolves in water, some of it reacts with the water to form carbonic acid (H_2CO_3). Because carbon dioxide breaks down into a weak acid, it affects the blood pH. An excessive amount of carbon dioxide in the blood due to decreased ventilation can lower the pH of blood; this is called **respiratory acidosis** (ah-sih-dō-sihs). If carbon dioxide levels are abnormally low due to increased ventilation, **respiratory alkalosis** (ahl-kah-lō-sihs) may result. Changes in blood pH also can result from metabolic factors (such as vomiting and renal disease) and are then called **metabolic acidosis** or **metabolic alkalosis** (Figure 9–9 and Table 9–2).

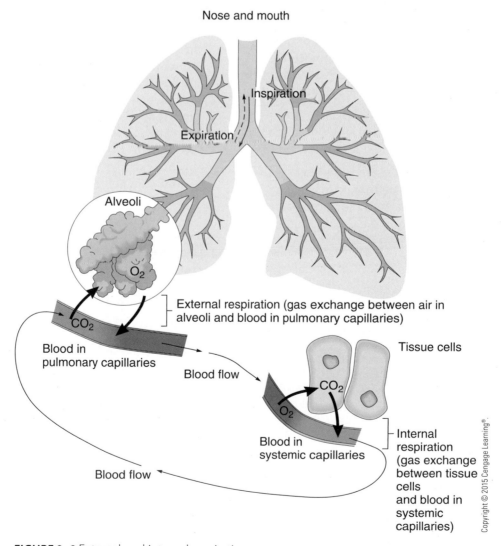

FIGURE 9–9 External and internal respiration.

TABLE 9–2 Lung Volume Terminology

tidal volume	amount of air exchanged during normal respiration (air inhaled and exhaled in one breath)
inspiratory reserve volume or **complemental air**	amount of air inspired over the tidal volume (extra amount that could be inhaled after normal inspiration)
expiratory reserve volume or **supplemental air**	amount of air expired over the tidal volume (extra amount that could be exhaled after normal expiration)
residual volume	air remaining in the lungs after a forced expiration (amount of air trapped in alveoli)
dead space	air in the pathway of the respiratory system (termed *dead* because this air is not currently participating in gas exchange)
minimal volume	amount of air left in alveoli after the lung collapses
vital capacity	largest amount of air that can be moved in the lung (tidal volume + inspiratory and expiratory reserve volumes)

TEST ME: RESPIRATORY SYSTEM

Diagnostic procedures performed on the respiratory system include the following:

- **arterial blood gases** = laboratory test in which pressure levels of oxygen (O_2) and carbon dioxide (CO_2) are measured in arterial blood; abbreviated ABG.
- **auscultation** (aws-kuhl-tā-shuhn) = act of listening. The respiratory tract is auscultated with a **stethoscope** (stehth-ō-skōp). Respiratory rhythm, rate, and sound are evaluated upon auscultation. Pathologic respiratory sounds are called **adventitious** (ahd-vehn-tish-uhs) **sounds**. Things to listen for include:
  ◦ **bubbling** = sound of popping bubbles that suggests fluid accumulation.

◦ **crepitation** (krehp-ih-tā-shuhn) = fine or coarse interrupted crackling noises coming from collapsed or fluid-filled alveoli during inspiration; also called **rales** (rahlz) or **crackles**.
  ◦ **decreased lung sounds** = less or no sound of air movement, suggesting consolidation of lung tissue. **Consolidation** is a condition in which lung tissue becomes firm and solid rather than elastic and air-filled because it has accumulated fluids and tissue debris.
- **respiratory rate** (RR) = number of respirations per minute. One inspiration and one expiration form a single respiration. RR varies with species.
  ◦ **rhonchi** (rohn-kī) = abnormal, continuous, musical, high-pitched whistling sounds heard during inspiration; also called **wheezes** (wē-zehz).
  ◦ **stridor** (strī-dōr) = snoring, squeaking, or whistling that suggests airway narrowing.
  ◦ **vesicular** (vehs-ihck-yoo-lahr) **sounds** = sound resulting from air passing through small bronchi and alveoli.
- **bronchoalveolar lavage** (brohng-kō-ahl-vē-ō-lahr lah-vahj) = collection of fluid or mucus from the bronchi and/or alveoli via an endoscope or through an endotracheal tube inserted as far down the trachea caudally as possible before infusing fluid and aspirating a sample. Fluid may be used for cytologic examination.
- **bronchoscopy** (brohng-kohs-kō-pē) = visual examination of the bronchus. Bronchoscopy may be used to examine the bronchi for disease or foreign objects. A **bronchoscope** (brohng-kō-skōp) is an instrument used to visually examine the bronchus.
- **laryngoscopy** (lahr-ihng-gohs-kō-pē) = visual examination of the voice box. Laryngoscopy is used to examine the larynx for disease, tissue repair, or foreign objects. A **laryngoscope** (lahr-ihn-gō-skōp) is an instrument used to visually examine the voice box.
- **percussion** (pər-kuhsh-uhn) = diagnostic procedure used to determine density in which sound is produced by tapping various body surfaces with the finger or an instrument (Figure 9–10). The sound produced over the chest where air is present differs from that of an area where fluid is present.
- **phlegm** (flehm) = thick mucus secreted by the respiratory lining. Mucus secretion from the lower respiratory tract is called **sputum** (spyoo-tuhm). Sputum can be used for cytologic examination.

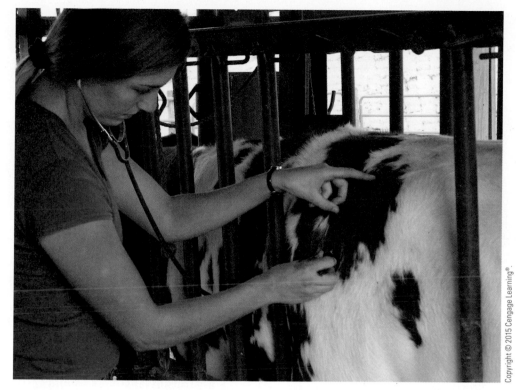

FIGURE 9–10 Percussion is tapping various body surfaces to determine density. Percussion of the caudal thoracic region of a cow to determine if she has a left displaced abomasm.

- **radiography** (rā-dē-ohg-rah-fē) = image of internal structures created by exposure of sensitized film to X-rays (Figure 9–11a and b). Ultrasound does not work well for the respiratory system because the ultrasound beam cannot pass through a gas-containing structure to provide information about the internal structures.

- **spirometer** (spər oh mē-tər) = instrument used to measure air taken in and out of the lungs. It measures the lungs' volume and capacity.
- **sputum culture** (spyoo-tuhm kuhl-tuhr) = procedure in which mucus from the lower respiratory tract is collected and placed in predetermined media (nutrient source) to allow microbes to reproduce.

(a)

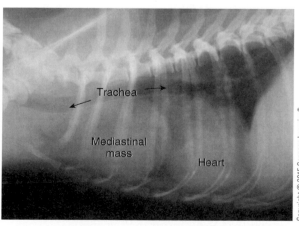

(b)

FIGURE 9–11 (a) Thoracic radiograph in a cat. This lateral thoracic radiograph reveals normal thoracic structures. (b) Thoracic radiograph of a dog with a mediastinal mass.

Courtesy of Kimberly Kruse Sprecher, CVT

FIGURE 9–12 Thoracocentesis in a dog. Thoracocentesis is used to collect fluid or gas from the chest cavity. Fluid is collected ventrally, and gas is aspirated dorsally.

- **thoracocentesis** (thō-rah-kō-sehn-tē-sihs) = puncture of the chest wall with a needle to obtain fluid from the pleural cavity. This fluid may be used for cytologic and microbiologic examination. Thoracocentesis also may be performed to drain pleural effusions or to reexpand a collapsed lung. Thoracocentesis also is called **thoracentesis** (thō-rah-sehn-tē-sihs) (Figure 9–12).
- **tracheal** (trā-kē-ahl) **wash** = collection of fluid or mucus from the trachea via an endotracheal tube to assess respiratory disease. Fluid may be used for cytologic and microbiologic examination.
- **transtracheal** (trahnz-trā-kē-ahl) **wash** = sterile collection of fluid or mucus from the trachea via a catheter inserted through the skin into the trachea to assess respiratory disease; abbreviated TTW. Fluid may be used for cytologic and microbiologic examination (Figure 9–13).
- **trephination** (trē-fīn-ā-shuhn) = insertion of a hole-boring instrument (trephine) into a sinus to establish fluid drainage or to allow access to the roots of teeth.

PATHOLOGY: RESPIRATORY SYSTEM

Pathologic conditions of the respiratory tract include the following:

- **anoxia** (ā-nohck-sē-ah) = absence of oxygen (almost complete lack of oxygen).
- **asphyxiation** (ahs-fihck-sē-ā-shuhn) = interruption of breathing resulting in lack of oxygen; also called suffocation.

Courtesy of Teri Raffel, CVT

FIGURE 9–13 Transtracheal wash in a dog. A TTW is initiated with insertion of a catheter into the trachea of a dog. A syringe filled with saline is attached to the opposite end of the catheter so that a small volume of saline may be introduced into the trachea and then quickly aspirated along with mucus. This sample is then ready for cytologic and microbiologic assessment.

- **aspiration** (ahs-pih-rā-shuhn) = inhalation of a foreign substance into the upper respiratory tract.
- **asthma** (ahz-mah) = chronic allergic disorder.
- **atelectasis** (aht-eh-lehck-tah-sihs) = incomplete expansion of the alveoli; also may mean collapse of a lung.
- **bronchiectasis** (brohng-kē-ehck-tah-sihs) = dilation of the bronchi. Bronchiectasis may be a **sequela** (sē-kwehl-ah) of inflammation or obstruction. Sequela is a condition following as a consequence of a disease.
- **bronchitis** (brohng-kī-tihs) = inflammation of the bronchi. Bronchitis may be **acute** (ah-kūt), which means occurring over a short course with a sudden onset, or **chronic** (krohn-ihck), which means occurring over a long course with a progressive onset.
- **bronchopneumonia** (brohng-kō-nū-mō-nē-ah) = abnormal condition of the bronchi and lung.
- **chronic obstructive pulmonary disease** (krohn-ihck ohb-struhck-tihv puhl-mah-nār-ē dih-zēz) = general term for abnormal conditions in equine species in which expiratory flow is slowed; commonly called heaves and abbreviated COPD. Horses with heaves may have a heave line, which is increased abdominal musculature associated with increased expiratory effort in horses with COPD.
- **cyanosis** (sī-ah-nō-sihs) = abnormal condition of blue discoloration. Cyanosis is caused by inadequate oxygen levels.
- **diaphragmatic hernia** (dī-ah-frahg-mah-tihck hər-nē-ah) = abnormal opening in the diaphragm

that allows part of the abdominal organs to migrate into the chest cavity (Figure 9–8).

- **emphysema** (ehm-fih-zē-mah) = chronic lung disease caused by enlargement of the alveoli or changes in the alveolar wall.
- **epistaxis** (ehp-ih-stahck-sihs) = nosebleed.
- **equine laryngeal hemiplegia** (ē-kwīn lahr-ihn-jē-ahl hehm-ih-plē-jē-ah) = disorder of horses that is characterized by abnormal inspiratory noise during exercise associated with degeneration of the left recurrent laryngeal nerve and atrophy of the laryngeal muscles; also called **left laryngeal hemiplegia** or **roaring**. The suffix **–plegia** means paralysis.
- **hemoptysis** (hē-mohp-tih-sihs) = spitting of blood from the lower respiratory tract.
- **hemothorax** (hē-mō-thō-rahcks) = accumulation of blood in the chest cavity.
- **inflammation** (ihn-flah-mā-shuhn) = localized protective response to destroy, dilute, or wall off injury; classic signs are heat, redness, swelling, pain, and loss of function.
- **laryngitis** (lahr-ihn-jı-tihs) = inflammation of the voice box.
- **laryngoplegia** (lahr-ihng-gō-plē-jē-ah) = paralysis of the voice box.
- **laryngospasm** (lah-rihng-ō-spahzm) = sudden fluttering or closure of the voice box.
- **pharyngitis** (fār-ihn-jī-tihs) = inflammation of the throat.
- **phonation** (fō-nā-shuhn) = act of producing sound. **Aphonation** (ā-foh-nā-shuhn) is the inability to produce sound.
- **pleural effusion** (ploor-ahl eh-fū-shuhn) = abnormal accumulation of fluid in the pleural space. **Effusion** (eh-fū-shuhn) is fluid escaping from blood or lymphatic vessels into tissues or spaces. A small amount of lubricating fluid in the pleural space is normal.
- **pleurisy** (ploor-ih-sē) = inflammation of the pleura; also called **pleuritis** (ploor-ī-tihs).
- **pleuropneumonia** (ploor-ō-nū-mō-nē-ah) = abnormal condition of the pleura and the lung (usually involves inflammation and congestion).
- **pneumonia** (nū-mō-nē-ah) = abnormal condition of the lung that usually involves inflammation and congestion of the lung (Figure 9–14 a, b, and c). **Congestion** (kohn-jehs-zhuhn) is the abnormal accumulation of fluid. **Interstitial** (ihn-tər-stih-shahl) pertains to the area between the cells; **interstitial pneumonia** is an abnormal lung condition with increased fluid between the alveoli and a decrease in lung function.

(a)

(b)

(c)

FIGURE 9–14 (a) Heifer suffering from pneumonia. This heifer has a thick nasal discharge and dyspnea. (b) Bovine lung with pneumonia. (c) Photomicrograph of a bovine lung with pneumonia. Note there are large numbers of white blood cells in the airways.

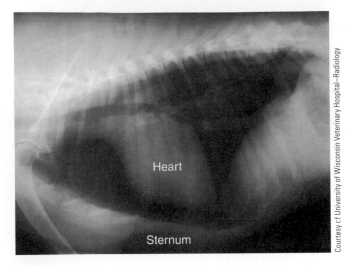

Heart

Sternum

FIGURE 9–15 Pneumothorax in a dog. Lateral projection of a pneumothorax demonstrates dorsal displacement of the heart from the sternum.

- **pneumothorax** (nū-mō-thōr-ahckz) = abnormal accumulation of air or gas in the chest cavity (Figure 9–15).
- **polyp** (pohl-uhp) = growth or mass protruding from a mucous membrane (usually benign).
- **pulmonary edema** (puhl-mohn-ār-ē eh-dē-mah) = accumulation of fluid in the lung tissue. **Edema** (eh-dē-mah) is abnormally large amounts of fluid in the intercellular tissue spaces.
- **pulmonary fibrosis** (puhl-ōm-nār-ē fī-brō-sihs) = abnormal formation of fibers in the alveolar walls.
- **pyothorax** (pī-ō-thō-rahcks) = accumulation of pus in the chest cavity. **Pus** is a fluid product of inflammation composed of leukocytes, **exudate** (ehcks-yoo-dāt) (high-protein fluid), and cell debris.
- **rhinitis** (rī-nī-tihs) = inflammation of the nasal mucous membranes. Nasal discharge can be described by its appearance, such as mucopurulent (myoo-kō-pər-ū-lehnt), which means containing mucus and pus.
- **rhinopneumonitis** (rī-nō-nū-moh-nī-tihs) = inflammation of the nasal mucous membranes and lungs.
- **rhinorrhea** (rī-nō-rē-ah) = nasal discharge.
- **sinusitis** (sī-nuh-sī-tihs) = inflammation of a sinus.
- **snuffles** (snuhf-uhlz) = common term for upper respiratory disease of rabbits caused by the bacterium *Pasteurella multocida*.
- **stenotic nares** (stehn-ah-tihck nār-ēz) = narrowed nostrils that reduce airway flow.

- **tracheitis** (trā-kē-ī-tihs) = inflammation of the windpipe.
- **tracheobronchitis** (trā-kē-ō-brohng-kī-tihs) = inflammation of the trachea and bronchi.
- **upper respiratory infection** = invasion of the nose, mouth, pharynx, epiglottis, or larynx (or trachea) by pathogenic organisms; abbreviated URI. Signs of URI include cough, nasal and ocular discharge, dyspnea, and respiratory noise (Figure 9–16). A **cough** (kowf) is a sudden, noisy expulsion of air from the lungs. Coughs may be **paroxysmal** (pahr-ohck-sihz-mahl), which means spasmlike and sudden. **Tuss/i** is the combining form for cough.

Drugs used on the respiratory system include **bronchoconstrictors** (brohng-kō-kohn-strihck-tərz), which are substances that narrow the openings into the lung, and **bronchodilators** (brohng-kō-dī-lā-tərz), which are substances that expand the openings into the lung. **Mucolytics** (mū-kō-lih-tihckz) are substances used to break down (**-lysis** means break down or separate) mucus. **Antitussives** (ahn-tih-tuhs-ihvz) are substances used to control or prevent coughing. The prefix **anti-** means against, and the combining form **tuss/i** means cough.

FIGURE 9–16 Young kitten with an upper respiratory tract infection. Note the nasal and ocular discharge.

PROCEDURES: RESPIRATORY SYSTEM

Procedures performed on the respiratory system include the following:

- **chest tube placement** = a chest tube is a hollow device inserted into the thoracic cavity to remove fluid or gas. Chest tubes are passed when animals are severely dyspneic because of pressure on the lungs (Figure 9–17).
- **endotracheal intubation** (ehn-dō-trā-kē-ahl ihn-too-bā-shuhn) = passage of a tube through the oral cavity, pharynx, and larynx into the windpipe. An endotracheal tube provides a **patent** (pā-tehnt) airway for administration of anesthetics or for critical care patients. *Patent* means open, unobstructed, or not closed.
- **laryngectomy** (lār-ihn-jehck-tō-mē) = surgical removal of the voice box.
- **laryngoplasty** (lah-rihng-ō-plahs-tē) = surgical repair of the voice box.
- **lobectomy** (lō-behck-tō-mē) = surgical removal of a lobe.
- **pharyngoplasty** (fār-rihng-ō-plahs-tē) = surgical repair of the throat.
- **pharyngostomy** (fār-ihng-ohs-tō-mē) = surgical creation of an opening into the throat. A **stoma** (stō-mah) is an opening on a body surface that may occur naturally or may be created surgically.

FIGURE 9–18 Tracheostomy tube in a dog.

- **pharyngotomy** (fār-ihng-oht-ō-mē) = surgical incision into the throat.
- **pleurectomy** (ploor-ehck-tō-mē) = surgical removal of all or part of the pleura.
- **pneumonectomy** (nū-mō-nehck-tō-mē) = surgical removal of lung tissue.
- **sinusotomy** (sī-nuhs-oht-ō-mē) = surgical incision into a sinus.
- **thoracotomy** (thō-rah-koht-ō-mē) = surgical incision into the chest wall.
- **tracheoplasty** (trā-kē-ō-plahs-tē) = surgical repair of the windpipe.
- **tracheostomy** (trā-kē-ohs-tō-mē) = surgical creation of an opening into the windpipe (usually involves insertion and placement of a tube) (Figure 9–18).
- **tracheotomy** (trā-kē-oht-ō-mē) = surgical incision into the windpipe.

FIGURE 9–17 A chest tube is a hollow device inserted into the thoracic cavity to remove fluid or gas. This dog is having blood removed from the thoracic cavity via a chest tube.

ABBREVIATIONS RELATED TO THE RESPIRATORY SYSTEM

- ABG = arterial blood gas
- CO_2 = carbon dioxide
- COPD = chronic obstructive pulmonary disease
- CPR = cardiopulmonary resuscitation
- ET = endotracheal
- O_2 = oxygen
- PEEP = positive end expiratory pressure
- RR = respiration rate
- TTA = transtracheal aspiration
- TTW = transtracheal wash
- URI = upper respiratory infection

REVIEW EXERCISES

Multiple Choice

Choose the correct answer.

1. The wall that divides the nasal cavity is called the
   a. nasodivision
   b. nares
   c. nasal septum
   d. nasal meatus

2. Parts of the respiratory tract contain thin hairs called
   a. flagella
   b. naris
   c. surfactant
   d. cilia

3. An abnormal condition of blue discoloration is called
   a. bluing
   b. cyanosis
   c. xanthochromia
   d. erythemia

4. Inhaling a foreign substance into the upper respiratory tract is called
   a. asphyxiation
   b. effusion
   c. atelectasis
   d. aspiration

5. Hypoxia is
   a. below-normal levels of oxygen
   b. above-normal levels of oxygen
   c. below-normal levels of carbon dioxide
   d. below-normal levels of carbon dioxide and oxygen

6. Liquid that reduces alveolar surface tension is called
   a. surfactant
   b. mucus
   c. rhinorrhea
   d. mucorrhea

7. A condition following as a consequence of disease is a(n)
   a. chronic condition
   b. acute condition
   c. sequela
   d. consequensosis

8. A substance that works against, controls, or stops a cough is a(n)
   a. bronchoconstrictor
   b. bronchodilator
   c. mucolytic
   d. antitussive

9. Abnormal accumulation of blood in the chest cavity is
   a. hemothorax
   b. hemoptysis
   c. hemopleuritis
   d. hemopneumonia

10. A growth or mass protruding from a mucous membrane is a
    a. nasogastric
    b. polyp
    c. bifurcation
    d. stridor

11. Tubes that bifurcate from the trachea are known as
    a. alveoli
    b. bronchi
    c. bronchioles
    d. nares

12. A nosebleed is known as
    a. pleurisy
    b. atelectasis
    c. aspiration
    d. epistaxis

13. Snoring, squeaking, or whistling that suggests airway narrowing is known as
    a. sequela
    b. polyp
    c. rhonchi
    d. stridor

14. Difficult breathing is known as
    a. dysphonia
    b. dyspnea
    c. dysphagia
    d. dyspepsia

15. Pus in the chest cavity is called
    a. pyothorax
    b. polyp
    c. hemiplegia
    d. hemoptysis

16. Tapping various body surfaces with the finger or an instrument to determine sound density is known as
    a. vesicular sounds
    b. crepitation
    c. percussion
    d. rales

17. The potential space between the parietal and visceral pleura is the
    a. parenchyma
    b. stroma
    c. diaphragmatic space
    d. pleural space

18. The alveoli are
    a. branches of the bronchial tree
    b. flexible air sacs where gas exchange occurs
    c. divisions of the lung
    d. sacs that surround the lung

19. The epiglottis
    a. acts as a lid over the entrance to the esophagus
    b. acts as a lid over the entrance to the trachea
    c. is commonly known as the Adam's apple
    d. is also known as the voice box

20. Which term means an abnormally rapid respiration rate?
    a. apnea
    b. bradypnea
    c. dyspnea
    d. tachypnea

21. When a lung segment becomes airless as a result of fluid accumulation, the lung is said to have
    a. congestion
    b. consolidation
    c. constriction
    d. congenital

22. An animal with a hemothorax has what in its chest cavity?
    a. pus
    b. mucus
    c. blood
    d. water

23. Which of the following terms contains a word part that means puncture?
    a. lobectomy
    b. tracheitis
    c. laryngoplasty
    d. thoracocentesis

24. Rhonchi, stridor, and rales are all
    a. surgical instruments
    b. pathologic respiratory sounds
    c. lung volume terminology
    d. sinuses in the respiratory system

25. Which pair of terms contains word parts meaning lung?
    a. pneumothorax and pulmonary edema
    b. bronchitis and larygnectomy
    c. tracheobronchitis and pulmonary fibrosis
    d. anoxia and transtracheal wash

Matching

Match the common term in Column I with the anatomical term in Column II.

Column I

1. _____ windpipe

2. _____ throat

3. _____ voice box

4. _____ chest

5. _____ nostril

6. _____ cilia

7. _____ edema

8. _____ atelectasis

9. _____ sequela

10. _____ anoxia

11. _____ antitussive

12. _____ hemoptysis

13. _____ acute

14. _____ chronic

15. _____ auscultation

Column II

a. pharynx

b. thorax

c. larynx

d. trachea

e. naris

f. abnormally large amounts of fluid in the intercellular tissue spaces

g. spitting up of blood from the lower respiratory tract

h. incomplete expansion of the alveoli

i. tiny hairs

j. occurring over a short course

k. occurring over a long course

l. condition following as a consequence of a disease

m. act of listening

n. substance that controls or prevents coughing

o. absence of oxygen

Fill in the Blanks

Write the correct term in the space provided.

1. Ox/i, ox/o, and ox/y refer to _____.

2. Nas/o and rhin/o mean _____.

3. Pulm/o and pulmon/o mean _____.

4. Pneum/o, pneumon/o, and pneu mean _____.

5. Ventilation and breathing mean _____.

True or False

If the statement is true, write T on the line. If the statement is false, write F on the line.

1. _____ Inflammation of the voice box is laryngitis.

2. _____ Foreign material comes into the body during internal respiration.

3. _____ Pleural effusion is a type of neoplasia.

4. _____ Atelectasis is another term for nosebleed.

5. _____ A paroxysmal cough is spasmlike and sudden.

6. _____ Stenotic nares can cause dyspnea.

7. _____ Hypopnea is abnormally slow or shallow respiration.

8. _____ Decreased levels of cellular carbon dioxide may result in hyperventilation.

9. _____ Adventitious sounds are pathologic respiratory sounds.

10. _____ The amount of air exchanged during normal respiration is vital capacity.

Spell Check

Circle the term that is spelled correctly.

1. abnormal condition of the lung that usually involves inflammation and congestion of the lung: pneumonia   pneumonia   pnuemolnia

2. interruption of breathing resulting in lack of oxygen: asfyxiation   asphyxiation   asphixiation

3. slimelike substance that is composed of glandular secretion, salts, cells, and leukocytes: mukous   mukus   mucus

4. chronic allergic disorder: astmah   asthma   asmah

5. muscle that separates the thoracic and abdominal cavities: diafragm   diaphragm   diaphram

Crossword Puzzle

Respiratory Terms

Supply the correct term in the appropriate space for the definition listed.

Across

1 abnormal condition of the pleura and the lung

3 incomplete expansion of the alveoli

6 absence of oxygen (almost complete lack of oxygen)

8 nasal discharge

15 abnormal accumulation of air or gas in the chest cavity

18 accumulation of blood in the chest cavity

19 inflammation of the nasal mucous membranes

20 interruption of breathing resulting in lack of oxygen

21 spitting of blood from the lower respiratory tract

Down

1 accumulation of pus in the chest cavity

2 inflammation of the pleura

4 inflammation of the trachea and bronchi

5 inflammation of the voice box

7 inhalation of a foreign substance into the upper respiratory tract

9 abnormal condition of the bronchi and lung

10 accumulation of fluid in the lung tissue

11 localized protective response to destroy, dilute, or wall off injury

12 abnormal condition of the lung that usually involves inflammation and congestion of the lung

13 abnormal condition of blue discoloration

14 nosebleed

16 inflammation of the bronchi

17 act of producing sound

Label the Diagrams

Label the diagram in Figure 9–19.

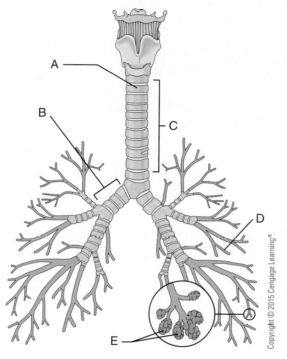

FIGURE 9–19 Respiratory tract. Label the parts of the lower respiratory tract as indicated by arrows. Provide the combining form for each part.

Case Studies

Define the underlined terms in each case study.

An 8-mo-old F DSH was presented to the clinic for routine surgery. The cat was given an IV drug to sedate it so that an <u>endotracheal tube</u> could be placed. The endotracheal tube (ET tube) (Figure 9–20) would serve as a delivery route for inhalant anesthesia. Upon attempts to <u>intubate</u> the cat, the larynx started to swell and spasm because of trauma. The cat had <u>inspiratory dyspnea</u>, a severe <u>cough, cyanotic</u> mucous membranes, and an elevated pulse rate. The <u>diagnosis</u> of <u>laryngeal spasm</u> was made. A drug (lidocaine) was placed in the laryngeal area to control the spasms, and endotracheal intubation was again attempted. After a few attempts, the ET tube was inserted, and the cat's breathing and mucous membrane color improved. If the ET tube could not have been placed, a <u>tracheotomy</u> would have been performed.

1. endotracheal tube _____

2. intubate _____

3. inspiratory dyspnea _____

4. cough _____

5. cyanotic _____

6. diagnosis _____

7. laryngeal spasm _____

8. tracheotomy _____

Copyright © 2015 Cengage Learning®.

FIGURE 9–20 Endotracheal tubes and a laryngoscope. Endotracheal tubes come in varying diameters (3 orange ones are in this image). Note the inflated cuff on the middle tube, which is used to seal the trachea. The laryngoscope provides a light to aide in visualization of the pharynx and is used to move the epiglottis to provide a view of the laryngeal folds.

A group of beef cattle was moved into a feedlot about 2 wk ago. Some of the cattle are experiencing <u>anorexia</u>, are <u>pyrexic</u>, and have <u>mucopurulent nasal discharge</u>. Some of the cattle have rapid shallow breathing and a cough. <u>Auscultation</u> of the lungs revealed moist <u>rales</u>. Because of the history and respiratory signs, the veterinarian suspected <u>bovine pneumonic</u> pasteurellosis, or shipping fever, a severe respiratory disease seen in younger animals after shipping or stress. The affected cattle were isolated and treated with antibiotics. Management practices such as immunization and stress reduction were discussed with the owner.

9. anorexia _____

10. pyrexic _____

11. mucopurulent nasal discharge _____

12. auscultation _____

13. rales _____

14. bovine _____

15. pneumonic _____

Critical Thinking Exercise

The following story and questions contain words and concepts from this chapter and are designed to stimulate critical thinking through class discussion, Internet searches, or brief writing responses.

Ms. Zimmerman called the veterinary clinic because the 4-year-old mustang female horse she adopted 6 months ago has crackles in her chest that seems to be getting worse. The horse has difficulty breathing (often lifting her head twice on expiration), a rapid respiratory rate, green discharge from her nose, and a consistent, deep cough. The horse is not eating because her hay is untouched when the owner comes to check on her. The owner has a horse-owner friend who told her that horses sometimes develop a condition called heaves. You tell the owner that her horse needs to be examined by a veterinarian and you schedule a call to her farm. After ending the phone conversation, you begin to document the horse's history in her medical record. Fill in the blanks of this record as you would enter it into the horse's medical record.

Phone conversation (9/15/XXXX): Ms. Zimmerman reported her 4-yr-old mustang _____ (adopted 6 mos ago) has _____ that are worsening. Horse has _____, _____, _____, and a consistent, deep cough. Horse is _____ ; does not touch her hay. Owner concerned that her horse has _____ (heaves). Appointment scheduled with Dr. Jones.

Dr. Jones went to Ms. Zimmerman's farm and examined her horse. Dr. Jones is concerned the horse may have COPD.

1. COPD is an abbreviation for what disease?

2. What makes a condition chronic?

3. What causes the sound the owner described as crackles?

4. What sign indicates that the horse may have an infection?

5. Horses with COPD may have weakened muscles. How might weakened muscles affect respiration?

Chapter 10
SKIN DEEP

Objectives

Upon completion of this chapter, the reader should be able to:

- Identify the structures of the integumentary system

- Describe the functions of the integumentary system

- Differentiate between surface lesions, fluid-filled lesions, and erosive lesions

- Recognize, define, spell, and pronounce the terms used to describe the diagnosis, pathology, and treatment of the integumentary system

- Interpret the meaning of abbreviations of the integumentary system

FUNCTIONS OF THE INTEGUMENTARY SYSTEM

The **integumentary** (ihn-tehg-yoo-mehn-tah-rē) **system** consists of skin and its appendages. (Appendages include glands, hair, fur, wool, feathers, scales, claws, beaks, horns, hooves, and nails.) One of the largest organ systems in the body, the integumentary system is involved in many processes.

Skin is a physical barrier that protects animals from microbial infection, injury, ultraviolet rays, chemicals, and toxins. Skin also plays a role in waterproofing the body, preventing fluid loss, providing temperature regulation through vasodilation and vasoconstriction, excreting wastes, providing species-specific coloration, sensation, and providing a site for vitamin D synthesis. Exocrine glands, both sebaceous and sweat, are located in the integumentary system. Sebaceous glands lubricate the skin and discourage bacterial growth on the skin. Sweat glands regulate body temperature and excrete wastes through sweat. Hair and nails are other components of the integumentary system. Hair helps control body heat loss and is a sense receptor. Nails protect the dorsal surface of the distal phalanx.

STRUCTURES OF THE INTEGUMENTARY SYSTEM

The integumentary system is on the outside of the body and is resilient and versatile because it is subjected to a variety of insults such as trauma, toxic chemicals, and environmental conditions.

Skin

Skin covers the external surfaces of the body. The skin is composed of **epithelial** (ehp-ih-thē-lē-ahl) **tissue** and is sometimes called the **epithelium** (ehp-ih-thē-lē-uhm). The combining forms for skin are **cutane/o**, **derm/o**, and **dermat/o**; the suffix **-derma** means skin. **Dermatology** (dər-mah-tohl-ō-jē) is the study of skin.

Skin Stratification

The skin is made up of three layers: the epidermis, dermis, and subcutaneous layer (Figure 10–1). The outermost, or most superficial, layer of skin is the **epidermis** (ehp-ih-dər-mihs). The prefix **epi-** means above, and **dermis** means skin. The epidermis is several layers thick and does not contain blood vessels. The epidermis is sometimes called the avascular layer because it lacks blood vessels. The epidermis depends on the deeper layers for nourishment.

The thickness of the epidermis varies greatly from region to region in all animals. The thickest layers of the epidermis are found in the areas of greatest exposure, such as the foot pads and teats.

The epidermis is made up of squamous epithelium and the basal layer (Table 10–1). **Squamous epithelium** (skwā-muhs ehp-ih-thē-lē-uhm) is composed of flat, platelike cells. Because these flat, platelike cells are arranged in many layers, this layer is called stratified squamous epithelium (Figure 10–2).

The basal layer is the deepest layer of the epidermis. Cells layer, multiply, and push upward into the basal layer. As the cells move superficially, they die and become filled with **keratin** (kehr-ah-tihn). Keratin is a protein that provides skin with its waterproofing properties. The combining form for keratin is **kerat/o**. Kerat/o also means horny or cornea. The basal layer also contains **melanocytes** (mehl-ah-nō-sītz), which produce and contain a black pigment. This black pigment is called **melanin** (mehl-ah-nihn). The combining form **melan/o** means black or dark. Melanin protects the skin from some of the harmful rays of the sun and is responsible for skin pigmentation. Animals that have dark skin produce more melanin. The absence of normal pigmentation is called **albinism** (ahl-bih-nihz-uhm); true albinism means that the hair, skin, and eyes have no pigmentation.

The **dermis** (dər-mihs) is the layer directly deep to the epidermis. The dermis also is called the **corium**

FIGURE 10–1 Skin layers.

Epidermis

Dermis

Subcutaneous layer

Hair shaft
Pore
Papilla
Capillary
Duct of sweat gland
Sebaceous gland
Nerve fiber
Sweat gland
Blood vessel
Adipose cells

TABLE 10–1 Layers of the Epidermis

The epidermis has five layers (from dermis to most superficial); *stratum* means layer or sheetlike mass.

stratum basale (strah-tuhm bā-sahl) or **stratum germinativum** (gər-mihn-ā-tihv-uhm) = deepest or basal layer that continually multiplies to replenish cells lost from the epidermal surface. **Cuboidal** (kyoo-boy-dahl) or cubelike cells are arranged in rows.

stratum spinosum (strah-tuhm spī-nō-suhm) = layer immediately superficial to the stratum basale, which is thickest in hairless regions and in areas of high wear and tear. **Keratinization** (kehr-ah-tihn-ah-zā-shuhn) and **desquamation** (deh-skwah-mā-shuhn) begin in this layer. Keratinization is the development of the hard, protein constituent of hair, nails, epidermis, horny structures, and tooth enamel. Desquamation is the process in which cell organelles gradually dissolve. The stratum spinosum also is called the **prickle** or **spinous layer**.

stratum granulosum (strah-tuhm grahn-yoo-lō-suhm) = layer immediately superficial to the stratum spinosum. Cells contain keratin granules in their cytoplasm.

stratum lucidum (strah-tuhm loo-sih-duhm) = layer immediately superficial to the stratum granulosum, which is clear because of the accumulation of keratin fibers in cell cytoplasm. This layer is not present in all species, but when present, it is found in areas of high wear and tear such as the foot pads.

stratum corneum (strah-tuhm kohr-nē-uhm) = most superficial layer of the epidermis, which consists of layers of dead, highly keratinized, and flattened cells; also called the **horny layer**.

Photomicrographs courtesy of William J. Bacha, PhD, and Linda M. Bacha, MS, VMD

FIGURE 10–2 Photomicrograph of the specialized skin found in the foot pad of a cat.

(kō-rē-uhm). The dermis is composed of blood and lymph vessels, nerve fibers, and the accessory organs of the skin. Sensory nerve endings in the dermis are the sensory receptors for touch, temperature, pain, and pressure. **Perception** (pər-sehp-shuhn) is the ability to recognize sensory stimuli. **Tactile** (tahck-tīl) perception is the ability to recognize touch sensation. The dermis also contains connective tissue, which is composed of the following cells:

- **fibroblasts** (fī-brō-blahsts) = fiber-producing cells. Collagen is the major fiber in the dermis.
- **collagen** (kohl-ah-jehn) = tough, flexible, fibrous protein found in skin, bone, cartilage, tendons, and ligaments. *Kolla* in Greek means glue, and **-gen** means to produce.
- **histiocytes** (hihs-tē-ō-sīts) = phagocytic cells that engulf foreign substances; also called tissue **macrophages** (mahck-rō-fājs-ehs or mahck-rō-fahjs-ehs).
- **mast cells** = cells that respond to insult by producing and releasing histamine and heparin. **Histamine** (hihs-tah-mēn) is a chemical released in response to allergens that causes itching. **Heparin** (hehp-ah-rihn) is an anticoagulant chemical released in response to injury.

The **subcutaneous** (suhb-kyoo-tahn-ē-uhs) **layer,** or **hypodermis** (hī-pō-dər-mihs), is located deep to or under the dermis and is composed of connective tissue. The subcutaneous layer contains a large amount of **fat,** or **lipid** (lihp-ihd). Adipocytes (ahd-ih-pō-sīts)

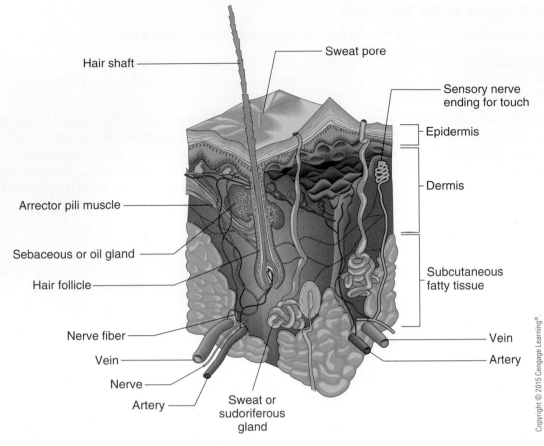

FIGURE 10–3 Skin structures.

are fat cells that produce lipid. **Adip/o** is the combining form for fat (Figure 10–3).

Skin Association

Appendages or structures associated with the skin include glands, hair, fur, wool, feathers, scales, claws, beaks, horns, hooves, and nails.

Glands

There are two main categories of skin glands: sebaceous and sweat glands. **Sebaceous** (seh-bā-shuhs) **glands**, or **oil glands**, secrete an oily substance called **sebum** (sē-buhm). **Seb/o** is the combining form that means sebum or oily substance. Sebaceous glands are located in the dermis and are closely associated with hair follicles. Sebum is released from its gland through **ducts** (duhckts) that open into the hair follicles. Ducts are tubelike passages; tiny ducts are called **ductules**. Sebum moves from the hair follicle to the skin surface, where it lubricates the skin. Sebum is slightly acidic and retards bacterial growth on the skin. Sebaceous glands are considered **holocrine** (hō-lō-krihn) **glands**

because the secreting cells and their secretions make up the discharge produced. Sebaceous glands are found in the anal sacs, glands that produce musk, and circumoral and supracaudal glands, which cats use to mark territory when they groom and rub their tail, respectively.

Sweat (sweht), or **sudoriferous** (soo-dohr-ihf-ohr-uhs), **glands** are aggregations of cells located in the dermis. Sweat glands are divided into **eccrine** (ē-krihn) **glands** and **apocrine** (ahp-ō-krihn) **glands**. Eccrine sweat glands produce and secrete water, salt, and waste (sweat) and are located in various regions of the body depending on the species. Eccrine sweat glands are tiny, coiled glands that have ducts that open directly onto the skin surface through pores. Apocrine glands produce and secrete a strong-smelling substance into the hair follicles. Apocrine glands are found throughout the body. Apocrine glands get their name from the fact that the free end or apical end of the cell is cast off along with the secretory products. Sweat glands help regulate body temperature against **hyperthermia** (hī-pər-thər-mē-ah), or high body temperature, and **hypothermia** (hī-pō-thər-mē-ah), or low body temperature.

BIOLOGY BONUS

How an animal repairs a wound varies with the severity of the lesion. When an animal injures itself and produces a superficial laceration in the epidermis, epithelial cells along the edges of the laceration divide rapidly and fill in the gap to heal the wound. When an animal injures itself and produces a wound extending into the dermis or when a surgeon makes an incision, blood vessels in the dermis break and blood escapes into the wound, which eventually forms a clot (a clot consists of the protein fibrin, platelets, blood cells, and dried tissue fluids). Phagocytic cells come into the wound to digest and clean up the tissue debris; meanwhile, the surface of the clot dries and hardens in the air to form a scab that seals and protects the wound from becoming infected. A few days after the injury the clot is invaded by new capillaries from the surrounding dermis, and other cells migrate into the wound to form new collagen fibers that pull the wound together. This soft tissue formed by collagen fibers in the wound is called granulation tissue, which is later replaced by a scar (Figure 10–4).

Copyright © 2015 Cengage Learning®.

FIGURE 10–4 Granulation tissue is formed on this dog's limb as its wound begins to heal.

Hidrosis (hī-drō-sihs) is the production and excretion of sweat. The combining form **hidr/o** means sweat. **Anhidrosis** (ahn-hī-drō-sihs) is the abnormal reduction of sweating; **hyperhidrosis** (hī-pər-hī-drō-sihs) is excessive sweating.

Ceruminous (seh-roo-mihn-uhs) glands are modified sweat glands that are located in the ear canal. The ceruminous glands secrete **cerumen** (seh-roo-mehn), a waxy substance of varying colors depending on the species; cerumen is commonly called **earwax**.

Hair

Hair is rodlike fibers made of dead protein cells filled with keratin. The combining forms for hair are **pil/i**, **pil/o**, and **trich/o**. The hair shaft is the portion of hair extending beyond the skin surface and is composed of the **cuticle** (kū-tih-kuhl), **cortex** (kōr-tehckz), and **medulla** (meh-doo-lah). The cuticle is one cell layer thick and appears scaly. The cortex is the main component of the hair shaft, is several layers thick, and is responsible for coat color. The medulla is the innermost component of the hair shaft. The **root** is the portion of the hair below the skin surface and is surrounded by a hair follicle. The shaft and root are continuous and together make up the hair, which is produced by the hair follicle.

Hair **follicles** are tubes that hold the hair fibers. In the dermis, the follicle expands to form a hair bulb. The cells at the base of the follicle divide rapidly and push the old cells toward the skin surface where they harden and undergo pigmentation. The **arrector pili** (ah-rehck-tər pī-lī) is a tiny muscle attached to the hair follicle that causes the hair to stand erect in response to cold temperatures or stress. When a dog contracts the arrector pili along the dorsal side of the neck and down the spine, it is called "raising the hackles." **Piloerection** (pī-lō-ē-rehck-shuhn) is the condition of the hair standing straight up.

Feathers and scales are discussed in Chapters 22 and 23, respectively.

Nails, Claws, and Hooves

The distal phalanx of animals is covered by nails, claws, or hooves. Nails, claws, and hooves all have a wall, sole, and pad, although they may be called

What Is Hair?

Animals have many different types of hair; therefore, many different terms describe the types of hair animals may have.

- **fur** = short, fine, soft hair.
- **pelt** = skin in addition to fur or hair.
- **guard hairs** = long, straight, stiff hairs that form the outer coat; also called **primary hairs** or **topcoat**. Guard hairs include tail and mane hair, bristly hair of swine, and most of the fur hair.
- **secondary hairs** = finer, softer, and wavy hair; also called the **undercoat**. Secondary hairs include wool and wavy hair located near the skin of rabbits.
- **tactile** (tahck-tīl) **hair** = long, brittle, extremely sensitive hairs usually located on the face; also called **vibrissae** (vī-brihs-ā), which are technically longer tactile hairs. An example of vibrissae are cat whiskers.
- **cilia** (sihl-ē-ah) = thin, short hairs. An example of cilia are the eyelashes.
- **simple pattern hair growth** = guard hairs that grow from separate follicular openings, as in cattle.
- **compound pattern hair growth** = multiple guard hairs that grow from single follicles, as in dogs.
- **shedding** = normal hair loss caused by temperature, hormones, photoperiod (light), nutrition, and other nondisease causes.

Things tend to be named for a reason. Sometimes bacteria, viruses, and fungi are named for the size, shape, or organism they affect. One example is the fungal genus *Trichophyton*. **Trich/o** means hair, and **phyt/o** means plant. **Fungi** are vegetative organisms (once considered to be plants) that grow as branched, hairlike filaments (trich/o).

different things. **Walls** usually are located dorsal and lateral to the distal phalanx. The **sole** is located ventral to the distal phalanx and usually is flaky. **Foot pads**, or **tori** (tohr-ē), provide cushioning and protection for the bones of the foot. Pads usually are thick and composed of keratinized epithelium. The pad has a subcutaneous layer that contains a large number of adipose cells and elastic connective tissue. Sweat glands are also found in most mammalian foot pads (Figure 10–5).

Dogs and cats have **digital pads** on the palmar and plantar surfaces of the phalanges. **Metacarpal** and **metatarsal pads** are singular pads located on the palmar and plantar surfaces of the metacarpal and metatarsal areas, respectively.

Carpal pads are located on the palmar surface of each carpus. Carpal pads do not bear weight when the animal is standing. Dogs and cats are called **digitigrade** (dihg-iht-ih-grād) animals because they walk on their toes (digits), with only the digital and metacarpal and metatarsal pads making contact with the ground. **Plantigrade** (plahnt-ih-grād) animals have well-developed foot pads, such as those in primates. Plantigrade animals walk with phalanges, metacarpals and metatarsals, and carpal and tarsal bones making contact with the ground. The suffix **–grade** means to go.

In dogs and cats, **nails** and **claws** are keratin plates covering the dorsal surface of the distal phalanx. The dorsal and lateral surface of the claw is covered by the wall, and the ventral surface is the sole. Beneath the wall and sole is the connective tissue

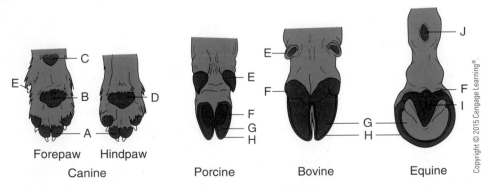

Forepaw   Hindpaw
Canine   Porcine   Bovine   Equine

FIGURE 10–5 Comparison of animal feet. A = digital pad, B = metacarpal pad, C = carpal pad, D = metatarsal pad, E = dewclaw, F = bulb or heel, G = sole, H = wall, I = frog, and J = ergot.

The equine hoof is divided into various regions: the coronary band, periople, wall, bars, sole, bulb, and frog (Figure 10–6).

- **coronary** (kohr-ō-nār-ē) **band** = region where hoof meets the skin; analogous to the cuticle of the human nail. The coronary band is the site of hoof wall growth; also called the **coronet**.
- **periople** (pehr-ē-ō-puhl) = flaky tissue band located at the junction of the coronary band and the hoof wall and extends distally. The periople widens at the heel to cover the bulbs of the heels.
- **wall** = epidermal tissue that includes the toe (front), quarters (sides), and heels (back).
- **bars** (bahrz) = raised V-shaped structure on ventral surface of hoof. Bars are located on either side of the frog.
- **sole** (sōl) = softer hoof tissue located on the ventral surface of the hoof (bottom of the hoof).
- **frog** (frohg) = V-shaped pad of soft horn located in the central region of the ventral hoof surface of equine (located between the bars). When weight is put on the frog, blood is forced out of the foot to promote circulation of blood throughout the foot.
- **bulbs** (buhlbz) **of heel** = upward thickening of the frog above the heels of the wall.

1 Horny bulbs of the heels
2 Middle cleft of the frog
3 Branches of the frog
4 Base of the frog
5 Collateral sulcus (cleft)
6 Buttress
7 Bars
8 Sole
9 White line
10 Wall
11 Apex of the frog

■ Quarter
□ Toe
▨ Heel

FIGURE 10–6 Parts of an equine hoof.

dermis, which contains numerous blood vessels and nerve endings. This sensitive tissue is known as the **quick** (Figure 10–7).

Quicking is the term used to describe trimming the nail or claw to the level of the dermis. Quicking results in bleeding and pain. The combining form for claw is **onych/o**.

Hooves are the horny covering of the distal phalanx in **ungulates** (uhng-yoo-lātz), or **hooved animals**, such as equine, ruminants, and swine. Some ungulates have a solid hoof, as in equine, and some have cloven or split hooves, as in ruminants and swine. The ventral surface of the hoof is the sole, which is large in equine and smaller in ruminants and swine. The combining form for hoof is **ungul/o**.

The pads of ungulates vary with the species. In ruminants and swine, the foot pad is called the bulb or heel. The pad of equine is called the frog and along with the bulb provides shock absorption. The **corium** (kōr-ē-uhm) is the dermis of the hoof and is located under the epidermal surface of the hoof wall, sole, and frog. The corium corresponds to the quick.

Vestigial Structures

Vestigial, or rudimentary, structures of the integumentary system include dewclaws, chestnuts, and ergots.

Quick

FIGURE 10–7 Quick of a cat nail.

Dewclaws (doo-klawz) are rudimentary bones. The dewclaw in dogs is the first digit, whereas in cloven-hoofed animals, the dewclaws are digits II and V. Dewclaws in dogs usually are found in the forepaw (although they occasionally are seen in the hindpaw) and may be removed within a few days of birth to avoid trauma.

Chestnuts and ergots are vestigial pads in equine. **Chestnuts** are located on the medial surface of the leg; in the front leg, they are located above the knee, and in the hind leg, they are located below the hock (Figure 10–8a and b). Chestnuts correspond to carpal pads in the dog. No two chestnuts are alike, and they do not change in size or shape throughout an equine's life. **Ergots** are located in a tuft of hair on the fetlock joint. Ergots correspond to metacarpal and metatarsal pads in the dog. Digital pads of dogs are replaced by the bulbs of the heel (and the frog in equine) in ungulates.

TEST ME: INTEGUMENTARY SYSTEM

Diagnostic procedures performed on the integumentary system include the following:

- **biopsy** (bī-ohp-sē) = removal of living tissue for examination of life. The combining form **bi/o** means life; the suffix **-opsy** means view of. An **incisional biopsy** (ihn-sih-shuhn-ahl bī-ohp-sē) is the removal of a piece of a tumor or lesion for examination (Figure 10–10). An **incision** (ihn-sih-shuhn) is a cut into tissue. An **excisional biopsy** (ehcks-sih-shuhn-ahl bī-ohp-sē) is the removal of an entire tumor or lesion in addition to a margin of surrounding tissue for examination. An **excision** (ehcks-sih-shuhn) is a cut out of tissue. A needle

What Is the Difference Between Horns and Antlers?

Horns and antlers are protective structures located in the head region of animals. **Horns** are permanent structures that grow continuously after birth. Horns grow from the frontal skull bones and originate from keratinized epithelium. **Cornification** (kohr-nih-fih-kā-shuhn) is the conversion of epithelium into keratin or horn. Horns may be located in different positions, as can be seen in the different ruminant species. Breeds that are naturally hornless are called **polled** (Figure 10–9a and b).

Antlers are not permanent structures and are shed and regrown annually. Antlers grow from the skull, as do horns. Antlers initially are covered with skin called **velvet**, which the animal rubs off after the skin dies. When the animal rubs off the velvet, the bone is exposed, the antlers lose their blood supply, and the antlers are eventually shed.

biopsy is the insertion of a sharp instrument (needle) into a tissue for examination. Examination of biopsies involves the use of a microscope.

- **culture** = diagnostic or research procedure used to allow microbes to reproduce in predetermined media (nutrient source). Fungal and bacterial cultures are performed using media that contains specific nutrients necessary for optimal growth of these types of microbes. **Tissue culture** takes epithelial cells and grows them in a medium so that intracellular microbes such as viruses can replicate.

(a)

Chestnut

Courtesy of Laura Lien, CVT, VTS (LAIM)

(b)

Ergot

Courtesy of Laura Lien, CVT, VTS (LAIM)

FIGURE 10–8 (a) Chestnut of a horse. (b) Ergot of a horse.

(a)

(b)

FIGURE 10–9 (a) Polled Hereford. (b) Unpolled Hereford.

FIGURE 10–10 Punch biopsy of the skin. A punch biopsy is a type of incisional biopsy commonly used to obtain full-thickness skin samples. The punch has a circular opening that comes in different sizes.

- **exfoliative cytology** (ehcks-fōl-ē-ah-tihv sī-tohl-ō-jē) = scraping of cells from tissue and examination under a microscope. **Exfoliative** means falling off.
- **intradermal** (ihn-trah-dər-mahl) **skin testing** = injection of test substances into the skin layer to observe a reaction (Figure 10–11a); used for diagnosis of **atopy** (ah-tō-pē) with the injection of multiple allergens or for tuberculosis testing by injecting tuberculin into the skin layer and observing the injection site for a 24-, 36-, and 72-hour postinjection reaction. (Tuberculosis testing is called **purified protein derivative,** or PPD, testing (Figure 10–11b)).
- **skin scrape** = microscopic examination of skin for the presence of mites; skin is sampled by scraping a scalpel blade across an area that is squeezed or raised so that the sample contains a deep skin sample (Figure 10–12).

(a)

(b)

FIGURE 10–11 (a) Intradermal skin testing. Intradermal skin testing is used to assess atopy. (b) Tuberculosis caudal fold skin test in a bovine.

(a)

(b)

Courtesy of Ron Fabrizius, DVM, Diplomate ACT

FIGURE 10–12 Skin scrape. Skin scrapings are used to detect mites.

(a)

Copyright © 2015 Cengage Learning®

(b)

Courtesy of Jordan Applied Technology Center, West Jordan, Utah

FIGURE 10–13 (a) Draining an abscess on the side of the face of an anesthetized cat. (b) Caseous abscess in a rodent.

PATHOLOGY: INTEGUMENTARY SYSTEM

Pathologic conditions of the integumentary system include the following:

- **abrasion** (ah-brā-shuhn) = injury in which superficial layers of skin are scraped.
- **abscess** (ahb-sehsz) = localized collection of pus (Figure 10–13a). The fluid from an abscess may be described as **caseous** (kā-sē-uhs) or cheese-like (Figure 10–13b).
- **acne** (ahck-nē) = skin inflammation caused by plugged sebaceous glands and comedone development from papules and pustules. **Chin acne** is a common condition in cats in which acne develops on the chin and lip area.
- **acute moist dermatitis** (ah-kūt moyst dər-mah-tī-tihs) = bacterial skin disease that is worsened by licking and scratching; also called **hot spot**.

- **alopecia** (ahl-ō-pē-shah) = hair loss resulting in hairless patches or complete lack of hair. The word root **alopec/o** means baldness (Figure 10–14). **Shedding** is normal hair loss due to various causes.
- **atopy** (ah-tō-pē) = hypersensitivity reaction in animals involving pruritus with secondary dermatitis; commonly called **allergies** or **allergic dermatitis**. **Hypersensitization** is an increased response to an allergen. **Hyposensitization** is a decreased response to an allergen. Animals with atopy may undergo a series of hyposensitization injections to decrease their response to a specific allergen.
- **bullae** (buhl-ā) = multiple contained skin elevations filled with fluid that are greater than 0.5 cm in diameter (Figure 10–19). The singular form is **bulla** (buhl-ah).
- **burn** = tissue injury caused by heat, flame, electricity, chemicals, or radiation.
- **carbuncle** (kahr-buhng-kuhl) = cluster of furuncles.
- **carcinoma** (kahr-sih-nō-mah) = malignant neoplasm of epithelial tissue. **Carcin/o** is the combining form for cancerous.

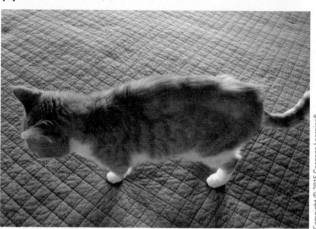

Courtesy of Ron Fabrizius, DVM, Diplomate ACT

Copyright © 2015 Cengage Learning®

FIGURE 10–14 (a) Alopecia in a pig with sarcoptic mange. (b) Asymmetrical alopecia in a cat due to excessive grooming. Skin surface injuries due to self-trauma are called excoriations.

Copyright © 2015 Cengage Learning®

FIGURE 10–15 Cellulitis in cat paw (right side of image).

- **cellulitis** (sehl-yoo-lī-tihs) = inflammation of connective tissue (Figure 10–15). Inflammation may be **diffuse** (dih-fuhs), meaning widespread, or **localized** (lō-kahl-īzd), meaning within a well-defined area.
- **comedo** (kōm-eh-dō) = blackhead or buildup of sebum and keratin in a pore. Plural is **comedones** (kōm-eh-dō-nehz).

- **contusion** (kohn-too-shuhn) = injury that does not break the skin; characterized by pain, swelling, and discoloration.
- **crust** = collection of dried sebum and cellular debris (Figure 10–19).
- **cyst** (sihst) = closed sac or pouch containing fluid or semisolid material (Figure 10–19).
- **dermatitis** (dər-mah-tī-tihs) = inflammation of the skin. **Contact dermatitis** is inflammation of the skin caused by touching an irritant.
- **dermatocellulitis** (dər-mah-tō-sehl-yoo-lī-tihs) = inflammation of the skin and connective tissue.
- **dermatomycosis** (dər-mah-tō-mī-kō-sihs) = abnormal skin condition caused by superficial fungus; also called **dermatophytosis** (dər-mah-tō-fī-tō-sihs). **Dermatophytes** (dər-mah-tō-fītz) are superficial fungi that are found on the skin. An example of a dermatophyte is the fungus that causes ringworm (Figure 10–16a and b).

Copyright © 2015 Cengage Learning®

Copyright © 2015 Cengage Learning®

FIGURE 10–16 (a) Holstein heifer with ringworm. (b) Ringworm lesion on the abdomen of a dog.

- **dermatosis** (dər-mah-tō-sihs) = abnormal skin condition. Plural is **dermatoses** (der-mah-tō-sēz).
- **discoid lupus erythematosus** (dihs-koyd loo-pihs eh-rih-thehm-ah-tō-sihs) = canine autoimmune disease in which the bridge of the nose (and sometimes the face and lips) exhibit depigmentation, erythema, scaling, and erosions; abbreviated DLE; may have been called collie nose or solar dermatitis in the past.
- **dyskeratosis** (dihs-kehr-ah-tō-sihs) = abnormal alteration in keratinization.
- **ecchymosis** (ehck-ih-mō-sihs) = purplish nonelevated patch of bleeding into the skin; also called a **bruise**; plural is **ecchymoses** (ehck-ih-mō-sēz).
- **ecthyma** (ehck-thih-mah) = skin infection with shallow eruptions caused by a pox virus (Figure 10–17); also known as **soremouth**.
- **eczema** (ehcks-zeh-mah) = general term for inflammatory skin disease characterized by erythema, papules, vesicles, crusts, and scabs either alone or in combination.
- **eosinophilic granuloma** (ē-ō-sihn-ō-fihl-ihck grahn-yoo-lō-mah) **complex** = collective name for autoimmune lesion of eosinophilic ulcer, eosinophilic plaque, and linear granuloma found in cats and rarely in dogs. This complex of diseases affects the skin, mucocutaneous junctions, and oral mucosa of cats, involving raised, ulcerated plaques (Figure 10–18). These lesions are named for their location: **eosinophilic ulcer**, or **rodent ulcer**, is located on the lip and oral mucosa of cats; **eosinophilic plaques** are raised pruritic lesions on the ventral abdomen of cats; and **linear granulomas** are located in a line usually on the caudal aspect of the hindlimb of cats.

FIGURE 10–18 Eosinophilic ulcer in a cat.

- **erythema** (ehr-ih-thē-mah) = skin redness. **Erythematous** (ehr-ih-thehm-ah-tuhs) means pertaining to redness.
- **erythroderma** (eh-rihth-rō-dər-mah) = abnormal redness of skin occurring over a widespread area. Combining forms for red are **erythr/o**, **erythem/o**, and **erythemat/o**.
- **exanthema** (ehcks-ahn-thē-mah) = cutaneous rash caused by fever or disease. Singular is **exanthem** (ehcks-ahn-thuhm).
- **excoriation** (ehcks-kohr-ē-ā-shuhn) = skin surface injury caused by self-trauma (Figure 10–14b).
- **feline miliary dermatitis** (mihl-ē-ahr-ē dər-mah-tī-tihs) = skin disease of cats in which multiple crusts and bumps are present predominantly on the dorsum; the disease can be associated with many causes.
- **fissure** (fihs-sər) = cracklike sore (Figure 10–19).
- **fistula** (fihs-tyoo-lah) = abnormal passage from an internal organ to the body surface or between two internal organs. Plural is **fistulae** (fihs-tyoo-lā). The word root **fistul/o** means tube or pipe.
- **flea allergy dermatitis** (dər-mah-tī-tihs) = inflammation of the skin caused by an allergic reaction to flea saliva; abbreviated FAD. An **allergen** (ahl-ər-jehn) is a substance that produces an allergic response.
- **footrot** = bacterial (*Fusobacterium* sp.) hoof disease that spreads from the interdigital skin to the deeper foot structures.
- **frostbite** = tissue damage caused by extreme cold or contact with chemicals with extreme temperature (e.g., liquid nitrogen).
- **furuncle** (fyoo-ruhng-kuhl) = localized skin infection in a gland or hair follicle; also called a **boil**. **Furunculosis** (fyoo-ruhng-kuh-lō-sihs) is the abnormal condition of persistent boils over a period of time.

FIGURE 10–17 Contagious ecthyma (soremouth) in a sheep.

A **papule** is a small solid raised lesion that is less than 0.5 cm in diameter.

A **plaque** is a solid raised lesion that is greater than 0.5 cm in diameter.

A **macule** is a flat discolored lesion that is less than 1 cm in diameter.

A **patch** is a flat discolored lesion that is greater than 1 cm in diameter.

A **scale** is a flaking or dry patch made up of excess dead epidermal cells.

A **crust** is a collection of dried serum and cellular debris.

A **wheal** is a smooth, slightly elevated swollen area that is redder or paler than the surrounding skin. It is usually accompanied by itching.

(a) Surface lesions

A **cyst** is a closed sac or pouch containing fluid or semisolid material.

A **pustule** is a small circumscribed elevation of the skin containing pus.

A **vesicle** is a circumscribed elevation of skin containing fluid that is less than 0.5 cm in diameter.

A **bulla** is a large vesicle that is more than 0.5 cm in diameter.

(b) Fluid-filled lesions

An **ulcer** is an open sore or erosion of the skin or mucous membrane resulting in tissue loss.

A **fissure** of the skin is a groove or crack-like sore.

(c) Erosive lesions

Copyright © 2015 Cengage Learning®.

FIGURE 10–19 Skin lesions. Skin lesions may be (a) raised and discolored, (b) fluid-filled, or (c) erosive in nature.

- **gangrene** (gahng-grēn) = necrosis associated with loss of circulation. **Necrosis** (neh-krō-sihs) is condition of dead tissue; **necrotic** (neh-kroh-tihck) means pertaining to dead tissue. Decay that produces a foul smell is called **putrefaction** (pyoo-treh-fahck-shuhn).
- **granuloma** (grahn-yoo-lō-mah) = small area of healing tissue.
- **hemangioma** (hē-mahn-jē-ō-mah) = benign tumor composed of newly formed blood vessels.
- **hyperkeratosis** (hī-pər-kehr-ah-tō-sihs) = abnormal condition of increased growth of the horny layer of the skin (Figure 10–20); also called **acanthokeratodermia** (ā-kahn-thō-kehr-ah-tō-dər-mah).
- **infestations** (ihn-fehs-tā-shuhns) = occupation and dwelling of a parasite on the external surface of tissue. **Ectoparasites** (ehck-tō-pahr-ah-sīts) live on the external surface; **ecto-** means outside (Figure 10–21a, b, c, d, e, and f). A **louse** (lows) is a wingless parasitic insect; plural is **lice** (līs). **Pediculosis** (pehd-ih-koo-lō-sihs) is lice infestation. A **mite** is an insect with a hard exoskeleton and paired, jointed legs. **Mange** (mānj) is a common term for skin disease caused by mites. There are different types of mange, such as **sarcoptic** (sahr-kohp-tihck) and **demodectic** (deh-mō-dehck-tihck), depending on the type of mite involved (Figure 10–22a and b). **Chiggers** (chihg-gərs) is infestation by mite larvae that results in severe pruritus. **Acariasis** (ahck-ah-rī-ah-sihs) is infestation with ticks or mites. **Maggots** (mah-gohts) are insect larvae found especially in dead or decaying tissue (Figure 10–23). **Myiasis** (mī-ī-ah-sihs) is infestation by fly larvae.
- **keratosis** (kehr-ah-tō-sihs) = abnormal condition of epidermal overgrowth and thickening. Plural is **keratoses** (kehr-ah-tō-sēs).

(a)　(b)

(c)　(d)

(e)　(f)

FIGURE 10–21 Ectoparasites. (a) *Cheyletiella* sp. (b) *Demodex canis*. (c) *Sarcoptes scabiei*. (d) *Otodectes cynotis*. (e) *Linognathus* sp. (sucking louse). (f) *Trichodectes* sp. (biting louse).

- **laceration** (lahs-ər-ā-shuhn) = accidental cut into the skin.
- **lesion** (lē-shuhn) = pathologic change of tissue; used to describe abnormalities in many locations.
- **lipoma** (lī-pō-mah) = benign growth of fat cells (Figure 10–24); also called **fatty tumor**; commonly seen in older dogs.
- **macule** (mahck-yool) = flat, discolored lesion less than 1 cm in diameter; also called **macula** (mahck-yoo-lah) (Figure 10–19).
- **melanoma** (mehl-ah-nō-mah) = tumor or growth of pigmented skin cells (Figure 10–25). **Malignant melanoma** is the term used to describe cancer of the pigmented skin cells. One form of melanoma is **amelanotic melanoma**, which is an unpigmented malignant melanoma.
- **nodule** (nohd-yoo-uhl) = small knot protruding above the skin.
- **onychomycosis** (ohn-kē-ō-mī-kō-sihs) = superficial fungal infection of the claw.
- **pallor** (pahl-ohr) = skin paleness.
- **papilloma** (pahp-ih-lō-mah) = benign epithelial growth that is lobed. The wired root **papill/o** or **papill/i** means nipplelike.

FIGURE 10–20 Hyperkeratosis of the foot pads of a dog.

(a)

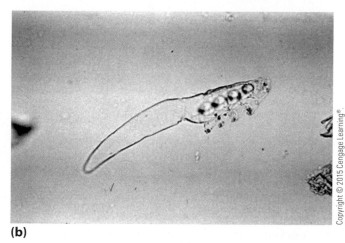

(b)

FIGURE 10–22 (a) Demodectic mange in a dog. (b) *Demodex canis.*

FIGURE 10–23 Maggots in a dog's wound.

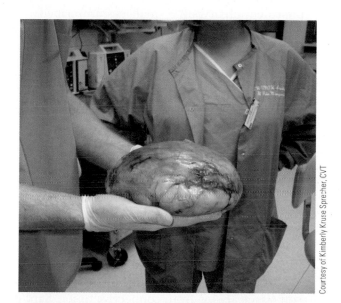

FIGURE 10–24 Lipoma that has been surgically removed from a dog.

- **papule** (pahp-yool) = small, raised skin lesion less than 0.5 cm in diameter (Figure 10–19).
- **parakeratosis** (pahr-ah-kehr-ah-tō-sihs) = lesion characterized by thick scales, cracking, and red raw surface caused by the persistence of keratinocyte nuclei in the horny layer of skin.
- **paronychia** (pahr-ohn-kē-ah) = bacterial or viral infection of the claw.
- **patch** = localized skin color change greater than 1 cm in diameter (Figure 10–19).
- **pemphigus** (pehm-fih-guhs) = group of immune mediated skin diseases characterized by vesicles, bullae, and ulcers. The most common form is **pemphigus vulgaris** (pehm-fih-guhs vuhl-gahr-ihs), which consists of shallow ulcerations frequently involving the oral mucosa and mucocutaneous junctions. *Pemphix* is Greek for blister.

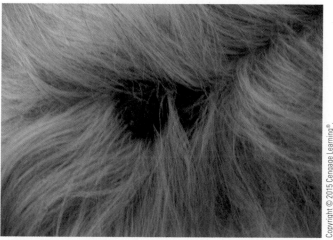

FIGURE 10–25 Melanoma in a dog.

- **petechiae** (peh-tē-kē-ā) = small, pinpoint hemorrhages (Figure 10-26). Singular is **petechia** (peh-tē-kē-ah).
- **plaque** (plahck) = solid raised lesion greater than 0.5 cm in diameter (Figure 10-19).
- **pododermatitis** (pō-dō- dər-mah-tī-tihs) = foot infection in animals, particularly birds (from walking on hard surfaces), small mammals, or rodents (from standing on wire or mesh flooring that has not been recently cleaned) (Figure 10-27); also called **bumblefoot** or **sore hocks**.
- **polyp** (pohl-uhp) = growth from mucous membranes.
- **pruritus** (proo-rī-tuhs) = itching.
- **purpura** (pər-pə-rah) = condition characterized by hemorrhage into the skin that causes bruising. The two types of purpura are ecchymosis and petechia.
- **pustule** (puhs-tyool) = small, circumscribed, pus-filled skin elevation (Figure 10-19). **Circumscribed** (sehr-kuhm-skrībd) means contained in a limited area.
- **pyoderma** (pī-ō-dər-mah) = skin disease containing pus (Figure 10-28). **Py/o** is the combining form meaning pus. **Pus** (puhs) is an inflammatory product made up of leukocytes, cell debris, and fluid. **Purulent** (pər-ū-lehnt) means containing or producing pus. **Puppy pyoderma** is a skin disease in puppies characterized by pus-containing lesions. **Juvenile pyoderma** is a skin disease in puppies that progresses to a systemic disease characterized by fever, anorexia, and enlarged and abscessing lymph nodes; juvenile pyoderma also is called **puppy strangles**.
- **sarcoma** (sahr-kō-mah) = malignant neoplasm of soft tissue arising from connective tissue. **Sarc/o** is the combining form for flesh (connective tissue) or malignancy of connective tissue.
- **scale** = flake (Figure 10-19).

Courtesy of Jordan Applied Technology Center, West Jordan, Utah

FIGURE 10-27 Pododermatitis in a chinchilla. Pododermatitis is caused by pressure on hairless or unprotected feet on hard surfaces (like wire cages) and poor husbandry practices such as not cleaning properly.

- **scar** (skahr) = mark left by a healing lesion where excess collagen was produced to replace injured tissue; also called **cicatrix** (sihck-ah-trihcks) or **cicatrices** (sihck-ah-trih-sēz), which are multiple scars.
- **sebaceous cyst** (seh-bā-shuhs sihst) = closed sac of yellow fatty material.
- **seborrhea** (sehb-ō-rē-ah) = skin condition characterized by overproduction of sebum (oil).
- **skin tag** = small growth that hangs from the body by stalks.

Courtesy of Kimberly Kruse Sprecher, CVT

FIGURE 10-26 Petechiae in a dog.

Courtesy of Kimberly Kruse Sprecher, CVT

FIGURE 10-28 Pyoderma in a dog.

- **ulcer** (uhl-sihr) = erosion of skin or mucous membrane (Figure 10–19). **Decubital ulcers** (dē-kyoo-bih-tahl uhl-sihrz) are erosions of skin or mucous membranes as a result of prolonged pressure; also called **bedsores**.
- **urticaria** (ər-tih-kā-rē-ah) = localized areas of swelling that itch; also called **hives**.
- **verrucae** (veh-roo-sē) = warts.
- **vesicle** (vehs-ih-kuhl) = contained skin elevation filled with fluid that is greater than 0.5 cm in diameter; also called a **blister**, **bulla**, (buhl-ah), or **bleb** (Figure 10–19).
- **wheal** (whēl) = smooth, slightly raised swollen area that itches (Figure 10–19).

(a)

PROCEDURES: INTEGUMENTARY SYSTEM

Procedures performed on the integumentary system include the following:

- **cauterization** (kaw-tər-ī-zā-shuhn) = destruction of tissue using electric current, heat, or chemicals (Figure 10–29).
- **cryosurgery** (krī-ō-sihr-jər-ē) = destruction of tissue using extreme cold.
- **debridement** (dē-brīd-mehnt) = removal of tissue and foreign material to aid healing.
- **lance** (lahnz) = to open or pierce with a lancet (scalpel blade) to allow drainage; abscesses are lanced to drain the pus present in an area.
- **laser** (lā-zər) = device that transfers light into an intense beam for various purposes; acronym for *light amplification by stimulated emission of radiation* (Figure 10–30a, b, and c).

(b)

FIGURE 10–29 Cauterization destroys tissue through the use of electric current, heat, or chemicals.

(c)

FIGURE 10–30 (a) A laser unit display screen. (b) A laser feline declaw surgery. (c) A laser treatment to reduce inflammation.

ABBREVIATIONS RELATED TO THE INTEGUMENTARY SYSTEM

- bx = biopsy
- DLE = discoid lupus erythematosus
- FAD = flea allergy dermatitis
- ID = intradermal
- PPD = purified protein derivative
- SLE = systemic lupus erythematosus
- SQ, SC, subq, subcu = subcutaneous`

REVIEW EXERCISES

Multiple Choice

Choose the correct answer.

1. Pruritus is commonly called
   - a. hair loss
   - b. dry skin
   - c. itching
   - d. pus

2. Skin redness is called
   - a. cellulitis
   - b. erythema
   - c. scleroderma
   - d. scarring

3. Hypersensitivity reaction in animals involving pruritus with secondary dermatitis is called
   - a. atrophy
   - b. allergen
   - c. antigen
   - d. atopy

4. Hair loss resulting in hairless patches or complete lack of hair is called
   - a. shedding
   - b. lesion
   - c. alopecia
   - d. plaque

5. Occupation and dwelling of parasites on the external skin surfaces is called
   - a. parasitism
   - b. ectoparasites
   - c. infestation
   - d. myiasis

6. A skin disease containing pus is
   - a. pyometra
   - b. pyoderma
   - c. pyoerythema
   - d. pyosis

7. Producing or containing pus is called
   - a. abscess
   - b. purulent
   - c. mucocutaneous
   - d. polyp

8. Inflammation of connective tissue is
   - a. connectitis
   - b. dermatitis
   - c. cutaneitis
   - d. cellulitis

9. Putrefaction is
   - a. foul-smelling decay
   - b. homogenation
   - c. granulomatous
   - d. cellulitis

10. Large tactile hair is
    - a. sensogenic
    - b. plantigrade
    - c. vibrissa
    - d. cerumen

11. A skin condition characterized by overproduction of sebum (oil) is known as
    - a. alopecia
    - b. seborrhea
    - c. hemangioma
    - d. abscess

12. The term for a skin flake is
    - a. patch
    - b. crust
    - c. scale
    - d. wheal

13. Which term means skin paleness?
    a. petechiae
    b. pallor
    c. pustule
    d. purpura

14. The protein that provides skin with its water-proofing properties is known as
    a. dermis
    b. squamous
    c. sebum
    d. keratin

15. Removal of tissue and foreign material to aid healing is
    a. cauterization
    b. cryosurgery
    c. debridement
    d. laser treatment

16. A localized collection of pus is a(n)
    a. bleb
    b. abscess
    c. nodule
    d. vesicle

17. A benign growth of fat cells is known as a(n)
    a. sarcoma
    b. carcinoma
    c. lipoma
    d. adoma

18. Which term means pertaining to dead tissue?
    a. abrasion
    b. carbuncle
    c. purulent
    d. necrotic

19. Which term means erosion of skin or mucous membrane?
    a. fistula
    b. ulcer
    c. fissure
    d. urticaria

20. A tumor or growth of pigmented skin is a(n)
    a. melanoma
    b. ecthyma
    c. exanthema
    d. eczema

21. Superficial fungi found on skin are called
    a. pediculoses
    b. dermatophytes
    c. sarcoptic
    d. demodectic

22. The conversion of epithelium into keratin or horn is
    a. velvet
    b. polled
    c. cornification
    d. macule

23. Necrosis associated with loss of circulation is
    a. gangrene
    b. infestation
    c. fissure
    d. eczema

24. An ecchymosis is a/an
    a. bruise
    b. ulcer
    c. comedo
    d. carbuncle

25. Which three terms represent diagnostic tests?
    a. atopy, alopecia, and furuncle
    b. cauterization, debridement, and cryosurgery
    c. vesicle, laser, and sarcoma
    d. cytology, culture, and skin scrape

Matching

Match the common dermatologic term in Column I with the medical term in Column II.

Column I	Column II
1. _____ hive	a. polled
2. _____ hot spot	b. cicatrix
3. _____ fatty tumor	c. furuncle
4. _____ pale	d. urticaria
5. _____ blister	e. abrasion
6. _____ scar	f. acute moist dermatitis
7. _____ oil	g. pruritus
8. _____ scrape	h. atopy
9. _____ boil	i. lipoma
10. _____ redness	j. scale
11. _____ itching	k. erythema
12. _____ flake	l. pallor
13. _____ warts	m. sebum
14. _____ allergic dermatitis	n. verrucae
15. _____ hornless	o. fissure
16. _____ crack	p. vesicle
17. _____ adipocytes	q. malignant neoplasm of epithelial tissue
18. _____ dermis	r. most superficial layer of skin
19. _____ epidermis	s. protein material found in skin, hair, and nails
20. _____ subcutaneous	t. sweat glands
21. _____ collagen	u. fat cells
22. _____ pus	v. small area of healing tissue
23. _____ granuloma	w. middle layer of skin
24. _____ carcinoma	x. deepest layer of the skin
25. _____ sudoriferous glands	y. oil glands
26. _____ sebaceous glands	z. inflammatory product made up of leukocytes, cell debris, and fluid

Fill in the Blanks

Write the correct term in the space provided.

1. Cutane/o, derm/o, dermat/o, and -derma mean _____.

2. Pil/i, pil/o, and trich/o mean _____.

3. The dermis and corium are names for _____.

4. The subcutaneous and hypodermis are names for _____.

5. A cyst, vesicle, pustule, and bulla are all _____.

True or False

If the statement is true, write T on the line. If the statement is false, write F on the line.

1. _____Pruritus may be present with many skin diseases.

2. _____Decubital ulcers are caused by prolonged pressure on an area of skin or mucous membrane.

3. _____A sarcoma is a malignant growth.

4. _____Keratoses is the plural form of keratosis.

5. _____Decay that does not have a foul odor is called putrefaction.

6. _____An incisional biopsy is the removal of an entire tumor or lesion.

7. _____Intradermal skin testing is used to diagnosis carcinomas.

8. _____Hypersensitization is a decreased response to an allergen.

9. _____Hair standing straight up is called piloerection.

10. _____Injury to the deep layers of skin is an abrasion.

Spell Check

Circle the term that is spelled correctly.

1. itching:	pruritis	pruritus	puritis
2. buildup of sebum and keratin in a pore:	komedo	comedo	comedoe
3. hair loss resulting in hairless areas:	alopecia	ahlopesia	alopecea
4. production and secretion of sweat:	hidrosis	hydrosis	hihdrosis
5. common term for skin disease caused by mites:	manje	mainge	mange

Crossword Puzzle

Skin Terms

Supply the correct term in the appropriate space for the definition listed.

Across

2 accidental cut into the skin

4 removal of a piece of a tumor or lesion for examination

6 growth from mucous membrane

9 dermis under the nail or hoof that has a blood supply and is sensitive

11 small, pinpoint hemorrhage

13 closed sac containing fluid or semisolid material

14 earwax

15 pertaining to falling off

20 contained within a limited area

21 normal hair loss

22 localized collection of pus

23 permanent structure originating from the skull

24 secondary hairs that are soft, thin, and wavy

25 superficial fungal infection of the claw

Down

1 small area of healing tissue

3 fat cell

5 act of hair standing upright

7 condition characterized by hemorrhage into the skin that causes bruising

8 decreased response to an allergen

10 destruction of tissue using electric current, heat, or chemicals

12 widespread

16 within a well-defined area

17 pertaining to touch

18 injury that does not break the skin

19 abnormal condition of epidermal overgrowth and thickening

Label the Diagrams

Label the diagram in Figure 10–31.

FIGURE 10–31 Skin layers. Label the skin layers and structures as indicated by the arrows.

Case Studies

Define the underlined terms in each case study.

A 3-yr-old F/S golden retriever was presented with clinical signs of <u>pruritus</u>, abdominal <u>dermatitis</u>, and <u>otitis</u>. <u>Skin scrapes</u> were negative for external parasites. Ear cytology revealed a large number of yeast. The dog was referred to a <u>dermatologist</u>, who diagnosed <u>atopy</u> via <u>intradermal skin testing</u>. The dog was put on a <u>hypoallergenic</u> diet and was given <u>hyposensitization</u> injections. Medications were prescribed to control the pruritus and secondary <u>pyoderma</u>.

1. pruritus   _____

2. dermatitis   _____

3. otitis   _____

4. skin scrapes   _____

5. dermatologist   _____

6. atopy   _____

7. intradermal skin testing   _____

8. hypoallergenic   _____

9. hyposensitization   _____

10. pyoderma   _____

The veterinarian was on a farm performing a routine herd check when the farmer pointed out a cow to the veterinarian. The cow was a 2-yr-old Holstein cow that was treated 5 days ago with an <u>IM</u> injection of long-acting penicillin for mastitis (inflammation of the mammary gland). The cow's mastitis improved, and she was eating and <u>ruminating</u> normally. However, the cow now had an enlarged swelling on her <u>R</u> hind limb (Figure 10–32). On <u>PE</u>, the cow's <u>HR</u> and <u>RR</u> were normal and the temperature was elevated. The cow was not painful upon <u>palpation</u> of the swelling. The swelling had a soft center that appeared to be fluid-filled. The veterinarian diagnosed an <u>abscess</u> and treated the cow by cleaning the wound area, <u>lancing</u> the abscess, draining the pus from the swelling, and flushing the abscess with <u>antiseptic</u> solution. The farmer was advised to monitor the cow to make sure the swelling does not recur.

11. IM   _____

12. ruminating   _____

13. R   _____

14. PE   _____

15. HR   _____

16. RR   _____

17. palpation   _____

18. abscess   _____

19. lancing   _____

20. antiseptic   _____

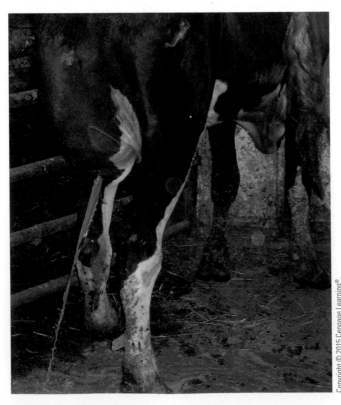

FIGURE 10–32 Leg abscess in a Holstein.

Critical Thinking Exercise

The following story and questions contain words and concepts from this chapter and are designed to stimulate critical thinking through class discussion, Internet searches, or brief writing responses.

Bailey, a 5-yr-old F/S DLH cat, was in a house fire. When the firefighters brought her to the veterinary clinic, her initial treatment included large volumes of IV fluids.

1. What does the abbreviation IV stand for?

2. Why does Bailey need large volumes of IV fluids?

The veterinarian explains to the owners that there are different levels of burns. First-degree burns redden the skin, but the skin layers are intact. Fur may or may not be missing and there is minor pain or discomfort. Second-degree burns have blisters in addition to redness, which indicates that several skin layers have been damaged. Second-degree burns are more painful than first-degree burns. Third-degree burns damage the full thickness of the skin and damage the tissue under the skin. The skin edges may be blackened with third-degree burns. Third-degree burns are more painful than second- and first-degree burns.

3. What is the medical term for redness of the skin?

4. What is the medical term for blister?

5. Because Bailey needed IV fluids upon arrival at the veterinary clinic, what type of burn did she have?

6. What do you think burn patients like Bailey are at risk of acquiring?

7. Why might the veterinary staff want to isolate Bailey from other patients?

8. What type of procedure may the veterinarian recommend to remove dead tissue to help heal Bailey's wounds?

Chapter 11
THE GREAT COMMUNICATOR

Objectives

Upon completion of this chapter, the reader should be able to:

- Identify the major structures of the endocrine system

- Describe the functions of the endocrine system

- Describe the role of the hypothalamus and pituitary gland in hormone secretion

- Recognize, define, spell, and pronounce terms related to the diagnosis, pathology, and treatment of the endocrine system

- Interpret the meaning of abbreviations of the endocrine system

FUNCTIONS OF THE ENDOCRINE SYSTEM

The endocrine (ehn-dō-krihn) system is composed of ductless glands that secrete chemical messengers, called hormones, into the bloodstream. The prefix **endo-** means within, and the suffix **-crine** means to secrete or separate. Hormones enter the bloodstream and are carried throughout the body, affecting a variety of tissues and organs. Tissues and organs on which the hormones act are called **target organs**. **Hormone** (hōr-mōn) is a Greek term meaning *impulse* or *to set in motion*; hormones may excite or inhibit a motion or an action.

STRUCTURES OF THE ENDOCRINE SYSTEM

The glands of the normal endocrine system include the following (Figure 11–1):

- One pituitary gland (with two lobes)
- One thyroid gland (right and left lobes fused ventrally)
- Parathyroid glands (four in most species)
- Two adrenal glands
- One pancreas
- One thymus
- One pineal gland
- Two gonads (ovaries in females, testes in males)

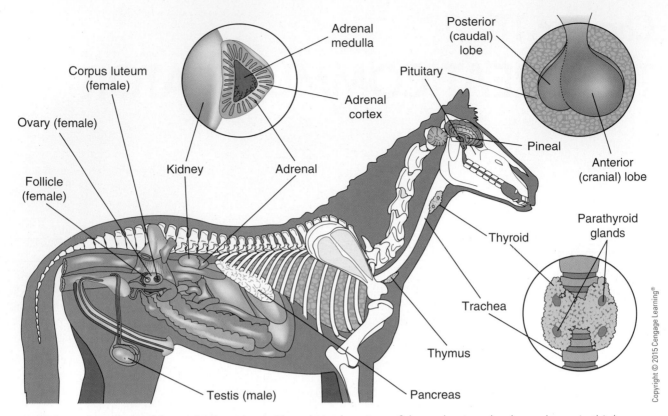

FIGURE 11–1 Locations of the endocrine glands. The relative locations of the endocrine glands are shown in this horse.

The Pituitary Gland

The **pituitary** (pih-too-ih-tār-ē) **gland** is located at the base of the brain just below the hypothalamus. The pituitary gland is also called the **hypophysis** (hī-poh-fī-sihs) because it is a growth located beneath or ventral to the hypothalamus (part of the brain). (*Physis* is Greek for growth.) The pituitary gland is known as the master gland because it secretes many hormones that control or master other endocrine glands. The combining form for pituitary gland is **pituit/o**.

The pituitary gland acts in response to stimuli from the **hypothalamus** (hī-pō-thahl-ah-muhs). The hypothalamus is located below the thalamus in the brain and secretes releasing and inhibiting factors that affect the release of substances from the pituitary gland (Figure 11–2). The hypothalamus also produces antidiuretic hormone and oxytocin, which are released from the posterior pituitary gland. The hypothalamus is connected to the pituitary gland via a stalk called the **infundibulum** (ihn-fuhn-dihb-yoo-luhm). *Infundibulum* means funnel-shaped passage or opening.

The pituitary gland is made up of two lobes: the anterior (cranial) and posterior (caudal). The anterior lobe is also known as the **adenohypophysis** (ahd-ehn-ō-hī-pohf-ih-sihs) because it produces hormones (that

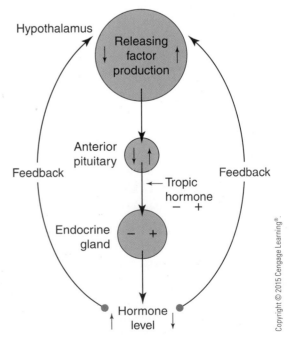

FIGURE 11–2 Feedback control mechanisms. Negative feedback mechanisms control the levels of most hormones in the blood through the secretion of releasing or inhibiting factors that affect hormone release.

is, has glandular function; **aden/o** = gland). Hormones released from the anterior pituitary gland sometimes are called **indirect-acting hormones** because they cause their target organ to produce a second hormone.

The posterior lobe is also known as the **neurohypophysis** (nū-rō-hī-pohf-ih-sihs) because it responds to a neurologic stimulus and does not produce hormones, but stores and secretes them. Hormones released from the posterior pituitary gland sometimes are called **direct-acting hormones** because they produce the desired effect directly in the target organ (Figure 11–3).

Secretions of the Anterior Pituitary Gland

- **thyroid** (thī-royd) **-stimulating hormone** = augments growth and secretions of the thyroid gland; abbreviated TSH.
- **adrenocorticotropic** (ahd-rēn-ō-kōr-tih-kō-trō-pihck) **hormone** = augments the growth and secretions of the adrenal cortex; abbreviated ACTH.

- **follicle** (fohl-lihck-kuhl) **-stimulating hormone** = augments the secretion of estrogen and growth of eggs in the ovaries (female) and the production of sperm in the testes (male); abbreviated FSH. FSH is a type of **gonadotropic** (gō-nahd-ō-trō-pihck) **hormone**. Gonadotropic can be divided into **gonad/o**, which means gamete-producing gland (ovary or testes), and **-tropic**, which means having an affinity for.
- **luteinizing** (loo-tehn-īz-ihng) **hormone** = augments ovulation and aids in the maintenance of pregnancy in females; abbreviated LH. **Lute/o** is the combining form for yellow. Luteinizing hormone transforms an ovarian follicle into a corpus luteum, or yellow body. LH is a type of gonadotropic hormone.
- **interstitial** (ihn-tər-stihsh-ahl) **cell-stimulating hormone** = stimulates testosterone secretion in males; abbreviated ICSH. ICSH is now considered to be LH.
- **prolactin** (prō-lahck-tihn) = augments milk secretion and influences maternal behavior; also known as **lactogenic hormone** or

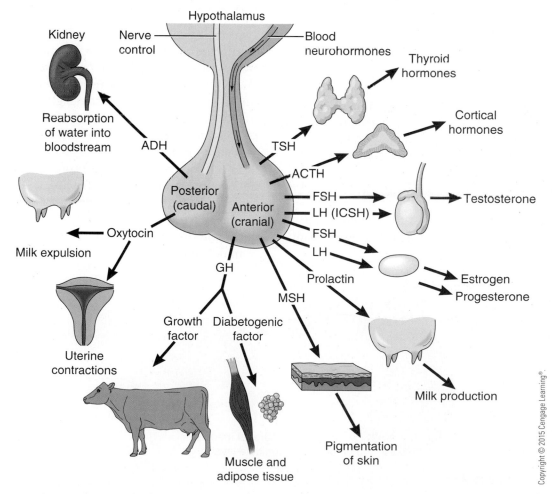

FIGURE 11–3 Secretions of the pituitary gland.

luteotropin (the combining form **lact/o** means milk); sometimes abbreviated LTH.
- **growth hormone** = accelerates body growth; abbreviated GH; also known as **somatotropin** (the combining form **somat/o** means body); sometimes abbreviated STH.
- **melanocyte** (mehl-ah-nō-sīt) **-stimulating hormone** = augments skin pigmentation; abbreviated MSH.

Secretions of the Posterior Pituitary Gland

- **antidiuretic** (ahn-tih-dī-yoo-reht-ihck) **hormone** = maintains water balance in the body by augmenting water reabsorption in the kidneys (**anti-** means against; **diuretic** means pertaining to increased urine secretion); abbreviated ADH; also known as **vasopressin** (**vas/o** is the combining form for vessel, and **press/i** is the combining form for tension; vasopressin affects blood pressure).
- **oxytocin** (ohcks-ē-tō-sihn) = stimulates uterine contractions during parturition and milk letdown from the mammary ducts.

Thyroid Gland

The **thyroid** (thī-royd) **gland** is a butterfly-shaped gland, with the right and left lobes fused ventrally by an isthmus. (The isthmus may be rudimentary in horses and dogs.) The thyroid gland is located on either side of the larynx. The thyroid gland regulates metabolism, iodine uptake, and blood calcium levels. The combining forms for thyroid are **thyr/o** and **thyroid/o**.

Secretions of the Thyroid Gland

- **triiodothyronine** (trī-ī-ō-dō-thī-rō-nēn) = one of the thyroid hormones that regulates metabolism; abbreviated T_3.

- **thyroxine** (thī-rohcks-ihn) = one of the thyroid hormones that regulates metabolism; abbreviated T_4.
- **calcitonin** (kahl-sih-tō-nihn) = thyroid hormone that promotes the absorption of calcium from blood into bones (lowers blood calcium and raises bone calcium levels).

Parathyroid Glands

The **parathyroid** (pahr-ah-thī-royd) **glands** are four glands located on the surface of the thyroid gland that secrete parathyroid hormone (or **parathormone** [pahr-ah-thōr-mōn] or PTH). The prefix **para-** means near or before. PTH helps regulate blood calcium and phosphorus levels. PTH increases blood calcium levels by reducing bone calcium levels. Calcium is regulated in the body by the antagonistic actions of PTH and calcitonin.

PTH also regulates phosphorus content of blood and bones. The combining form for the parathyroid glands is **parathyroid/o**.

Adrenal Glands

Two small glands, known as the **adrenal** (ahd-rē-nahl) or **suprarenal** (soop-rah-rē-nahl) glands, are located cranial to each kidney. The prefix **ad-** means toward, the combining form **ren/o** means kidney, and the prefix **supra-** means above. The adrenal gland regulates electrolytes, metabolism, sexual functions, and the body's response to injury. The combining forms **adren/o** and **adrenal/o** refer to the adrenal glands.

Each adrenal gland consists of two parts: the **adrenal cortex** (ahd-rē-nahl kōr-tehcks), or outer portion, and the **adrenal medulla** (ahd-rē-nahl meh-doo-lah), or inner portion. The combining form **cortic/o** means outer, and the combining form **medull/o** means inner or middle portion (Figure 11–4).

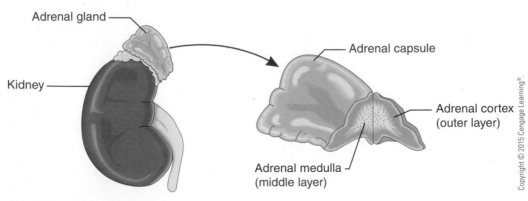

FIGURE 11–4 Portions of the adrenal gland.

Secretions of the Adrenal Cortex

The adrenal cortex hormones are classified as steroids. A **steroid** (stehr-oyd) is a substance that has a specific chemical structure of carbon atoms in four interlocking rings. **Corticosteroids** (kōr-tih-kō-stehr-oydz) are steroids produced by the adrenal cortex.

- **mineralocorticoids** (mihn-ər-ahl-ō-kōr-tih-koydz) = group of corticosteroids that regulates electrolyte and water balance by affecting ion transport in the kidney. The principal mineralocorticoid is **aldosterone** (ahl-dohs-tər-ōn), which is a hormone that regulates sodium and potassium.
- **glucocorticoids** (gloo-kō-kōr-tih-koydz) = group of corticosteroids that regulates carbohydrate, fat, and protein metabolism; resistance to stress; and immunologic functioning. An example of a glucocorticoid is **hydrocortisone** (hī-drō-kōr-tih-zōn), or **cortisol** (kōr-tih-zohl), which regulates carbohydrate, fat, and protein metabolism.
- **androgens** (ahn-drō-jehnz) = group of corticosteroids that aids in the development and maintenance of male sex characteristics. The combining form **andr/o** means male, and the suffix **-gen** means producing. **Anabolic** (ahn-ah-bohl-ihck) **steroids** are synthetic medications similar in structure to testosterone that are used to increase strength and muscle mass. The prefix **ana-** means up or excessive, and the combining form **bol/o** means throwing. Anabolic means pertaining to building up. The opposite of anabolism is catabolism. The prefix **cata-** means down; therefore, **catabolism** (kaht-ah-bōl-ihz-uhm) means breaking down.

Secretions of the Adrenal Medulla

- **epinephrine** (ehp-ih-nehf-rihn) = **catecholamine** (kaht-ih-kōl-ih-mēn) that stimulates the sympathetic nervous system (the fight-or-flight system) and increases blood pressure, heart rate, and blood glucose. (Catecholamines are nitrogen-containing compounds that act as hormones or neurotransmitters.) Epinephrine is also known as **adrenaline** (ahd-rehn-ah-lihn). Because epinephrine increases blood pressure, it is known as a vasopressor. A **vasopressor** (vahs-ō-prehs-ōr) is a substance that stimulates blood vessel contraction and increases blood pressure.

- **norepinephrine** (nōr-ehp-ih-nehf-rihn) = catecholamine that stimulates the sympathetic nervous system (the fight-or-flight system) and increases blood pressure, heart rate, and blood glucose. Norepinephrine also is known as **noradrenaline.**

What Is in a Name?

It may seem as though medical terminology is a bunch of long words strung together with no relationship to each other. However, consider the names of the hormones released from the adrenal medulla:

- **epinephrine** = the prefix **epi-** means above. The combining form **nephr/o** means kidney. Its name indicates that epinephrine is made in the gland above the kidney, which is known as the adrenal gland.
- **adrenaline** = the prefix **ad-** means toward. The combining form **ren/o** means kidney. Its name indicates that adrenaline is made in the gland toward the kidney, which is known as the adrenal gland.
- **norepinephrine** = the prefix **nor-** means normal and denotes the parent compound in a pair of related substances. Its name indicates that norepinephrine is a substance very similar in chemical structure and function to epinephrine.

Pancreas

The **pancreas** (pahn-krē-ahs) is an aggregation of cells located near the proximal duodenum that has both exocrine and endocrine functions. The exocrine function of the pancreas involves the secretion of digestive enzymes, which is covered in Chapter 6. The endocrine function of the pancreas involves the secretion of blood glucose–regulating hormones. (Blood glucose is also regulated in conjunction with other hormones such as thyroid hormones.) The combining form for pancreas is **pancreat/o**.

Specialized cells in the pancreas called the **islets of Langerhans** (ī-lehts ohf lahng-ər-hahnz) secrete the hormones that help regulate blood glucose (Figure 11–5).

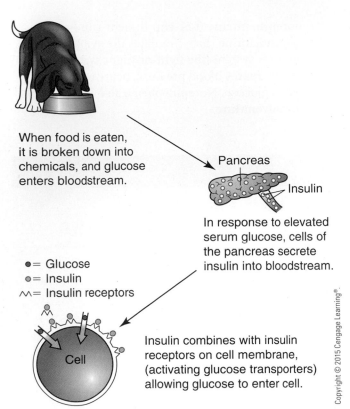

When food is eaten, it is broken down into chemicals, and glucose enters bloodstream.

Pancreas

Insulin

In response to elevated serum glucose, cells of the pancreas secrete insulin into bloodstream.

● = Glucose
● = Insulin
⋏ = Insulin receptors

Cell

Insulin combines with insulin receptors on cell membrane, (activating glucose transporters) allowing glucose to enter cell.

FIGURE 11–5 How insulin works.

Endocrine Secretions of the Pancreas

- **insulin** (ihn-suh-lihn) = hormone that decreases blood glucose levels by transporting blood glucose into body cells or into storage as glycogen (glī-kō-jehn). Glycogen is the main carbohydrate storage unit in animals. Insulin is secreted by the beta cells of the islets of Langerhans in response to high blood glucose.
- **glucagon** (gloo-kah-gohn) = hormone that increases blood glucose levels by breaking down glycogen to glucose. Glucagon is secreted by the alpha cells of the islets of Langerhans in response to low blood glucose.

Thymus

The **thymus** (thī-muhs) is a gland predominant in young animals located near midline in the cranioventral portion of the thoracic cavity. The thymus gland has an immunologic function through its role in the maturation of T-lymphocytes. The combining form for the thymus is **thym/o.** Thym/o also refers to the soul; some people believed that the thymus gland was the seat of the soul because it was so close to the heart.

Secretion of the Thymus

- **thymosin** (thī-mō-sihn) = augments maturation of T-lymphocytes; therefore, it plays a role in the immune system, which is discussed in Chapter 15.

Pineal Gland

The **pineal** (pih-nē-ahl or pī-nē-ahl) **gland,** or **pineal body**, is an aggregation of cells located in the central portion of the brain. The pineal gland's functions are not fully understood, but one of its functions is secretion of serotonin, which is converted to melatonin, a hormone that affects circadian rhythm. The combining form **pineal/o** refers to the pineal gland.

Secretion of the Pineal Gland

- **melatonin** (mehl-ah-tōn-ihn) = controls circadian rhythm and reproductive timing. **Circadian** can be divided into **circa** (around) and **diem** (day). *Circadian* refers to the events occurring within a 24-hour period.

Gonads

The **gonads** (gō-nahdz) are gamete-producing glands. A **gamete** (gahm-ēt) is a sex cell. Gamete-producing glands are the **ovaries** in females and **testes** in males. **Gonad/o** is the combining form for gonad or gamete-producing gland. **Gonadotropic hormones** stimulate the gonads.

Secretions of the Ovary

Secretions of the ovary are stimulated by **human chorionic** (kōr-ē-ohn-ihck) **gonadotropin,** or **hCG**. hCG is secreted by the embryo and by the placenta after implantation has occurred. Other gonadotropins also influence ovarian secretion.

- **estrogen** (ehs-trō-jehn) = hormone that aids in the development of secondary sex characteristics (an example is mammary gland development) and regulates ovulation in females.
- **progesterone** (prō-jēhs-tər-own) = hormone that aids in the maintenance of pregnancy. Progesterone is also secreted from the corpus luteum and placenta.

Secretion of the Testes

- **testosterone** (tehs-tohs-tər-own) = augments the development of secondary sex characteristics. Examples of secondary sex characteristics in animals include horns in rams (not horns in all species), boar tusks, and shoulder girth in cattle and horses. Testosterone is also thought to be secreted from the ovaries and adrenal cortex, although those secretions are in very small amounts.

TEST ME: ENDOCRINE SYSTEM

Diagnostic procedures performed on the endocrine system include the following:

- **ACTH stimulation test** = blood analysis for cortisol levels after administration of synthetic adrenocorticotropic hormone; used to differentiate pituitary-dependent hyperadrenocorticism from adrenal-dependent hyperadrenocorticism.
- **assays** (ahs-āz) = laboratory technique used to determine the amount of a particular substance in a sample. Assays are discussed in Chapter 16.
- **dexamethasone** (dehcks-ah-mehth ah-zōn) **suppression test** = blood analysis for cortisol levels after administration of synthetic glucocorticoid (dexamethasone); used to differentiate pituitary-dependent hyperadrenocorticism from adrenal-dependent hyperadrenocorticism. Also called **dex suppression test.** Dex suppression tests may be high-dose or low-dose.
- **radioactive** (rā-dē-ō-ahck-tihv) **iodine uptake test** = analysis of thyroid function after radioactive iodine has been given orally or intravenously. Absorption of the radioactive iodine is measured with a counter for a specific time period.
- **thyroid stimulation test** = blood analysis for thyroid hormone levels after administration of synthetic thyroid-stimulating hormone; used to differentiate pituitary-dependent from thyroid-dependent dysfunction. **Synthetic** (sihn-theh-tihck) means pertaining to artificial production. TSH levels can also be measured before and after administration of thyrotropin-releasing hormone.

PATHOLOGY: ENDOCRINE SYSTEM

Pathologic conditions of the endocrine system include the following:

- **acromegaly** (ahck-rō-mehg-ah-lē) = enlargement of the extremities caused by excessive secretion of growth hormone after puberty. **Acr/o** means extremities.
- **adrenopathy** (ahd-rēn-ohp-ah-thē) = disease of the adrenal glands (Figure 11–6).
- **aldosteronism** (ahl-doh-stər-ōn-ihzm) = disorder caused by excessive secretion of aldosterone by the adrenal cortex, resulting in electrolyte imbalance. An **electrolyte** (ē-lehck-trō-līt) is a charged substance found in blood.
- **diabetes insipidus** (dī-ah-bē-tēz ihn-sihp-ih-duhs) = insufficient antidiuretic hormone production or the inability of the kidneys to respond to ADH stimuli; abbreviated DI. A **stimulus** (stihm-yoo-luhs) is an agent, act, or influence that produces a reaction. The plural of *stimulus* is **stimuli.**
- **diabetes mellitus** (dī-ah-bē-tez mehl-ih-tuhs) = metabolic disorder of inadequate secretion of insulin or recognition of insulin by the body, resulting in increased urination, drinking, and weight loss; abbreviated DM. The combining form **mellit/o** means honey, which describes the "sweetness" of blood due to hyperglycemia in patients with diabetes mellitus. Severe insulin deficiency may result in **ketoacidosis** (kē-tō-ah-sih-dō-sihs), which is an abnormal condition of low pH accompanied by ketones (by-products of fat metabolism). The combining form **ket/o** means breakdown product of fat. **Acidosis** (ah-sih-dō-sihs) is an abnormal condition of low pH.
- **endocrinopathy** (ehn-dō-krih-nohp-ah-thē) = disease of the hormone-producing system.

BIOLOGY BONUS

Homeostasis and the response to stress are controlled by two distinct but interconnected body systems: the nervous and endocrine systems. The nervous system (covered in Chapter 13) creates an immediate yet short-lived response while the endocrine system has a slightly slower onset and a longer duration of action. The endocrine system maintains homeostasis when certain endocrine glands are stimulated to secrete hormones in response to other hormones, thereby keeping the animal's body in balance. Diseases of the endocrine system typically involve an abnormal increase or decrease in the secretion of hormones. Clinical signs vary with the degree of increase or decrease in hormone secretion and age of the patient. Treatment aims to either decrease hypersecretion of a hormone or supplement hyposecretion of a hormone.

(a)

Courtesy of Eric Klaphake, DVM

(a)

Courtesy of Dr. Mark Jackson, University of Glasgow

(b)

Courtesy of Eric Klaphake, DVM

(b)

Courtesy of Lynn Paron

FIGURE 11–6 Adrenopathy or adrenal disease in a ferret. (a) Untreated adrenal gland disease in ferrets may result in alopecia with erythematosis. (b) A typical pattern of alopecia and edema occurs in jills and hobs with adrenal gland disease.

FIGURE 11–7 (a) Dog with hyperadrenocorticism. Hyperadrenocorticism or Cushing's disease results in polyuria, polydipsia, and redistribution of body fat. (b) Hyperadrenocorticism in horses presents as a failure to shed out the coat in spring, making the coat look long, thick, and matted; excessive sweating; and loss of muscle mass causing a dipped back and pendulous abdomen.

- **gynecomastia** (gī-neh-kō-mahs-tē-ah) = condition of excessive mammary development in males. The combining form **gynec/o** means woman and **mast/o** means mammary gland.
- **hyperadrenocorticism** (hī-pər-ahd-rēn-ō-kōr-tih-kihz-uhm) = disorder caused by excessive adrenal cortex production of glucocorticoid, resulting in increased urination, drinking, and redistribution of body fat; also called **Cushing's** (kuhsh-ihngz) **disease** (Figure 11–7a and b).
- **hypercrinism** (hī-pər-krī-nihzm) = condition of excessive gland secretion.
- **hyperglycemia** (hī-pər-glī-sē-mē-ah) = abnormally elevated blood glucose.
- **hypergonadism** (hī-pər-gō-nahd-ihzm) = abnormal condition of excessive hormone secretion by the sex glands (ovaries in females; testes in males).

- **hyperinsulinism** (hī-pər-ihn-suh-lin-ihzm) = disorder of excessive hormone that transports blood glucose to body cells.
- **hyperparathyroidism** (hī-pər-pahr-ah-thī-royd-ihzm) = abnormal condition of excessive parathyroid hormone secretion resulting in hypercalcemia.
- **hyperpituitarism** (hī-pər-pih-too-ih-tahr-ihzm) = condition of excessive secretion of the pituitary gland.
- **hyperthyroidism** (hī-pər-thī-royd-ihzm) = condition of excessive thyroid hormone. Signs of hyperthyroidism include increased metabolic rate, weight loss, polyuria, and polydipsia (Figure 11–8).

FIGURE 11–8 Cat with endocrine disease. Increased metabolic rate, weight loss, polyuria, and polydipsia may indicate hyperthyroidism or diabetes mellitus in cats.

FIGURE 11–9 Dog with hypothyroidism.

- **hypoadrenocorticism** (hī-pō-ahd-rĕn-ō-kōr-tih-kihz-uhm) = disorder caused by deficient adrenal cortex production of glucocorticoid, resulting in weakness, vomiting, and weight loss; also called **Addison's** (ahd-ih-sohnz) **disease**.
- **hypocrinism** (hī-pō-krī-nihzm) = condition of deficient gland secretion.
- **hypoglycemia** (hī-pō-glī-sē-mē-ah) = abnormally low blood glucose. The brain is the first organ affected by hypoglycemia because metabolism in the brain depends primarily on glucose. Signs of hypoglycemia include shakiness, anxiety, confusion, convulsions, seizures, and, if untreated, loss of consciousness.
- **hypogonadism** (hī-pō-gō-nahd-ihzm) = abnormal condition of deficient hormone secretion by the sex glands (ovaries in females; testes in males).
- **hypoparathyroidism** (hī-pō-pahr-ah-thī-royd-ihzm) = abnormal condition of deficient parathyroid hormone secretion resulting in hypocalcemia. **Hypocalcemia** (hī-pō-kahl-sē-mē-ah) is abnormally low levels of blood calcium. **Hypo-** is the prefix for deficient, **calc/i** is the combining form for calcium, and **-emia** is the suffix for blood condition. Hypercalcemia is abnormally high blood calcium levels because **hyper-** is the prefix meaning excessive.
- **hypophysitis** (hī-pō-fī-sī-tihs) = inflammation of the pituitary gland.
- **hypothyroidism** (hī-pō-thī-royd-ihzm) = condition of thyroid hormone deficiency. Signs of hypothyroidism include decreased metabolic rate, poor hair coat, lethargy, and increased sensitivity to cold (Figure 11–9). **Euthyroidism** (yoo-thī-royd-ihzm) is the condition of normal thyroid function. The prefix **eu-** means good, well, or easily.

- **insulinoma** (ihn-suh-lihn-ō-mah) = tumor of the islet of Langerhans of the pancreas.
- **pancreatitis** (pahn-krē-ah-tī-tihs) = inflammation of the pancreas (Figure 11–10a and b).
- **pheochromocytoma** (fē-ō-krō-mō-sī-tō-mah) = tumor of the adrenal medulla resulting in increased secretion of epinephrine and norepinephrine. **Phe/o** is the combining form for dusky; **chrom/o** is the combining form for color; **cyt/o** is the combining form for cell; **-oma** is the suffix meaning tumor. A pheochromocytoma is a tumor that takes on a dark (dusky) color because it is composed of chromaffin (or colored) cells.
- **pinealopathy** (pīn-ē-ah-lohp-ah-thē) = disorder of the pineal gland.
- **pituitarism** (pih-too-ih-tār-ihzm) = any disorder of the pituitary gland.
- **thymoma** (thī-mō-mah) = tumor of the thymus.
- **thyroiditis** (thī-roy-dī-tihs) = inflammation of the thyroid gland.
- **thyromegaly** (thī-rō-mehg-ah-lē) = enlargement of the thyroid gland.
- **thyrotoxicosis** (thī-rō-tohck-sih-kō-sihs) = abnormal life-threatening condition of excessive, poisonous quantities of thyroid hormones. The combining form **toxic/o** means poison.

PROCEDURES: ENDOCRINE SYSTEM

Procedures performed on the endocrine system include the following:

- **adrenalectomy** (ahd-rē-nahl-ehck-tō-mē) = surgical removal of one or both adrenal glands.

(a)

(b)

FIGURE 11–10 (a) Normal pancreas in a puppy. (b) Pancreas of a dog with diabetes mellitus.

- **chemical thyroidectomy** (thī-royd-ehck-tō-mē) = administration of radioactive iodine to suppress thyroid function; also called **radioactive iodine therapy**.
- **hypophysectomy** (hī-pohf-ih-sehck-tō-mē) = surgical removal of the pituitary gland.
- **lobectomy** (lō-behck-tō-mē) = surgical removal of a lobe or a well-defined portion of an organ.
- **pancreatectomy** (pahn-krē-ah-tehck-tō-mē) = surgical removal of the pancreas.
- **pancreatotomy** (pahn-krē-ah-toht-ō-mē) = surgical incision into the pancreas.
- **parathyroidectomy** (pahr-ah-thī-royd-ehck-tō-mē) = surgical removal of one or more parathyroid glands.
- **pinealectomy** (pīn-ē-ahl-ehck-tō-mē) = surgical removal of the pineal gland.
- **thymectomy** (thī-mehck-tō-mē) = surgical removal of the thymus.
- **thyroidectomy** (thī-royd-ehck-tō-mē) = surgical removal of all or part of the thyroid gland.

ABBREVIATIONS RELATED TO THE ENDOCRINE SYSTEM

- ACTH = adrenocorticotropic hormone
- ADH = antidiuretic hormone
- DI = diabetes insipidus
- DM = diabetes mellitus
- FSH = follicle-stimulating hormone
- GH = growth hormone
- GLU = glucose
- hCG = human chorionic gonadotropin
- ICSH = interstitial cell-stimulating hormone
- LH = luteinizing hormone
- LTH= luteotropin hormone
- MSH = melanocyte-stimulating hormone
- PTH = parathyroid hormone
- STH = somatotropin hormone
- T_3 = triiodothyronine
- T_4 = thyroxine
- TSH = thyroid-stimulating hormone

REVIEW EXERCISES

Multiple Choice

Choose the correct answer.

1. The gland known as the master gland that helps maintain the appropriate levels of hormone in the body is the
   a. hypothalamus
   b. pituitary gland
   c. thyroid gland
   d. pancreas

2. The chemical substance secreted by the posterior pituitary gland that stimulates uterine contractions during parturition is
   a. prolactin
   b. luteinizing hormone
   c. ACTH
   d. oxytocin

3. The regulator of the endocrine system is the
   a. thyroid gland
   b. calcitonin
   c. hypothalamus
   d. parathyroid gland

4. Thyromegaly is
   a. enlargement of the thyroid gland
   b. augmentation of the thymus
   c. dissolution of the parathyroid glands
   d. radioactive iodine treatment of the thyroid gland

5. Surgical removal of a well-defined portion of an organ is a
   a. sacculectomy
   b. lumpectomy
   c. lobectomy
   d. cystectomy

6. An aggregation of cells specialized to secrete or excrete materials not related to their function is a
   a. hormone
   b. gland
   c. hypoadenum
   d. hyperadenum

7. Hypoadrenocorticism, a disorder caused by deficient adrenal cortex production of glucocorticoid, also is known as
   a. Cushing's disease
   b. Graves' disease
   c. Addison's disease
   d. Langerhans' disease

8. A tumor of the islets of Langerhans of the pancreas is called
   a. diabetes mellitus
   b. diabetes insipidus
   c. ketoacidosis
   d. insulinoma

9. The chemical substance that helps control circadian rhythm is
   a. circadianin
   b. pinealin
   c. melatonin
   d. thymin

10. Excessive mammary development in males is called
    a. feminum
    b. gynecomastia
    c. gyneconium
    d. feminomastia

11. Which of the following is a function of the thyroid gland?
    a. secretes thymosin
    b. secretes thyroid-stimulating hormone
    c. secretes corticosteroids
    d. secretes triiodothyronin, thyroxine, and calcitonin

12. What is another name for the anterior lobe of the pituitary gland?
    a. hypophysis
    b. epiphysis
    c. adenohypophysis
    d. neurohypophysis

13. Which hormone maintains water balance in the body by increasing water reabsorption in the kidney?
    a. insulin
    b. epinephrine
    c. antidiuretic hormone
    d. FSH

14. What is the name of a laboratory technique used to determine the amount of a particular substance in a sample?
    a. stimulation test
    b. synthetic test
    c. uptake test
    d. assay

15. An abnormally low blood glucose level is known as
    a. hyperglycemia
    b. hypoglycemia
    c. glucosemia
    d. diabetes mellitus

16. The condition of normal thyroid function is known as
    a. thyrotoxicosis
    b. thyromegaly
    c. euthyroidism
    d. thyroiditis

17. Which hormone transports blood glucose to the cells?
    a. glucagon
    b. insulin
    c. vasopressin
    d. melatonin

18. Glucocorticoids are secreted by the
    a. adrenal cortex
    b. adrenal medulla
    c. parathyroid gland
    d. pancreas

19. The pituitary gland acts in response to stimuli from the
    a. hypothalamus
    b. thymus
    c. parathyroid glands
    d. thyroid gland

20. The word part that means to secrete or separate is
    a. endo
    b. -crine
    c. hormon/o
    d. aden/o

21. Which is the only hormone that can lower blood glucose?
    a. GH
    b. insulin
    c. glucagon
    d. oxytocin

22. In the term *hypoglycemia*, the prefix hypo means
    a. excessive
    b. beside
    c. between
    d. below

23. Which hormone reduces the volume of urine produced by the kidneys and is also known as vasopressin?
    a. FSH
    b. LH
    c. ADH
    d. ACTH

24. Which term has a word part that means black?
    a. oxytocin
    b. melatonin
    c. pineal
    d. acromegaly

25. An agent, act, or influence that produces a reaction is a(n)
    a. stimulus
    b. electrolyte
    c. synthetic
    d. gamete

Matching
Match the endocrine term in Column I with the correct description in Column II.

Column I

1. _____ adrenals
2. _____ thyroid
3. _____ thymus
4. _____ pancreas
5. _____ pituitary
6. _____ parathyroid
7. _____ pineal
8. _____ gonads
9. _____ thyroidectomy
10. _____ hyperthyroidism
11. _____ diabetes mellitus
12. _____ diabetes insipidus
13. _____ adrenopathy
14. _____ acidosis
15. _____ glucagon
16. _____ hypocalcemia
17. _____ insulinoma
18. _____ hypercrinism

Column II

a. butterfly-shaped gland on either side of the larynx

b. gland located dorsal to the sternum

c. contains specialized cells secrete hormones affecting sugar and starch metabolism

d. secretes melatonin

e. gamete-producing glands

f. small gland at the base of the brain

g. secretes hormone that reduces bone calcium levels and regulates phosphorus

h. two small glands located on top of each kidney

i. tumor of the islet of Langerhans cells of the pancreas

j. abnormally low blood calcium levels

k. insufficient antidiuretic hormone

l. insufficient secretion of insulin or recognition of insulin

m. surgical removal of the thyroid gland

n. condition of excessive gland secretion

o. condition of excessive thyroid hormone

p. disease of the adrenal gland

q. abnormal condition of low pH

r. hormone that increases blood glucose

Fill in the Blanks
Write the correct term in the space provided.

1. Thyr/o and thyroid/o mean _____.

2. Adren/o and adrenal/o mean _____.

3. -tropic means _____.

4. Lute/o means _____.

5. Somat/o means _____.

True or False

If the statement is true, write T on the line. If the statement is false, write F on the line.

1. _____Epinephrine decreases blood pressure, heart rate, and blood glucose.

2. _____The thyroid gland is a gland predominant in young animals.

3. _____The "S" in MSH stands for stimulating.

4. _____Insulin and glucagon have opposite effects on blood glucose levels.

5. _____Gonadotropic hormones affect the ovaries or testes.

6. _____Circadian refers to events occurring within a 24-hour period.

7. _____Acidosis is an abnormal condition of high pH.

8. _____Synthetic pertains to artificial production.

9. _____A laboratory technique used to determine the amount of a particular substance in a sample is a suppression test.

10. _____An example of a test to analyze thyroid function is a radioactive iodine uptake test.

Spell Check

Circle the term that is spelled correctly.

1. gland that has both exocrine and endocrine function:	pancrease	pancreus	pancreas
2. substance that increases urine secretion:	dyuretic	diyuretic	diuretic
3. gland located at the base of the brain:	pitwoitary	pituitary	pitootary
4. part of the brain that controls secretions of the pituitary gland:	hypothalamus	hypothalmus	hypothalmis
5. gamete-producing glands:	gonids	gonads	goknads

Word Scramble

Use the definitions to unscramble the terms relating to the endocrine system.

1. condition of building up   mlsioaanb _____

2. condition of breaking down   mlsioaactb _____

3. abnormally elevated blood glucose   aieyeyhprglcm _____

4. inflammation of the pancreas   iittseaacrnp _____

5. gamete-producing   naodg _____

6. surgical removal of the thyroid gland   dyorihtetcmoy _____

7. study of the hormonal system   glnooyndeocri _____

8. production of new glucose   ssgiluncoeogeen _____

9. artificially produced   nttheicys _____

10. abnormal condition of deficient thyroid hormone secretion   smidioryhtophy _____

Crossword Puzzle

Endocrine Terms

Supply the correct term in the appropriate space for the definition listed.

Across

1 condition of normal thyroid function

2 pertaining to artificial production

7 surgical incision into the pancreas

9 prefix meaning excessive, above normal, or elevated

10 combining form for gland

11 sex cell

14 abnormally low blood calcium levels

17 chemical substance produced by an organ that is transported by the bloodstream to regulate another organ

19 steroid produced by the adrenal cortex

21 prefix meaning deficient, less than normal, or decreases

22 agent, act, or influence that produces a reaction

23 abnormal condition of low pH

24 substance that stimulates blood vessel contraction and increases blood pressure

Down

1 disease of hormone-producing system

3 funnel-shaped passage or opening

4 condition of building up

5 drowsiness or indifference

6 condition of excessive gland secretion

8 tumor of the adrenal medulla

9 surgical removal of the pituitary gland

12 suffix meaning affinity for

13 another term for hyperadrenocorticism

15 condition of breaking down

18 suffix meaning to secrete or separate

20 combining form for body

Label the Diagrams

Label the diagram in Figure 11–11.

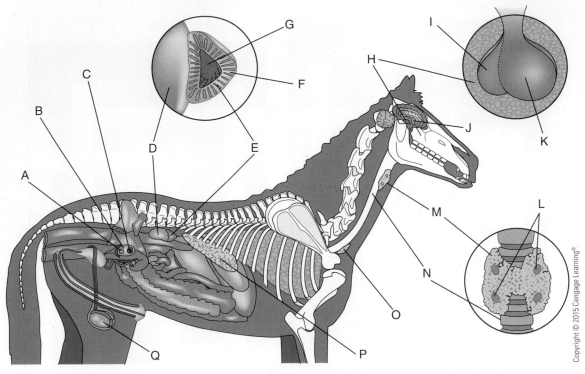

FIGURE 11–11 Endocrine organs. Label the endocrine organs and related structures. Provide the combining form for each endocrine gland.

Case Studies

Define the underlined terms in each case study.

A 7-yr-old M/N golden retriever, Goldie, was presented to the clinic for bite wounds from another dog in the household. After the dog's vaccination status was checked, he was examined. The dog's vital signs included T = 103°F, HR = 120 BPM, RR = 42 breaths/min. Upon examination, it was noted that the dog weighed 85#, the haircoat had areas of alopecia, and the dog was lethargic. The wounds from the dog fight consisted of 2 lacerations; 1 laceration was 4 cm long located on the lateral side of the L forelimb proximal to the elbow, and the other laceration was 2 cm long located on the medial surface of the L hock. The owner was questioned as to whether these two dogs usually got along in the house, and the owner stated that they had before Goldie's recent history of lethargy and unwillingness to go out for walks during the past month. The owner thought that Goldie did not want to go out on walks because of the change of seasons and the cold, damp weather. Goldie was admitted to the clinic and anesthetized, and the lacerations were cleaned and sutured. Antibiotics were dispensed 2 T BID for 7 d. Goldie came back for suture removal in 10 days. The lacerations had healed; but hair had not begun to regrow in the areas of the lacerations, and more areas of alopecia were noted. A thyroid panel was recommended to assess Goldie's thyroid function. Blood was drawn, and a thyroid panel was run. The thyroid panel revealed that Goldie was hypothyroid, and thyroid supplementation was begun.

1. yr   _____

2. M/N   _____

3. T   _____

4. HR   _____

5. BPM _____

6. # _____

7. alopecia _____

8. lethargic _____

9. cm _____

10. lateral _____

11. proximal _____

12. elbow _____

13. medial _____

14. hock _____

15. anesthetized _____

16. BID _____

17. d _____

18. hypothyroid _____

A 15-yr-old F/S DSH was brought to the clinic for weight loss. Examination revealed that the cat was tachycardic, emaciated, and lethargic. Vital signs were T = 102°F, HR = 260 BPM, RR = 60 breaths/min. Abdominal palpation revealed a moderately full urinary bladder, slightly small kidneys, and gas-filled loops of bowel. Palpation of the neck revealed enlargement of the thyroid glands. Blood was collected from the jugular vein for a CBC, chem screen, and T_4 level. Blood work revealed a slightly elevated BUN and creatinine, hypercholesterolemia, and elevated T_4 levels. It was recommended that the cat receive radioactive iodine treatments; however, the options of thyroidectomy and pharmaceutical management also were discussed. The cat was scheduled for radiation treatments at a referral center.

19. yr _____

20. F/S _____

21. DSH _____

22. tachycardic _____

23. emaciated _____

24. lethargic _____

25. T _____

26. HR _____

27. BPM _____

28. RR _____

29. palpation _____

30. vein _____

31. CBC _____

32. chem screen _____

33. T$_4$ _____

34. BUN _____

35. creatinine _____

36. hypercholesterolemia _____

37. radioactive iodine treatments _____

38. thyroidectomy _____

Critical Thinking Exercise

The following story and questions contain words and concepts from this chapter and are designed to stimulate critical thinking through class discussion, Internet searches, or brief writing responses.

Snickers, a 12-yr-old F/S DLH cat, presents to the veterinary clinic with a voracious appetite, weight loss, and PU/PD. Within the past 12 hours Snickers has started vomiting. On PE, she was tachycardic, panting, and appeared emaciated (even though her owner reported that her appetite was increased over normal). The veterinarian ordered a CBC, chemistry panel, and UA to determine the cause of Snicker's illness.

Define the following terms or abbreviations:

1. Snickers is described as "emaciated." Describe how she would look.

2. List Snickers' clinical signs. Provide definitions of any abbreviations used.

3. What does *voracious* mean?

4. Based on the meaning of voracious, what does not make sense about Snickers' weight loss?

5. Examine the following UA results (Figure 11–12). List any abnormalities from this report.

Urinalysis

Specimen	Free catch
Color	Yellow
Appearance	Cloudy
Specific gravity	1.043
pH	7.0
Protein	Negative
Glucose	3+
Ketones	1+
Bilirubin	Negative
Blood	Negative
Urobilinogen	Negative
WBCs/hpf	0–2
RBCs/hpf	0–2
Epithelial cells/hpf	0–2
Other	Amorphous debris

Copyright © 2015 Cengage Learning®

FIGURE 11–12 Urinalysis report.

The CBC (complete blood count) was normal and the chemistry panel showed that the cat had hyperglycemia.

6. Based on the UA and chemistry panel results, what may be a cause of this cat's clinical signs?

7. In the condition listed for question 6, do you think this cat's insulin levels would be high or low? Why?

8. Why would you think this cat may be more prone to developing a UTI?

Chapter 12
1 + 1 = 3 (OR MORE)

Objectives

Upon completion of this chapter, the reader should be able to:

- Identify the structures of the male and female reproductive systems

- Describe the functions of the male and female reproductive systems

- Use medical terminology to describe the estrous cycle in females

- Recognize, define, spell, and pronounce the terms related to the diagnosis, pathology, and treatment of the reproductive systems

- Interpret the meaning of abbreviations of the male and female reproductive systems

THE FUNCTIONS OF THE REPRODUCTIVE SYSTEM

The **reproductive** (rē-prō-duhck-tihv) **system** is responsible for producing offspring. The reproductive system needs both male- and female-specific organs to complete offspring production. The term **theriogenology** (thēr-ē-ō-jehn-ohl-ō-jē) is used to describe animal reproduction or the study of producing beasts. **Theri/o** is the combining form for beast, **gen/o** is the combining form for producing, and **-logy** is the suffix meaning to study.

The reproductive organs, whether male or female, are called the **genitals** (jehn-ih-tahlz), or **genitalia** (jehn-ih-tā-lē-ah). The combining form **genit/o** refers to the organs of reproduction. The genitalia include external and internal organs.

The male reproductive system will be described first, followed by the female reproductive system.

FUNCTIONS OF THE MALE REPRODUCTIVE SYSTEM

The functions of the male reproductive system are to produce and deliver sperm to the egg to create life.

STRUCTURES OF THE MALE REPRODUCTIVE SYSTEM

The structures of the male reproductive system produce sperm, transport sperm out of the body, and produce hormones.

Scrotum

The **scrotum** (skrō-tuhm), or **scrotal sac**, is the external pouch that encloses and supports the testes. The scrotum encloses the testes outside the body so that the testes are at a temperature lower than body temperature. This lower temperature is needed for sperm development. *Scrotum* is the Latin term for bag. The combining form **scrot/o** means scrotum.

The area between the scrotum and the anus in males is the **perineum** (pehr-ih-nē-uhm). **Perine/o** is the combining form for the area between the scrotum (or vulva in females) and anus.

Testes

The **testes** (tehs-tēz), or **testicles** (tehs-tih-kuhlz), are the male sex glands that produce spermatozoa. Testes refer to glands, and **testis** (tehs-tihs) refers to a single gland.

The orientation of the testes in the scrotum varies between species. Some species have open inguinal rings that allow the testes to be withdrawn from the scrotum and into the abdomen. Sex glands are called **gonads** (gō-nahdz). The combining forms for testes are **orch/o**, **orchi/o**, **orchid/o**, **test/o**, and **testicul/o** (Figure 12–1).

The testes develop in the fetal abdomen and descend into the scrotum before birth. The testes are suspended in the scrotum by the spermatic cord. The testicle is divided into compartments that contain coiled tubes called the **seminiferous tubules** (seh-mih-nihf-ər-uhs too-buhlz), and cells between the spaces are called **interstitial** (ihn-tər-stih-shahl) cells. The interstitial cells of the testes are called the **Leydig's** (lih-dihgz) **cells**. Leydig's cells have endocrine function. **Sertoli** (sihr-tō-lē) **cells** are specialized cells in the testes that support and nourish sperm growth. The seminiferous tubules are channels in the testes in which sperm are produced and through which the sperm leave the testes. **Sperm** (spərm), or **spermatozoa** (spər-mah-tō-zō-ah), are the male gametes, or sex cells. **Spermatozoon** (spər-mah-tō-zō-uhn) is one gamete. The combining forms for spermatozoa are **sperm/o** and **spermat/o**.

Ejaculated semen occasionally is evaluated microscopically to determine whether the spermatozoa have normal morphology and motility and whether they are present in adequate numbers (Figure 12–2a and b). A spermatozoon has a head, midpiece, and tail (Figure 12–3). The head contains the nucleus. At the top of the head is a structure called the **acrosome** (ahck-rō-zōm), which contains enzymes that allow the spermatozoon to penetrate the ovum. The midpiece contains mitochondria to provide energy to the sperm. The tail is actually a flagellum, providing movement for the spermatozoon to reach the ovum. **Spermatogenesis** (spər-mah-tō-jehn-eh-sihs) is production of male gametes.

Epididymis

The seminiferous tubules join together to form a cluster. Ducts emerge from this cluster and enter the epididymis. The **epididymis** (ehp-ih-dihd-ih-mihs)

Gamete comes from the Greek terms *gametes*, which means husband, and *gamete*, which means wife. *Gamein* is Greek for to marry. Johann Gregor Mendel was the first to apply the term *gamete* in biology to mean sex cells. Gametes are the sperm in males and ovum (egg) in females.

FIGURE 12–1 Male genital organs. (a) Right testis and epididymis, (b) ductus deferens, (c) urinary bladder, (d) ureter, (e) penis and extrapelvic urethra, (f) pelvic symphysis, and (g) sigmoid flexure.

Male Genital Organs

☐ Ampulla*
☐ Prostate Gland
☐ Vesicular Gland**
■ Bulbourethral Gland

* glandular part of ductus deferens in pig
** seminal vesicle in horse

(a)　　**(b)**

FIGURE 12–2 Sperm with abnormal shapes are not desirable. (a) Normal. (b) Defective sperm with bent tails.

FIGURE 12–3 Parts of sperm.

Acrosome

Head

Midpiece

Tail

is the tube at the upper part of each testis that secretes part of the semen, stores semen before ejaculation, and provides a passageway for sperm. The epididymis is divided into head (or caput), body, and tail portions. The epididymis runs down the length of the testicle, turns upward, and becomes a narrower tube called the ductus deferens (Figure 12–4). Sperm are collected in the epididymis, where they become **motile** (mō-tihl), or capable of spontaneous motion. The combining form for epididymis is **epididym/o**.

Ductus Deferens

The **ductus deferens** (duhck-tuhs dehf-ər-ehnz) is a tube connected to the epididymis that carries sperm into the pelvic region toward the urethra. Each ductus deferens is encased by the spermatic cord. (The spermatic cord also encases nerves, blood and lymph vessels, and the cremaster muscle along with the ductus deferens.) The ductus deferens is the excretory duct of the testes. The ductus deferens in swine has a glandular portion called the **ampulla** (ahmp-yoo-lah). An ampulla is an enlarged part of a tube or canal.

Accessory Sex Glands

The male accessory sex glands include the seminal vesicles, prostate gland, and bulbourethral glands. Not all glands are present in all species (Table 12–1). The accessory sex glands add secretions to the sperm and flush urine from the urethra before sperm enter it.

The **seminal vesicles** (sehm-ih-nahl vehs-ih-kuhls), or **vesicular** (vehs-ih-koo-lahr) **glands**, are two glands that open into the ductus deferens where it joins the

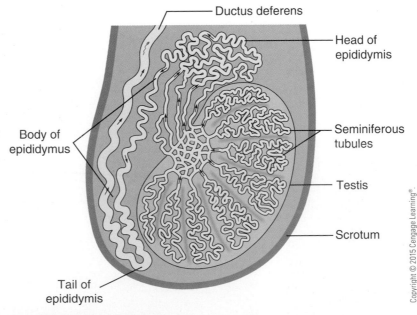

Ductus deferens

Head of epididymis

Seminiferous tubules

Testis

Scrotum

Body of epididymus

Tail of epididymis

FIGURE 12–4 Cross section of scrotal structures.

TABLE 12–1 Male Accessory Glands of Different Species

	Prostate	Vesicular Gland or Seminal Vesicle	Bulbourethral
dog	+	–	–
cat	+	–	+
pig	+	+	+
ruminants	+	+	+
horse	+	+	+

Note: In some species, the ampulla of the ductus deferens is glandular (as in horses).

Copyright © 2015 Cengage Learning®.

FIGURE 12–5 The prepuce is the retractable fold of skin covering the glans penis.

Courtesy of Terri Raffel, CVT

urethra. The seminal vesicles secrete a thick, yellow substance that nourishes sperm and adds volume to the ejaculated semen. **Semen** (sē-mehn) is the ejaculatory fluid that contains sperm and the secretions of the accessory sex glands. **Semin/i** is the combining form for semen.

The **ejaculatory** (ē-jahck-yoo-lā-tōr-ē) **duct** is formed by the union of the ductus deferens and the duct from the seminal vesicles. The ejaculatory duct passes through the prostate and enters the urethra.

The **prostate** (proh-stāt) **gland** is a single gland that surrounds or is near the urethra and may be well defined or diffuse depending on the species. The prostate gland secretes a thick fluid that aids in the motility of sperm. The combining form **prostat/o** means prostate gland.

The **bulbourethral** (buhl-bō-yoo-rē-thrahl) **glands** are two glands located on either side of the urethra. The bulbourethral glands secrete a thick mucus that acts as a lubricant for sperm. These glands are called **Cowper's** (cow-pərz) **glands** in humans.

Urethra

The **urethra** (yoo-rē-thrah) is a tube passing through the penis to the outside of the body; it serves both

Seminal vesicles is the term used in horses; *vesicular gland* is used in the other species when the gland is present.

reproductive and urinary systems. The combining form for urethra is **urethr/o**.

Penis

The **penis** (pē-nihs) is the male sex organ that carries reproductive and urinary products out of the body. The glans penis is the distal part of the penis on which the urethra opens. The **prepuce** (prē-pyoos) is the retractable fold of skin covering the glans penis (Figure 12–5). The prepuce is sometimes called the **foreskin**. Dogs have an **os penis** (ohs pē-nihs), which is a bone encased in the penile tissue. All species except the cat have a cranioventrally directed penis. The combining forms **pen/i** and **priap/o** mean penis.

The penis is composed of erectile tissue that upon sexual stimulation fills with blood (under high pressure) and causes an erection. Some species, such as ruminants and swine, achieve an erection by straightening of the **sigmoid flexure** (sihg-moyd flehck-shər), an S-shaped bend in the penis. Other species, such as equine and canine, have a penis with almost all erectile tissue. Erection in these animals is caused by blood engorgement of the erectile tissue (Figure 12–6).

FUNCTIONS OF THE FEMALE REPRODUCTIVE SYSTEM

The functions of the female reproductive system are to create and support new life.

Colon

Prostate gland

Seminal vesicle

Urinary bladder

Kidney

Bulbourethral gland

Ureter

Testicular artery, vein, nerve

Ductus deferens

Body of epididymis

Glans penis

Prepucial fold

Retractor penis muscle

Urethra

Tail of epididymis

Scrotum

Head of epididymis

Testis

Urethral process

Copyright © 2015 Cengage Learning®

FIGURE 12–6 Reproductive tract of a stallion.

STRUCTURES OF THE FEMALE REPRODUCTIVE SYSTEM

The structures of the female reproductive system produce eggs, transport eggs for fertilization and implantation, house the embryo, deliver the fetus, and produce hormones.

Ovaries

The **ovaries** (ō-vah-rēz) are a small pair of organs located in the caudal abdomen (Figure 12–7). An ovary is the female gonad that produces estrogen; progesterone; and **ova** (ō-vah), or **eggs**. The ovaries contain many small sacs called **graafian follicles** (grahf-ē-ahn fohl-lihck-kuhlz) (Figure 12–8). Each graafian follicle contains an ovum. The ova develop in the ovaries and are expelled (ovulated) when the egg matures. The combining forms **ovari/o** and **oophor/o** mean female gonad. The combining forms **oo/o**, **ov/i**, and **ov/o** mean egg. An egg cell is an **oocyte** (ō-ō-sīt).

Uterine Tubes

The **uterine tubes** are paired tubes that extend from the cranial portion of the uterus to the ovary (although they are not attached to the ovary). The uterine tubes

are also called **oviducts** (ō-vih-duhckts) and **fallopian** (fah-lō-pē-ahn) **tubes**. The combining form for uterine tubes is **salping/o**, which means tube.

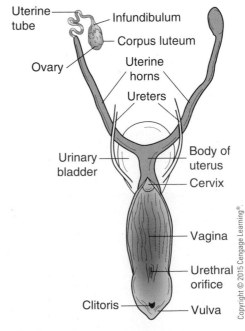

Uterine tube

Infundibulum

Corpus luteum

Ovary

Uterine horns

Ureters

Urinary bladder

Body of uterus

Cervix

Vagina

Urethral orifice

Clitoris

Vulva

Copyright © 2015 Cengage Learning®

FIGURE 12–7 Reproductive tract of a bitch.

FIGURE 12–8 The follicle, which appears as a clear blister on the surface of the ovary, secretes a hormone called estrogen.

FIGURE 12–9 Bovine reproductive tract.

The distal end of each uterine tube is a funnel-shaped opening called the **infundibulum** (ihn-fuhn-dihb-yoo-luhm). The infundibulum contains fringed extensions called **fimbriae** (fihm-brē-ā) that catch ova when they leave the ovary. The fimbriae are not attached to the ovaries.

The proximal end of each uterine tube is connected to the uterine horns. The uterine tubes carry ova from the ovary to the uterus. The uterine tubes also transport sperm traveling up from the vagina and uterus. **Fertilization** (fər-tihl-ih-zā-shuhn), or egg and sperm union, usually occurs in the uterine tube.

Uterus

The **uterus** (yoo-tər-uhs) is a thick-walled, hollow organ with muscular walls and a mucous membrane lining that houses the developing embryo in pregnant females (Figure 12–9). The uterus is situated dorsal to the urinary bladder and ventral to the rectum. The combining forms **hyster/o**, **metri/o**, **metri/o**, and **uter/o** mean uterus.

The uterus consists of three parts:

- **cornus** (kōr-nuhs) = horn. The cranial end of the uterus has two horns that travel toward the uterine tubes. **Cornu** means horn. Some animals are **bicornuate** (bī-kōrn-yoo-āt), which means having two large, well-defined uterine horns.
- **corpus** (kōr-puhs) = body. The middle portion of the uterus. **Corpu** means body.
- **cervix** (sihr-vihckz) = neck. The caudal portion of the uterus that extends into the vagina. **Cervic/o** means neck.

The uterus is composed of three major tissue types:

- **perimetrium** (pehr-ih-mē-trē-uhm) = membranous outer layer of the uterus. **Peri-** is the prefix for surrounding.
- **myometrium** (mī-ō-mē-trē-uhm) = muscular middle layer of the uterus. **My/o** is the combining form for muscle.
- **endometrium** (ehn-dō-mē-trē-uhm) = inner layer of the uterus. **Endo-** is the prefix meaning within.

Cervix

The **cervix** (sihr-vihckz) is the caudal continuation of the uterus and marks the cranial extent of the vagina. The cervix contains ringlike smooth muscle called **sphincters**. The main function of the cervix is to prevent foreign substances from entering the uterus. The cervix usually is closed tightly except during estrus, when it relaxes to allow entry of sperm. During pregnancy, the cervix is closed with a mucous plug. The mucous plug is released near parturition to allow fetal passage. The combining form **cervic/o** means neck or necklike structure.

Vagina

The **vagina** (vah-jī-nah) is the muscular tube lined with mucosa that extends from the cervix to the outside of the body. The vagina accepts the penis during copulation and serves as a passage for semen into the body and excretions and offspring out of the body. The combining forms **colp/o** and **vagin/o** mean vagina.

A membranous fold of tissue may partially or completely cover the external vaginal orifice. This fold is called the **hymen** (hī-mehn). An **orifice** (ōr-ih-fihs) is an entrance or outlet from a body cavity.

Vulva

The **vulva** (vuhl-vah), also known as the female external genitalia, or **pudendum** (pyoo-dehn-duhm), is the external opening to the urogenital tract and consists of the vaginal orifice, vestibular glands, clitoris, hymen, and urethral orifice. The combining forms for the vulva are **vulv/o** and **episi/o**. The **perineum** (pehr-ih-nē-uhm) is the region between the vaginal orifice and anus in females.

The **labia** (lā-bē-ah) are the fleshy borders or edges of the vulva and are occasionally called the **lips.** In animals, the vulva contains simple lips, whereas humans have major and minor labia. The **vaginal orifice** is the entrance from the vagina to the outside of the body. The **vestibular glands** (also known as **Bartholin's glands** in primates) are found in bovine, feline, and occasionally ovine species. The vestibular glands secrete mucus to lubricate the vagina. The **clitoris** (kliht-ə-rihs) is the sensitive erectile tissue of females located in the ventral portion of the vulva. The clitoris is the analog of the glans penis of the male. The **urethral orifice** is found where the vagina and vulva join and is sometimes associated with a vestigial hymen (Figure 12–10).

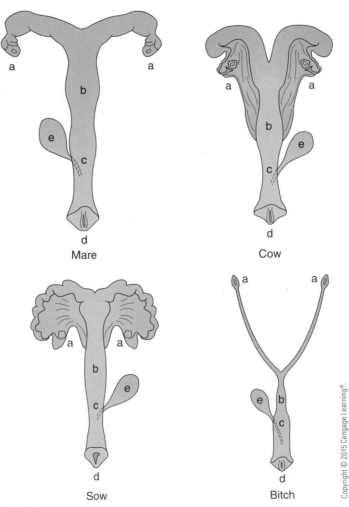

FIGURE 12–10 Female genital organs and urinary bladder: (a) ovaries, (b) uterus, (c) vagina, (d) vulva, and (e) urinary bladder.

Mammary Glands

The **mammary** (mahm-mah-rē) **glands** are milk-producing glands in females. The number of mammary glands varies with the species: the mare, ewe, and doe (goat) have two; cows have four; sows have six or more pairs; and bitches and queens have four or more pairs. In litter-bearing species, the glandular structures usually are paired, located on the ventral surface, and called mammary glands, or **mammae** (mahm-ā). Each singular gland (mamma) is associated with one nipple. In large animals, the mammary gland is called an **udder** (uh-dər), is located in the inguinal area, and has two or four functional teats. The nipple area is called a **teat** (tēt) in large animals. In cows, the four mammae are called **quarters** (kwahr-tərz) (Figure 12–11).

Mammary glands are composed of connective and adipose tissue organized into lobes and lobules that contain milk-secreting sacs called **alveoli** (ahl-vē-ō-lī). Each lobe drains toward the teat or papilla via a **lactiferous** (lahck-tihf-ər-uhs) **duct**. **Lact/i** means milk. The lactiferous ducts come together to form the lactiferous, or teat, sinus. The **lactiferous sinus** is composed of the gland cistern (within the gland) and the teat cistern (within the teat). Milk travels from the gland cistern into the teat cistern. From the teat cistern, milk empties into the **papillary duct**, which is commonly called the **streak canal**.

The combining forms for mammary glands are **mamm/o** and **mast/o**.

THE ESTROUS CYCLE

The **estrous** (ehs-truhs) **cycle**, sometimes called the heat cycle, occurs at the onset of puberty and continues throughout an animal's life. The ability to sexually reproduce begins at puberty and varies among species. The estrous cycle prepares the uterus to accept a fertilized ovum.

The female reproductive system functions on cyclic intervals. Hormones secreted from the anterior pituitary gland and ovary control the estrous cycle. Although there is species variation during estrous cycle phases, the basic patterns are the same.

The estrous cycle starts when ova develop in ovarian follicles. One or more follicles continue to develop until they reach the ripened follicle size (called the graafian follicle). The graafian follicle ruptures, a process called **ovulation** (ohv-yoo-lā-shuhn). The ovum is expelled from the ovary into the uterine tube. The ruptured follicle continues to grow and becomes filled with a yellow substance. The yellow ruptured follicle is called the **corpus**

(a)

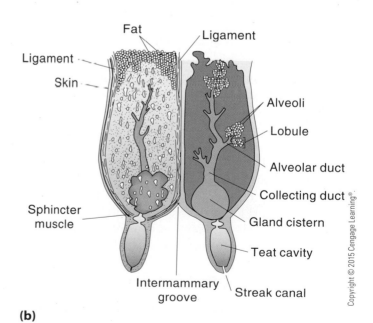

(b)

FIGURE 12–11 (a) Cross section of cow reproductive system lateral view. (b) Cross section of an udder cranial view.

BIOLOGY BONUS

When an animal produces milk, the demand for calcium is great and at times may exceed the amount that a female can mobilize from calcium reserves. At or near parturition, the onset of lactation results in the sudden loss of calcium from blood into milk. The sudden decline of serum calcium levels, especially in high-producing dairy cows, can interfere with muscle and nerve function, causing weakness (difficulty standing), anorexia, ataxia, and eventually heart failure. This condition is called parturient paresis, more commonly known as milk fever. "Fever" is a misnomer since the animal's temperature during this disease is usually below normal. Milk fever can be avoided with nutritional management of calcium prior to parturition.

luteum (kōr-puhs loo-tē-uhm), or yellow body. The corpus luteum is abbreviated CL. The CL secretes progesterone. If the ovum is fertilized, the CL will continue to secrete progesterone to prevent future estrous cycling. If the ovum is not fertilized, the CL will shrink and reduce its progesterone secretion, and a new estrous cycle begins. The stage of the estrous cycle in which the graafian follicle is present is called the **follicular** (fohl-ihck-yoo-lahr) **phase**. Estrogen is the predominant hormone during the follicular phase. The stage of the estrous cycle in which the corpus luteum is present is called the **luteal** (loo-tē-ahl) **phase**. Progesterone is the predominant hormone during the luteal phase.

The estrous cycle is also divided into phases. The phases of the estrous cycle are as follows:

- **proestrus** (prō-ehs-truhs) = period of the estrous cycle before sexual receptivity. The prefix **pro-** means before. Proestrus involves the secretion of follicle-stimulating hormone (FSH) by the anterior pituitary gland, which causes the follicles to develop in the ovary. FSH stimulates ovarian release of estrogen, which helps prepare the reproductive tract for pregnancy.
- **estrus** (ehs-truhs) = period of the estrous cycle in which the female is receptive to the male. During estrus, FSH levels decrease and luteinizing hormone (LH) levels increase, causing the graafian follicle to rupture and release its egg (ovulation). **Ovulation** (ohv-yoo-lā-shuhn) occurs, and the animal is said to be in **heat**. This also is called **standing heat**.
- **metestrus** (meht-ehs-truhs) = short period of the estrous cycle after sexual receptivity. The CL forms and produces progesterone during this phase. Progesterone ensures proper implantation and maintenance of pregnancy. If an animal is not pregnant, the CL will decrease in size and

become a **corpus albicans** (kōr-puhs ahl-bih-kahnz), or white body. Metestrus may be followed by diestrus, estrus, pregnancy, or false pregnancy. Metestrus is more commonly used to describe the estrous cycle of cattle; the term is rarely used in dogs, cats, and horses.

- **diestrus** (dī-ehs-truhs) = period of the estrous cycle after metestrus. This short phase of inactivity and quietness is seen in polyestrous animals before the onset of proestrus.
- **anestrus** (ahn-ehs-truhs) = period of the estrous cycle when the animal is sexually quiet. This long phase of quietness is seen in seasonally polyestrous and seasonally monestrous animals.

What do the Following Terms Mean?

- **monestrous** (mohn-ehs-truhs) = having one estrous or heat cycle per year.
- **polyestrous** (pohl-ē-ehs-truhs) = having more than one estrous or heat cycle per year.
- **spontaneous ovulators** (spohn-tā-nē-uhs oh-vū-lā-tōrz) = animals in which ovum release occurs cyclically.
- **induced ovulators** (ihn-doosd oh-vū-lā-tōrz) = animals in which the ovum is released only after copulation; also called **reflex ovulators**. Examples include cats, rabbits, ferrets, llamas, and mink.
- **seasonally** = pertaining to a specific time of year. For example, queens are seasonally polyestrous. The queen will have multiple cycles from January through October (varies depending on photoperiod and geographic location).

MATING, PREGNANCY, AND BIRTH

Mating

For reproduction to occur, the male and female of the species must **copulate** (kohp-yoo-lāt) to allow the sperm from the male to be transferred into the female. Copulation and **coitus** (kō-ih-tuhs) are terms that mean sexual intercourse. The combining forms **copulat/o** and **coit/o** mean joining together.

Intromission (ihn-trō-mihs-shuhn) is insertion of the penis into the vagina. During coitus, the male **ejaculates** (ē-jahck-yoo-lātz) into the female's vagina. Ejaculate is to release semen during copulation. **Ejaculat/o** means to throw or hurl out. Sperm travel through the vagina, into the uterus, and into the uterine tube. When a sperm penetrates the ovum that is descending down the uterine tube, fertilization occurs. **Fertilization** (fər-tihl-ih-zā-shuhn) is the union of ovum and sperm (Figure 12–12). If more than one ovum is passing down the uterine tube when sperm are present, multiple fertilizations may take place.

Afterbirth

The **placenta** (plah-sehn-tah) is the female organ of mammals that develops during pregnancy and joins mother and offspring for exchange of nutrients,

Copulation and **coitus** are two words used to describe sexual intercourse. The following terms also are used in reference to mating:

- **mount** (mownt) = preparatory step to animal mating that involves one animal climbing on top of another animal or object; used as an indicator of heat (Figure 12–13).
- **tie** (tī) = period of copulation between a male and female canine during which the two animals are locked together by penile erectile tissue.
- **conception** (kohn-sehp-shuhn) = beginning of a new individual resulting from fertilization. The fertilized egg is called a **zygote** (zī-gōt). Division of sex cells in which the cell receives half the chromosomes from each parent is called **meiosis** (mī-ō-sihs). **Implantation** (ihm-plahn-tā-shuhn) is the attachment and embedding of the zygote in the uterus. The developing zygote after implantation is called an **embryo** (ehm-brē-ō). An unborn animal is called a **fetus** (fē-tuhs); this term is used more toward the end of pregnancy. Pregnancy is the time period between conception and parturition.

oxygen, and waste products. The placenta also is called the **afterbirth**. The **umbilical** (uhm-bihl-ih-kuhl) **cord** is the structure that forms where the fetus communicates with the placenta. The **umbilicus** (uhm-bihl-ih-kuhs) is the structure that forms on the abdominal wall where the umbilical cord was connected to the fetus.

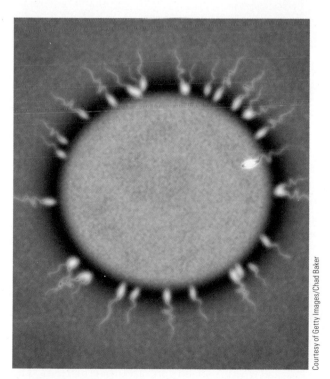

FIGURE 12–12 The outer membrane of the egg must be dissolved before the sperm can enter the egg.

FIGURE 12–13 Cattle mounting.

FIGURE 12–14 The relationship of a fetus with the umbilical cord leading to the placenta.

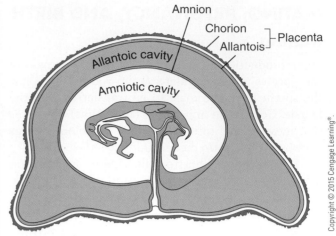

(a)

The umbilicus also is called the **navel** (nā-vuhl) (Figure 12–14). **Umbilic/o** is the combining form for navel.

The placenta and its associated structures are called the fetal membranes. The innermost membrane enveloping the embryo in the uterus is called the **amnion** (ahm-nē-ohn). The amnion forms the amniotic cavity and protects the fetus by engulfing it in amniotic fluid. The amnion may be called the **amniotic sac** or **bag of waters**. The **allantois** (ahl-ahn-tō-ihs) is the innermost layer of the placenta. It forms a sac between itself and the amnion, where fetal waste products accumulate. The **chorion** (kōr-ē-ohn) is the outermost layer of the placenta (Figure 12–15).

The ruminant placenta has elevations on it that are located on the maternal or fetal surface. The **cotyledon** (koht-eh-lē-dohn) is the elevation of the ruminant

(b)

FIGURE 12–15 (a) Equine placenta and fetus. (b) Equine fetus within placenta.

placenta that is on the fetal surface and adheres to the maternal caruncle. Cotyledons are also called **buttons**. The **caruncle** (kahr-uhnck-uhl) is the fleshy mass on the maternal ruminant placenta that attaches to the fetal cotyledon. (Think c**a**runcle = m**a**ternal; cotyl**e**don = f**e**tus.) Together the caruncle and cotyledon form the **placentome** (plahs-ehn-tohm) (Figures 12–16 and 12–17; also see Figure 12–25b).

Pregnancy

Pregnancy (prehg-nahn-sē) is the condition of having a developing fetus in the uterus and is the time period between conception and parturition. The

Twins

Twins are two offspring born from the same gestational period. Fraternal twins are two offspring born during the same labor resulting from fertilization of separate ova by separate sperm. Maternal twins are two offspring born during the same labor resulting from fertilization of a single ovum by a single sperm. (The fertilized egg separates into two parts.) The term *twin* is not used in litter-bearing species, even if only two offspring are produced.

FIGURE 12–16 Bovine placenta. Caruncles are present on the maternal aspect of this bovine placenta.

Courtesy of Ron Fabrizius, DVM, Diplomate ACT

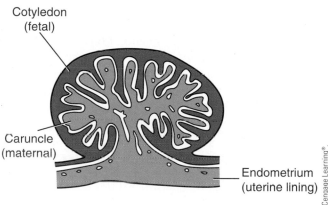

Cotyledon (fetal)

Caruncle (maternal)

Endometrium (uterine lining)

Placentome
1) Fetal attachment to uterus
2) Nutrients for fetus
3) Waste disposal for fetus

FIGURE 12–17 A placentome is the site of attachment between the fetal placenta and the maternal uterus in the cow. The blood supply of the two animals does not mix.

Copyright © 2015 Cengage Learning®.

combining form for pregnancy is **pregn/o**. **Cyesis** (sī-eh-sihs) also means pregnancy. **Gestation** (jehs-tā-shuhn) is the period of development of the fetus in the uterus from conception to parturition and is the term more commonly used in reference to animals. The combining forms for gestation are **gest/o** and **gestat/o**. Gestation periods vary in length between animal species. A fetus is said to be **viable** (vī-ah-buhl) when it is capable of living outside the mother. Viability depends on the species and the age and weight of the fetus.

The embryo has distinct layers that give rise to various tissue types. Layers of the embryo include the **ectoderm** (ehck-tō-dərm), or outer layer of the embryo; the **meso-derm** (meh-sō-dərm), the middle layer of the embryo; and the **endoderm** (ehn-dō-dərm), the inner layer of the embryo.

Terms Related to Gestational and Parturitional Status

- **gravid/o** (grahv-ih-dō) = combining form for pregnant.
- **-para** (pahr-ah) = suffix meaning to bear or bring forth.
- **nulligravida** (nuhl-ih-grahv-ih-dah) = one who has never been pregnant; **nulli-** means none; **nullipara** (nuhl-ih-pahr-ah) = female who has never borne a viable fetus.
- **primigravida** (prī-mih-grahv-ih-dah) = female during first pregnancy; **primi-** means first; **primipara** (prī-mihp-ah-rah) = female who has borne one offspring.
- **multigravida** (muhl-tih-grahv-ih-dah) = one who has had multiple pregnancies; **multi-** means many.
- **multiparous** (muhl-tihp-ah-ruhs) = female who has borne multiple offspring during the same gestation. A **litter** (liht-tər) is a group of offspring born during the same labor.
- **viviparous** (vī-vihp-ahr-uhs) = bearing live young. **Vivi-** is the prefix for live.
- **oviparous** (ō-vihp-ahr-uhs) = bearing eggs. **Ovi** means egg.
- **ovovivparous** (ō-vō-vī-vihp-ahr-uhs) = producing eggs that are hatched within the body, so that young are born alive but without placental attachment such as snakes.

Birth

The act of giving birth is called **parturition** (pahr-tyoo-rihsh-uhn) (Figure 12–18). Parturition is also called **labor**. **Part/o** is the combining form for giving birth. The period before the onset of labor is the **antepartum** (ahn-tē-pahr-tuhm) period; the period immediately after labor is the **postpartum** (pōst-pahr-tuhm) period.

FIGURE 12–18 Parturition in a Hereford cow. In cattle, the most common presentation of the calf is the cranial presentation in which the calf's head is tucked between its front legs.

© iStockphoto/emholk

PRESENTATION

Presentation is the orientation of the fetus before delivery. Presentation varies with species. For example, in cattle and sheep, the fetus adopts a **cranial** (anterior) **presentation** in which the legs and head are directed toward the cervix. In swine, cranial and caudal presentations are considered normal. In a **caudal** (posterior) **presentation**, the pelvis and rear legs are directed toward the cervix. **Transverse presentation** involves the fetus lying across the cervix, and normal parturition is not achieved. In a **breech presentation**, the tail of the fetus is presented first and delivery may or may not be obstructed.

Parturition is divided into stages. The first stage of labor involves **dilation** (dī-lā-shuhn) of the cervix. Dilation is the act of stretching. The second stage of labor involves uterine contractions of increasing frequency and strength and expulsion of the fetus. Expulsion of the fetus is called **delivery** (deh-lihv-ər-ē) (Figure 12–19). The third stage of labor involves separation of the placenta from the uterus.

The postpartum period begins after delivery of the fetus. The newborn is called a **neonate** (nē-ō-nāt). The neonatal period varies from species to species but usually is less than 4 weeks. The first stool of a newborn that consists of material collected in the intestine of the fetus is called the **meconium** (meh-kō-nē-uhm).

After delivery of the fetus, the mother's uterus returns to its normal size. The process of the uterus returning to normal size is called **uterine involution** (yoo-tər-ihn ihn-vō-loo-shuhn). The mammary glands of the mother secrete **colostrum** (kuh-lohs-truhm), which is a thick fluid that contains nutrients and antibodies needed by the neonate. **Lactation** (lahck-tā-shuhn) is the process of forming and secreting milk.

TEST ME: REPRODUCTIVE SYSTEM

Diagnostic procedures performed on the reproductive system include the following:

- **amniocentesis** (ahm-nē-ō-sehn-tē-sihs) = surgical puncture with a needle through the abdominal and uterine walls to obtain amniotic fluid to evaluate the fetus.
- **radiography** (rā-dē-ohg-rah-fē) = procedure in which film is exposed as ionizing radiation passes through the patient and shows the internal body structures in profile.
- **ultrasound** (uhl-trah-sownd) = diagnostic test using high-frequency waves to evaluate internal structures. Ultrasonography works well in evaluating the uterus during pregnancy because the fluid present in the uterus helps define structures.

PATHOLOGY: REPRODUCTIVE SYSTEM

Pathologic conditions of the reproductive system include the following:

- **abortion** (ah-bōr-shuhn) = termination of pregnancy. The combining form **abort/o** means premature expulsion of a nonviable fetus.
- **azoospermia** (ā-zō-ō-spər-mē-ah) = absence of sperm in the semen.

Copyright © 2015 Cengage Learning®

FIGURE 12–19 A Chihuahua giving birth. A fetus in the placenta is being presented and two whelps are nursing.

- **benign prostatic hypertrophy** (beh-nīn prohs-tah-tihck hī-pər-trō-fē) = abnormal noncancerous enlargement of the prostate; also called **prostato-megaly** or **enlarged prostate**; abbreviated BPH.
- **cervicitis** (sihr-vih-sī-tihs) = inflammation of the neck of the uterus.
- **cryptorchidism** (krihp-tōr-kih-dihzm) = developmental defect in which one or both testes fail to descend into the scrotum; also called **undescended testicle(s)**. Animals may be unilaterally or bilaterally cryptorchid. Unilaterally cryptorchid is sometimes called **monorchid** (mohn-ōr-kihd). The combining form **crypt/o** means hidden.
- **dystocia** (dihs-tōs-ah) = difficult birth; the female is having difficulty expelling the fetus. The suffix –**tocia** means birth or labor.
- **ectopic pregnancy** (ehck-tohp-ihck prehg-nahn-sē) = fertilized ovum implanted outside the uterus.
- **epididymitis** (ehp-ih-dihd-ih-mī-tihs) = inflammation of the epididymis.
- **fetal defects** = abnormalities that occur in the development of the fetus (Figure 12–20). **Teratogens** (tər-ah-tō-jehnz) are substances that produce defects in the fetus. The combining form **terat/o** means malformed fetus. **Mutagens** (mū-tah-jehnz) are substances that produce change or that create genetic abnormalities. The combining form **mut/a** means genetic change.
- **fibroid** (fī-broyd) = benign tumor arising from the smooth muscle of the uterus; also called **leiomyoma** (lī-ō-mī-ō-mah).

- **hermaphroditism** (hər-mahf-rō-dih-tihzm) = condition of having both ovarian and testicular tissue. **Pseudohermaphroditism** (soo-dō-hər-mahf-rō-dih-tihzm) is the condition of having gonads of one sex but the physical characteristics of both sexes.
- **mastitis** (mahs-tī-tihs) = inflammation of the mammary gland(s) (Figure 12–21a and b).
- **metritis** (meh-trī-tihs) = inflammation of the uterus.
- **oligospermia** (ohl-ih-gō-spər-mē-ah) = deficient amount of sperm in semen; **oligo-** is the prefix for scant or few.
- **orchitis** (ōr-kī-tihs) = inflammation of the gonads of the male; also called **testitis**.
- **ovarian cyst** (ō-vahr-ē-ahn sihst) = collection of fluid or solid material in the female gonad.

(a)

(b)

FIGURE 12–20 A calf born with a fetal defect (a portion of two heads).

FIGURE 12–21 (a) Mastitis in the rear quarter of a cow udder. (b) Normal white milk is shown in the left tube. Abnormal secretion from a cow with mastitis is shown in the right tube.

FIGURE 12–22 Paraphimosis in a horse.

- **paraphimosis** (pahr-ah-fih-mō-sihs) = retraction of the skin of the prepuce causing a painful swelling of the glans penis that prevents the penis from being retracted; the penis is extruded from the prepuce but cannot be returned to its normal position (Figure 12–22). The combining form **phim/o** means muzzling or constriction of an orifice.
- **phimosis** (fih-mō-sihs) = narrowing of the skin of the prepuce so that it cannot be retracted to expose the glans penis; the penis cannot be extruded from the prepuce due to the small orifice.
- **pneumovagina** (nū-mō-vah-jī-nah) = conformational defect in the perineum of cows and mares that allows air to enter the vagina; also called **windsuckers**.
- **priapism** (prī-ahp-ihzm) = persistent penile erection not associated with sexual excitement.
- **prostatitis** (prohs-tah-tī-tihs) = inflammation of the prostate.
- **pseudocyesis** (soo-dō-sī-ē-sihs) = false pregnancy; also called **pseudopregnancy**. A behavioral and physical syndrome (most commonly seen in bitches 2–3 months after estrus) in which mammary glands develop, lactation occurs, and mothering behaviors occur. **Cyesis** means pregnancy; **pseudo-** is the prefix for false.
- **pyometra** (pī-ō-mē-trah) = pus in the uterus (Figure 12–23).
- **retained placenta** (rē-tānd plah-sehn-tah) = nonpassage of the placenta after delivery of the fetus; abbreviated RP. A retained placenta can lead to metritis and infertility in the female.
- **scrotal hydrocele** (skrō-tahl hī-drō-sēl) = hernia of fluid in the testes or along the spermatic cord. **Hydro-** is the prefix for water; **-cele** is the suffix meaning hernia (Figure 12–24).

- **sterility** (stər-ihl-ih-tē) = inability to reproduce.
- **supernumerary** (soo-pər-nū-mahr-ē) = more than the normal number. Supernumerary teats is a condition in which an animal has more than the normal number of nipples (commonly seen in ruminants).
- **transmissible venereal tumor** (trahnz-mihs-ih-buhl vehn-ēr-ē-ahl too-mər) = naturally occurring, sexually transmitted tumor of dogs that affects the external genitalia and other mucous membranes; abbreviated TVT.
- **uterine prolapse** (yoo-tər-ihn prō-lahps) = protrusion of the uterus through the vaginal orifice (Figure 12–25a and b). Commonly called **cast her withers**.
- **vaginal prolapse** (vah-jih-nahl prō-lahps) = protrusion of the vagina through the vaginal wall or vaginal orifice (Figure 12–26).
- **vaginitis** (vahj-ih-nī-tihs) = inflammation of the vagina.

FIGURE 12–23 Pyometra in a bitch.

FIGURE 12–24 Scrotal hydrocele in a ram.

(a)

(b)

FIGURE 12–25 (a) Cow with uterine prolapse. (b) Close-up view of prolapsed uterus showing placentomes.

FIGURE 12–26 Vaginal prolapse in a ewe.

PROCEDURES: REPRODUCTIVE SYSTEM

Procedures performed on the reproductive system include the following:

- **assisted delivery** = manual use of hands or equipment to aid in delivery of a fetus. In cattle, obstetric chains may be placed around the calf's legs and then force is applied to help extract the fetus (Figure 12–27).
- **cesarean section** (sē-sā-rē-ahn sehck-shuhn) = delivery of offspring through an incision in the maternal abdominal and uterine walls; also called a **C-section** (Figure 12–28).
- **electroejaculation** (ē-lehck-trō-ē-jahck-yoo-lā-shuhn) = method of collecting semen for artificial insemination or examination in which electrical stimulation is applied to the nerves

FIGURE 12–27 Assisted delivery can be accomplished in cows with a fetal extractor (calf jack), which applies pressure to the hind limbs of the cow as traction is applied to the calf.

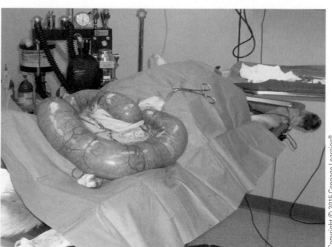

FIGURE 12–28 C-section in a bitch.

Courtesy of ABS Global, Inc.

FIGURE 12–29 Artificial insemination of a mare.

Courtesy of Lodi Veterinary Hospital, SC

FIGURE 12–30 Canine castration.

to promote ejaculation. Electroejaculation is achieved by use of an **electroejaculator** (ē-lehck-trō-ē-jahck-yool-ā-tər), which is a probe and power source used to apply current to the nerves that promote ejaculation. Artificial insemination, abbreviated AI, is a breeding method in which semen is collected, stored, and deposited in the uterus or vagina without copulation taking place (Figure 12–29). AI is also described in Chapter 20.

- **episiotomy** (eh-pihz-ē-oht-ō-mē) = surgical incision of the perineum and vagina to facilitate delivery of the fetus and to prevent damage to maternal structures.
- **fetotomy** (fē-toh-tō-mē) = cutting apart of a fetus to enable removal from the uterus; also called **embryotomy** (ehm-brē-ah-tō-mē). The combining form **fet/o** means fetus.
- **hysterectomy** (hihs-tər-ehck-tō-mē) = surgical removal of the uterus.
- **mastectomy** (mahs-tehck-tō-mē) = surgical removal of the mammary gland or breast.
- **neuter** (nū-tər) = to sexually alter; usually used to describe the sexual altering of males. An animal that is not neutered is **intact** (ihn-tahckt), or has reproductive capability.
- **oophorectomy** (ō-ohf-ō-rehck-tō-mē) = surgical removal of the ovary (ovaries).
- **orchidectomy** (ōr-kih-dehck-tō-mē) = surgical removal of the testis (testes); also known as

orchectomy (ōr-kehck-tō-mē), **orchiectomy** (ōr-kē-ehck-tō-mē), and **castration** (kahs-trā-shuhn) (Figure 12–30).
- **ovariohysterectomy** (ō-vahr-ē-ō-hihs-tər-ehck-tō-mē) = surgical removal of the ovaries, uterine tubes, and uterus; also called a **spay**; abbreviated OHE or OVH.
- **vasectomy** (vah-sehck-tō-mē) = sterilization of a male in which a portion of the ductus deferens is surgically removed, yet the animal may retain its libido. **Libido** (lih-bē-dō) is sexual desire. The combining form **vas/o** means vas deferens (the human term for ductus deferens), vessel, or duct.

ABBREVIATIONS RELATED TO THE REPRODUCTIVE SYSTEM

- ACT = American College of Theriogenologists
- AI = artificial insemination
- BPH = benign prostatic hypertrophy
- C = castrated
- C-section = cesarean section
- CL = corpus luteum
- CMT = California Mastitis Test
- N = neutered
- OB = obstetrics
- OHE = ovariohysterectomy
- OVH = ovariohysterectomy
- pg = pregnant
- RP = retained placenta
- S = spayed
- TVT = transmissible venereal tumor

REVIEW EXERCISES

Multiple Choice

Choose the correct answer.

1. The inner layer of the uterus is called the
   a. endohysteria
   b. myometrium
   c. perimetrium
   d. endometrium

2. The area between the vaginal orifice or scrotum and the anus is called the
   a. clitoris
   b. perineum
   c. vulva
   d. inguinal area

3. Copulation also is called
   a. coitus
   b. impotence
   c. sterility
   d. zygote

4. The act of giving birth is
   a. freshening
   b. calving
   c. gestation
   d. parturition

5. A difficult birth is known as
   a. dystocia
   b. dyshernia
   c. dyspartia
   d. dyslaboratum

6. A false pregnancy also is called
   a. pseudo
   b. pseudopara
   c. pseudocyesis
   d. pseudogestia

7. A condition of an individual having both ovarian and testicular tissue is called
   a. hemisexual
   b. hermaphroditism
   c. supernumerary
   d. orchioovaris

8. Pyometra is
   a. pus in the uterus
   b. increased temperature of the uterus
   c. tumors in the uterus
   d. necrosis of the uterus

9. The innermost membrane enveloping the embryo in the uterus is the
   a. allantois
   b. umbilicus
   c. amnion
   d. chorion

10. Attachment and embedding of the zygote in the uterus is
    a. zygotion
    b. conception
    c. fertilization
    d. implantation

11. The ovum is the
    a. female gonad
    b. male gonad
    c. female gamete
    d. male gamete

12. The heat cycle in females is known as the
    a. estrus cycle
    b. estrous cycle
    c. lactogenic cycle
    d. follicular cycle

13. In large animals, the mammary gland is called the
    a. teat
    b. mammae
    c. lactiferous duct
    d. udder

14. The female organ of mammals that develops during pregnancy and joins the mother and offspring for exchange of nutrients, oxygen, and waste products is known as the
    a. umbilicus
    b. placenta
    c. mount
    d. navel

15. Substances that produce change or that create genetic abnormalities are known as
    a. teratogens
    b. priapisms
    c. dystociae
    d. mutagens

16. Reproductive organs, whether male or female, are called the
    a. theriogens
    b. genitals
    c. gametes
    d. perineum

17. An enlarged part of a tube or canal is called a(n)
    a. vesicle
    b. bulla
    c. ampulla
    d. motile

18. The term for surgical incision of the perineum and vagina to facilitate delivery of the fetus and to prevent damage to maternal structures is
    a. mastectomy
    b. cesarean section
    c. episiotomy
    d. oophorectomy

19. An animal that has not been neutered is referred to as
    a. sterile
    b. gravid
    c. intact
    d. cyesis

20. Another term for spay is
    a. orchidectomy
    b. ovariohysterectomy
    c. C-section
    d. hysterectomy

21. Which of the following will fix a cryptorchid?
    a. fetotomy
    b. hysterectomy
    c. episiotomy
    d. orchidectomy

22. The term for orientation of the fetus before delivery is
    a. presentation
    b. phimosis
    c. primigravida
    d. partum

23. The terms *cornus*, *corpus*, and *cervix* are associated with which organ?
    a. ovary
    b. uterus
    c. vagina
    d. vulva

24. Which term has a suffix that means condition?
    a. pyometra
    b. vaginitis
    c. phimosis
    d. fibroid

25. The "H" in OHE stands for
    a. hyper
    b. hypo
    c. hyster
    d. holo

Matching

Match the term in Column I with the definition in Column II.

Column I

1. _____ gestation
2. _____ sterility
3. _____ dystocia
4. _____ viviparous
5. _____ oviparous
6. _____ allantois
7. _____ chorion
8. _____ parturition
9. _____ colostrum
10. _____ abortion

Column II

a. act of giving birth

b. termination of pregnancy

c. thick fluid that contains nutrients and antibodies needed by neonates, which is secreted by the mother's mammary glands

d. period of pregnancy

e. outermost layer of the placenta

f. innermost layer of the placenta

g. bearing live young

h. inability to reproduce

i. bearing eggs

j. difficult birth

Fill in the Blanks

Write the correct term in the space provided.

1. Colop/o and vagin/o mean _____.

2. Hyster/o, metr/o, metri/o, and uter/o mean _____.

3. Pen/i and priap/o mean _____.

4. Orch/o, orchi/o, orchid/o, test/o, and testicul/o mean _____.

5. Sperm/o and spermat/o mean _____.

True or False

If the statement is true, write T on the line. If the statement is false, write F on the line.

1. _____A fertilized egg is called a zygote.

2. _____The labia are entrances or outlets from a body cavity.

3. _____Another name for the prepuce is foreskin.

4. _____Acrosome is capable of spontaneous movement.

5. _____Cotyledons are also called buttons.

6. _____A bull with oligospermia has less than the normal amount of semen.

7. _____The "V" in TVT stands for vagina.

8. _____Expulsion of the fetus is delivery.

9. _____When a cow has cast her withers, she has a breech presentation.

10. _____Psuedopregnancy is spelled correctly.

Spell Check

Circle the term that is spelled correctly.

1. developmental defect in which one or both testes fail to descend into the scrotum: cryptorchidism criptorchidism cryptorchydism

2. inflammation of the mammary glands: mastis mastitis masitits

3. fertilized egg: zigoat zygoat zygote

4. pregnancy: syesis cyesis ciesis

5. to sexually alter: neuter nueter newter

Word Building

Build the following terms using word parts.

1. pus in the uterus:

    word part for pus _____

    word part for uterus _____

    term for pus in the uterus _____

2. near the ovary:

    word part for near _____

    word part for ovary _____

    term for near the ovary _____

3. pertaining to the urinary and reproductive systems:

    word part for urinary _____

    word part for reproductive system _____

    word part for pertaining to _____

    term for pertaining to the urinary and reproductive systems _____

4. capable of stimulating milk production:

    word part for milk _____

    word part for producing _____

    word part for pertaining to _____

    term for capable of stimulating milk production _____

5. surgical removal of the uterus:

    word part for uterus _____

    word part for surgical removal _____

    term for surgical removal of the uterus _____

Crossword Puzzle

Reproductive Terms

Supply the correct term in the appropriate space for the definition listed.

Across

1 eggs

6 muscular middle layer of the uterus

10 persistent penile erection not associated with sexual excitement

11 retractable fold of skin covering the glans penis

12 female external genitalia

13 fleshy borders or edges

14 inflammation of the gland that surrounds the urethra that secretes a thick fluid that aids in motility of sperm

18 to sexually alter (either sex)

19 capable of spontaneous motion

20 process of egg maturation and release

21 bearing eggs

22 common name for placenta

Down

2 bearing live young

3 structure that forms where the fetus communicates with the placenta

4 medical term for spay

5 condition of undescended testicles

7 ejaculatory fluid that contains sperm plus secretions from the secondary sex glands

8 preparatory step to animal mating that involves one animal jumping on top of another animal or object

9 union of ovum and sperm

10 retraction of the skin of the prepuce causing a painful swelling of the glans penis that prevents the penis from being retracted

15 entrance or outlet from a body cavity

16 funnel-shaped opening

17 male sex cells

Case Studies

Define the underlined terms in each case study.

A 12-yr-old F miniature poodle was presented with lethargy, anorexia, and PU/PD. The bitch had been in proestrus 1 mo before presentation. On PE, the dog was pyrexic and tachypneic. Abdominal palpation yielded an enlarged uterus, and a purulent vaginal discharge was noted. Blood was collected for a CBC and chem panel. The CBC results included leukocytosis with a left shift. (A left shift is an increase in banded neutrophils in peripheral blood due to many causes including bacterial infection.) The dx of pyometra was made, antibiotics were started, and the dog was scheduled for an emergency OHE.

1. yr _____

2. F _____

3. lethargy _____

4. anorexia _____

5. PU/PD _____

6. bitch _____

7. proestrus _____

8. PE _____

9. pyrexic _____

10. tachypneic _____

11. abdominal _____

12. palpation _____

13. uterus _____

14. purulent _____

15. vaginal _____

16. CBC _____

17. chem panel _____

18. leukocytosis _____

19. dx _____

20. pyometra _____

21. OHE _____

A 10-yr-old intact M German shepherd was presented with stranguria and hematuria. PE revealed pyrexia, anorexia, and a stiff gait. On rectal palpation, the veterinarian discovered that the prostate gland was bilaterally enlarged. Upon review of the record, it was noted that this dog has had a hx of recurrent UTIs . Radiographs were taken, and prostatomegaly was noted. After a dx of prostatitis was made, the dog was scheduled to be neutered the next day.

22. yr _____

23. intact _____

24. M _____

25. stranguria _____

26. hematuria _____

27. PE _____

28. pyrexia _____

29. anorexia _____

30. gait _____

31. rectal palpation _____

32. prostate gland _____

33. bilaterally _____

34. hx _____

35. recurrent _____

36. UTI _____

37. radiographs _____

38. prostatomegaly _____

39. dx _____

40. prostatitis _____

41. neutered _____

42. another term for neutering in male dogs _____

A 2-yr-old Holstein cow was examined because the farmer noticed that she was off feed. PE revealed that the cow had a slightly elevated rectal temperature. The farmer told the veterinarian that this cow had stepped on her teat previously, but it had appeared to be healing. The udder was palpated, and it was not warm to the touch or swollen. Milk was expressed from each quarter, and the milk appeared more watery than normal. A CMT paddle test was performed on the milk, and moderate precipitation was noted (Figure 12–31). The diagnosis of mastitis was made, and milk samples were taken for culture. Antibiotic treatment was started pending culture results, and milking hygiene was discussed with the farmer.

43. cow _____

44. off feed _____

45. medical term for off feed _____

46. teat _____

47. udder   _____

48. quarter   _____

49. CMT   _____

50. diagnosis   _____

51. mastitis   _____

52. culture   _____

FIGURE 12–31 CMT paddle. The CMT (California Mastitis Test) is a cow side indicator of somatic cell count used to detect mastitis (see Chapter 20). Using a four-well plastic paddle (one well for each quarter), a little milk is drawn into each well (after the foremilk is discarded). An equal volume of test reagent is added, the paddle is gently agitated, and a reaction is scored on a scale of 0 (mixture remains unchanged) to 3 (an almost solid gel forms). A score of 2 or 3 is considered positive. If a cow has mastitis, the milk cannot be introduced into the bulk tank or sold.

A 5-yr-old quarter horse <u>mare</u> had <u>foaled</u> 10 hours earlier. The foal appeared normal, and the mare was letting the foal suckle. The mare had not moved since foaling, and the owner was concerned because the mare seemed quieter than normal. PE revealed a mildly elevated rectal temperature, <u>vaginal discharge</u> was noted, milk production seemed adequate, and the milk appeared normal. The owner was asked whether the mare had passed its <u>placenta</u>, and the owner did not know whether she had. The stall was examined for remnants of the placenta, and none were found. The veterinarian was concerned that the mare had a <u>retained placenta</u> and that an infection was starting. The equine placenta normally is passed within a few hours of foaling, and because it was at least 10 hours <u>postpartum</u>, an injection of oxytocin was given by slow <u>IV</u>. Blood was drawn for a <u>CBC</u> to assess <u>leukocyte</u> numbers. The owner was advised to watch for the passing of the placenta, and general hygiene was discussed with the client.

53. mare   _____

54. foaled   _____

55. vaginal discharge   _____

56. placenta   _____

57. retained placenta   _____

58. postpartum   _____

59. IV   _____

60. CBC   _____

61. leukocyte   _____

Critical Thinking Exercise

The following story and questions contain words and concepts from this chapter and are designed to stimulate critical thinking through class discussion, Internet searches, or brief writing responses.

A client comes into the clinic with a male boxer puppy. During your discussion of puppy care, the client asks you about using her dog for breeding and any tests that need to be performed to "clear" her dog as a stud. You note this question on the record for the veterinarian to discuss with the client; however, you notice while taking the dog's history and performing a brief examination of the pup that he is a unilateral cryptorchid.

1. What does cryptorchid mean?

2. What does unilateral mean?

After discussing cryptorchidism with the client, you go back to the treatment area and look up information on cryptorchidism. You read that it is the most common disorder of sexual development in dogs and predisposing factors include testicular hypoplasia, breech presentation that compromises blood supply to the testes, and delayed closure of the umbilicus resulting in an inability to increase abdominal pressure.

3. What is testicular hypoplasia?

4. What is a breech presentation?

5. What is the umbilicus?

Based on what you read, you notice that unilateral cryptorchidism is more common than bilateral cryptorchidism. The article also says that the undescended testicle may be located anywhere caudal to the kidney to within the inguinal canal. Abdominal testicles produce male hormones; therefore, cryptorchid animals will develop secondary sex characteristics. It is also known that cryptorchidism is inherited and that affected animals have a higher incidence of cancer in retained testicles.

6. Can the dog produce offspring if he is a unilateral cryptorchid?

7. Could the dog produce offspring if he was a bilateral cryptorchid?

8. What would you tell the owner about her puppy since she is interested in using him as a breeding dog?

Chapter 13
NERVES OF STEEL

Objectives

Upon completion of this chapter, the reader should be able to:

- Identify the structures of the nervous system

- Describe the functions of the nervous system

- Identify the divisions of the nervous system and describe the structures and functions of each

- Recognize, define, spell, and pronounce terms related to the functions, disorders, and treatment of the nervous system

- Interpret the meaning of abbreviations of the nervous system

FUNCTIONS OF THE NERVOUS SYSTEM

The **nervous** (nər-vuhs) **system** coordinates and controls body activity. The three basic types of the nervous system response are sensory, integrative, and motor. The sensory role of the nervous system is to detect and process internal and external stimuli. The integrative role of the nervous system is to process sensory information. The motor role of the nervous system is to formulate appropriate responses by either causing muscle contraction or glandular secretion.

FIGURE 13–1 Divisions of the nervous system.

FIGURE 13–2 Photomicrograph of multiple motor neurons from the spinal cord of a bovine.

STRUCTURES OF THE NERVOUS SYSTEM

The major structures of the nervous system are the brain, spinal cord, peripheral nerves, and sensory organs. The two major divisions of the nervous system are as follows:

- **central nervous system** (sehn-trahl nər-vuhs sihs-tehm) = portion of the nervous system that consists of the brain and spinal cord; abbreviated CNS.
- **peripheral nervous system** (pehr-ihf-ər-ahl nər-vuhs sihs-tehm) = portion of the nervous system that consists of the cranial and spinal nerves, autonomic nervous system, and ganglia; abbreviated PNS (Figure 13–1).

Back to Basics

The basic unit of the nervous system is the **neuron** (nū-rohn) (Figure 13–2). There are three types of neurons, based on their function.

- **sensory neurons** (sehn-sōr-ē nū-rohnz) = nerves that carry sensory impulses toward the CNS; also called **afferent** (ahf-fər-ahnt) or **ascending tracts** because they carry information toward the CNS. Sensory information such as sound and light is converted into electrical impulses so that the nerves can transport it.
- **associative neurons** (ahs-ō-shē-ah-tihv nū-rohnz) = nerves that carry impulses from one neuron to another; also called **connecting neurons**.

- **motor neurons** (mō-tər nū-rohnz) = nerves that carry impulses away from the CNS and toward the muscles and glands; also called **efferent** (ē-fər-ahnt) or **descending tracts** because they carry information away from the CNS.

The parts of a neuron are the cell body, one or more dendrites, one axon, and terminal end fibers. The cell body, or **soma** (sō-mah), has a nucleus and is responsible for maintaining the life of the neuron. The **dendrites** (dehn-drīts) are rootlike structures that receive impulses and conduct them toward the cell body. The combining form for dendrite is **dendr/o**. The **axon** (ahcks-ohn) is a single process that extends away from the cell body and conducts impulses away from the cell body. The combining form **ax/o** means axis or main stem. The terminal end fibers are the branching fibers that lead the impulse away from the axon and toward the synapse.

The dendrites and axons also are called **nerve fibers**. Bundles of nerve fibers bound together by specialized tissues are called **nerves** or **nerve trunks**. Nerve fibers are covered with a tubelike membrane called the **neurolemma** (nū-rō-lehm-ah) or **neurilemma** (nū-rih-lehm-ah). Neuron cell bodies grouped together within the CNS are called **nuclei** (nū-klē-ī), and those outside the CNS are called **ganglia** (gahng-glē-ah) (Figure 13–3).

The Gap

The junction between two neurons or between a neuron and a receptor is the **synapse** (sihn-ahps) (Figure 13–4). The combining forms for this space or point of contact are **synaps/o** and **synapt/o**. A synapse is the junction where neural impulses cause a release of a chemical substance called a **neurotransmitter** (nū-rō-trahnz-miht-ər) that allows the signal to move from one neuron to another. There are different neurotransmitters, each with specific functions.

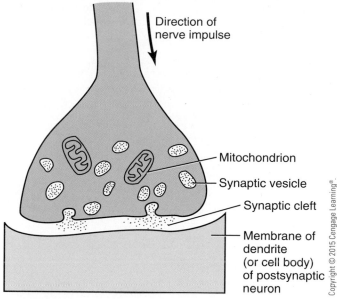

FIGURE 13–4 A synapse. Nerve impulse from the axon stimulates the release of a neurotransmitter.

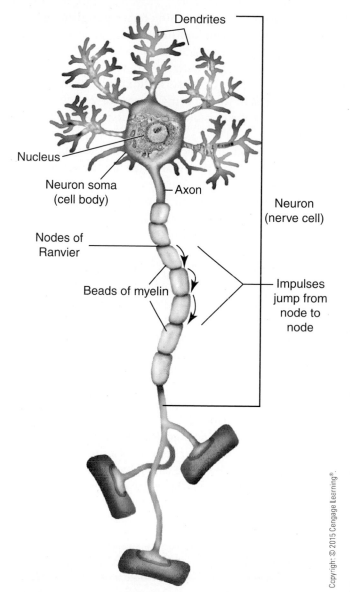

FIGURE 13–3 Structures of a neuron. Dendrites conduct impulses toward the neuron cell body, and axons conduct impulses away from the cell body.

Supporting Role

The **neuroglia** (nū-rohg-lē-ah or nū-rō-glē-ah), or **glial** (glē-ahl) **cells**, are the supportive cells of the nervous system. Glial cells consist of **astrocytes** (ahs-trō-sītz), **microglia** (mī-krō-glē-ah), **oligodendrocytes** (ohl-ih-gō-dehnd-rō-sītz), and **Schwann** (shwahn) **cells**. The combining form **gli/o** means glue, which helps explain the function of glial cells.

- **astr/o** means star. Astrocytes are star-shaped, cover the capillary surface of the brain, and help form the blood-brain barrier in the CNS. The blood-brain barrier separates the nervous tissue from blood and consists of capillary walls in the CNS without

fenestrations (perforations) and the surrounding astrocytes. The blood brain barrier prevents or slows the passage of drugs, chemicals, and some microbes from entering the CNS; abbreviated BBB.

- **micro-** means small. Microglia are small phagocytic cells that help fight infection in the CNS.
- **oligo-** means few, **dendr/o** means branching, and **-cyte** means cell. Oligodendrocytes are cells with few branches that hold the nerve fibers together and help form myelin in the CNS.
- Schwann cells (named after a German anatomist) help form myelin in the PNS.

Surrounding Tissue

Myelin (mī-eh-lihn) is a protective covering over some nerve cells, including parts of the spinal cord, white matter of the brain, and most peripheral nerves. Myelin also is called the **myelin sheath**. Myelin serves as an electrical insulator. Nerves that are described as **myelinated** (mī-lihn-āt-ehd) are surrounded by myelin and transport a signal much faster than nonmyelinated nerves. Myelin gives nerve fibers a white color, and myelinated nerves are called **white matter**. The **gray matter** does not contain myelinated fibers, so it is darker in color (Figure 13–5a and b). The gray matter is composed of cell bodies, branching dendrites, and neuroglia.

Myelin is interrupted at regular intervals along the length of a fiber by gaps called **nodes of Ranvier** (nōdz of rohn-vē-ā). Ionic exchange takes place at the nodes of Ranvier. Some nerve fibers have a very thin layer of myelin and are referred to as **nonmyelinated**. The autonomic nervous system contains nonmyelinated fibers.

Gray matter

White matter (myelin sheath)

(a)

(b)

FIGURE 13–5 (a) Cross section of the spinal cord shows white matter and gray matter. The orientation of the white and gray matter in the spinal cord is opposite of that in the brain. (b) Cross section of the brain shows the white matter and gray matter.

Carry an Impulse

A **nerve** (nərv) is one or more bundles of impulse-carrying fibers that connect the CNS to the other parts of the body. **Neur/i** and **neur/o** are combining forms for nerve or nerve tissue. Terms used in reference to nerves include the following:

- A **tract** (trahckt) is a group of nerve fibers located in the CNS. Ascending tracts (groups going *up*, or cranially) carry nerve impulses toward the brain, whereas descending tracts (groups going *down*, or caudally) carry nerve impulses away from the brain.
- A **ganglion** (gahng-glē-ohn) is a knotlike mass of neuron cell bodies located *outside* the CNS. More than one ganglion are called **ganglia**, or less often

ganglions. The combining forms **gangli/o** and **ganglion/o** mean knotlike masses of nerve cell bodies located outside the CNS.

- A **plexus** (plehck-suhs) is a network of intersecting nerves or vessels. **Plexi** (plehck-sī) are networks of intersecting nerves or vessels. **Plex/o** is the combining form for a plexus, or network.
- **Innervation** (ihn-nər-vā-shuhn) is the supply or stimulation of a body part through the action of nerves.
- **Receptors** (rē-sehp-tərz) are sensory organs that receive external stimulation and transmit that information to the sensory neurons. There are different types of receptors. For example, **nociceptive** (nō-sih-sehp-tihv) receptors are pain receptors, whereas **proprioceptive** (prō-prē-ō-sehp-tihv) receptors are spatial orientation or perception of movement receptors.
- A **stimulus** (stihm-yoo-luhs) is something that excites or activates. Multiple excitations or activations are called **stimuli** (stihm-yoo-lī).
- An **impulse** (ihm-puhlz) is a wave of excitation transmitted through nervous tissue.
- A **reflex** (rē-flehcks) is an automatic, involuntary response to change. Reflex actions include the patellar and ulnar reflexes (Figure 13–6). Decreased reflexes are called **hyporeflexive**, while exaggerated reflexes are called **hyperreflexive**.

FIGURE 13–6 A veterinarian tests the patellar reflex on a dog. The patellar reflex is tested by stimulating the patellar tendon with a neurologic hammer. A normal reaction is extension of the stifle because the femoral nerve (with nerve roots between L4 and L6 in small animals; L4 and L5 in large animals) is stimulated.

CENTRAL NERVOUS SYSTEM

The central nervous system (or cerebrospinal system or CNS) is made up of the brain and the spinal cord. The combining form for brain is **encephal/o**, and the combining form for spinal cord is **myel/o**. (Remember, myel/o also means bone marrow.) The CNS contains both white and gray matter; white matter contains myelinated fibers, and gray matter consists of nerve cell bodies.

Membrane

The brain and spinal cord are encased in connective tissue called the **meninx** (meh-nihncks). Because this connective tissue has three layers, it more commonly is referred to by its plural form of **meninges** (meh-nihn-jēz) (Figure 13–7).

Mening/o and **meningi/o** are combining forms for the layers of connective tissue enclosing the CNS.

The three layers of the meninges are as follows:

- **dura mater** (doo-rah mah-tər) = thick, tough, outermost layer of the meninges. **Dura** means tough, and this layer of the meninges is tough and strong. **Dur/o** is the combining form for dura mater. The dura mater also is called **pachymeninx** (pahck-ē-meh-nihcks); the prefix **pachy-** means thick.
- **arachnoid** (ah-rahck-noyd) **membrane** = second layer of the meninges. **Arachn/o** means spider, and the arachnoid membrane resembles a spider web. The arachnoid membrane is loosely attached to the other layers of the meninges to allow space between the layers.

Meninges →

FIGURE 13–7 Meninges of the sheep brain.

- **pia mater** (pē-ah mah-tər) = third and deepest layer of the meninges. **Pia** means soft or tender, and this layer of the meninges is soft with a rich supply of blood vessels. The pia adheres to the CNS. The pia mater and arachnoid membranes collectively are called the **leptomeninges** (lehp-tō-meh-nihn-jēz) (Figure 13–8). **Lept/o** means thin.

Certain terms are used to describe the location of structures in reference to the meninges. **Epidural** (ehp-ih-doo-rahl) means located above or superficial to the dura mater. (**Epi-** is the prefix meaning above or upon.) The **subdural** (suhb-doo-rahl) **space** is the area located deep to (below) the dura mater and superficial to (above) the arachnoid membrane. (The prefix **sub-** means below.) The **subarachnoid** (suhb-ah-rahck-noyd) **space** is the area located deep to (below) the arachnoid membrane and superficial to (above) the pia mater. The subarachnoid space contains cerebrospinal fluid.

Fluid

Cerebrospinal fluid (sər-ē-brō-spīn-ahl flū-ihd) is the clear, colorless ultrafiltrate that nourishes, cools, and cushions the CNS. Cerebrospinal fluid is abbreviated CSF. CSF is produced by special capillaries in the ventricles of the brain. The ventricles of the brain are cavities. Vascular folds of the pia mater in the ventricles called the **choroid plexus** (kōr-oyd plehck-suhs) secrete CSF.

Brain

The brain is the enlarged and highly developed portion of the CNS that lies in the skull and is the main site of nervous control. The portion of the skull that encloses and protects the brain is called the **cranium** (krā-nē-uhm). **Crani/o** is the combining form for skull. **Intracranial** (ihn-trah-krā-nē-ahl) means within the cranium.

The brain is commonly divided into parts based on functional group or on location. **Encephal/o** is the combining form for brain; however, other combining forms that refer to specific regions of the brain also are used.

The divisions of the brain (Figure 13–9) based on functional group include the following:

- **cerebrum** (sər-ē-bruhm). The combining form for cerebrum is **cerebr/o**. The cerebrum, the largest part of the brain, is responsible for receiving and processing stimuli, initiating voluntary movement, and storing information. The **cerebral**

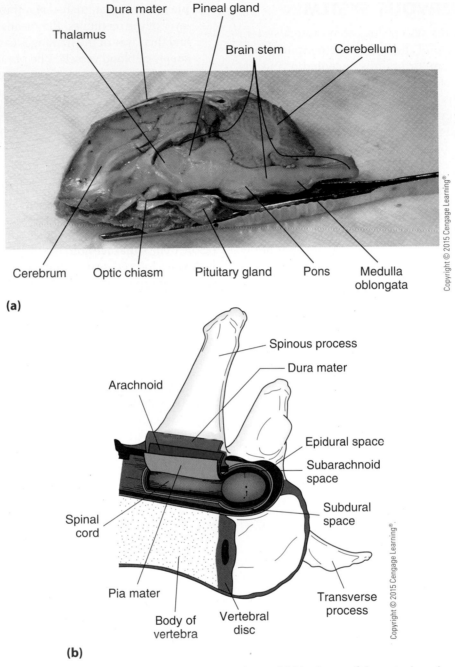

Dura mater   Pineal gland

Thalamus

Brain stem   Cerebellum

Cerebrum   Optic chiasm   Pituitary gland   Pons   Medulla oblongata

(a)

Spinous process

Dura mater

Arachnoid

Epidural space

Subarachnoid space

Subdural space

Spinal cord

Pia mater

Body of vertebra   Vertebral disc   Transverse process

(b)

FIGURE 13–8 (a) Sagittal view of the sheep's brain. (b) Meninges of the spinal cord.

cortex (outer region) is made up of gray matter and is arranged in folds (Figure 13–10). The elevated portions of the cerebral cortex are known as **gyri** (jī-rī). The combining form **convolut/o** means coiled, and **gyr/o** means folding. The grooves of the cerebral cortex are called fissures or **sulci** (suhl-kī). The combining form **sulc/o** means groove. The medullary substance of the cerebrum is made up of white matter. The brain also has small cavities called **ventricles** (vehn-trih-kuhlz) (Figure 13–11). There are four ventricles of the brain: two lateral ventricles, a third

ventricle, and a fourth ventricle. The ventricles of the brain (and central canal of the spinal cord) are lined with a membrane called the **ependyma** (eh-pehn-dih-mah).

- **cerebellum** (sehr-eh-behl-uhm). The combining form for cerebellum is **cerebell/o**. The cerebellum is the second largest part of the brain, and it coordinates muscle activity for smooth movement. The cerebellum has an inner portion, called the **vermis** (vər-mihs) because it is wormlike, and other portions divided into right and left cerebellar hemispheres.

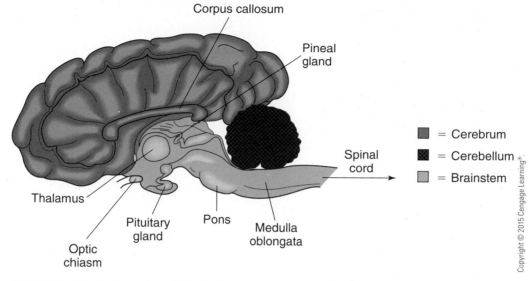

FIGURE 13–9 Sagittal section of the brain.

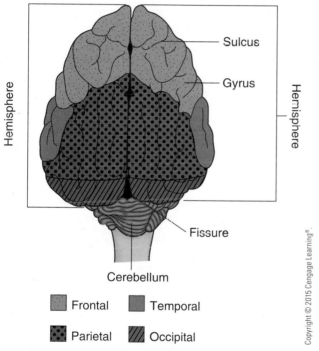

FIGURE 13–10 Dorsal view of the cerebral cortex. The cerebrum is divided into right and left hemispheres. Each hemisphere is further divided into lobes, and each lobe is named for the bone plate covering it: **frontal** (frohn-tahl) lobe = most cranial lobe that controls motor function; **parietal** (pahr-ī-ih-tahl) lobe = receives and interprets sensory nerve impulses; **occipital** (ohcks-ihp-ih-tahl) lobe = most caudal lobe that controls vision; **temporal** (tehmp-ruhl) lobe = laterally located lobe that controls hearing and smell.

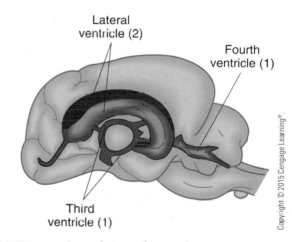

FIGURE 13–11 Lateral view of ventricles.

- **brainstem** (brān-stehm). The brainstem is the stalklike portion of the brain that connects the cerebral hemispheres with the spinal cord. The brainstem consists of the pons, medulla oblongata, midbrain, and interbrain. The **interbrain** contains structures such as the pituitary gland, hypothalamus, and thalamus. These structures are responsible for endocrine activity, regulation of thirst and water balance, and regulation of body temperature. The **midbrain** contains structures responsible for visual and auditory reflexes, posture, and muscle control. The **pons** (pohnz) is the bridge at the base of the brain that allows nerves to cross over so that one side of the brain controls the opposite side of the body. The **medulla oblongata** (meh-duhl-ah ohb-lohng-gah-tah) is the cranial continuation of the spinal cord that controls basic life functions. Divisions of the brain also are based on location (Figure 13–12 and Table 13–1).

FIGURE 13–12 Brain divisions based on location.

TABLE 13–1 Brain Divisions Based on Location

Brain Part	Division	Some Components
forebrain; prosencephalon (prōs-ehn-sehf-ah-lohn)	**telencephalon** (tēl-ehn-sehf-ah-lohn)	cerebral cortex and olfactory brain, **limbic** (lihm-bihck) system (affects emotion and behavior); limbic means border
	diencephalon (dī-ehn-sehf-ah-lohn)	thalamus, epithalamus, and hypothalamus; limbic system; optic chiasm (crossing of the vision nerves); **chiasm** (kī-ahs-uhm) means crossing
midbrain; mesencephalon (mēz-ehn-sehf-ah-lohn)	**mesencephalon**	vision and hearing bodies, posture, and muscle control; limbic system
hindbrain; rhombencephalon (rohmb-ehn-sehf-ah-lohn)	**metencephalon** (meht-ehn-sehf-ah-lohn)	cerebellum and pons
	myelencephalon (mī-lehn-sehf-ah-lohn)	medulla oblongata

Some parts of the brain are named not based on location or division, but rather on how they look. The vermis of the cerebellum is one example. Another example is the hippocampus, a portion of the limbic system that involves memory. The **hippocampus** (hihp-ō-kahm-puhs) is shaped like a seahorse, which exists in mythology as a sea monster with the head of a horse and the tail of a fish and as an actual sea creature. The name *hippocampus* comes from the Greek *hippos,* meaning horse, and *kampos,* meaning sea monster. The arbor vitae (tree of life) of the cerebellum are the tree-like outlines seen on sagittal views of the cerebellum.

Spinal Cord

The **spinal cord** is the caudal continuation of the medulla oblongata. The spinal cord passes through an opening in the occipital bone called the **foramen magnum** (fər-ā-mehn mahg-nuhm). **Foramen** means passage or hole, and **magnum** means great. The spinal cord carries all of the tracts that influence the innervation of the limbs and lower part of the body. In addition, the spinal cord is the pathway for impulses going to and from the brain. **Myel/o** is the combining form for spinal cord. (Remember that myel/o also is the combining form for bone marrow.)

The spinal cord is part of the central nervous system, which means that it is surrounded by the meninges and is protected by cerebrospinal fluid. The gray matter of the spinal cord is located in the internal portion and is not protected by myelin. The white matter of the spinal cord is located in the external portion and is myelinated.

The spinal cord contains two areas of swelling. The medical term for swelling (normal or abnormal) is **intumescence** (ihn-too-meh-sehns). The swelling is caused by an increase in white matter and cell bodies that are associated with the innervation of the limbs. One swelling occurs in the area of C6–T2, which is known as the **cervical intumescence**. The other swelling occurs in the area of the L4–caudal segment, which is known as the **lumbosacral intumescence.** The cranial parts of the spinal cord have tracts with fibers from the cranial and caudal portions, but the caudal parts of the spinal cord have only tracts with fibers from the caudal portions. Therefore, as the spinal cord proceeds caudally, its cross-sectional area decreases. At the level of the cranial lumbar vertebrae, the spinal cord becomes cone-shaped. This cone-shaped segment is called the **conus medullaris** (kō-nuhs mehd-yoo-lahr-ihs). **Conus** means cone. At the caudal end of the spinal cord (caudal lumbar, sacral, and coccygeal nerve segments), the tracts terminate and the spinal nerves fan outward and backward, giving the appearance of a horse's tail. This collection of spinal roots at the caudal part of the spinal cord is the **cauda equina** (kaw-dah ē-kwī-nah). The cauda equina includes the conus medullaris to the caudal vertebrae. The threadlike tapering section of the cauda equina is known as the **filum terminale** (fī-luhm tər-mih-nahl). **Filum** means threadlike structure. The filum terminale attaches the conus medullaris to the caudal vertebrae.

Cushions

The spinal cord is housed within the vertebrae to protect it from injury. The vertebrae are protected from each other by **intervertebral** (ihn-tər-vər-tə-brahl) discs that are located between vertebra (Figure 13–13). Intervertebral discs are layers of fibrocartilage that form pads separating and cushioning the vertebrae from each other. The center of the intervertebral disc is the gelatinous **nucleus pulposus** (nū-klē-uhs puhl-pō-suhs), and the outer layer is the fibrous **annulus fibrosis** (ahn-yoo-luhs fī-brō-sihs).

PERIPHERAL NERVOUS SYSTEM

The peripheral nervous system (PNS) consists of the cranial nerves, the autonomic nervous system, the spinal nerves, and ganglia.

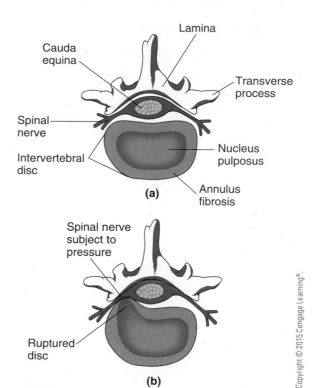

FIGURE 13–13 Intervertebral disc. (a) Normal intervertebral disc. (b) Ruptured disc.

Cranial Nerves

Twelve pairs of **cranial nerves** originate from the undersurface of the brain. The cranial nerves generally are named for the area or function they serve and are represented by Roman numerals (Figure 13–14 and Table 13–2).

Spinal Nerves

The **spinal nerves** arise from the spinal cord. With the exception of some cervical and coccygeal nerves, the spinal nerves are paired and emerge caudal to

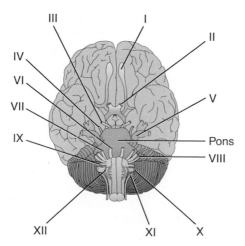

I. Olfactory
II. Optic Nerve
III. Oculomotor Nerve
IV. Trochlear Nerve
V. Trigeminal Nerve
VI. Abducent Nerve
VII. Facial Nerve
VIII. Vestibulocochlear Nerve
IX. Glossopharyngeal Nerve
X. Vagus Nerve
XI. Accessory Nerve
XII. Hypoglossal Nerve

FIGURE 13–14 Cranial nerves and ventral surface of the brain.

TABLE 13–2 Cranial Nerves

Cranial Nerve	Name	Function
I	olfactory	conducts sensory impulses from the nose to the brain (smell)
II	optic	conducts sensory impulses from the eyes to the brain (vision)
III	oculomotor	sends motor impulses to the external eye muscles (dorsal, medial, and ventral rectus; ventral oblique; and levator superioris) and to some internal eye muscles
IV	trochlear	sends motor impulses to one external eye muscle (dorsal oblique)
V	trigeminal	three branches: ophthalmic = sensory to cornea; maxillary = motor to upper jaw; mandibular = motor to lower jaw
VI	abducent	motor innervation to two muscles of the eye (retractor bulbi and lateral rectus)
VII	facial	motor to facial muscles, salivary glands, and lacrimal glands and taste sensation to anterior two-thirds of tongue
VIII	acoustic or vestibulocochlear	two branches: cochlear = sense of hearing; vestibular = sense of balance
IX	glossopharyngeal	motor to the parotid glands and pharyngeal muscles, taste sensation to caudal third of tongue, and sensory to the pharyngeal mucosa
X	vagus	sensory to part of the pharynx and larynx and parts of thoracic and abdominal viscera; motor for swallowing and voice production
XI	accessory	accessory motor to shoulder muscles
XII	hypoglossal	motor to the muscles that control tongue movement

the vertebra of the same number and name. The first cervical vertebra (C1, or the atlas) has a pair of spinal nerves emerging cranially and caudally to it; therefore, there are eight cervical spinal nerves. C1 exits the spinal canal in a foramen in the wing of the atlas, and C2 emerges at the C1–C2 intervertebral foramen. C8 emerges caudal to vertebra C7, and T1 emerges caudal to T1 vertebra. The coccygeal vertebrae usually have fewer pairs of nerves than the number of vertebrae (Figure 13–15).

Spinal nerves have dorsal and ventral roots. The **dorsal root** enters the dorsal portion of the spinal cord and carries afferent or sensory impulses from the periphery to the spinal cord. The **ventral root** emerges from the ventral portion of the spinal cord and carries efferent or motor impulses from the spinal cord to muscle fibers or glands (Figure 13–16).

Spinal nerves supply sensory and motor fibers to the body region associated with their emergence from the spinal cord. After spinal nerves exit the spinal cord, they branch to form the peripheral nerves of the trunk and limbs. Several spinal nerves may join together to form a single peripheral nerve. This network of intersecting nerves is called a **plexus**. Each appendage is innervated by a plexus. Each forelimb is supplied from nerves that arise from the **brachial** (brā-kē-ahl) **plexus** (C6–T2) (Figure 13–17), and each hindlimb is supplied from nerves that arise from the **lumbosacral plexus** (L4–S3). Brachial means the arm, and lumbosacral means the loin and sacrum.

Spinal nerves are named for where they arise from the spinal cord. *C* represents cervical, *T* represents thoracic, *L* represents lumbar, *S* represents sacral, and *Co* and *Cy* represent coccygeal (or *Cd* represents caudal). These letter abbreviations are followed by numbers to represent the vertebral area from which the nerve exits the spinal cord. C4 represents cervical spinal nerve 4, T11 represents thoracic spinal nerve 11, and so on.

Autonomic Nervous System

The **autonomic nervous system** (aw-tō-nah-mihck nər-vuhs sihs-tehm), or ANS, is the part of the peripheral nervous system that innervates smooth muscle, cardiac muscle, and glands. The two divisions of the autonomic nervous system are the sympathetic nervous system (sihm-pah-theh-tihck nər-vuhs sihs-tehm) and the **parasympathetic nervous system** (pahr-ah-sihm-pah-theh-tihck nər-vuhs sihs-tehm). The two divisions of the autonomic nervous system work together to maintain homeostasis within the body. **Homeostasis** (hō-mē-ō-stā-sihs) is the process of maintaining a stable internal body environment. The combining form **home/o** means unchanging.

sympathetic	provides emergency and stress response; "fight or flight"	↑ heart rate, and blood flow to muscles; ↓ gastrointestinal function; pupil dilation
parasympathetic	returns body to normal after stressful response; maintains normal body function	returns heart rate, respiratory rate, and blood flow to normal levels; returns normal gastrointestinal function; constricts pupil size to normal

levels of consciousness = descriptive terms used to describe mentation

- **BAR** = bright, alert, and responsive.
- **coma** (kō-mah) = deep state of unconsciousness.
- **conscious** (kohn-shuhs) = awake, aware, and responsive; also known as **alert**.
- **disorientation** (dihs-ōr-ē-ehn-tā-shuhn) = condition in which the animal appears mentally confused.
- **lethargy** (lehth-ahr-jē) = drowsiness, indifference, and listlessness.
- **obtunded** (ohb-tuhn-dehd) = depressed.
- **stupor** (stoo-pər) = impaired consciousness with unresponsiveness to stimuli.

Cauda equina

C1 C2 C3 C4 C5 C6 C7 T1 T2 T3 T4 T5 T6 T7 T8 T9 T10 T11 T12 T13 L1 L2 L3 L4 L5 L6 L7 S Cd1 Cd2 Cd3 Cd4 Cd5 Cd6

FIGURE 13–15 Naming scheme for spinal nerves.

FIGURE 13–16 Spinal nerve.

FIGURE 13–17 Intraoperative photograph of the brachial plexus showing nerves entering the leg and the underside of the scapula.

- **discography** (dihs-kō-grah-fē) = radiographic study of an intervertebral disc after injection of contrast material into the disc, also spelled *disk-ography*. The record of this procedure is called a **discogram** (Figure 13–19).
- **electroencephalography** (ē-lehck-trō-ehn-sehf-ah-lohg-rah-fē) = process of recording electrical activity of the brain; abbreviated EEG. An **electroencephalograph** (ē-lehck-trō-ehn-sehf-ah-lō-grahf) is the instrument used to record the electrical activity of the brain, and an **electroencephalogram** (ē-lehck-trō-ehn-sehf-ah-lō-grahm) is the record of the electrical activity of the brain.

TEST ME: NERVOUS SYSTEM

Diagnostic procedures performed on the nervous system include the following:

- **cerebrospinal fluid tap** (sər-ē-brō-spī-nahl flū-ihd tahp) = removal of cerebrospinal fluid; also called a CSF tap. CSF is obtained by inserting a needle or catheter into the **cisterna magna** (sihs-tər-nah mahg-nah) or lumbosacral area. The cisterna magna is the subarachnoid space located between the caudal surface of the cerebellum and the dorsal surface of the medulla oblongata. Intracranial pressures also may be measured before removal of CSF (Figure 13–18).

FIGURE 13–18 Lumbar CSF tap in a lion.

- **magnetic resonance imaging** and **computed axial tomography** are covered in Chapter 16.
- **myelography** (mī-eh-lohg-rah-fē) = diagnostic study of the spinal cord after injection of contrast material. A **myelogram** (mī-eh-lō-grahm) is the record of the spinal cord after injection of contrast material (Figure 13–20).
- **pupillary light reflex** = response of pupil to a bright light source; abbreviated PLR. Light is shone in one eye, and that eye (direct) and the opposite eye (consensual) should constrict. It is used to assess neurologic damage.

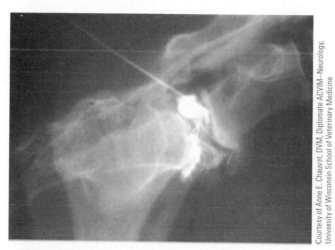

FIGURE 13–19 Discogram in a dog.

(a)

(b)

FIGURE 13–20 (a) Normal myelogram in a Labrador retriever. (b) Myelogram demonstrating disc disease. (Note that the dye does not flow continuously.)

PATHOLOGY: NERVOUS SYSTEM

Pathologic conditions of the nervous system include the following:

- **amnesia** (ahm-nē-zē-ah) = memory loss.
- **astrocytoma** (ahs-trō-sī-tō-mah) = malignant intracranial tumor composed of astrocytes.
- **ataxia** (ā-tahck-sē-ah) = without coordination; "stumbling" (Figure 13–21).
- **catalepsy** (kaht-ah-lehp-sē) = waxing rigidity of muscles accompanied by a trancelike state.
- **cataplexy** (kaht-ah-plehck-sē) = sudden attacks of muscular weakness and hypotonia triggered by an emotional response.
- **cerebellar hypoplasia** (sehr-eh-behl-ahr hī-pō-plā-zē-ah) = smaller-than-normal cerebellum; seen in cats secondary to feline panleukopenia virus, which leads to incoordination.
- **cervical vertebral malformation** = abnormal formation or instability of the caudal cervical vertebrae that causes ataxia and incoordination; seen more often in horses and dogs; also called **wobbler's syndrome**.
- **chorea** (kōr-ē-ah) = repetitive, rhythmic contraction of limb or facial muscles; also called **myoclonus** (mī-ō-klō-nuhs); usually the result of distemper viral infection in dogs.

FIGURE 13–21 Ataxia in a cat.

- **choriomeningitis** (kōr-ē-ō-meh-nihn-jī-tihs) = inflammation of the choroid plexus and meninges.
- **concussion** (kohn-kuhsh-uhn) = shaking of the brain caused by injury. The combining form **concuss/o** means shaken together.
- **conscious proprioceptive deficit** (kohn-shuhs prō-prih-ō-sehp-tihv dehf-ih-siht) = neurologic defect in which the animal appears not to know where its limbs are; abbreviated CP deficit, "knuckling" (Figure 13–22).
- **contusion** (kohn-too-zhuhn) = bruising. The combining form **contus/o** means bruise.
- **decerebration** (dē-sər-ē-brā-shuhn) = condition of loss of mental functions caused by damage to the midbrain.
- **demyelination** (dē-mī-eh-lih-nā-shuhn) = destruction or loss of myelin.
- **discospondylitis** (dihs-kō-spohn-dih-lī-tihs) = destructive inflammatory disorder that involves the intervertebral discs, vertebral end-plates, and vertebral bodies.
- **encephalitis** (ehn-sehf-ah-lī-tihs) = inflammation of the brain.
- **encephalocele** (ehn-sehf-ah-lō-sēl) = herniation of the brain through a gap in the skull.
- **encephalomalacia** (ehn-sehf-ah-lō-mah-lā-shē-ah) = abnormal softening of the brain.
- **encephalomyelitis** (ehn-sehf-ah-lō-mī-ih-lī-tihs) = inflammation of the brain and spinal cord.
- **encephalopathy** (ehn-sehf-ah-lohp-ah-thē) = any disease of the brain.
- **epilepsy** (ehp-ih-lehp-sē) = recurrent seizures of nonsystemic origin or of intracranial disease.

Epilepsy may be described as **idiopathic** (ihd-ē-ō-pahth-ihck). *Idiopathic* means unknown cause or disease of an individual. **Idio-** is the prefix meaning individual.

- **hallucination** (hah-loo-sehn-ā-shuhn) = false sensory perception.
- **hematoma** (hē-mah-tō-mah) = mass or collection of blood. In the nervous system, a hematoma usually is described by the area where it is found. An **epidural hematoma** (ehp-ih-doo-rahl hē-mah-tō-mah) is a collection of blood above or superficial to the dura mater. A **subdural hematoma** (suhb-doo-rahl hē-mah-tō-mah) is a collection of blood deep to (below) the dura mater and superficial to (above) the arachnoid membrane.
- **hemiplegia** (hehm-ih-plē-jē-ah) = paralysis of one side of the body. The suffix **-plegia** means paralysis.
- **Horner's syndrome** (hōr-nərz sihn-drōm) = collection of signs relating to injury of the cervical sympathetic innervation to the eye; signs include sinking of the eyeball (enophthalmus), ptosis of the upper eyelid, pupil constriction, and prolapse of the third eyelid (Figure 13–23).
- **hydrocephalus** (hī-drō-sehf-ah-luhs) = abnormally elevated amount of cerebrospinal fluid in the ventricles of the brain; "water on the brain."
- **hyperesthesia** (hī-pər-ehs-thē-zē-ah) = excessive sensitivity. The suffix **-esthesia** means sensation or feeling.
- **hyperkinesis** (hī-pər-kihn-ē-sihs) = increased motor function or activity.
- **hypnosis** (hihp-nō-sihs) = condition of altered awareness; trancelike state.

FIGURE 13–22 CP deficit in a dog.

FIGURE 13–23 Horner's syndrome in a cat.

- **intervertebral** (ihn-tər-vər-tə-brahl) **disc disease** = condition of pain and neurologic deficits resulting from the displacement of part or all of the material in the disc located between the vertebrae (see Figure 13–13); abbreviated IVDD.
- **leukoencephalomalacia** (loo-kō-ehn-sehf-ah-lō-mah-lā-shah) = abnormal softening of the white matter of the brain.
- **macrocephaly** (mahck-rō-sehf-ah-lē) = abnormally large skull.
- **meningioma** (meh-nihn-jē-ō-mah) = benign tumor of the meninges (Figure 13–24).
- **meningitis** (mehn-ihn-jī-tihs) = inflammation of the meninges.
- **meningocele** (meh-nihng-gō-sēl) = protrusion of the meninges through a defect in the skull or vertebrae.
- **meningoencephalitis** (meh-nihn-go-ehn-sehf-ah-lī-tihs) = inflammation of the meninges and brain.
- **meningoencephalomyelitis** (meh-nihn-gō-ehn-sehf-ah-lō-mī-eh-lī-tihs) = inflammation of the meninges, brain, and spinal cord.
- **microcephaly** (mī-krō-sehf-ah-lē) = abnormally small skull.
- **monoplegia** (mohn-ō-plē-jē-ah) = paralysis of one limb.
- **myelitis** (mī-eh-lī-tihs) = inflammation of the spinal cord (or bone marrow).
- **myelopathy** (mī-eh-lah-pahth-ē) = disease of the spinal cord (or bone marrow).

- **myoparesis** (mī-ō-pahr-ē-sihs) = weakness of muscles. The suffix **-paresis** means weakness. As is the case with **-plegia**, the suffix *-paresis* is modified to describe the area of weakness. **Hemiparesis** (hehm-ih-pahr-ē-sihs) is weakness on one side of the body; **paraparesis** (pahr-ah-pahr-ē-sihs) is weakness of the lower body in bipeds or of hindlimbs in quadrupeds. In reference to the nervous system, the prefix **para-** means hind or lower portion.
- **narcolepsy** (nahr-kō-lehp-sē) = syndrome of recurrent uncontrollable sleep episodes. The combining form **narc/o** means stupor, and the suffix **-lepsy** means seizure (episode).
- **neuralgia** (nū-rahl-jē-ah) = nerve pain.
- **neuritis** (nū-rī-tihs) = inflammation of the nerves.
- **opisthotonos** (ohp-ihs-thoht-ō-nohs) = tetanic spasm in which the head and tail are bent dorsally and the back is arched. The combining form **opisth/o** means backward.
- **paralysis** (pahr-ahl-ih-sihs) = loss of voluntary movement or immobility. Paralysis is further described by the areas it involves.
- **paraplegia** (pahr-ah-plē-jē-ah) = paralysis of the lower body in bipeds or of hindlimbs in quadrupeds (Figure 13–25).
- **paresthesia** (pahr-ehs-thē-zē-ah) = abnormal sensation. Abnormal sensations may include tingling, numbness, or burning and may be difficult to assess in animals. The combining forms for burning are **caus/o** and **caust/o**. Caustic substances are capable of burning.

FIGURE 13–24 Magnetic resonance image of a meningioma.

FIGURE 13–25 Paraplegia in a rabbit.

FIGURE 13-26 Polioencephalomalacia in a sheep. Opisthotonos is one sign of PEM.

- **polioencephalomalacia** (pō-lē-ō-ehn-sehf-ah-lō-mah-lā-shah) = abnormal softening of the gray matter of the brain; abbreviated PEM (Figure 13–26). The combining form **poli/o** means gray.
- **polioencephalomyelitis** (pō-lē-ō-ehn-sehf-ah-lō-mī-eh-lī-tihs) = inflammation of the gray matter of the brain and spinal cord.
- **poliomyelitis** (pō-lē-ō-mī-eh-lī-tihs) = inflammation of the gray matter of the spinal cord. The combining form **poli/o** means gray.
- **polyneuritis** (poh-lē-nū-rī-tihs) = inflammation of many nerves.
- **polyradiculoneuritis** (poh-lē-rah-dihck-yoo-lō-nū-rī-tihs) = inflammation of many peripheral nerves and spinal nerve roots that may lead to progressive paralysis; commonly called **coonhound paralysis** in coonhound dogs and **idiopathic polyradiculoneuropathy** in other dogs.
- **ptosis** (tō-sihs) = prolapse or drooping. The suffix -**ptosis** means prolapse, drooping, or falling downward; refers specifically to the upper eyelid.
- **radiculitis** (rah-dihck-yoo-lī-tihs) = inflammation of the root of a spinal nerve. **Radicul/o** is the combining form for root.
- **roaring** = noisy respiration caused by air passing through a narrowed larynx in horses; common term for **equine laryngeal hemiplegia** because of nerve fiber degeneration of the left recurrent laryngeal nerve (Figure 13–27).
- **seizure** (sē-zhər) = sudden, involuntary contraction of some muscles caused by a brain disturbance; also called **convulsions**. The most common type of seizure in animals is **grand mal**

FIGURE 13-27 Laryngeal area of a horse with equine laryngeal hemiplegia (roaring).

(grahnd mahl), in which the animal experiences loss of consciousness and muscle contractions. Other types of seizures include partial, which have a seizure focus that does not spread, and petit mal, which is a mild generalized seizure in which loss of consciousness and generalized loss of muscle tone occur.

What Side Are You On?

Pathologic conditions of the nervous system may involve lesions that cause abnormal clinical signs on the same side or opposite side that the lesion occurs. In describing lesions of the nervous system, the terms *ipsilateral* and *contralateral* are used. **Ipsi-** is the prefix meaning the same, and **contra-** is the prefix meaning opposite. **Ipsilateral** (ihp-sē-laht-ər-ahl) means on the same side, and **contralateral** (kohn-trah-laht-ər-ahl) means on the opposite side.

FIGURE 13–28 A disc fenestration involves removal of intervertebral disc material.

Seizure Stages

Seizures are divided into stages to aid in determining when they may start or end. These stages are as follows:

- **preictal** (prē-ihck-tahl) = period before a seizure; also called the **aura** (aw-rah). An animal may pace, excessively lick, fly bite, or seem anxious during this stage.
- **ictus** (ihck-tuhs) = attack or actual seizure. An animal may convulse, lose control of excretory functions, shake, and appear confused during this stage.
- **postictal** (pōst-ihck-tahl) = period after a seizure. An animal may appear obtunded, tired, fearful, or anxious during this stage.

- **spasticity** (spahs-tihs-ih-tē) = state of increased muscular tone.
- **spina bifida** (spī-nah bihf-ih-dah) = congenital anomaly in which the spinal canal does not close over the spinal cord. The combining form **bifid/o** means split or cleft.
- **syncope** (sihn-kō-pē) = fainting; sudden fall in blood pressure or cardiac systole resulting in cerebral anemia and loss of consciousness. The combining form syncop/o means to cut short or to cut off.
- **tetraplegia** (teht-rah-plē-jē-ah) = paralysis of all four limbs; also called **quadriplegia** (kwohd-rih-plē-je-ah).
- **tremor** (treh-mər) = involuntary trembling.
- **vestibular disease** (vehs-tihb-yoo-lahr dih-zēz) = neurologic disorder characterized by head tilt, nystagmus, rolling, falling, and circling. **Nystagmus** (nī-stahg-muhs) is involuntary, rhythmic movement of the eye and is discussed in Chapter 14.

PROCEDURES: NERVOUS SYSTEM

Procedures performed on the nervous system include the following:

- **analgesia** (ahn-ahl-jēz-ē-ah) = without pain. *Analgesia* is used to describe pain relief, which is different from *anesthesia* (absence of sensation). **Endorphins** (ehn-dōr-fihnz) are natural, opioid-like chemicals that are produced in the brain and raise the pain threshold.
- **anesthesia** (ahn-ehs-thē-zē-ah) = absence of sensation. An **anesthetic** (ahn-ehs-theht-ihck) is a substance used to induce anesthesia. There are different types of anesthesia. Some types include **topical anesthesia** = absence of sensation after a substance has been applied to the skin or external surface; **local anesthesia** = absence of sensation after chemical injection to an adjacent area; **epidural anesthesia** = absence of sensation to a region after injection of a chemical into the epidural space; **general anesthesia** = absence of sensation and consciousness.
- **disc fenestration** (dihsk fehn-ih-strā-shuhn) = removal of intervertebral disc material by perforating and scraping out its contents (Figure 13–28).
- **dysesthesia** (dihs-eh-stēsh-ah) = impaired sensation.
- **laminectomy** (lahm-ihn-ehck-tō-mē) = surgical removal of the lamina of the vertebra to relieve pressure on the spinal cord.
- **neurectomy** (nū-rehck-tō-mē) = surgical removal of a nerve.
- **neuroanastomosis** (nū-rō-ahn-ahs-tō-mō-sihs) = connecting nerves together.
- **neuroplasty** (nū-rō-plahs-tē) = surgical repair of a nerve.
- **neurorrhaphy** (nū-rōr-ah-fē) = suturing the ends of a severed nerve.
- **neurotomy** (nū-roht-ō-mē) = surgical incision or dissection of a nerve.

ABBREVIATIONS RELATED TO THE NERVOUS SYSTEM

- ACh = acetylcholine
- AChE = acetylcholinesterase
- ANS = autonomic nervous system
- BAR = bright, alert, and responsive
- BBB = blood–brain barrier
- ChE = cholinesterase
- CNS = central nervous system
- CP = conscious proprioception
- CSF = cerebrospinal fluid
- CT scan = computed axial tomography; also abbreviated CAT scan
- EEG = electroencephalography
- IVDD = intervertebral disc disease
- LOC = level of consciousness
- MRI = magnetic resonance imaging
- PEM = polioencephalomalacia
- PLR = pupillary light reflex
- PNS = peripheral nervous system

REVIEW EXERCISES

Multiple Choice

Choose the correct answer.

1. The space between two neurons or between a neuron and a receptor is a(n)
   a. synapse
   b. ganglion
   c. axon
   d. dendrite

2. Maintaining a constant internal environment is called
   a. osmosis
   b. homeosmosis
   c. homeostasis
   d. ipsistasis

3. Inflammation of the root of a spinal nerve is
   a. myelitis
   b. radiculitis
   c. polyneuritis
   d. poliomyelitis

4. The three-layered membrane lining the CNS is called the
   a. hippocampus
   b. pons
   c. myelin
   d. meninges

5. The protective sheath that covers some nerve cells of the spinal cord, white matter of the brain, and most peripheral nerves is called
   a. pia
   b. dura
   c. glia
   d. myelin

6. The division of the autonomic nervous system that is concerned with body functions under emergency or stress is the
   a. peripheral
   b. central
   c. sympathetic
   d. parasympathetic

7. A network of intersecting nerves is a
   a. bundle
   b. trunk
   c. tract
   d. plexus

8. What type of neuron carries impulses away from the CNS and toward the muscles?
   a. afferent (sensory)
   b. efferent (motor)
   c. sympathetic
   d. parasympathetic

9. An automatic, involuntary response to change is called a(n)
   a. impulse
   b. stimulus
   c. reflex
   d. receptor

10. The largest portion of the brain that is involved with thought and memory is the
    a. cerebrum
    b. cerebellum
    c. brainstem
    d. spinal cord

11. Elevated portions of the cerebral cortex are
    a. sulci
    b. plexuses
    c. gyri
    d. hemispheres

12. The term meaning without pain is
    a. analgesia
    b. endorphin
    c. idiopathic
    d. ptosis

13. Inflammation of the gray matter of the spinal cord is known as
    a. myelitis
    b. poliomyelitis
    c. leukomyelitis
    d. neuritis

14. The term for without coordination or "stumbling" is
    a. ataxia
    b. spasticity
    c. seizure
    d. epilepsy

15. A knotlike mass of neuron cell bodies in the peripheral nervous system is known as a
    a. microglia
    b. synapse
    c. receptor
    d. ganglion

16. A "depressed" animal is referred to as
    a. conscious
    b. disoriented
    c. obtunded
    d. depressed

17. Inflammation of the membranes surrounding the brain and spinal cord is known as
    a. meningitis
    b. encephalopathy
    c. encephalitis
    d. chorea

18. Conscious proprioceptive deficit is sometimes called
    a. nystagmus
    b. roaring
    c. knuckling
    d. wobbler's syndrome

19. The medical term for crossing is
    a. intumescence
    b. plexus
    c. chiasm
    d. filum

20. Small cavities in the brain are known as
    a. gyri
    b. sulci
    c. ventricles
    d. ependymal

21. The opposite of afferent is
    a. convergent
    b. recurrent
    c. efferent
    d. divergent

22. How many combining forms are in the term meningoencephalomyelitis?
    a. 1
    b. 2
    c. 3
    d. 4

23. The "P" in PNS stands for
    a. paralysis
    b. peripheral
    c. paralysis
    d. poly

24. The "V" in IVDD stands for
    a. ventral
    b. vestibular
    c. vertebral
    d. vertical

25. A dog in the postictal state would have had a
    a. laminectomy
    b. concussion
    c. contusion
    d. seizure

Matching

Match the term in Column I with the definition in Column II.

Column I

1. _____ conscious
2. _____ BAR
3. _____ coma
4. _____ lethargy
5. _____ obtunded
6. _____ disorientation
7. _____ stupor

Column II

a. depressed

b. impaired consciousness with unresponsiveness to stimuli

c. bright, alert, and responsive

d. deep state of unconsciousness

e. awake, aware, and responsive; also known as alert

f. condition in which the animal appears mentally confused

g. drowsiness, indifference, and listlessness

Match the term in Column I with the definition in Column II.

Column I

8. _____ homeostasis
9. _____ ganglia
10. _____ stimulus
11. _____ reflex
12. _____ synapse
13. _____ soma
14. _____ impulse
15. _____ neuroglia
16. _____ myelin
17. _____ myoclonus

Column II

a. supportive cells of the nervous system

b. protective covering over some nerve cells that serves as an electrical insulator

c. wave of excitation transmitted through nervous tissue

d. neuron cell bodies grouped together outside the CNS

e. repetitive, rhythmic contraction of limb or facial muscles

f. something that excites or activates

g. automatic, involuntary response to change

h. space between two neurons or between a neuron and a receptor

i. cell body

j. process of maintaining a constant internal body environment

Fill in the Blanks

Write the correct term in the space provided.

1. Neur/i and neur/o mean _____.

2. Mening/o and meningi/o mean _____.

3. Pachy- means _____.

4. Gangli/o and ganglion/o mean _____.

5. Gli/o means _____.

True or False

If the statement is true, write T on the line. If the statement is false, write F on the line.

1. _____ Receptors are sensory organs that receive external stimuli.

2. _____ **S**ensory nerves are **a**fferent, while **m**otor nerves are **e**fferent. (Remember the acronym SAME.)

3. _____ The bony structure that protects the brain is the cranium.

4. _____ Neuron structures that conduct nerve impulses toward the cell body are called axons.

5. _____ Ventricles hold CNS.

6. _____ The part of the brain with two hemispheres is the cerebellum.

7. _____ Schwann cells help form myelin in the CNS.

8. _____ Astrocytes are star-shaped cells that help form the blood–brain barrier.

9. _____ The spinal cord passes through an opening in the occipital bone called the foramen magnum.

10. _____ Vermis is the medical term for swelling.

Spell Check

Circle the term that is spelled correctly.

1. fainting:	sincope	syncopy	syncope
2. substance that produces absence of sensation:	anestetic	anesthetic	anesthetick
3. loss of voluntary movement:	paralysis	paralisis	parlysis
4. convulsions:	seezure	seizure	seizure
5. prolapse or drooping:	tosis	ptosis	ptoesis

Word Scramble

Use the definitions to unscramble the terms relating to the nervous system.

1. incision into a nerve   tronumoye _____

2. period before a seizure   uaar _____

3. disease of the spinal cord (or bone marrow)   pthayyelom _____

4. passage or hole   fmnraoe _____

5. paralysis of the lower limbs in bipeds or of hindlimbs in quadrupeds   ppaaliger _____

6. opposite   rcoatn _____

7. recurrent seizures of nonsystemic origin   yspelipe _____

Crossword Puzzle

Nervous System Terms

Supply the correct term in the appropriate space for the definition listed.

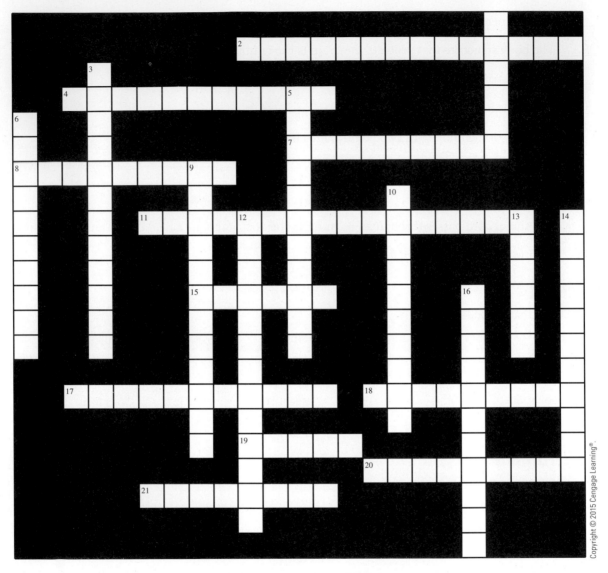

Across

2 spatial orientation or perception of movement

4 diagnostic radiographic study of the spinal cord after injection of contrast material into the subarachnoid space

7 repetitive, rhythmic contraction of skeletal muscle

8 surgical incision or dissection of a nerve

11 abnormal softening of the brain

15 crossing

17 the same side

18 star-shaped cell

19 period of an actual seizure

20 nerve pain

21 inflammation of the spinal cord (or bone marrow)

Down

1 prolapse or drooping

3 increased motor function or activity

5 weakness of one side of the body

6 shaking of brain caused by injury

9 abnormally small skull

10 syndrome of recurrent uncontrollable sleep episodes

12 inflammation of the gray matter of the spinal cord

13 without coordination or stumbling

14 abnormal sensation

16 surgical repair of a nerve

Case Studies

Define the underlined terms in each case study.

A 3-yr-old F mixed breed dog was presented to the clinic for <u>convulsions</u> (Figure 13–29). A thorough history was taken, including a description of the convulsion episodes, possible toxin exposure, and eating history. PE revealed that the animal was <u>obtunded</u> but otherwise in normal health. The <u>neurologic</u> examination was normal. Blood was drawn for a CBC and chem panel. The CBC was normal, and the chem panel did not show evidence of <u>renal</u> disease, <u>hepatic</u> disease, or <u>hypoglycemia</u>. Urine was collected for <u>UA</u> via <u>cystocentesis</u>. The results of the UA were normal. The dog's history was examined, and it was noted that the dog's vaccinations were current. The <u>signalment</u>, history, and <u>clinical</u> signs were used to <u>diagnose idiopathic epilepsy</u>. Diagnostic tests, including <u>CSF tap</u> and <u>EEG</u>, were recommended to the owners but were declined at this time. The owners were advised to have the dog <u>spayed</u> (hormone levels may contribute to <u>seizure</u> activity) and to monitor the animal's activity for recurrence of seizures. If the seizures recur and persist, <u>anticonvulsant</u> medication will be used in an attempt to control them.

1. convulsions _____
2. obtunded _____
3. neurologic _____
4. renal _____
5. hepatic _____
6. hypoglycemia _____
7. UA _____
8. cystocentesis _____
9. signalment _____
10. clinical _____
11. diagnose _____
12. idiopathic _____
13. epilepsy _____
14. CSF tap _____
15. EEG _____
16. spayed _____
17. seizure _____
18. anticonvulsant _____

FIGURE 13–29 A seizuring dog. During a grand mal seizure, dogs typically are unresponsive, lying on their sides, and paddling their legs.

A 5-yr-old <u>M/N</u> dachshund was presented to the clinic with a history of not being able to walk up stairs or jump on the bed. The owner stated that the dog had decreased its eating and was more <u>lethargic</u> than usual. PE revealed an <u>obese</u> dog that had normal <u>vital signs</u>. The neurologic examination revealed normal <u>cranial nerves</u> and <u>CP deficit</u> present on both hindlimbs. <u>Patellar reflexes</u> were <u>hyporeflexive</u>, with the right side worse than the left. <u>Anal</u> tone was adequate. <u>Radiographs</u> were recommended to assess whether the dog had calcified or herniated <u>intervertebral discs</u>. Abnormal discs were noted in the <u>lumbar</u> region. The dog was referred for <u>myelography</u>. The <u>myelogram</u> confirmed that the dog had <u>herniated</u> discs in the lumbar region, and surgery was recommended. <u>Disc fenestration</u> surgery was performed the following day. The dog went home with orders for strict cage rest and rechecks by the referring veterinarian.

19. M/N _____

20. lethargic _____

21. obese _____

22. vital signs _____

23. cranial nerves _____

24. CP deficit _____

25. patellar reflexes _____

26. hyporeflexive _____

27. anal _____

28. radiographs _____

29. intervertebral discs _____

30. lumbar _____

31. myelography _____

32. myelogram _____

33. herniated _____

34. disc fenestration _____

Critical Thinking Exercise

The following story and questions contain words and concepts from this chapter and are designed to stimulate critical thinking through class discussion, Internet searches, or brief writing responses.

You are reading a continuing education article about drugs that affect the nervous system and their clinical use in animals. The following is from the journal article:

The autonomic nervous system consists of two divisions working together to ensure that the body responds appropriately to different situations. The ANS controls blood pressure, heart and breathing rates, body temperature, digestion, metabolism, the production of body fluids (saliva, sweat, and tears), urination, and defecation. Many organs are controlled primarily by either the sympathetic or the parasympathetic division; however, there are times that the two divisions have opposite effects on the same organ. For example, both systems affect blood pressure.

1. How does the sympathetic nervous system affect blood pressure?

2. How does the parasympathetic nervous system affect blood pressure?

Generally, the sympathetic division prepares the body for stressful or emergency situations—fight or flight. The parasympathetic division controls body process during ordinary situations.

3. How does the sympathetic nervous system affect the following?
   - HR

   - Bronchiole diameter

   - BP

   - Defecation

   - Urination

   - Pupil diameter

4. How does the parasympathetic nervous system affect the following?
   - HR

   - Bronchiole diameter

   - BP

   - Defecation

   - Urination

   - Pupil diameter

Two chemical messengers (neurotransmitters), acetylcholine and norepinephrine, are used to communicate within the autonomic nervous system. Nerve fibers that secrete acetylcholine are called cholinergic fibers. Fibers that secrete norepinephrine are called adrenergic fibers. Generally, acetylcholine has parasympathetic (inhibiting) effects and norepinephrine has sympathetic (stimulating) effects.

5. What types of action would an anti-cholinergic drug have?

6. What types of action would a cholinergic drug have?

Chapter 14
SEEING AND HEARING

Objectives

Upon completion of this chapter, the reader should be able to:

- Identify the structures of the eyes and ears

- Describe the functions of the eyes and ears

- Describe how light enters the eye and is processed for vision

- Describe how animals hear

- Describe how animals maintain equilibrium

- Recognize, define, spell, and pronounce terms related to the diagnosis, pathology, and treatment of eye and ear disorders

- Interpret the meaning of abbreviations of the ocular and auditory systems

FUNCTIONS OF THE EYE

The **ocular** (ohck-yoo-lahr) **system** is responsible for vision. The **eyes** are the receptor organs for sight. Combining forms for the eye or sight are **opt/i**, **opt/o**, **optic/o**, **ocul/o**, and **ophthalm/o**. **Extraocular** (ehcks-trah-ohck-yoo-lahr) means outside the eyeball, and **intraocular** (ihn-trah-ohck-yoo-lahr) means within the eyeball. **Periocular** (pehr-ē-ohck-yoo-lahr) means around the eyeball.

STRUCTURES OF THE EYE

The structures of the eye include the accessory structures and the eyeball.

Accessories

The accessory structures of an organ are called **adnexa** (ahd-nehck-sah). **Stroma** (strō-mah) is another term used to describe the supporting tissue of an organ. The adnexa of the eye include the orbit, eye muscles, eyelids, eyelashes, conjunctiva, and lacrimal apparatus (Figure 14–1).

- **orbit** (ohr-biht) = bony cavity of the skull that contains the eyeball. The term **periorbita** (pehr-ih-ōr-bih-tah) means eye socket.
- **eye muscles** = seven major muscles attached to each eye that make a range of movement possible (two oblique muscles, four rectus muscles, and the retractor bulbi) (Figure 14–2). The muscles of both eyes work together in coordinated movements to make normal binocular vision possible. **Binocular** (bī-nohck-yoo-lahr) means both eyes.

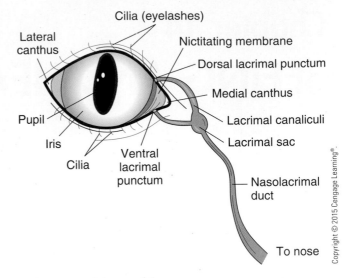

FIGURE 14–1 Adnexa of the eye.

The **extrinsic muscles** (ehcks-trihn-sihck muhs-uhlz) are six muscles that attach the outside of the eyeball to the bones of the orbit. The combining form **extrins/o** means contained outside. The **levator palpebrae muscles** (lē-vā-tər pahl-pē-brā muhs-uhlz) are muscles that raise the upper eyelid.

- **eyelids** = each eye has an upper and lower eyelid to protect the eye from injury, foreign material, and excessive light. The combining form **blephar/o** (blehf-ah-rō) means eyelid. **Palpebra** (pahl-pē-brah) is another term used for the eyelid; the plural form is **palpebrae** (pal-pē-brā). **Palpebral** (pahl-pē-brahl) means pertaining to the eyelid.

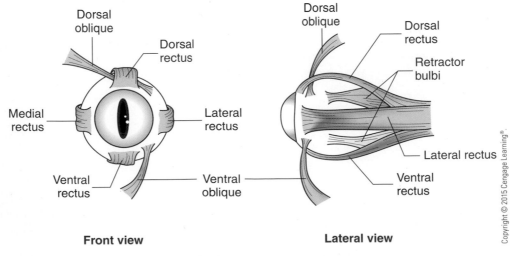

Front view

Lateral view

FIGURE 14–2 Extrinsic eye muscle of a dog.

- The angle where the upper and lower eyelids meet is called the **canthus** (kahn-thuhs). The combining form **canth/o** means corner of the eye. The medial canthus is the corner of the eye nearer to the nose (also called the inner canthus). The lateral canthus is the corner of the eye farther away from the nose (also called the outer canthus).
- The **tarsal** (tahr-sahl) **plate** (or tarsus) is the platelike framework within the upper and lower eyelids that provides stiffness and shape. The combining form **tars/o** means edge of the eyelid or "ankle" joint.
- **Meibomian** (mī-bō-mē-ahn) **glands** are the sebaceous glands on the margins of each eyelid; also called **tarsal glands**.
- **eyelashes** = the edge of each eyelid has hairlike structures called **cilia** (sihl-ē-ah). The cilia, or eyelashes, protect the eye from foreign material.
- **conjunctiva** (kohn-juhnck-tī-vah) = mucous membrane that lines the underside of each eyelid. The conjunctiva forms a protective covering of the exposed surface of the eyeball when the eyelids are closed. The combining form **conjunctiv/o** means conjunctiva. The **nictitating membrane** (nihck-tih-tāt-ihng mehm-brān) is the conjunctival fold attached at the medial canthus that moves across the cornea when the eyelids close; it is also called the **third eyelid** or **nictitans** (nihck-tih-tahnz) or **haws** (hawz) (Figure 14–3).
- **lacrimal apparatus** (lahck-rih-mahl ahp-ah-raht-tuhs) = structures that produce, store, and remove tears. **Lacrimation** (lahck-rih-mā-shuhn) is the condition of normal tear secretion. The combining forms **lacrim/o** and **dacry/o** mean teardrop, tear duct, or lacrimal duct. The lacrimal glands are glands that secrete tears. The **lacrimal**

canaliculi (kahn-ah-lihck-yoo-lī) are the ducts at the medial canthus that collect tears and drain them into the lacrimal sac. The **lacrimal sac**, or **dacryocyst** (dahck-rē-ō-sihst), is the enlargement that collects tears at the upper portion of the tear duct. The **nasolacrimal** (nā-sō-lahck-rih-mahl) **duct** is the passageway that drains tears into the nose. The **dorsal punctum** (dōr-sahl puhnck-tuhm) is the small spot near the upper medial canthus where the nasolacrimal duct begins; the **ventral punctum** (vehn-trahl puhnck-tuhm) is the small spot near the lower medial canthus where the nasolacrimal duct begins. A **punctum** (puhnck-tuhm) is a point or small spot.

Eyeball

The **eyeball**, or **globe**, is a sphere with multilayered walls. These walls are the sclera, choroid, and retina. Another term for the eyeball is **orb** (ōrb) (Figure 14–4).

Sclera

The **sclera** (sklehr-ah) is the fibrous outer layer of the eye that maintains the shape of the eye. It is sometimes called the **white of the eye**. The combining form **scler/o** means sclera, or hard. The thin membrane covering the sclera is the **episclera** (ehp-ih-sklehr-ah).

The anterior portion of the sclera is transparent and is called the **cornea** (kōr-nē-ah). The cornea provides most of the focusing power of the eye. The combining forms **corne/o** and **kerat/o** mean cornea. **Descemet's membrane** (dehs-eh-māz mehm-brān) is the innermost or deepest layer of the cornea.

Choroid

The **choroid** (kō-royd) is the opaque middle layer of the eyeball that contains blood vessels and supplies blood for the entire eye. **Opaque** (ō-pāk) means that light cannot pass through. The **tapetum lucidum** (tah-pē-duhm loo-sehd-uhm) is the brightly colored iridescent reflecting tissue layer of the choroid of most species. The tapetum lucidum is also called **choroid tapetum**. The **tapetum nigrum** (tah-pē-duhm nī-gruhm) is the black pigmented tissue layer of the choroid in some species. *Tapetum* is the medical term for a layer of cells. The combining form **choroid/o** means choroid. Associated with the choroid are the iris, pupil, lens, and ciliary muscles. The **iris** (ī-rihs) is the pigmented muscular diaphragm of the choroid that surrounds the pupil. The iris is composed of muscle fiber rings that contract or relax to change the size of the

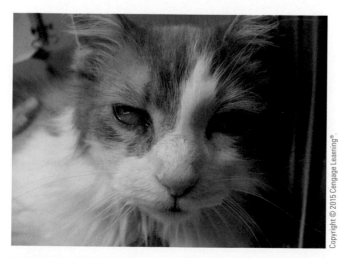

FIGURE 14–3 Third eyelid, or nictitans, in a cat.

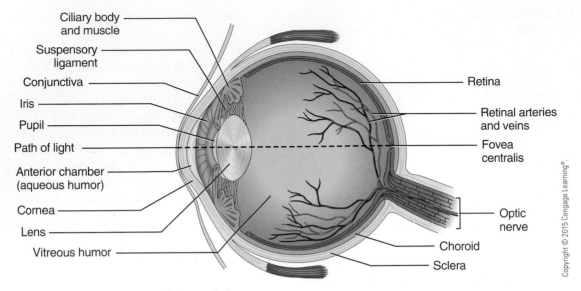

FIGURE 14–4 Cross section of the eyeball.

Labels (clockwise from top left):
Ciliary body and muscle
Suspensory ligament
Conjunctiva
Iris
Pupil
Path of light
Anterior chamber (aqueous humor)
Cornea
Lens
Vitreous humor
Retina
Retinal arteries and veins
Fovea centralis
Optic nerve
Choroid
Sclera

pupil and thus regulate the amount of light entering the lens. The **corpora nigra** (kōr-pōr-ah nī-grah) is the black pigmentation at the edge of the iris in equine and ruminants. The combining forms **ir/i, ir/o, irid/o,** and **irit/o** refer to the iris of the eye.

The **pupil** (pū-pihl) is the circular opening in the center of the iris. The combining forms **pupill/o** and **core/o** mean pupil. Muscles in the iris control the amount of light entering the pupil. To decrease the amount of light entering the eye, the iris muscles contract and make the opening smaller. Making the opening smaller is called **constriction**; when used in reference to pupillary constriction, it is called **miosis** (mī-ō-sihs). To increase the amount of light entering the eye, the iris muscles relax and make the opening larger. Making the opening larger is called **dilation**; when used in reference to pupillary dilation, it is called **mydriasis** (mih-drī-ah-sihs).

The **lens** is the clear, flexible, avascular, curved capsule located posterior to (behind) the iris and pupil. The shape of the lens is altered by the ciliary muscles. Adjusting the shape of the lens to improve near or far vision is known as accommodation. The varying shape of the lens affects the angle at which light rays enter the retina. The combining form **phac/o** means lens of the eye.

The **ciliary** (sihl-ē-ər-ē) **body** is the thickened extension of the choroid that assists in accommodation or adjustment of the lens. The **ciliary muscles**, located in the ciliary body, are muscles that adjust the shape and thickness of the lens. These adjustments make it possible for the lens to refine the focus of light rays on the retina.

Working Together

Parts of the sclera and choroid sometimes are referred to together. Examples of these terms are as follows:

- **iridocorneal** (ihr-ihd-ō-kōr-nē-ahl) = pertaining to the iris and cornea.
- **uvea** (yoo-vē-ah) = term used to describe the iris, ciliary body, and choroid.
- **limbus** (lihm-buhs) = term used for the corneoscleral junction.

Retina

The **retina** (reht-ih-nah) is the nervous tissue layer of the eye that receives images and lines the posterior chamber of the eye. The combining form **retin/o** means retina.

The retina contains specialized cells called rods and cones that convert visual images to nerve impulses that travel from the eye to the brain via the optic nerve. **Rods** are specialized cells of the retina that react to light, and **cones** are specialized cells of the retina that react to color and fine detail.

The **optic** (ohp-tihck) **disk** is the region of the eye where nerve endings of the retina gather to form the optic nerve. It is also called the **blind spot** because it does not contain any rods or cones.

The **macula lutea** (mahck-yoo-lah lū-tē-ah) is a centrally depressed, clearly defined yellow area in the center of the retina. The macula lutea surrounds a small depression called the fovea centralis. The **fovea centralis** (fō-vē-ah sehn-trah-lihs) contains the greatest concentration of cones in the retina (no rods) and the area of sharpest vision. The combining form **macul/o** means spot; **lute/o** is the combining form for yellow; the combining form for pit is **fove/o**. The term *macula* is used in other parts of the body, such as the ear and kidney.

The **fundus** (fuhn-duhs) is the part of the retina opposite the lens that includes the retina, optic disc, macula lutea, and fovea centralis (Figure 14–5).

Eye Chambers

The eye is divided into parts to make identification and location of structures easier. The **anterior segment** (also known as the aqueous chamber) is the cranial one-third of the eyeball and is divided into anterior and posterior chambers. The **anterior chamber** is the eye cavity located between the caudal surface of the cornea and the cranial surface of the iris. The **posterior chamber** is the eye cavity located between the caudal surface of the iris and the cranial surface of the lens.

The anterior and posterior chambers of the eye are filled with a watery fluid called **aqueous humor** (ah-kwē-uhs hū-mər). Aqueous humor is the anterior segment fluid that nourishes the intraocular structures. The combining form **aque/o** means water. Humor is any clear body fluid.

The caudal two-thirds of the eyeball is called the **vitreous chamber**. **Vitreous humor** (viht-rē-uhs hū-mər), or vitreous, is the soft, clear, jellylike mass that fills the vitreous chamber. The combining form **vitre/o** means glassy (Figure 14–6).

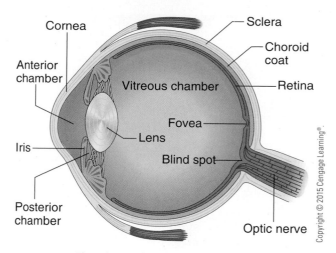

FIGURE 14–6 Chambers of the eye. The anterior and posterior chambers of the eye contain aqueous humor. The vitreous chamber of the eye contains vitreous humor.

VISION

Vision occurs when light rays enter the eye through the cornea, pass through the lens, and travel to the retina. The image is focused on the retina and is then transmitted to the optic nerve. Stimulations are transmitted from the optic nerve, to the optic chiasm, to the midbrain, and to the visual cortex of the occipital lobe of the cerebrum.

Accommodation (ah-kohm-ō-dā-shuhn) is the process of eye adjustments for seeing objects at various distances. These adjustments are accomplished through changes in lens shape.

Refraction (rē-frahck-shuhn) is the process of the lens bending the light rays to help them focus on the retina. Refraction also is called **focusing**.

Convergence (kohn-vər-jehns) is simultaneous inward movement of both eyes. Convergence usually occurs in an effort to maintain single binocular vision as an object approaches.

Acuity (ah-kū-ih-tē) means sharpness or acuteness, usually used in reference to vision.

TEST ME: EYES

Diagnostic tests performed on the eyes include the following:

- **conjunctival** (kohn-juhnck-tī-vahl) **scrape** = diagnostic test using an instrument to peel cells from the conjunctiva so that they can be viewed microscopically.
- **electroretinography** (ē-lehck-trō-reh-tihn-ohg-rah-fē) = procedure of recording the electrical activity of the retina. An **electroretinogram** (ē-lehck-trō-reh-tihn-ō-grahm) is the record of electrical activity of the retina; abbreviated ERG.

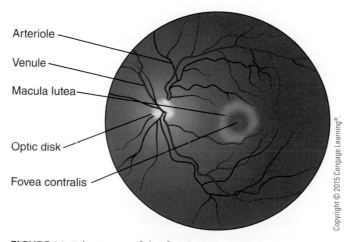

FIGURE 14–5 Anatomy of the fundus.

FIGURE 14–7 Fluorescein dye stain is used to detect whether corneal defects are present in this dog.

- **fluorescein** (fluhr-ō-sēn) **dye stain** = diagnostic test to detect corneal injury by placing dye on the surface of the cornea (Figure 14–7).
- **goniometry** (gō-nē-ah-meh-trē) = procedure to measure the drainage angle of the eye. **Gon/i** is the combining form for angle or seed.
- **menace response** (mehn-ahs rē-spohns) = diagnostic test to detect vision in which movement is made toward the animal to test whether it will see movement and try to close its eyelids.
- **ophthalmoscope** (ohp-thahl-mō-skōp) = instrument used for ophthalmoscopy (Figure 14–8).
- **ophthalmoscopy** (ohp-thahl-mohs-kō-pē) = procedure used to examine the interior eye structures; may be direct or indirect (Figure 14–9); also known as **funduscopy** (fuhn-duhs-kō-pē).
- **palpebral** (pahl-pē-brahl) **reflex** = diagnostic test in which the eye should blink in response to touch

FIGURE 14–8 An ophthalmoscope is used to examine the internal structures of the eye.

(a)

(b)

FIGURE 14–9 Ophthalmoscopy. Two methods of ophthalmoscopy are (a) direct and (b) indirect.

to the medial canthus of the eye. This test is used to make neurologic assessment of cranial nerves V and VII and to assess depth of anesthesia.

- **pupillary light reflex** (pū-puhl-ār-ē līt rē-flehcks) = response of pupil to light; abbreviated PLR. When light is shown in the pupil, constriction should take place.
- **Schirmer** (shər-mər) **tear test** = diagnostic test using a graded paper strip to measure tear production (Figure 14–10).
- **slit lamp examination** = visual testing of the cornea, lens, fluids, and membranes of the interior of the eye using a narrow beam of light.
- **tonometry** (tō-nohm-eh-trē) = procedure using an instrument to measure intraocular pressure

FIGURE 14–10 The Schirmer tear test is used to measure tear production in this dog.

(a)

(b)

FIGURE 14–11 Detection of intraocular pressure by tonometry. (a) A Schiotz tonometer, an example of an applanation tonometer, measures corneal resistance to an applied force. (b) A pneumatic tonometer detects intraocular pressure by measuring corneal resistance to a puff of air.

indirectly. **Intraocular pressure**, abbreviated IOP, is determined by the resistance of the eyeball to indentation by an applied force. Tonometry can be applanation, in which the instrument and weights are placed on the cornea to measure resistance, or pneumatic, in which a puff of air is blown against the cornea to flatten it slightly to measure resistance. A **Schiotz** (shē-ohtz) **tonometer** is an example of an applanation tonometer (Figure 14–11).

In a description of the eyes, the abbreviations *OD, OS,* and *OU* are used. OD means right eye (oculus dexter), OS means left eye (oculus sinister), and OU means both eyes (oculus uterque). Similarly AD, AS, and AU are used to describe the right, left, and both ears, respectively. *A* stands for auris.

Examination of the eyes may entail treatment with chemicals. Anesthetics may be used so that tonometers can be placed on the cornea and intraocular pressure can be measured. Retinal examination may be aided with the use of cycloplegics or mydriatics. **Cycloplegics** (sī-klō-plē-jihcks) cause paralysis of the ciliary muscle that may aid in dilation of the pupil and ease the pain of ciliary muscle spasms. **Mydriatics** (mihd-rē-ah-tihcks) are agents that dilate the pupil.

- **anisocoria** (ahn-ih-sō-kō-rē-ah) = condition of unequal pupil size. **Anis/o** is the combining form meaning unequal (**an-** is not; **iso-** is equal) (Figure 14–12).
- **anophthalmos** (ahn-ohp-thahl-mōs) = without development of one or both eyes.
- **aphakia** (ah-fahk-ē-ah) = absence of the lens.
- **blepharitis** (blehf-ah-rī-tihs) = inflammation of the eyelid.

PATHOLOGY: EYES

Pathologic conditions of the eyes include the following:

- **amblyopia** (ahm-blē-ō-pē-ah) = dimness or loss of sight without detectable eye disease. The combining form **ambly/o** means dim.

FIGURE 14–12 Anisocoria.

- **blepharoptosis** (blehf-ah-rō-tō-sihs) = drooping of the upper eyelid.
- **blepharospasm** (blehf-rō-spahzm) = rapid, involuntary contractions of the eyelid.
- **blindness** (blīnd-nehs) = inability to see.
- **buphthalmos** (boof-thahl-muhs) = abnormal enlargement of the eye.
- **cataract** (kaht-ah-rahckt) = cloudiness or opacity of the lens (Figure 14–13).
- **chalazion** (kah-lā-zē-ohn) = localized swelling of the eyelid resulting from the obstruction of a sebaceous gland of the eyelid.
- **conjunctivitis** (kohn-juhnck-tih-vī-tihs) = inflammation of the conjunctiva.
- **corneal ulceration** (kōr-nē-ahl uhl-sihr-ah-shuhn) = surface depression on the cornea (Figure 14–14).
- **cyclopia** (sī-klō-pē-ah) = congenital anomaly characterized by a single orbit.
- **dacryoadenitis** (dahck-rē-ō-ahd-ehn-ī-tihs) = inflammation of the lacrimal gland.
- **dacryocystitis** (dahck-rē-ō-sihs-tī-tihs) = inflammation of the lacrimal sac and abnormal tear drainage.
- **diplopia** (dih-plō-pē-ah) = double vision. **Dipl/o** is the combining form for double; **-opia** is the suffix meaning vision.
- **distichiasis** (dihs-tē-kē-ī-ah-sihs) = abnormal condition of a double row of eyelashes that usually result in conjunctival injury. **Distichia** (dihs-tēck-ē-ah) is a double row of eyelashes.
- **ectropion** (ehck-trō-pē-ohn) = eversion, or turning outward, of the eyelid.
- **entropion** (ehn-trō-pē-ohn) = inversion, or turning inward, of the eyelid (Figure 14–15).

FIGURE 14–14 Corneal ulceration in a cat secondary to keratoconjunctivitis sicca.

(a)

FIGURE 14–13 Cataract in a dog.

(b)

FIGURE 14–15 (a) Entropion in a lamb. (b) Note the severe conjunctivitis associated with entropion.

Copyright © 2015 Cengage Learning®.

Courtesy of Ron Fabrizius, DVM, Diplomate ACT

Courtesy of Ron Fabrizius, DVM, Diplomate ACT

Courtesy of Dr. Mark Jackson, University of Glasgow

- **epiphora** (ē-pihf-ōr-ah) = excessive tear production.
- **episcleritis** (ehp-ih-sklehr-ī-tihs) = inflammation of the tissue of the cornea.
- **exophthalmos** (ehcks-ohp-thahl-mōs) = abnormal protrusion of the eyeball; sometimes referred to as **exophthalmosis** (ehcks-ohp-thahl-mō-sihs).
- **floaters** (flō-tərz) = particles suspended in the vitreous fluid that cast shadows on the retina; also called **vitreous floaters**.
- **glaucoma** (glaw-kō-mah) = group of disorders resulting from elevated intraocular pressure (Figure 14–16).
- **hordeolum** (hōr-dē-ō-luhm) = infection of one or more glands of the eyelid; also called a **stye** (stī).
- **hypertropia** (hī-pər-trō-pē-ah) = deviation of one eye upward.
- **hypopyon** (hī-pō-pē-ohn) = pus in the anterior chamber of the eye.
- **hypotropia** (hī-pō-trō-pē-ah) = deviation of one eye downward.
- **iritis** (ī-rī-tihs) = inflammation of the iris.
- **keratitis** (kehr-ah-tī-tihs) = inflammation of the cornea (Figure 14–17).
- **keratoconjunctivitis** (kehr-ah-tō-kohn-juhnck-tih-vī-tihs) = inflammation of the cornea and conjunctiva. **Infectious keratoconjunctivitis** is caused by a variety of microbes and is commonly called pinkeye (Figure 14–18). **Keratoconjunctivitis sicca** (sihka) is inflammation of the cornea and conjunctiva due to a tear deficiency and can lead to persistent, mucopurulent discharge and corneal ulceration and scarring;

FIGURE 14–17 Ulcerative keratitis in a guinea pig.

FIGURE 14–18 A cow with pinkeye, which may be spread by flies in livestock.

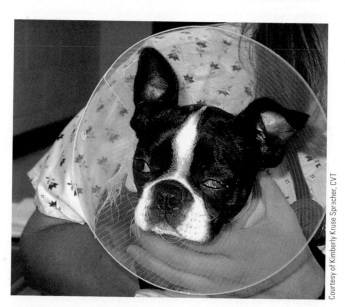

FIGURE 14–16 Glaucoma in a dog.

also called **dry eye** and abbreviated KCS. KCS occurs in dogs, cats, and horses (see Figure 14–15).
- **macular degeneration** (mahck-yoo-lahr dē-jehn-ər-ā-shuhn) = condition of central vision loss.
- **microphthalmia** (mī-krohf-thahl-mē-ah) = abnormally small eyes; also called **microphthalmos** (mī-krohp-thahl-mōs).
- **monochromatism** (mohn-ō-krō-mah-tihzm) = lack of ability to distinguish colors; also called **color blindness**. **Mono-** is the prefix meaning one; **chrom/o** is the combining form for color.
- **nictitating gland prolapse** (nihck-tih-tāt-ihng glahnd prō-lahpz) = hypertrophy and protrusion of the conjunctival fold of the medial canthus; occurs in puppies and more commonly in some breeds of dog (Figure 14–19). Also called **cherry eye** or **third eyelid prolapse**.

FIGURE 14–19 Cherry eye in a puppy. In the acute stage, the red glandular mass swells and protrudes over the edge of the nictitating membrane, and there may be a mucopurulent discharge. The swelling may recede for short periods; it often remains prolapsed. Since this structure contains a tear gland, it should be preserved if possible; however, surgical removal is often needed. Surgical removal may lead to keratoconjunctivitis sicca (KCS).

- **nuclear sclerosis** (nū-klē-ahr sklehr-ō-sihs) = drying out of the lens with age.
- **nyctalopia** (nihck-tah-lō-pē-ah) = condition of inability or difficulty seeing at night; also called **night blindness**.
- **nystagmus** (nī-stahg-muhs) = involuntary, constant, rhythmic movement of the eye.
- **ophthalmoplegia** (ohp-thahl-mō-plē-jē-ah) = paralysis of eye muscles.
- **panophthalmitis** (pahn-ohp-thahl-mih-tī-tihs) = inflammation of all eye structures.
- **papilledema** (pahp-ehl-eh-dē-mah) = swelling of the optic disk.
- **photophobia** (fō-tō-fō-bē-ah) = fear or intolerance of light.
- **proptosis** (prohp-tō-sihs) = displacement of the eye from the orbit.
- **retinal detachment** (reht-ih-nahl dē-tahch-mehnt) = separation of the nervous layer of the eye from the choroid; also called **detached retina**.

- **retinopathy** (reht-ih-nohp-ah-thē) = any disorder of the retina.
- **scleral injection** (skleh-rahl ihn-jehck-shuhn) = dilation of blood vessels into the sclera (Figure 14–20).
- **scleritis** (skleh-rī-tihs) = inflammation of the sclera.
- **strabismus** (strah-bihz-muhs) = disorder in which the eyes are not directed in a parallel manner; deviation of one or both eyes (Figure 14–21). **Convergent strabismus** (kohn-vər-jehnt strah-bihz-muhs) is deviation of the eyes toward each other; also known as **crossed eyes** and **esotropia** (ehs-ō-trō-pē-ah). The suffix **-tropia** means turning; the prefix **eso-** means inward. **Divergent strabismus** (dī-vər-jehnt strah-bihz-muhs) is deviation of the eyes away from each other; also

FIGURE 14–20 Scleral injection in a dog.

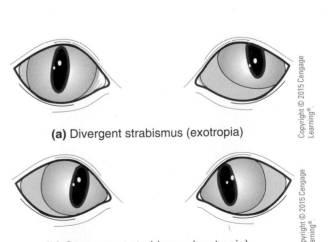

(a) Divergent strabismus (exotropia)

(b) Convergent strabismus (esotropia)

FIGURE 14–21 Strabismus. (a) Divergent strabismus (exotropia). (b) Convergent strabismus (esotropia).

called **exotropia** (ehck-sō-trō-pē-ah). The prefix **exo-** means outward.

- **synechia** (sī-nēk-ē-ah) = adhesion that binds the iris to an adjacent structure; plural is **synechiae** (sī-nēk-ē-ā).
- **uveitis** (yoo-vē-ī-tihs) = inflammation of the uvea.

PROCEDURES: EYES

Procedures performed on the eyes include the following:

- **blepharectomy** (blehf-ār-ehck-tō-mē) = surgical removal of all or part of the eyelid.
- **blepharoplasty** (blehf-ār-rō-plahs-tē) = surgical repair of the eyelid.
- **blepharorrhaphy** (blehf-ār-ōr-rah-fē) = suturing together of the eyelids; also called **tarsorrhaphy** (tahr-soh-rah-fē).
- **blepharotomy** (blehf-ār-ah-tō-mē) = incision of the eyelid; also called **tarsotomy** (tahr-soh-tō-mē).
- **canthectomy** (kahn-thehck-tō-mē) = surgical removal of the corner of the eyelid.
- **canthoplasty** (kahn-thō-plahs-tē) = surgical repair of the palpebral fissure.
- **canthotomy** (kahn-thoh-tō-mē) = incision into the corner of the eyelid.
- **conjunctivoplasty** (kohn-juhnck-tī-vō-plahs-tē) = surgical repair of the conjunctiva.
- **dacryocystectomy** (dahck-rē-ō-sihs-tehck-tō-mē) = surgical removal of the lacrimal sac.
- **dacryocystotomy** (dahck-rē-ō-sihs-tah-tō-mē) = incision into the lacrimal sac.
- **enucleation** (ē-nū-klē-ā-shuhn) = removal of the eyeball (Figure 14–22).

- **extracapsular extraction** (ehcks-trah-kahp-soo-lahr ehcks-trahck-shuhn) = removal of a cataract that leaves the posterior lens capsule intact.
- **goniotomy** (gō-nē-ah-tō-mē) = incision into the anterior chamber angle for treatment of glaucoma.
- **intracapsular extraction** (ihn-trah-kahp-soo-lahr ehcks-trahck-shuhn) = cataract removal that includes the surrounding capsule.
- **iridectomy** (ihr-ih-dehck-tō-mē) = surgical removal of a portion of the iris.
- **keratectomy** (kehr-ah-tehck-tō-mē) = surgical removal of part of the cornea.
- **keratocentesis** (kehr-ah-tō-sehn-tē-sihs) = puncture of the cornea to allow aspiration of aqueous humor.
- **keratoplasty** (kehr-ah-tō-plahs-tē) = surgical repair of the cornea (may include corneal transplant).
- **keratotomy** (kehr-ah-toh-tō-mē) = incision into the cornea.
- **lacromotomy** (lahck-rō-moh-tō-mē) = incision into the lacrimal gland or duct.
- **lensectomy** (lehn-sehck-tō-mē) = surgical removal of a lens (usually performed on cataracts).
- **tarsectomy** (tahr-sehck-tō-mē) = surgical removal of all or part of the tarsal plate of the third eyelid.
- **tarsorrhaphy** (tahr-sōr-ah-fē) = suturing together of the eyelids.

FUNCTIONS OF THE EAR

The **ear** is the sensory organ that enables hearing and helps to maintain balance. The combining forms **audit/o**, **aud/i**, and **ot/o** mean ear. **Acoust/o** and **acous/o** are combining forms for sound or hearing. Therefore, the term **auditory** pertains to the ear, and **acoustic** pertains to sound.

STRUCTURES OF THE EAR

The ear is divided into the outer, middle, and inner portions. Each part has its own unique structures (Figure 14–23).

Outer or External Ear

The **pinna** (pihn-ah) is the external portion of the ear that catches sound waves and transmits them to the external auditory canal. The pinna also is known as the **auricle** (awr-ihck-kuhl). The combining form

FIGURE 14–22 Enucleation.

Courtesy of Kimberly Kruse Sprecher, CVT

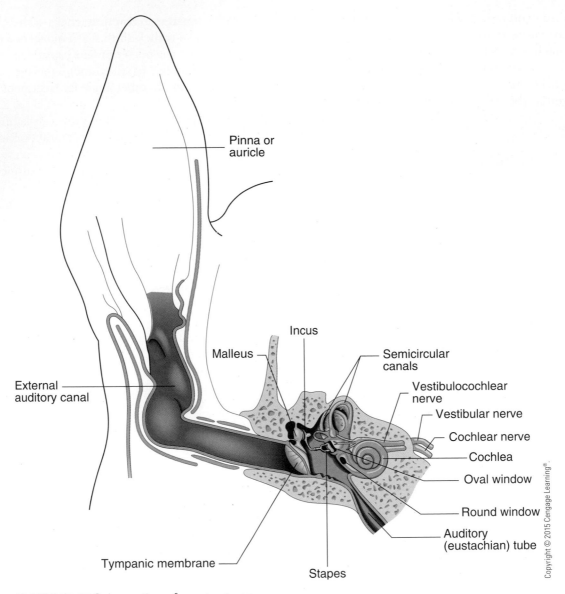

FIGURE 14–23 Cross section of ear structures.

pinn/i means external ear. The combining forms **aur/i** and **aur/o** mean external ear but are commonly used simply to mean ear.

The **external auditory canal** is the tube that transmits sound from the pinna to the tympanic membrane. The external auditory canal is also known as the **external auditory meatus**. Glands that line the external auditory canal secrete **cerumen** (seh-roo-mehn), which is commonly known as **earwax**.

Middle Ear

The middle ear begins with the **eardrum**. The medical term for eardrum is **tympanic membrane** (tihm-pahn-ihck mehm-brān). The tympanic membrane is the tissue that separates the external ear from the middle ear.

When sound waves reach the tympanic membrane, it transmits sounds to the ossicles. The combining forms **tympan/o** and **myring/o** mean eardrum.

The **auditory ossicles** (aw-dih-tōr-ē ohs-ih-kulz) are three little bones of the middle ear that transmit sound vibrations. The three bones are as follows:

- **malleus** (mahl-ē-uhs) = auditory ossicle known as the hammer.
- **incus** (ihng-kuhs) = auditory ossicle known as the anvil.
- **stapes** (stā-pēz) = auditory ossicle known as the stirrup.

The **eustachian tube** (yoo-stā-shuhn toob) follows the auditory ossicles. The eustachian tube, or **auditory tube**, is the narrow duct that leads from the middle ear to the nasopharynx. It helps equalize air pressure in the middle ear with that of the atmosphere. Equalizing air pressure on both sides of the eardrum is essential to hearing.

The **oval window** (ō-vahl wihn-dō), located at the base of the stapes, is the membrane that separates the middle and inner ear.

The **round window** (rownd wihn-dō) is the membrane that receives sound waves through fluid after they have passed through the cochlea.

The **tympanic bulla** (tihm-pahn-ihck buhl-ah) is the osseous chamber at the base of the skull. **Bulla** (buhl-ah) is the medical term for large vesicle.

Inner Ear

The inner ear contains sensory receptors for hearing and balance (Figure 14–24). The inner ear consists of three spaces in the temporal bone assembled in the **bony labyrinth** (lahb-ih-rihnth). The combining form **labyrinth/o** means maze, labyrinth, and inner ear. The bony labyrinth is filled with a waterlike fluid called **perilymph** (pehr-eh-lihmf). A membranous sac is suspended in the perilymph and follows the shape of the bony labyrinth. This membranous labyrinth is filled with a thicker fluid called **endolymph** (ehn-dō-lihmf).

The bony labyrinth is divided into three parts:

• **vestibule** = located adjacent to the oval window and between the semicircular canals and cochlea.

The vestibule (and semicircular canals) contain specialized mechanoreceptors for balance and position.

• **semicircular canals** = located adjacent to the vestibule. The semicircular canals are oriented at right angles to each other. The three canals are the **vestibular**, **tympanic**, and **cochlear**. Each canal has a dilated area called the **ampulla** that contains sensory cells with hairlike extensions. These sensory cells are suspended in endolymph, and when the head moves, the hairlike extensions bend. This bending generates nerve impulses for the regulation of position.

• **cochlea** (kōck-lē-ah) = spiral-shaped passage that leads from the oval window to the inner ear. The combining form **cochle/o** means snail or spiral. The following tube and organ are located in the cochlea:

  ◦ **cochlear duct** (kōck-lē-ahr duhckt) = membranous tube in the bony cochlea that is filled with endolymph. Endolymph vibrates when sound waves strike it.

  ◦ **organ of Corti** (ōr-gahn ohf kōr-tē) = spiral organ of hearing located in the cochlea that receives and relays vibrations. Specialized cells in the organ of Corti generate nerve impulses when they are bent by the movement of endolymph. Nerve impulses are relayed to the auditory nerve fibers that transmit them to the cerebral cortex.

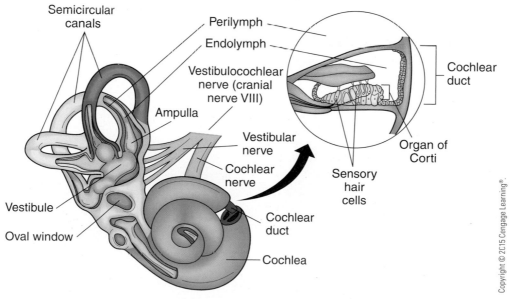

FIGURE 14–24 The inner ear. The bony labyrinth (semicircular canals, vestibule, and cochlea) is the hard outer wall of the entire inner ear. The membranous labyrinth is located in the bony labyrinth. The membranous labyrinth is surrounded by perilymph and is filled with endolymph.

MECHANISM OF HEARING

Sound waves enter the ear through the pinna, travel through the external auditory canal, and strike the tympanic membrane. This is called **air conduction**. As the tympanic membrane vibrates, it moves the ossicles. The ossicles conduct the sound waves through the middle ear. This is called **bone conduction**. Sound vibrations reach the inner ear via the round window. The structures of the inner ear respond to sound waves that displace fluid in the inner ear. Stimulation of hair cells in the organ of Corti initiates a nerve impulse that is transmitted to the vestibulocochlear nerve.

This nerve impulse is then relayed to the brain. This is called **sensorineural conduction** (Figure 14–25).

MECHANISM OF EQUILIBRIUM

In addition to the structures listed previously, the ear also has structures that maintain equilibrium. **Equilibrium** (ē-kwihl-ih-brē-uhm) is the state of balance. The sense of equilibrium is a combination of static equilibrium (maintaining the position of the head relative to gravity) and dynamic equilibrium (maintaining balance in response to rotational or angular movement).

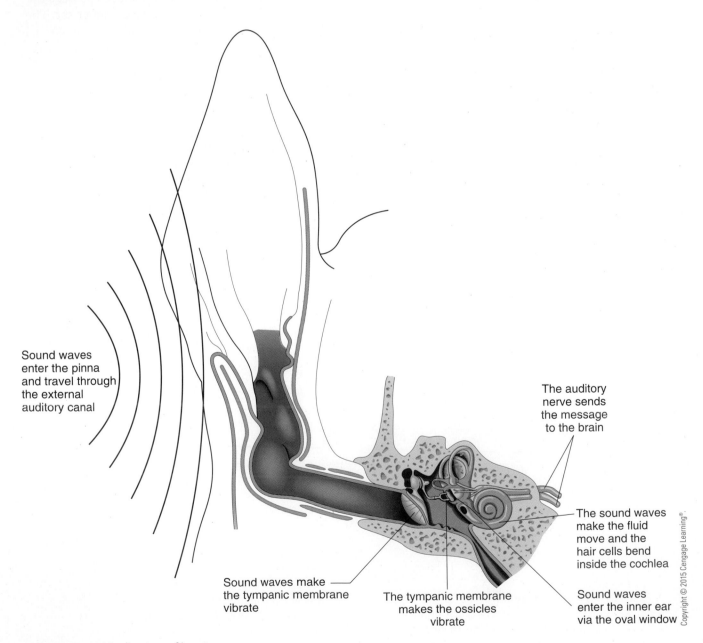

Sound waves enter the pinna and travel through the external auditory canal

The auditory nerve sends the message to the brain

The sound waves make the fluid move and the hair cells bend inside the cochlea

Sound waves make the tympanic membrane vibrate

The tympanic membrane makes the ossicles vibrate

Sound waves enter the inner ear via the oval window

FIGURE 14–25 Mechanism of hearing.

Do You Hear What I Hear?

Sound waves enter the pinna, travel through the external auditory canal, and strike the tympanic membrane. } **air conduction**

The tympanic membrane vibrates.
Ossicles conduct sound waves through the middle ear. } **bone conduction**

Sound vibrations enter the inner ear via the oval window.
Structures of the inner ear respond to sound waves.
Sound waves initiate nerve impulses that are relayed to the brain. } **sensorineural conduction**

(a)

(b)

Static equilibrium is controlled by organs in the vestibule of the inner ear. The membranous labyrinth inside the vestibule has a **saccule** (sahck-yoo-uhl) and **utricle** (yoo-trih-kuhl), which are small, hairlike sacs. The saccule and utricle contain a **macula** (mahck-yoo-lah), an organ consisting of hair cells covered by a gelatinous mass containing **otoliths** (ō-tō-lihthz), which are small stones. When the head is in an upright position, the hairs are straight. When the head tilts or bends, the otoliths and gelatinous mass move in response to gravity (Figure 14–26). As the gelatinous mass moves, it bends some of the hairs on the receptor cells, which in turn initiates an impulse that stimulates the vestibular branch of cranial nerve VIII. The impulse travels from the vestibular branch of cranial nerve VIII to the brain, which interprets the information and sends motor impulses to the appropriate muscles to maintain balance.

Dynamic equilibrium is controlled by the semicircular canals. Three semicircular canals in the inner ear lie in planes at right angles to each other. At the base of each canal is a swelling called the ampulla, which contains sensory organs called **cristae** (krihs-tā). Each crista contains sensory hair cells and a gelatinous mass. When the head turns rapidly, the semicircular canals move with the head, but the endolymph within the membrane of the semicircular canals remains stationary. The fluid in the semicircular canals pushes against the gelatinous mass, which bends some of the hairs on the hair cells. This triggers a sensory impulse to the

FIGURE 14–26 The role of the macula in maintaining static equilibrium. (a) When an animal's head is upright, gravity causes otoliths to press on hair cells, which initiate impulses to the brain. (b) When an animal's head is bent, the gelatinous mass in the macula moves and bends the hair cells. This sends an impulse to the vestibular branch of cranial nerve VIII, which in turn signals the brain to send motor impulses to the appropriate muscles to maintain balance.

vestibular branch of cranial nerve VIII. The impulse travels from the vestibular branch of cranial nerve VIII to the brain, which interprets the information and sends appropriate responses to maintain balance.

BIOLOGY BONUS

Motion sickness in a vehicle or airplane results from sudden changes in an animal's motion that occur when the organs of equilibrium are disrupted. Animals that experience motion sickness may feel dizziness and develop an upset stomach and drool or vomit. Some medications relieve the feeling of dizziness and stomach upset that accompany motion sickness.

TEST ME: EARS

Diagnostic procedures performed on the ear include the following:

- **cytology** (sī-tohl-ō-jē) = the study of cells; ear cytology involves taking a sample of exudate or cerumen from the horizontal ear canal with a cotton-tipped applicator, rolling it on a microscope slide, staining it, and examining it under a microscope; used to detect bacteria, yeast, leukocytes or abnormal cells (Figure 14–27).
- **otoscopy** (ō-tohs-kō-pē) = procedure used to examine the ear for parasites, irritation to the ear lining, discharge, and the integrity of the tympanic membrane. An **otoscope** (ō-tō-skōp) is the instrument used for otoscopy (Figure 14–28).

PATHOLOGY: EARS

Pathologic conditions of the ears include the following:

- **aural hematoma** (awr-ahl hē-mah-tō-mah) = collection or mass of blood on the outer ear. **Aural** means pertaining to the ear (external ear) (Figure 14–29).

© iStockphoto/zilli

FIGURE 14–28 An otoscope is used to examine the ear.

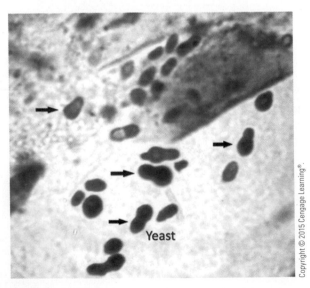

Copyright © 2015 Cengage Learning®.

FIGURE 14–27 Ear cytology of a dog with otitis externa. Yeast is found in the ear of this dog.

Copyright © 2015 Cengage Learning®.

FIGURE 14–29 Aural hematoma in a cat.

- **deafness** (dehf-nehs) = complete or partial hearing loss.
- **myringitis** (mihr-ihn-jī-tihs) = inflammation of the eardrum.
- **otalgia** (ō-tahl-jē-ah) = ear pain.
- **otitis** (ō-tī-tihs) = inflammation of the ear; usually has a second term that describes the location: **otitis externa** (ō-tī-tihs ehcks-tər-nah) = inflammation of the outer ear; **otitis media** (ō-tī-tihs mē-dē-ah) = inflammation of the middle ear; **otitis interna** (ō-tī-tihs ihn-tər-nah) = inflammation of the inner ear.
- **otomycosis** (ō-tō-mī-kō-sihs) = fungal infection of the ear.
- **otopathy** (ō-tohp-ah-thē) = disease of the ear.
- **otopyorrhea** (ō-tō-pī-ō-rē-ah) = pus discharge from the ear.
- **otorrhea** (ō-tō-rē-ah) = ear discharge.
- **panotitis** (pahn-ō-tī-tihs) = inflammation of all ear parts.
- **vertigo** (vər-tih-gō) = sense of dizziness.

What Is in a Name?

The names of parasites may seem long and cumbersome at first glance. However, many times the name of something can tell a great deal about what it does or where it lives. The scientific name of the common ear mite of dogs and cats is *Otodectes cynotis*. From the combining form **ot/o** in the name, it can be concluded that *Otodectes* are parasites that infiltrate the ear. **Dectes** means biter; therefore, *Otodectes* are biting parasites found in the ear (Figure 14–30).

FIGURE 14–30 *Otodectes* mite.

Copyright © 2015 Cengage Learning®

PROCEDURES: EARS

Procedures performed on the ears include the following:

- **ablation** (ah-blā-shuhn) = removal of a part (Figure 14–31).
- **myringectomy** (mihr-ihn-jehck-tō-mē) = surgical removal of all or part of the eardrum; also called tympanectomy (tihm-pahn-ehck-tō-mē).
- **otoplasty** (ō-tō-plahs-tē) = surgical repair of the ear.

ABBREVIATIONS RELATED TO THE OCULAR AND AUDITORY SYSTEMS

- AD = right ear (A stands for auris)
- AS = left ear
- AU = both ears
- ERG = electroretinogram
- IOP = intraocular pressure
- KCS = keratoconjunctivitis sicca
- OD = right eye (O stands for oculus; also abbreviation for overdose)
- OS = left eye
- OU = both eyes
- PLR = pupillary light response

FIGURE 14–31 Ear ablation surgery in a dog.

Courtesy of Kimberly Kruse Sprecher, CVT

REVIEW EXERCISES

Multiple Choice
Choose the correct answer.

1. The state of balance is
   a. vertigo
   b. hemostasis
   c. vestibular
   d. equilibrium

2. The outer or external ear is separated from the middle ear by the
   a. oval window
   b. round window
   c. tympanic membrane
   d. pinna

3. Another term for earwax is
   a. pinna
   b. auricle
   c. cerumen
   d. corti

4. The fibrous tissue that maintains the shape of the eye is the
   a. choroid
   b. sclera
   c. white of the eye
   d. b and c

5. The term for corner of the eye is
   a. canthus
   b. cilia
   c. cerumen
   d. cornus

6. The colored muscular layer of the eye that surrounds the pupil is the
   a. cornea
   b. choroid
   c. lens
   d. iris

7. Involuntary, constant, rhythmic movement of the eyeball is called
   a. ectropion
   b. nystagmus
   c. strabismus
   d. entropion

8. A group of eye disorders resulting from increased intraocular pressure is
   a. ophthalmopathy
   b. glaucoma
   c. floaters
   d. hypertension

9. Opacity of the lens is called
   a. opaque
   b. turgid
   c. cataract
   d. diplopia

10. The process of the lens bending the light ray to help focus the rays on the retina is called
    a. convergence
    b. refraction
    c. humo
    d. fovea

11. The fibrous layer of clear tissue that extends over the anterior portion of the eye and is continuous with the white of the eye is the
    a. ciliary body
    b. pupil
    c. cornea
    d. iris

12. The region of the eye where nerve endings of the retina gather to form the optic nerve is called the
    a. optic disk
    b. posterior chamber
    c. sclera
    d. choroid

13. What eye structure is transparent and focuses light on the retina?
    a. conjunctiva
    b. lens
    c. vitreous humor
    d. aqueous humor

14. The meaning of palpebr/o is
    a. eyelid
    b. cornea
    c. conjunctiva
    d. eyelashes

15. Glaucoma usually is diagnosed by
    a. ophthalmoscopy
    b. slit lamp examination
    c. Schirmer tear test
    d. tonometry

16. The term for removal of a part is
    a. otalgia
    b. vertigo
    c. ablation
    d. acoustic

17. The term for pupillary dilation is
    a. miosis
    b. meiosis
    c. macula
    d. mydriasis

18. The condition of normal tear production is
    a. lacrimation
    b. nictitans
    c. vitreous
    d. acuity

19. The instrument used to visually examine the ear
    (the ear lining, presence or absence of discharge,
    and integrity of the tympanic membrane) is
    known as a(n)
    a. myringoscope
    b. mycoscope
    c. otoscope
    d. dectescope

20. What is the name of the spiral-shaped
    passage that leads from the oval window to
    the inner ear?
    a. organ of Corti
    b. semicircular canals
    c. ampulla
    d. cochlea

21. The area of sharpest vision in the retina is the
    a. cones
    b. rods
    c. macula lutea
    d. fovea centralis

22. The posterior chamber of the eye contains
    a. vitreous humor
    b. aqueous humor
    c. fovea centralis
    d. macula lutea

23. Sharpness of vision is
    a. acuity
    b. accommodation
    c. convergence
    d. refraction

24. The "A" in AD stands for
    a. acuity
    b. accurate
    c. auris
    d. aqueous

25. The abbreviation OD stands for
    a. left ear
    b. right ear
    c. left eye
    d. right eye

Matching

Match the ocular term in Column I with the definition in Column II.

Column I	Column II
1. _____ palpebra	a. iris, ciliary body, and choroid
2. _____ orbit	b. platelike frame within the upper and lower eyelids
3. _____ cilia	c. eyelid
4. _____ cornea	d. eyelashes
5. _____ conjunctiva	e. bony cavity of the skull that contains the eyeball
6. _____ tarsus	f. transparent anterior portion of the sclera
7. _____ uvea	g. mucous membrane that lines the underside of each eyelid

Match the auditory term in Column I with the definition in Column II.

Column I	Column II
8. _____ pinna	a. narrow duct that leads from the middle ear to the nasopharynx
9. _____ cerumen	b. auditory ossicle known as the hammer
10. _____ ampulla	c. auditory ossicle known as the anvil
11. _____ otoliths	d. auditory ossicle known as the stirrup
12. _____ tympanic bulla	e. earwax
13. _____ incus	f. external portion of the ear
14. _____ stapes	g. osseous chamber at the base of the skull
15. _____ malleus	h. tube that transmits sound from the pinna to the eardrum
16. _____ eustachian tube	i. dilated area in the semicircular canals
17. _____ external auditory canal	j. small stones in the saccule and utricle

Fill in the Blanks

Write the correct term in the space provided.

1. Lacrim/o and dacry/o mean _____.

2. Tympan/o and myring/o mean _____.

3. Pinn/i, aur/i, and aur/o mean _____.

4. Opt/i, opt/o, optic/o, ocul/o, and ophthalm/o mean _____.

5. Acoust/o and acous/o mean _____.

True or False

If the statement is true, write T on the line. If the statement is false, write F on the line.

1. _____The aqueous humor is a thick, gelatinous substance.

2. _____Semicircular canals in the ears are important for hearing.

3. _____Refraction is focusing on distant objects.

4. _____Tonometry measures intraocular pressure.

5. _____The lens of the eye is avascular because it has no blood supply.

6. _____Periorbital structures are located around the eye.

7. _____The medical term for pinkeye is blepharoptosis.

8. _____The ossicles of the ear are the malleus, incus, and stapes.

9. _____The "S" in KCS stands for sclera.

10. _____A slit lamp examination is used to visualize the ear.

Spell Check

Circle the term that is spelled correctly.

1. double row of eyelashes: dystichiasys   distechiasis   distichiasis

2. groups of disorders resulting from elevated   glaukoma   glawcoma   glaucoma
intraocular pressure:

3. inflammation of the iris, ciliary body, and   uveitis   uveititis   uveytis
choroid:

4. blocking the passage of light: opague   opaque   opaiue

5. narrow duct that leads from the middle ear   eustayshian tube   eustaichian tube   eustachian tube
to the nasopharynx:

Crossword Puzzles

Eye Terms

Supply the correct term in the appropriate space for the definition listed.

Across

3 eversion of the eyelid

6 inflammation of the eyelid

9 condition of unequal pupil size

12 removal of the eyeball

14 any disorder of the retina

16 incision into the lacrimal sac

17 disorder in which the eyes are not directed in a parallel manner

Down

1 another term for third eyelid

2 inflammation of the cornea and mucous membranes that line the eyelid

4 surface depression on the cornea

5 intolerance of light

7 inversion of the eyelid

8 paralysis of the eye muscles

10 corneoscleral junction

11 incision into the cornea

13 suturing together of the eyelids

15 displacement of the eye from the orbit

Ear Terms

Across

3 another term for hammer

7 external portion of the ear

10 another term for maze or inner ear

11 inflammation of the eardrum

Down

1 collection of blood on the outer ear

2 removal of a part by cutting

4 another term for stapes

5 pus discharge from the ear

6 inflammation of the ear

8 another term for anvil

9 another term for pinna

Case Studies
Define the underlined terms in each case study.

A 6-yr-old DSH M/N cat was presented for inappetence and blepharospasm. PE revealed T = 103.4°F, HR = 200 bpm, RR = 40breaths/min, mm = pink and dry, CRT = 1.5 sec. The conjunctiva was reddened, and the sclera was infected. Heart and lungs auscultated normally. Abdominal palpation was normal. Oral examination was normal. Ocular exam: Anterior chamber was cloudy. Iris appeared normal. Tonometer readings were OS 7 mm Hg, OD 10 mm Hg. Dx was uveitis—anterior chamber. Possible causes include infections (FIP, FeLV, toxoplasmosis), immune mediated, systemic disease, and trauma. Further diagnostic tests are being pursued by the veterinarian.

1. blepharospasm _____

2. conjunctiva _____

3. sclera _____

4. ocular _____

5. anterior chamber _____

6. iris _____

7. tonometer _____

8. OS _____

9. mm Hg _____

10. OD _____

11. uveitis _____

An 8-mo-old M black Labrador retriever was presented to the clinic for ocular discharge and rubbing at the eyes. Upon the animal's presentation to the clinic, the vital signs were normal, attitude was normal for a puppy, and bilateral mucopurulent ocular discharge was noted. Ophthalmic examination revealed normal Schirmer tear test values, a normal-looking retina, and no stain retention via fluorescence staining. Upon examination of the eyelids, it was noted that the dog had blepharospasm and entropion was noted. The entropion probably was the cause of the eye infection because the eyelashes were brushing against the cornea. Blepharoplasty was recommended to this owner. Topical antibiotics were dispensed pending surgery.

12. ocular _____

13. bilateral mucopurulent ocular discharge _____

14. ophthalmic _____

15. Schirmer tear test _____

16. retina _____

17. fluorescence staining _____

18. blepharospasm _____

19. entropion _____

20. cornea _____

21. blepharoplasty _____

22. topical _____

Buddy, a 3-yr-old M/N springer spaniel, was presented to the clinic with <u>recurrent</u> ear problems. Previously diagnosed <u>bilateral</u> ear problems responded well to treatment. Buddy had been seen 2 wk earlier for ear problems, and <u>otitis externa</u> was diagnosed. Ear <u>cytology</u> revealed that Buddy had a severe yeast infection of both ears, and an antifungal drug was prescribed. On today's examination, <u>mucopurulent</u> discharge was noted <u>AU.</u> <u>Otoscopic</u> examination revealed that <u>AD</u> was <u>hyperemic</u> and <u>hyperkeratotic,</u> and the ear was so swollen that the <u>tympanic membrane</u> could not be visualized. <u>AS</u> was hyperemic and hyperkeratotic, and the tympanic membrane was intact. Ear cytology revealed yeast, and the dog was getting worse. Oral antifungal drugs and anti-inflammatory drugs were prescribed. The owner was advised that if the problem did not resolve, <u>ear ablation</u> surgery may be warranted.

23. recurrent _____

24. bilateral _____

25. otitis externa _____

26. cytology _____

27. mucopurulent _____

28. AU _____

29. otoscopic _____

30. AD _____

31. hyperemic _____

32. hyperkeratotic _____

33. tympanic membrane _____

34. AS _____

35. ear ablation _____

Critical Thinking Exercise

The following story and questions contain words and concepts from this chapter and are designed to stimulate critical thinking through class discussion, Internet searches, or brief writing responses.

As a veterinary technician in a busy, small animal clinic, you are responsible for maintaining equipment in the examination rooms. You also want to make sure that procedures are done properly, so animals do not contract diseases while in the clinic and that owners comply with treating their animals based on veterinarian recommendations.

While cleaning the otoscope in the examination room, you begin to wonder if this basic piece of equipment is being used properly by the staff. What are some recommendations you want to make to the staff to make sure the otoscope is used properly in regards to the following:

1. Light source

2. Speculum (plastic piece that goes into the opening of the ear)

3. Technique

Chapter 15
FEED AND PROTECT ME

Objectives

Upon completion of this chapter, the reader should be able to:

- Identify the major structures of the hematologic, lymphatic, and immune systems

- Describe the functions of the hematologic, lymphatic, and immune systems

- Recognize, define, spell, and pronounce the major terms related to diagnosis, pathology, and treatment of the hematologic, immune, and lymphatic systems

- Describe how tumors affect an animal's body

- Recognize, define, spell, and pronounce terms related to diagnosis, pathology, and treatment of tumors

- Interpret the meaning of abbreviations of the hematologic, lymphatic, and immune systems

- Interpret the meaning of abbreviations in oncology

HEMATOLOGIC SYSTEM

The hematologic system encompasses the production of blood and the transport of blood throughout the body. Blood is a type of connective tissue and consists of cells contained in a liquid matrix. Blood is dependent on the heart to move it around the body.

Functions of Blood

Blood supplies body tissues with oxygen, nutrients (such as vitamins, minerals, glucose, and amino acids), and various chemicals such as hormones. Blood transports waste products to various organs for removal from the body as well as transports gases to and from the lungs and cells. Blood also helps maintain the body's homeostasis and plays an important role in the immune system.

Structures of Blood

Blood is composed of 55 percent liquid plasma and 45 percent formed elements. Formed elements include red blood cells, white blood cells, and clotting cells. The combining forms for blood are **hem/o** and **hemat/o**. The study of blood is **hematology** (hē-mah-tohl-ō-jē).

Blood is formed in the bone marrow. **Hematopoiesis** (hē-mah-tō-poy-ē-sihs) is the medical term for formation of blood. The suffix **-poiesis** means formation. The components of blood can be separated clinically and examined microscopically. A blood sample is collected with a needle and syringe. **Drawing blood** is a common term for collecting a blood sample. Blood can be collected in a tube that has an **anticoagulant** (ahn-tih-kō-ahg-yoo-lahnt). An anticoagulant is a substance that prevents clotting of blood. **EDTA** (ethylenediaminetetraacetic acid) and **heparin** (hehp-ahr-ihn) are types of anticoagulants found in blood tubes and are used clinically as drugs. **Coagulation** (kō-ahg-yoo-lā-shuhn) is the process of clotting. Sometimes coagulation of blood is desired after blood is placed in a collection tube. When blood coagulates, a layer of leukocytes and thrombocytes forms, which appears at the interface of the erythrocytes and plasma after blood is centrifuged. This layer that appears at the interface of the erythrocytes and plasma is called the **buffy coat** (buhf-ē kōt) (Figure 15–1).

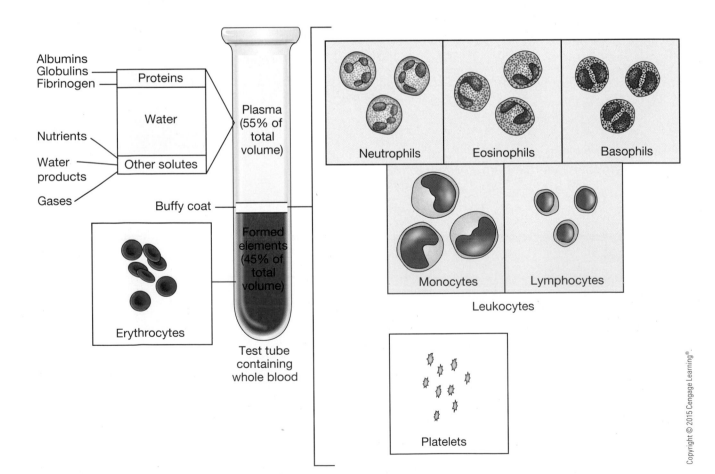

FIGURE 15–1 Blood components.

a liquid that contains suspended particles (in the case of plasma, most of the suspended particles are clotting proteins).

- Clotting proteins found in plasma include **fibrinogen** (fih-brihn-ō-jehn) and **prothrombin** (prō-throhm-bihn). The combining form **fibrin/o** means fibrin or threads of a clot, the prefix **pro-** means before, and the combining form **thromb/o** means clot. **Albumin** (ahl-byoo-mihn) is another example of a plasma protein.
- Fats also circulate in plasma. **Cholesterol** (kō-lehs-tər-ohl) and **triglyceride** (trī-glihs-ər-īd) are types of lipids that circulate in blood.

What Is Normal?

Blood is about 55 percent liquid and 45 percent cellular material, but what is the normal blood volume of an animal? Approximately 9 percent of an animal's body weight is blood.

Liquid

The liquid portion of blood consists of the following:

- **Serum** (sē-ruhm) is the liquid portion of blood with clotting proteins removed (Figure 15–2). When blood is allowed to clot, the liquid portion left is serum.
- **Plasma** (plahz-mah) is the straw-colored fluid portion of blood that transports nutrients, hormones, and waste products. Plasma contains clotting proteins. Plasma is a **colloid** (kohl-oyd),

Serum
Clot
Plasma
Sedimentation of blood

Copyright © 2015 Cengage Learning®

FIGURE 15–2 Serum versus plasma. Serum is the liquid portion of blood seen in the sample collected in the red-topped tube without anticoagulant (left), and plasma is the liquid portion of blood seen in the sample collected in the purple-topped tube containing anticoagulant (right). Although serum and plasma look similar, plasma contains clotting proteins while serum does not.

Formed Elements

Erythrocytes. An **erythrocyte** (eh-rihth-rō-sīt) is a mature red blood cell (oxygen-carrying cell) and is abbreviated RBC. The combining form **erythr/o** means red, and the suffix **-cyte** means cell. Mature mammalian erythrocytes do not contain nuclei while mature non-mammalian (birds, reptiles, amphibians, and fish) erythrocytes contain nuclei. Erythrocytes have a biconcave disc shape and contain **hemoglobin** (hē-mō-glō-bihn), a blood protein that transports oxygen. **Heme** (hēm) is the nonprotein, iron-containing portion of hemoglobin. The central pale area of an RBC that represents the thinnest part of the biconcave disc is the area of **central pallor** (sehn-trahl pahl-ohr). RBCs that have more central pallor than normal have decreased hemoglobin concentration in their cells. These RBCs are described as hypochromic and appear pale. It would seem logical then that RBCs that have less central pallor than normal would have increased hemoglobin concentration in their cells and would be described as hyperchromic; however, that term is misleading. RBCs can hold only so much hemoglobin before their cell membrane is altered. This alteration of the RBC cell membrane makes the membrane shrink so that the amount of hemoglobin is contained in a smaller space, making the cells appear darker (bluer) in color. These cells are more commonly called polychromatic. Polychromatic RBCs also may be young erythrocytes that have been released early from the bone marrow. They are typically larger and bluer in color than mature erythrocytes due to the presence of organelles such as ribosomes and mitochondria that are still present in immature red blood cells.

Erythrocytes are produced in the bone marrow. The combining form for bone marrow (and spinal cord) is **myel/o**. Erythrocytes vary in appearance from species to species.

A **reticulocyte** (reh-tihck-yoo-lō-sīt) is an immature, non-nucleated erythrocyte characterized by polychromasia (Wright's stain) or a meshlike pattern of threads (new methylene blue stain). Reticulocytes may increase as the bone marrow responds to blood loss; **reticulocytosis** (reh-tihck-yoo-lō-sī-tō-ihs) is seen with regenerative anemia.

When RBCs are no longer useful, they are destroyed by macrophages. A **macrophage** (mahck-rō-fahj or mahck-rō-fāj) is a large cell that destroys by eating (engulfing). The combining form **macr/o** means large, and the suffix **-phage** means eating. A **phagocyte** (fā-gō-sīt) is "a cell that eats." The formal definition of a phagocyte is a leukocyte that ingests foreign material.

Hematology (hē-mah-tah-lō-jē) is the study of blood. When blood cells are studied, it is important to note the cell's morphology. **Morphology** (mōr-fah-lō-jē) is the study of form or shape (Table 15–1 and Figure 15–3).

Leukocytes. A **leukocyte** (loo-kō-sīt) is a white blood cell and is abbreviated WBC. The combining form **leuk/o** means white. Leukocytes are produced in the bone marrow (and mature in other places) and function primarily in fighting disease in the body. **Leukocytopoiesis** (loo-kō-sīt-ō-poy-ē-sihs) is the production of white blood cells. The production of leukocytes is also called **leucopoiesis** (loo-kō-poy-ē-sihs).

A **granulocyte** (grahn-yoo-lō-sīt) is a cell that contains prominent grainlike structures in its cytoplasm; an **agranulocyte** (ā-grahn-yoo-lō-sīt) is a cell that does not contain prominent grainlike structures in its cytoplasm. Agranulocytes often are referred to as other leukocytes due to the fact that they have less prominent cytoplasmic granules and are not truly agranulocytic. Types of WBCs are summarized in Table 15–2 and Figure 15–3.

Some cells are described as basophilic or eosinophilic. **Basophilic** (bā-sō-fihl-ihck) means stained readily with basic, or blue, dyes in many commonly used stains such as hematoxylin and eosin (H&E), Giemsa, and Wright's. **Eosinophilic** (ē-ō-sihn-ō-fihl-ihck) means stained readily with acidic, or pink, dyes in many commonly used stains such as H&E, Giemsa, and Wright's.

TABLE 15–1 Terms Used to Describe Erythrocytes

Term	Pronunciation	Description
normocytic	(nohr-mō-siht-ihck)	normal cell size
macrocytic	(mahck-rō-siht-ihck)	larger-than-normal cell size
microcytic	(mī-krō-siht-ihck)	smaller-than-normal cell size
poikilocytosis	(poy-kē-loh-sī-tō-sihs)	condition of irregular cells; clinically means varied shapes of erythrocytes; **poikilo** means irregular
anisocytosis	(ahn-eh-sō-sī-tō-sihs)	condition of unequal cell size; excessive variation in RBC size
normochromic	(nohr-mō-krō-mihck)	normal RBC color (has area of central pallor of normal size)
hypochromic	(hī-pō-krō-mihck)	less-than-normal color (has enlarged area of central pallor)
hyperchromic	(hī-pər-krō-mihck)	more-than-normal color (term is not commonly used)
polychromasia	(poh-lē-krō-mah-zē-ah)	"condition of many colors" that appears as an overall blue tint of Wright-stained RBCs due to shrinking of the cell membrane; indicates slight immaturity of the erythrocyte
rouleaux	(roo-low)	RBCs that are arranged like stacks of coins on the peripheral blood smear; may be artifact or may be normal in species such as cats and horses
agglutination	(ah-gloo-tih-nā-shuhn)	clumping of RBCs due to the presence of an antibody directed against RBC surface antigens that forms a latticework that links them together

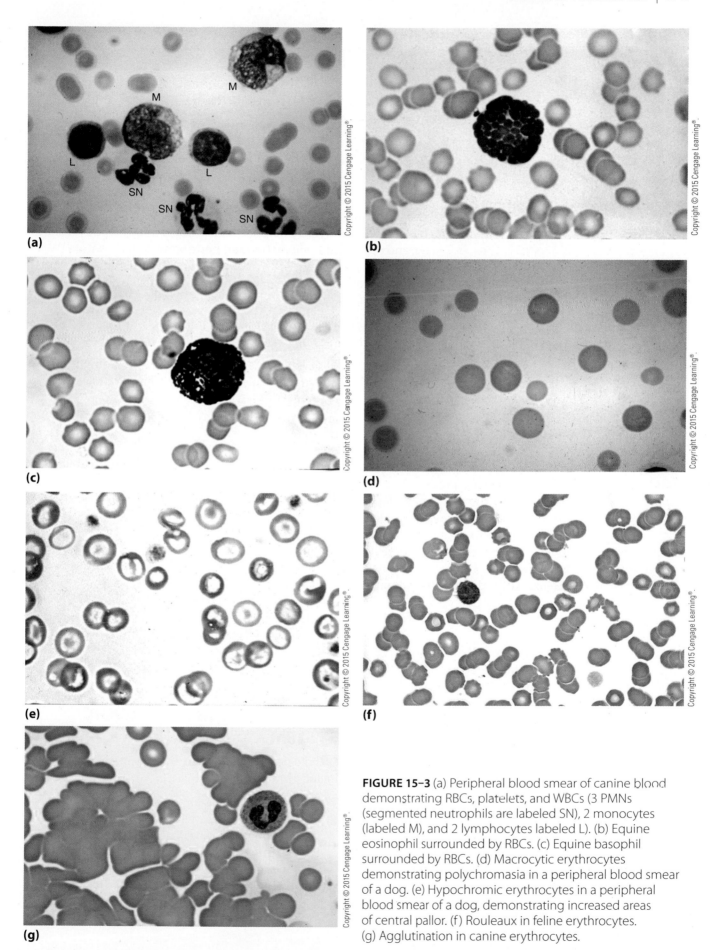

FIGURE 15–3 (a) Peripheral blood smear of canine blood demonstrating RBCs, platelets, and WBCs (3 PMNs (segmented neutrophils are labeled SN), 2 monocytes (labeled M), and 2 lymphocytes labeled L). (b) Equine eosinophil surrounded by RBCs. (c) Equine basophil surrounded by RBCs. (d) Macrocytic erythrocytes demonstrating polychromasia in a peripheral blood smear of a dog. (e) Hypochromic erythrocytes in a peripheral blood smear of a dog, demonstrating increased areas of central pallor. (f) Rouleaux in feline erythrocytes. (g) Agglutination in canine erythrocytes.

TABLE 15–2 Major Groups of Leukocytes

Type of Leukocyte	Pronunciation	Description
lymphocyte	(lihm-fō-sīt)	class of "agranulocytic" leukocyte that contains a diverse set of cells including those that can directly attack specific pathogens or produce antibodies
monocyte	(mohn-ō-sīt)	class of "agranulocytic" leukocyte that has a phagocytic function and participates in the inflammatory response
neutrophil	(nū-trō-fihl)	class of granulocytic leukocyte that has a phagocytic function (mainly against bacteria); a mature neutrophil is called a **segmented neutrophil** or **polymorphonuclear** (poh-lē-mōr-fō-nū-klē-ahr) **leukocyte**, or PMN; *polymorphonuclear* means multishaped nucleus; a **band cell** (bahnd sehl) or **banded neutrophil** is an immature PMN; in rabbits, birds, amphibians, and reptiles neutrophils are called **heterophils** (heht-ər- ō-fihls), which are large and stain a deep orange or red
eosinophil	(ē-ō-sihn-ō-fihl)	class of granulocytic leukocyte that detoxifies allergens and controls parasitic infections by damaging parasite membranes
basophil	(bā-sō-fihl)	class of granulocytic leukocyte that promotes the inflammatory response and contains histamine in its cytoplasmic granules

Clotting Cells. Clotting cells also are produced in the bone marrow and play a part in the clotting of blood. A **thrombocyte** (throhm-bō-sīt) is a nucleated clotting cell, and a **platelet** (plāt-leht) is an anucleated clotting cell. Occasionally, those terms are used interchangeably; however, they have different meanings. Non-mammalian vertebrates (birds, reptiles, amphibians, and fish) have nucleated clotting cells (thrombocytes) while mammals have anucleated clotting cells (platelets).

A **megakaryocyte** (mehg-ah-kahr-ē-ō-sīt) is a large, nucleated cell found in the bone marrow from which platelets are formed.

TEST ME: HEMATOLOGIC SYSTEM

Some diagnostic tests performed on the hematologic system are discussed in Chapter 8 (if the test involves assessing the heart and blood). Other diagnostic tests performed on the hematologic system include the following:

- **blood smear** (smēr) = blood specimen for microscopic examination in which blood is spread thinly across a microscope slide and typically stained (Figure 15–4a through f).

- **bone marrow biopsy** (bōn mahr-ō bī-ohp-sē) = sample of bone marrow obtained by needle aspiration for examination of cells. Bone marrow samples are taken from long bones, ribs, the pelvis, or the sternum.
- **laboratory tests**, which are covered in Chapter 16.
- **ultrasound** = diagnostic procedure to image the structures of blood vessels (diagnostic ultrasound) and flow of blood through these vessels (Doppler ultrasound). Ultrasound is discussed in Chapters 6 and 16.

PATHOLOGY: HEMATOLOGIC SYSTEM

Pathologic conditions of the hematologic system include the following:

- **anemia** (ah-nē-mē-ah) = blood condition of less-than-normal levels of red blood cells and/or hemoglobin.
- **basopenia** (bā-sō-pē-nē-ah) = deficiency in the number of basophils in the blood; **-penia** is a suffix that indicates a decrease in a particular type of cell.

FIGURE 15–4 Making a blood smear (a) dispense one drop of blood onto the end of a slide; (b) using a spreader slide, gently bring the back edge of the spreader slide into contact with the drop of blood; (c) and (d) quickly push the spreader slide forward to make the smear. (e) Blood smears are then stained with a quick stain kit. (f) A stained blood smear is then examined under a microscope to examine the types of cells present.

- **basophilia** (bā-sō-fihl-ē-ah) elevation in the number of basophils in the blood; **-philia** is a suffix that indicates an increase in a particular type of cell.
- **disseminated intravascular coagulation** = disorder in which the proteins that control blood clotting become abnormally active, producing clots that obstruct the blood vessels, cut off blood supply to various organs, and in time alter the functioning of these organs. The production of clots consumes coagulation factors and platelets and leads to bleeding; abbreviated DIC.

DIC may result from inflammation, infection, or cancer.
- **dyscrasia** (dihs-krā-zē-ah) = any abnormal condition of the blood.
- **edema** (eh-dē-mah) = accumulation of fluid in the intercellular space (Figure 15–5). **Edemateous** (eh-deh-mah-tuhs) is the adjective form of edema.
- **eosinopenia** (ē-ō-sihn-ō-pē-nē-ah) = deficiency in the number of eosinophils in the blood.
- **eosinophilia** (ē-ō-sihn-ō-fihl-ē-ah) = elevation in the number of eosinophils in the blood.

Courtesy of Kimberly Kruse Sprecher, CVT

FIGURE 15–5 Edema in the hind legs of a dog.

Copyright © 2015 Cengage Learning®.

FIGURE 15–6 Lipemic serum contains an excessive amount of fat and appears white (tube on the left), while **hemolytic** serum contains red blood cell components that are released when erythrocytes are damaged due to a variety of causes (such as improper specimen processing, collection, or transport) and appears pink or blood-tinged (second tube from the left). **Icteric** serum is yellow-tinged due to increased bilirubin levels (third tube from the left), while **normal** serum is clear (tube on the right).

- **erythrocytosis** (eh-rihth-rō-sī-tō-sihs) = abnormal increase in red blood cells. The suffix **-cytosis** means condition of cell but implies elevated cell numbers.
- **exudate** (ehcks-yoo-dāt) = material that has escaped from blood vessels and is high in protein, cells, or solid materials derived from cells (Table 15–3); exudates are typically inflammatory in nature.
- **hemolytic** (hē-mō-liht-ihck) = removing and destroying red blood cells. **Hemolytic anemia** is excessive RBC destruction, resulting in lower-than-normal levels of RBCs. **Hemolytic serum** contains red blood cell components that are released when erythrocytes are damaged due to a variety of causes (such as improper specimen processing, collection, or transport) and appears pink or blood-tinged (Figure 15–6). **Hemolysis** (hē-mohl-eh-sihs) is the breaking down of red blood cells. **Lysis** (lī-sihs) is the medical term for destruction or breakdown.

- **hemophilia** (hē-mō-fihl-ē-ah) = hereditary condition of deficient blood coagulation.
- **hemorrhage** (hehm-ōr-ihdj) = loss of blood (usually in a short period of time). **Hemostasis** (hē-mō-stā-sihs) is the act of controlling blood or bleeding. A **hemostat** (hē-mō-staht) is an instrument used to control bleeding.
- **hyperalbuminemia** (hī-pər-ahl-byoo-mih-nē-mē-ah) = blood condition of abnormally high albumin levels.
- **hyperemia** (hī-pər-ē-mē-ah) = excess blood in a part; engorgement.
- **hyperlipidemia** (hī-pər-lihp-ih-dē-mē-ah) = blood condition of abnormally high fat levels; more accurately means abnormally high fat levels caused by fat metabolism. **Lip/o** is the combining form for fat.
- **icteric** (ihck-tər-ihck) **serum** = serum that has yellow pigmentation that is suggestive of hyperbilirubinemia (Figure 15–6). Hyperbilirubinemia may be caused by increased erythrocyte damage, liver disease, and disruption of bile flow.

TABLE 15–3 Exudates Found in Blood

Fluid Type	Total Protein (g/dL)	Cell Count (cells/mL)
Transudate	<2.5	<1,500
Modified transudate	2.5–7.5	1,500–7,000
Exudate	>3.0	>7,000

- **left shift** (lehft shihft) = common term for an alteration in the distribution of leukocytes in which there are increases in banded neutrophils usually in response to severe bacterial infection.
- **leukemia** (loo-kē-mē-ah) = elevation in the number of malignant white blood cells; animals with leukemia have extremely elevated total leukocyte counts. Leukemias may be classified according to the concentration of neoplastic cells that are circulating in blood (such as granulocytic leukemia and lymphocytic leukemia). Leukemias are further classified as either acute or chronic based on the majority or degree of neoplastic cell differentiation and clinical course.
- **leukocytosis** (loo-kō-sī-tō-sihs) = elevation in the number of white blood cells.
- **leukopenia** (loo-kō-pē-nē-ah) = deficiency of white blood cells; sometimes called **leukocytopenia** (loo-kō-sīt-ō-pē-nē-ah).
- **lipemia** (lī-pē-mē-ah) = excessive amount of fats in the blood.
- **lipemic serum** (lī-pē-mihck sē-ruhm) – fats from blood that have settled in the serum. Clinically the serum appears cloudy and white (Figure 15–6).
- **lymphocytosis** (lihm-fō-sī-tō-sihs) = elevated numbers of lymphocytes in the blood.
- **lymphopenia** (lihm-fō-pē-nē-ah) = deficiency of lymphocytes in the blood.
- **modified transudate** (trahnz-yoo-dāt) = material that has cell numbers and distribution similar to transudates but the protein concentration is greater than 2.5 g/dL resulting from leakage of high protein lymph or occasionally inflammatory proteins; modified transudates are common in right sided heart failure.
- **monocytopenia** (moh-nō-sīt-ō-pē-nē-ah) = deficiency of monocytes in the blood.
- **monocytosis** (moh-nō-sī-tō-sihs) = elevated numbers of monocytes in the blood.
- **myelodysplasia** (mī-ehl-ō-dihs-plā-zē-ah) = hematologic disorder characterized clinically and morphologically by ineffective hematopoiesis that results in some form of cytopenia such as anemia, neutropenia, and/or thrombocytopenia.
- **neutropenia** (nū-trō-pē-nē-ah) = deficiency in the number of neutrophils in the blood.
- **neutrophilia** (nū-trō-fihl-ē-ah) = elevation in the number of neutrophils in the blood.
- **pancytopenia** (pahn-sīt-ō-pē-nē-ah) = deficiency of all types of blood cells.
- **phagocytosis** (fahg-ō-sī-tō-sihs) = condition of engulfing or eating cells.
- **polycythemia** (poh-lē-sī-thē-mē-ah) = condition of many cells; clinically means excessive erythrocytes.
- **septicemia** (sehp-tih-sē-mē-ah) = blood condition in which pathogenic microorganisms (bacteria) and their toxins are present. The suffix **-emia** means blood condition. **Pathogenic** (pahth-ō-jehn-ihck) means producing disease. **Bacteremia** (bahck-tər-ē-mē-ah) is the blood condition in which bacteria are present.
- **thrombocytopenia** (throhm-bō-sīt-ō-pē-nē-ah) = abnormal decrease in the number of clotting cells. The suffix **-penia** means less than normal or deficiency.
- **thrombocytosis** (throhm-bō-sī-tō-sihs) = elevation in the number of clotting cells.
- **transudate** (trahnz-yoo-dāt) = material that has passed through a membrane and is high in fluidity and low in protein, cells, or solid materials derived from cells; transudates are observed with severe hepatic disease, protein losing enteropathy, or protein losing nephropathy.

PROCEDURES: HEMATOLOGIC SYSTEM

Procedures performed on the hematologic system are discussed in Chapter 8.

LYMPHATIC SYSTEM

Functions of the Lymphatic System

The lymphatic system functions as part of the immune system by providing the location to gather and concentrate foreign substances in the body so the immune system can destroy and remove them from the body. The lymphatic system returns excess lymph to the blood, traps and filters cellular debris (such as cancer cells and microorganisms), reduces tissue edema by removing fluid from capillary beds, and absorbs fats and fat-soluble vitamins from the digestive system and transports them to cells. The combining form **lymph/o** means lymph fluid, lymph vessels, and lymph nodes. **Lymphoid** (lihm-foyd) pertains to lymph or tissue of the lymphatic system.

Structures of the Lymphatic System

The major structures of the lymphatic system include lymph vessels, lymph nodes, lymph fluid, tonsils, spleen, thymus, and lymphocytes.

Lymph Fluid

Interstitial fluid (ihn-tər-stihsh-ahl flū-ihd) is the clear, colorless tissue fluid that leaves the capillaries and flows in the spaces between the cells of a tissue or an organ. Interstitial fluid functions to bathe and nourish the cells. *Interstitial* pertains to the spaces in a tissue or an organ. **Lymph** (lihmf) is formed when interstitial fluid moves into the capillaries of the lymphatic system. Lymph brings nutrients and hormones to cells and carries waste products from tissue back to the bloodstream (Figure 15–7).

Lymph Vessels

Lymph is carried from the tissue space via thin-walled tubes called **lymph capillaries**. Lymph capillaries take the lymph to the lymphatic vessels. **Lymphatic vessels** (lihm-fah-tihck vehs-uhlz) are similar to veins in

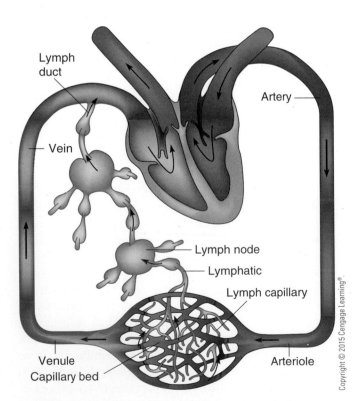

FIGURE 15–7 Fluids that leave circulation through the capillaries are returned to venous circulation by the lymphatic system.

Lymph duct

Vein

Artery

Lymph node

Lymphatic

Lymph capillary

Venule
Capillary bed

Arteriole

that they have valves to prevent the backflow of lymph. Unlike blood, lymph only flows in one direction, from its point of origin until its return to the venous circulation. Lymph always travels toward the thoracic cavity and as the vessels approach the heart, they carry more fluid and are larger in size. In the thoracic cavity, the right lymphatic duct and thoracic duct empty lymph into veins. Lymph ducts release lymph (and whatever is in lymph) into venous blood, where it is quickly passed to the lungs and then throughout the body. This mechanism can spread infection, cancers, and other diseases. The **cisterna chyli** (sihs-tər-nah kī-lē) is the origin of the thoracic duct and saclike structure for the lymph collection.

A comparison of blood versus lymph systems is summarized in Table 15–4. **Lacteals** (lahck-tē-ahls), located in the small intestine, are specialized lymph vessels that transport fats and fat-soluble vitamins (Figure 15–8).

Lymph Nodes

Lymph nodes (lihmf nōdz) are small bean-shaped structures that filter lymph and store B and T lymphocytes (Figure 15–9a). The primary function of lymph nodes is to filter lymph to remove harmful substances such as bacteria and viruses. Lymph nodes are located along the lymphatic vessels and by the time lymph reaches the thoracic cavity, it has been filtered many times. Because cells are destroyed in lymph nodes, swollen lymph nodes often are an indication of disease. Lymph nodes are described according to their location: Mandibular lymph nodes are located near the mandible, parotid lymph nodes are located near the ear (**para-** means near, and **otos** is Greek for ear), mesenteric lymph nodes are located in the mesentery (Figure 15–9b).

Tonsils

The **tonsils** (tohn-sahlz) are masses of lymphatic tissue that protect the nose and cranial (upper) throat. Tonsils are described according to their location: Lingual tonsils are located near the tongue, palatine tonsils are located near the palate or roof of the mouth, and pharyngeal tonsils are located near the throat. **Tonsill/o** is the combining form for tonsil.

Spleen

The **spleen** (splēn) is an organ located in the cranial abdomen that filters foreign material from the blood, stores erythrocytes, removes damaged or old erythrocytes, and maintains an appropriate balance of cells and plasma in the blood (Figure 15–10). The spleen is

TABLE 15-4 Comparison of Blood and Lymph Systems

Property	Blood	Lymph
Circulation	throughout entire body	in one direction from point of origin until its return to venous circulation
System type	open system in which blood leaves and re-enters the blood vessels through the capillaries	closed system in which lymph enters the lymphatic capillaries, but does not leave the lymphatic vessels until it returns to venous circulation
Fluid movement	pumped throughout the body by the heart	depends on movements of nearby muscles and blood vessels to aid its flow
Color	red, which makes arteries and veins readily visible	clear, which makes lymphatic vessels not visible
Filtering	filtered by kidneys and waste products excreted by the urinary system	filtered by lymph nodes located along the lymphatic vessels throughout the body

Copyright © 2015 Cengage Learning®.

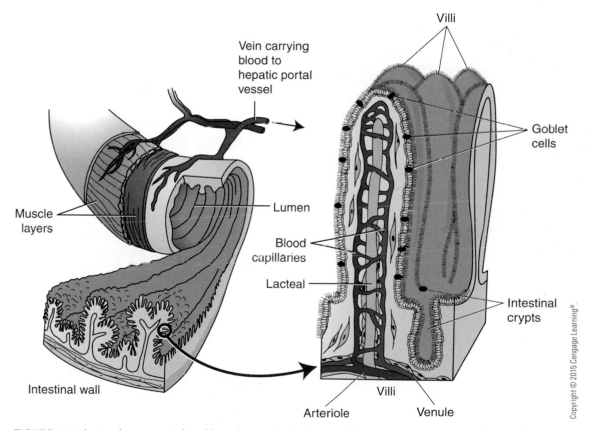

FIGURE 15-8 Lacteals are specialized lymph vessels in the small intestine that transport fats and fat-soluble vitamins.

also a secondary lymphoid tissue (as opposed to the primary lymphoid tissues, which are the thymus and bone marrow) where mature, differentiated B and T lymphocytes reside and wait for antigenic stimulation.

Upon antigenic stimulation, these lymphocytes proliferate. In addition to lymphocytes, macrophages are also found in the spleen. Macrophages line the sinusoids of the spleen (called sinusoidal lining cells) where they

(a)

(b)

FIGURE 15–9 (a) Location of lymph nodes that can be palpated on a dog. (b) Mesenteric lymph nodes are located in the mesentery.

FIGURE 15–10 The spleen is a mass of lymphatic tissue located in the cranial abdomen.

will phagocytize foreign material, break it down, and present antigenic parts on their surface for the helper T lymphocytes to recognize and initiate an immune response. The combining form **splen/o** means spleen.

Thymus

The **thymus** (thī-muhs) is a gland that has an immunologic function and is found predominantly in young animals. The thymus is located near midline in the cranioventral portion of the thoracic cavity. The immunologic role of the thymus is development of T cells.

Some of the lymphocytes formed in the bone marrow migrate to the thymus, where they multiply and mature into T cells. T cells play an important role in the immune response. The thymus also has an endocrine function. The combining form for thymus is **thym/o**.

IMMUNE SYSTEM

The immune system functions to protect the body from harmful substances and utilizes other systems to maintain its functions. The term **immunity** comes from the Latin term *immunitas*, which means exemption. Immunity was used to imply that an animal was exempt from or protected against foreign substances. The combining form **immun/o** means protected. **Immunology** (ihm-yoo-nohl-ō-jē) is the study of the immune system.

The immune system is not contained in one set of organs or in one area. Many structures from different body systems aid in protecting the body. The lymphatic system, respiratory tract, gastrointestinal tract, integumentary system, and others work together to prevent the body from being harmed from foreign invaders.

Specialized Cells

Some cells are specialized for immune reactions. The **lymphocyte** is a type of white blood cell that is involved in the immune response and works against specific antigens. Lymphocytes are formed in the bone marrow and mature in lymphatic tissue throughout the body, such as the spleen, thymus, and bone marrow. There are two subpopulations of lymphocytes—the

Anti- means against. The terms *antigen* and *antibody* must be distinguished from each other. **Antigen** (ahn-tih-jehn) is a substance that the body regards as foreign (such as a virus, bacterium, or toxin). **Antibody** (ahn-tih-boh-dē) is a disease-fighting protein produced by the body in response to the presence of a specific antigen.

T lymphocytes, which are responsible for cell-mediated immunity, and the B lymphocytes, which are responsible for humoral immunity (Figure 15–11). **B lymphocytes** are produced and mature in the bone marrow and are responsible for antibody-mediated or humoral immunity. In the presence of a specific antigen, B lymphocytes differentiate into plasma cells and memory B cells. Memory B cells remember a specific antigen and stimulate a faster and more intense response when that same antigen is presented to the body. A **plasma cell** (plahz-mah sehl) is an immune cell that produces and secretes a specific antibody for a specific antigen. Plasma cells are also called **plasmocytes** (plahz-mō-sītz). The antibodies made by plasma cells are called **immunoglobulins** (ihm-yoo-nō-glohb-yoo-lihnz). Immunoglobulin is abbreviated Ig. There are five distinct immunoglobulins (Table 15–5 and Figure 15–12).

T lymphocytes are small, circulating lymphocytes produced in the bone marrow. These cells mature in the thymus, so they are called **T cells**. The primary function of T lymphocytes is to coordinate immune defenses and kill organisms on contact. T cells are

Cell-Mediated and Humoral (Antibody-Mediated) Immunity	
Cell-Mediated Immunity	**Humoral (Antibody-Mediated) Immunity**
T cells are responsible for cell-mediated immunity.	**B cells** are responsible for humoral immunity.
T cells directly attack the invading antigen.	Differentiated B cells produce antibodies that react with the antigen or substances produced by the antigen.
Cell-mediated immunity is most effective against viruses that infect body cells, cancer cells, and foreign tissue cells.	Humoral immunity is most effective against bacteria, viruses that are outside body cells, and toxins. It is also involved in allergic reactions.

FIGURE 15–11 Comparison of cell-mediated and antibody-mediated immunity.

TABLE 15–5 Types of Immunoglobulin and Their Functions

IgA	found in the mucous membrane lining of intestines and bronchi, and secretions such as saliva, sweat, and tears; protects those areas by preventing attachment of bacteria and viruses to the epithelial surfaces that line most organs
IgD	found in large amounts on the surface of B cells; unknown function but is important in B cell activation
IgE	found bound to mast cells in lungs, skin, and cells of mucous membranes; provides defense against the environment and is involved in allergic reactions;
IgG	found in blood stream, interstitial fluid (tissue), and lymph (lymph vessels); synthesized in response to invading germs such as bacteria, fungi, and viruses; most abundant antibody and only class that can cross the placenta; found in high concentrations in colostrum of large domestic species (horses, cattle, pigs)
IgM	found in circulating fluids (due to its size it cannot exit the blood stream); first immunoglobulin produced in response to an initial exposure to an antigen; invades and provides protection in the earliest stages of infection

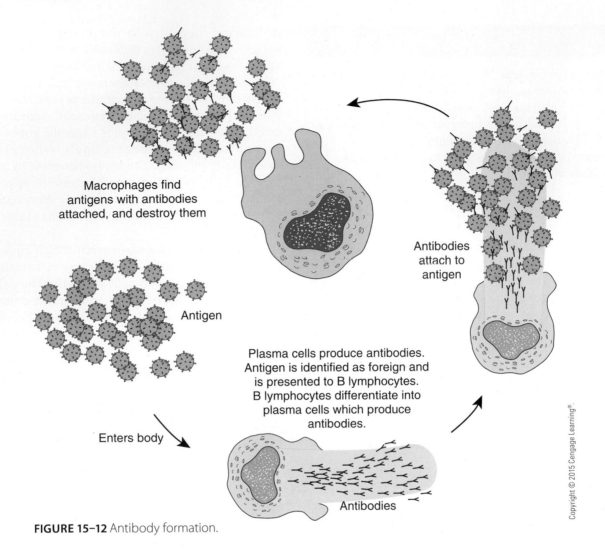

Macrophages find
antigens with antibodies
attached, and destroy them

Antigen

Antibodies
attach to
antigen

Enters body

Plasma cells produce antibodies.
Antigen is identified as foreign and
is presented to B lymphocytes.
B lymphocytes differentiate into
plasma cells which produce
antibodies.

Antibodies

FIGURE 15–12 Antibody formation.

involved in cell-mediated immunity. Cell-mediated immunity involves T cell activation and cellular secretions. Although cell-mediated immunity does not involve antibody production, the antibodies produced during humoral immunity may play a role in some cell-mediated responses.

There are different types of T lymphocytes. **Helper T cells** (also known as CD4 cells) secrete substances such as cytokines that stimulate the production of B lymphocytes and cytotoxic T cells. **Cytotoxic T cells** (also known as CD8 cells) destroy intracellular pathogens. **Suppressor T cells** stop B and T lymphocyte activity when this activity is no longer needed. **Memory T cells** remember a specific antigen and stimulate a faster and more intense response when that same antigen is presented to the body.

Monocytes are another type of leukocyte formed in the bone marrow and transported to other parts of the body. Monocytes migrate to tissues such as the spleen to become macrophages. A **macrophage** is a phagocytic cell that protects the body by engulfing invading cells and by interacting with other cells of the immune system. **Histiocytes** (hihs-tē-ō-sītz) are large macrophages found in loose connective tissue.

How Does the Immune System Work?

The body has many ways to protect itself from invading organisms and substances. The initial defenses against pathogen invasion are nonspecific. The first line of defense limits access to internal tissues and organs of the body. Anatomical barriers such as intact skin help protect the animal's body from infection. **Intact** (ihn-tahckt) means having no cuts, scrapes, openings, or alterations. Intact skin makes it more difficult for invading organisms and substances to obtain access to an animal. Oil secreted by sebaceous glands discourages bacterial growth on the skin.

BIOLOGY BONUS

Three characteristics (specificity, memory, and discrimination) distinguish immunity from nonspecific defense mechanisms. Even though these three characteristics are important in protecting animals from infection, they can also cause problems under certain circumstances. For example, if a bacterium changes a component of its genetic code due to mutation (**mut/a** means genetic change), it then becomes a new antigen. Since the immune system has specificity to only the original antigen, the animal's immune system does not recognize it. Another example is when the immune system's property of discrimination breaks down, causing the animal's immune system to not recognize "self" versus "non-self." These animals may develop autoimmune disease due to the immune system attacking itself.

The respiratory system has its own line of defense. Foreign material breathed in is trapped in the cilia of the nares and the moist mucous membrane that lines the respiratory tract. Mucus continually flushes away trapped debris. Coughing and sneezing also remove foreign material.

The digestive system destroys invading organisms that are swallowed by the acidic nature of the stomach.

An animal's health, age, and heredity also play roles in protecting the body. Some animals may be immunodeficient or hypersensitive.

Infectious agents that penetrate the first line of defense are usually destroyed by other nonspecific responses such as fever, inflammation, chemicals, and complement. The complement system is a nonspecific defense mechanism, and its activation can result in initiation of inflammation, activation of leukocytes, lysis of pathogens, and increased phagocytosis. **Complement** is a series of enzymatic proteins that are continually present in normal plasma. Complement is thus named because it is complementary to the immune system and it aids phagocytes in destroying antigens and causes cell lysis.

The immune system is activated when the previously listed defenses fail. Three characteristics distinguish immunity from the first two lines of defense: specificity, memory, and discrimination. Specificity means that the immune response is directed against a specific antigen. Immunity to one antigen does not confer immunity to other antigens. Memory helps the immune system recognize an antigen it has been previously exposed to and allows the body to respond quickly to that antigen. Discrimination helps the immune system differentiate between "self" and "non-self." Most of this recognition develops prior to birth. Activation of the immune system is outlined here:

Organisms invade the body. Macrophages ingest the invading organisms and present antigen fragments to its surface. Helper T cells are activated.

↓

Helper T cells multiply and produce cytokines that stimulate cytotoxic T cell production. Complement goes to the affected area. B cells sensitive to the organism multiply. B cells transform into plasma cells that produce antibody.

↓

Complement proteins lyse affected cells. Antibodies bind to organisms.

↓

If infection is contained, suppressor T cells stop the immune response. Memory B and T cells remain ready if this particular antigen enters the body again.

Immunity (ihm-yoo-nih-tē) is the state of being resistant to a specific disease. Different forms of immunity are obtained during life and may be described with the terms naturally acquired, artificially acquired, active, or passive. Immunity may be acquired either naturally or artificially; naturally means the animal was exposed to the disease during normal activity while artificially means the animal was intentionally exposed. Immunity may be developed actively or passively; active means the protection was induced in the host by an antigen and the host produced immunity while passive means activated T cells or antibodies were transferred to the animal. These terms can be combined to describe the types of immunity and are listed as follows:

- **naturally acquired passive immunity** = resistance to a specific infection by the passing of protection from mother to offspring before birth or through colostrum.

Words Used When Discussing the Immune System

herd immunity = resistance of a group (herd) to a microbe/infection because a large proportion of the group's members are immune; for example, if most animals are vaccinated to protect them from a certain disease, the unvaccinated ones receive the benefit of herd immunity because the microbe is not present in the group.

debilitated (dē-bihl-ih-tāt-ehd) = weakened.

heredity (hər-eh-dih-tē) = genetic transmission of characteristics from parent to offspring.

innate immunity (ihn-āt) = immunity determined by the genetic makeup of an individual; for example, dogs do not get human immunodeficiency virus (HIV) infection due to innate immunity; also called **inherent** or **genetic immunity**.

inhibit (ihn-hihb-iht) = to slow or stop.

multiplication (muhl-tih-plih-kā-shuhn) = reproduction.

opportunistic (ohp-ər-too-nihs-tihck) = able to cause disease (due to debilitation or alteration of the animal) when disease normally would not be produced; for example, a microbe is able to cause disease because it is given an "opportunity" when the population of normal bacteria in the gastrointestinal tract is altered.

resistant (rē-zihs-tahnt) = not susceptible.

titer (tī-tər) = quantity of a substance required to react with a given amount of another substance; an **antibody titer** measures the concentration of antibodies circulating in the bloodstream of an animal. A rising antibody titer usually indicates the animal's response to antigens associated with an active infection. Titers can be used to determine if an animal is protected against contracting a disease or if the antibody levels are high enough to offer protection without giving a booster vaccination.

vaccination (vahck-sihn-ā-shuhn) = administration of antigen (vaccine) to stimulate a protective immune response against a specific infectious agent; also called **immunization** (ihm-yoo-nih-zā-shuhn).

vaccine (vahck-sēn) = preparation of pathogen (live, weakened, or killed) or a portion of pathogen that is administered to stimulate a protective immune response against the pathogen.

- **naturally acquired active immunity** = resistance to a specific infection after the development of antibodies during the actual disease.
- **artificially acquired passive immunity** = resistance to a specific infection by receiving antiserum-containing antibodies from another host.
- **artificially acquired active immunity** = resistance to a specific infection through vaccination.

TEST ME: LYMPHATIC AND IMMUNE SYSTEMS

Diagnostic tests performed on the immune and lymphatic systems are described in Chapter 16.

PATHOLOGY: LYMPHATIC AND IMMUNE SYSTEMS

Pathologic conditions of the immune and lymphatic systems include the following:

- **allergy** (ahl-ər-jē) = overreaction by the body to a particular antigen; also called **hypersensitivity** (hī-pər-sehn-sih-tihv-ih-tē). An **allergen** (ahl-ər-jehn) is a substance capable of inducing an allergic reaction.
- **anaphylaxis** (ahn-ah-fih-lahck-sihs) = severe response to a foreign substance. Signs develop acutely and may include swelling, blockage of airways, tachycardia, and ptylism.
- **autoimmune disease** (aw-tō-ihm-yoon dih-zēz) = disorder in which the body makes antibodies directed against itself.
- **immunosuppression** (ihm-yoo-nō-suhp-prehsh-uhn) = reduction or decrease in the ability of the immune system to respond to stimulation by antigens. An **immunosuppressant** (ihm-yoo-nō-suh-prehs-ahnt) is a chemical that prevents or reduces the body's normal reaction to disease.
- **lymphadenitis** (lihm-fahd-eh-nī-tihs) = inflammation of the lymph nodes; also called **swollen glands**.
- **lymphadenopathy** (lihm-fahd-eh-nohp-ah-thē) = disease of the lymph nodes; an example of incorrect use of *aden/o* because nodes are not glands.
- **lymphangioma** (lihm-fahn-jē-ō-mah) = abnormal collection of lymphatic vessels forming a mass (usually benign).
- **splenomegaly** (splehn-ō-mehg-ah-lē) = enlargement of the spleen.
- **tonsillitis** (tohn-sih-lī-tihs) = inflammation of the tonsils.

PROCEDURES: LYMPHATIC AND IMMUNE SYSTEMS

Procedures performed on the immune and lymphatic systems include the following:

- **splenectomy** (splehn-ehck-tō-mē) = surgical removal of the spleen.
- **thymectomy** (thī-mehck-tō-mē) = surgical removal of the thymus.
- **tonsillectomy** (tohn-sih-lehck-tō-mē) = surgical removal of the tonsils.

ONCOLOGY

Oncology (ohng-kohl-ō-jē) is the study, diagnosis, and treatment of tumors. The combining form **onc/o** means tumor. The term **tumor** does not mean cancerous. When a tumor is cancerous, the term **malignant** is used to describe it. Nonmalignant tumors are called **benign.** The term *neoplasm* is another word used to describe abnormal growths. A **neoplasm** (nē-ō-plahz-uhm) is any abnormal new growth of tissue in which the

Tumors may be described by their appearance under the microscope using the following terms:

- **anaplastic** = reverting to a more immature form
- **carcinoma *in situ*** = contained at a site (not spreading)
- **diffuse** = evenly spreading
- **dysplastic** = abnormal cell development
- **follicular** = containing glandular sacs
- **hyperchromatic** = intensely colored
- **hyperplastic** = excessive cell numbers
- **hypoplastic** = deficient cell numbers
- **nodular** = tight cell clusters
- **pleomorphic** (plē-ō-mohr-fihck) = having many cell types
- **undifferentiated** = lacking a defined cell structure

Tumors may also be described by their appearance during visual examination using the following terms:

- **cystic** = fluid filled
- **inflammatory** = appearing red and swollen
- **medullary** = large and fleshy
- **necrotic** = containing dead tissue
- **ulcerating** = having depressed, open areas
- **verrucous** = having wartlike growths

multiplication of cells is uncontrolled, more rapid than normal, and progressive. The death of normal cells in a normal time cycle is called **apoptosis** (ahp-ohp-tō-sihs); this is also referred to as programmed cell death.

Tumors are often described by their appearance. **Pedunculated** (peh-duhnck-yoo-lā-tehd) means having a peduncle, or stalk. **Well-circumscribed** means the mass has well-defined borders. **Encapsulated** means the tumor is retained within a border of connective tissue. **Invasive** means that the mass does not have well-defined borders and is spreading.

Malignant growths tend to spread to distant body sites. **Metastasis** (meh-tahs-tah-sihs) is a pathogenic growth distant from the primary disease site. Metastasis literally means beyond control. The plural form of metastasis is **metastases** (meh-tahs-tah-sēz). The term **metastasize** (meh-tahs-tah-sīz) is used to describe invasion by the pathogenic growth to a point distant from the primary disease site.

Malignant growths also may be described by their tissue of origin. A **carcinoma** (kahr-sih-nō-mah) is a malignant growth of epithelial cells, whereas a **sarcoma** (sahr-kō-mah) is a malignant neoplasm arising from any type of connective tissue. The combining form **carcin/o** means cancer, and the combining form **sarc/o** means flesh. Connective tissue sometimes is referred to as fleshy. The combining form *carcin/o* is used to form the term *carcinogen*. A **carcinogen** is a substance that produces cancer. The suffix **-gen** means producing. Tumors appear for a number of reasons including exposure to carcinogens (cancer-causing agents) or inherited defects.

TEST ME: ONCOLOGY

Diagnostic tests performed on tumors include the following:

- **biopsies** = incisional and excisional biopsies are described in Chapter 10. A **brush biopsy** is the passing of a catheter with bristles on it into an organ to remove cells for examination. A **needle biopsy** is any biopsy in which cells are aspirated through a needle; also known as a **fine needle aspirate**; abbreviated FNA. An **exfoliative biopsy** is a biopsy in which cells are scraped off of the skin for evaluation.
- **histopathology** (hihs-tō-pahth-ohl-lō-jē) = the microscopic study of the structure, composition, and function of tissues.
- **radiographs** = records of ionizing radiation used to visualize internal body structures. Radiographs are taken in oncology patients to assess the extent of some tumors or to check for metastases (Figure 15–13).

Courtesy of Anne E. Chauvet, DVM, Diplomate ACVIM

FIGURE 15–13 Ventrodorsal radiograph of pulmonary metastases.

- **leiomyosarcoma** (lī-ō-mī-ō-sahr-kō-mah) = malignant tumor of smooth muscle.
- **lymphoma** (lihm-fō-mah) = general term for neoplasm composed of lymphoid tissue (usually malignant); also called **lymphosarcoma**; abbreviated LSA.
- **mast cell tumor** = malignant growth of tissue mast cells (cells that release histamine); abbreviated MCT. Mast cell tumors are associated with vomiting, anorexia, and various signs, depending on the tissue involved (Figure 15–15).

Copyright © 2015 Cengage Learning®.

FIGURE 15–14 Intraoperative photograph of a spleen with hemangiosarcoma.

- **touch preps** = collections of cells on a glass slide pressed against a part of the mass. The slide is then examined under a microscope.
- **ultrasound** = imaging of internal body structures by recording echoes of sound waves (see Chapter 6).

PATHOLOGY: ONCOLOGY

Tumors are named for the tissues involved. Some examples include the following:

- **adenocarcinoma** (ahd-eh-nō-kahr-sih-nō-mah) = malignant growth of epithelial glandular tissue.
- **blastoma** (blahs-tō-mah) = neoplasm composed of immature undifferentiated cells; **blast/o** means immature cell.
- **hemangioma** (hē-mahn-jē-ō-mah) = benign neoplasm composed of newly formed blood vessels.
- **hemangiosarcoma** (hē-mahn-jē-ō-sahr-kō-mah) = malignant tumor of vascular tissue (Figure 15–14).

Copyright © 2015 Cengage Learning®.

FIGURE 15–15 A mast cell tumor on a dog's leg.

- **melanoma** (mehl-ah-nō-mah) = neoplasm composed of melanin-pigmented cells (see Figure 10–25).
- **myeloma** (mī-eh-lō-mah) = malignant neoplasm composed of bone marrow.
- **myosarcoma** (mī-ō-sahr-kō-mah) = malignant neoplasm composed of muscle.
- **myxoma** (mihcks-ō-mah) = tumor of connective tissue.
- **neuroblastoma** (nū-rō-blahs-tō-mah) = malignant neoplasm of nervous tissue origin.
- **osteosarcoma** (ohs-tē-ō-sahr-kō-mah) = malignant neoplasm composed of bone (Figure 15–16).
- **rhabdomyosarcoma** (rahb-dō-mī-ō-sahr-kō-mah) = malignant tumor of striated muscle.
- **squamous cell carcinoma** (skwā-mohs sehl kahr-sih-nō-mah) = malignant tumor developed from squamous epithelial tissue; abbreviated SCC (Figure 15–17).

FIGURE 15–16 Osteosarcoma in a dog. Lateral radiograph of a dog with osteosarcoma of the distal tibia.

FIGURE 15–17 SCC of a bovine eye.

PROCEDURES: ONCOLOGY

Procedures performed on tumors include the following:

- **chemotherapy** (kē-mō-thehr-ah-pē) = treatment of neoplasm through the use of chemicals.
- **fulguration** (fuhl-guh-rā-shuhn) = destruction of tissue by high-frequency electrical current.
- **lymphadenectomy** (lihm-fahd-ehn-ehck-tō-mē) = surgical removal of a lymph node. The name for removal of any tissue that may have a mass or tumor can be derived by adding the combining form for the area being removed.
- **protocol** (prō-tō-kahl) = course of treatment.
- **radiation therapy** (rā-dē-ah-shuhn thehr-ah-pē) = treatment of neoplasm through the use of X-rays; abbreviated RT. Cells that are treated with high-dose radiation are irradiated. Radiation is measured in **rads** (radiation absorbed dose). A **radiosensitive** tumor absorbs radiation and responds by dying or shrinking. A **radioresistant** tumor is minimally affected by radiation. Some side effects of radiation include alopecia and **leukotrichia** (loo-kō-trihk-ih-ah), whitening of the hair.
- **surgical excision** = removal of the entire mass in addition to some normal tissue to ensure that the entire mass is removed.

ABBREVIATIONS RELATED TO THE HEMATOLOGIC, LYMPHATIC, AND IMMUNE SYSTEMS AS WELL AS ONCOLOGY

Hematologic, Lymphatic, and Immune System Abbreviations

- CMI = cell-mediated immunity
- DIC = disseminated intravascular coagulation
- EDTA = ethylenediaminetetraacetic acid
- H & E = hematoxylin and eosin stain
- PMN = polymorphonuclear leukocyte
- RBC = red blood cell
- WBC = white blood cell

Oncology Abbreviations

- FNA = fine needle aspirate
- LSA = lymphosarcoma
- MCT = mast cell tumor
- rad = radiation absorbed dose
- RT = radiation therapy
- SCC = squamous cell carcinoma

REVIEW EXERCISES

Multiple Choice

Choose the correct answer.

1. Red blood cells are called
   a. erythrocytes
   b. leukocytes
   c. thrombocytes
   d. platelets

2. White blood cells are called
   a. erythrocytes
   b. leukocytes
   c. thrombocytes
   d. platelets

3. An elevation in white blood cells is called
   a. leukopenia
   b. leukocytosis
   c. leukemia
   d. leukosis

4. An immature, non-nucleated erythrocyte that is characterized by polychromasia (Wright's stain) or a meshlike pattern of threads (new methylene blue stain) is called a
   a. macrophage
   b. prothrombocyte
   c. reticulocyte
   d. phagocyte

5. An immature polymorphonuclear leukocyte is called a
   a. baby PMN
   b. banded neutrophil
   c. left shift
   d. micro PMN

6. Lacteals are located in the
   a. groin
   b. loin
   c. small intestine
   d. stomach

7. Lymph, when it is present in tissues, is known as
   a. serum
   b. plasma
   c. interstitial fluid
   d. a and b

8. T cells are responsible for
   a. allergies
   b. humoral immunity
   c. cell-mediated immunity
   d. thymus production

9. B cells are responsible for
   a. allergies
   b. humoral immunity
   c. cell-mediated immunity
   d. thymus production

10. The spleen is
    a. hemolytic
    b. an organ where lymphocytes are differentiated
    c. a storage area for RBCs
    d. all of the above

11. An example of a malignant tumor that developed from epithelial tissue is known as a(n)
    a. squamous cell carcinoma
    b. myxoma
    c. osteosarcoma
    d. blastoma

12. What is the term for a malignant tumor of vascular tissue?
    a. hemangioma
    b. hemangiosarcoma
    c. neuroblastoma
    d. lymphoma

13. Malignant neoplasms arising from connective tissue are known as
    a. carcinomas
    b. sarcomas
    c. melomas
    d. myxomas

14. When a growth does not have well-defined borders, it is described as being
    a. circumscribed
    b. pedunculated
    c. invasive
    d. metastasized

15. A deficiency in white blood cells is called
    a. leukopenia
    b. leukocytosis
    c. erythrocytosis
    d. anemia

16. Hemolysis is
    a. increasing the RBC count
    b. decreasing the RBC count
    c. breaking down red blood cells
    d. excess amounts of fat in the blood

17. A substance that prevents clotting is known as a(n)
    a. hematopoietic agent
    b. anticoagulant
    c. anisocytosis
    d. poikiloctosis

18. Larger-than-normal RBCs are described as being
    a. hyperchromic
    b. hypochromic
    c. microcytic
    d. macrocytic

19. What term is used to describe the condition in which microorganisms and their toxins are present in the blood?
    a. septicemia
    b. bacteremia
    c. anemia
    d. edema

20. The name for a disease-fighting protein produced by the body in response to a particular foreign substance is
    a. antigen
    b. antibody
    c. antipathogen
    d. dysrasia

21. What do erythrocyte and leukocyte have in common?
    a. They are diagnoses.
    b. They are procedures.
    c. They are blood cells.
    d. They are colored stains.

22. Which set of terms only contains lymphatic tissue?
    a. spleen, tonsil, pancreas
    b. spleen, adrenal, pancreas
    c. tonsil, thymus, spleen
    d. tonsil, blood, lymph node

23. Which list of tumors involves muscle?
    a. melanoma and myeloma
    b. leiomyosarcoma and rhabdomyosarcoma
    c. lymphoma and blastoma
    d. adenocarcinoma and myxoma

24. The "N" in FNA stands for
    a. nocturia
    b. null
    c. neuro
    d. needle

25. Fluid with a protein <2.5 g/dL is classified as a(n)
    a. exudate
    b. modified transudate
    c. transudate
    d. dyscrasia

Matching

Match the immunologic term in Column I with the definition in Column II.

Column I	Column II
1. _____ resistant	a. weakened
2. _____ heredity	b. able to cause disease due to debilitation when disease would not be produced normally
3. _____ vaccination	c. administration of antigen to stimulate a protective immune response against a specific infectious agent
4. _____ immunization	d. reproduction
5. _____ vaccine	e. not susceptible
6. _____ multiplication	f. genetic transmission of characteristics from parent to offspring
7. _____ inhibit	g. to slow or stop
8. _____ opportunistic	h. preparation of pathogen (live, weakened, or killed) or a portion of pathogen that is administered to stimulate a protective immune response against the pathogen
9. _____ debilitated	

Match the term in Column I with the definition in Column II.

Column I	Column II
10. _____ benign	a. abnormal new growth of tissue
11. _____ malignant	b. cancerous
12. _____ neoplasm	c. enlarged spleen
13. _____ metastasis	d. condition of eating cells
14. _____ splenomegaly	e. breakdown
15. _____ edema	f. noncancerous
16. _____ phagocytosis	g. accumulation of fluid in the intercellular space
17. _____ lysis	h. pathogenic growth distant from its primary disease site
18. _____ leukocytosis	i. abnormal increase in WBCs
19. _____ leucopenia	j. abnormal decrease in WBCs
20. _____ agglutination	k. clumping of incompatible blood cells

Anatomy Application

After each of the following, write the letter of the blood component it most closely relates to, either (a) for red blood cell, (b) for white blood cell, or (c) for component of plasma.

1. albumin _____

2. lymphocytes _____

3. neutrophils _____

4. fibrinogen _____

5. leukocyte _____

6. hemoglobin _____

7. basophils _____

8. monocytes _____

9. histamine _____

10. eosinophils _____

Fill in the Blanks

Write the correct term in the space provided.

1. _____ means no cuts, scrapes, openings, or alterations.

2. The state of being resistant to a specific disease is _____.

3. A substance that the body regards as foreign is a(n) _____.

4. A deficiency of red blood cells or hemoglobin is called _____.

5. Excessive blood in a part is called _____.

6. _____ is a severe response to a foreign substance.

7. A substance capable of inducing an allergic reaction is called a(n) _____.

8. The study of tumors is _____.

9. A general term for a malignant neoplastic disorder of lymphoid tissue is _____.

10. A blood condition in which pathogenic microorganisms or their toxins are present is called _____.

11. -poiesis means _____.

12. Hem/o and hemat/o mean _____.

13. Immun/o means _____.

14. -emia means _____.

15. Lip/o means _____.

True or False

If the statement is true, write T on the line. If the statement is false, write F on the line.

1. _____Agglutination is the clumping of cells and particles in the blood.

2. _____Plasma is the liquid portion of clotted blood.

3. _____The fluid in the space between tissues is called interstitial fluid.

4. _____A carcinogen is a substance that prevents cancer.

5. _____A vaccine is an agent intended to cause disease.

6. _____An antibody is produced in response to an antigen.

7. _____An antibody titer measures the concentration of antibodies circulating in the bloodstream of an animal.

8. _____Immunity is the state of being protected.

9. _____Immunosuppression is an increase in the ability of the immune system to respond to stimulation by antigens.

10. _____Chemotherapy and radiation therapy are examples of protocols for treating cancer.

Spell Check

Circle the term that is spelled correctly.

1. mature red blood cell:	erithrocyte	erythrocyte	erythrosite
2. study of form:	morfology	morphology	morphlogy
3. condition of unequal cell size:	anistocytosis	anisocytosis	anystocytosis
4. tumor of connective tissue:	mixoma	myxtoma	myxoma
5. malignant neoplasm composed of bone:	ostesarcoma	osteosarkoma	osteosarcoma

Crossword Puzzle

Hematologic, Lymphatic, and Immune Terms

Supply the correct term in the appropriate space for the definition listed.

Across

4 abnormal increase in the number of malignant WBCs

7 malignant neoplasm of bone marrow

10 abnormal decrease in the number of WBCs

13 disease-fighting protein produced by the body in response to the presence of a foreign substance

14 destruction or breakdown

15 malignant neoplasm of bone

16 another name for polymorphonuclear leukocyte

17 enlargement of the spleen

Down

1 malignant neoplasm of vascular tissue

2 study of blood

3 another term for allergy

5 substance that prevents clotting

6 study of form (shape)

8 accumulation of fluid in the intercellular space

9 formation of blood

11 pertaining to the spaces within a tissue or organ

12 pathogenic growth distant from the primary disease site

Case Studies

Define the underlined terms in each case study.

A 9-yr-old M/N toy poodle was scheduled for an <u>excisional biopsy</u> of a skin mass that the owner first observed 2 weeks ago. The surgery went well and the veterinarian sent the biopsy to a pathologist to determine if the mass was <u>malignant</u> or <u>benign</u>. The pathologist prepared a microscope slide and the following is the pathology report: Microscopic Exam: A microscope slide was prepared from a skin biopsy of a mass that was centrally <u>ulcerated</u>.

The ulcerated area was surrounded by <u>squamous epithelium</u> and exhibited full-thickness <u>dysplasia</u>. The dysplastic change had cells that had <u>hyperchromic</u> <u>nuclei</u>. The <u>lesion</u> does not appear to invade the underlying tissue. Moderate dysplastic changes were seen <u>peripherally</u> and do not extend into the surgical margins.

1. excisional biopsy _____

2. malignant _____

3. benign _____

4. ulcerated _____

5. squamous epithelium _____

6. dysplasia _____

7. hyperchromic _____

8. nuclei _____

9. lesion _____

10. peripherally _____

Bailey, a 7-yr-old M/N Bassett hound presented to the clinic with an <u>intermittent</u> 3 week history of <u>anorexia</u>, diarrhea, and trouble <u>ambulating</u>. Bailey was current on his <u>vaccines</u> and was on flea, tick, heartworm, and intestinal parasite <u>prophylaxis</u>. On PE, TPR was normal; however, the veterinarian <u>palpated</u> a mass in the abdomen. A complete blood count (CBC), chemistry panel, Lyme test, and <u>urinalysis</u> were recommended for Bailey as well as abdominal and thoracic <u>radiographs</u>. She tested negative for Lyme disease. The CBC revealed a moderate <u>normochromic</u>, <u>normocytic</u> <u>anemia</u> that was categorized as regenerative anemia based on the observation of <u>polychromasia</u> and <u>reticulocytosis</u>. Mild <u>leukocytosis</u> with a mild <u>neutrophilia</u>, <u>monocytosis</u>, and <u>basophilia</u> was reported. <u>Platelets</u> were normal.

Radiographs of the thorax were normal (no <u>metastases</u>) while abdomen revealed <u>splenomegaly</u>. An abdominal <u>ultrasound</u> was recommended and revealed numerous lesions arising from the spleen. The veterinarian suspected a splenic tumor and recommended an exploratory <u>laparotomy</u>. The surgery revealed a large spleen with multiple lesions; therefore, a <u>splenectomy</u> was performed and multiple biopsies were taken from the spleen and submitted for <u>histopathology</u>. The histopathology report stated there was moderate cellular <u>pleomorphism</u>, <u>hemorrhage</u>, and <u>necrosis</u> present in the splenic tissue. A diagnosis of <u>hemangiosarcoma</u> was made and a <u>chemotherapy</u> <u>protocol</u> was instituted.

11. intermittent _____

12. anorexia _____

13. ambulating _____

14. vaccine _____

15. prophylaxis _____

16. palpated

_____

17. urinalysis

_____

18. radiographs

_____

19. normochromic

_____

20. normocytic

_____

21. anemia

_____

22. polychromasia

_____

23. reticulocytosis

_____

24. leukocytosis

_____

25. neutrophilia

_____

26. monocytosis

_____

27. basophilia

_____

28. platelets

_____

29. metastases

_____

30. splenomegaly

_____

31. ultrasound

_____

32. laparotomy

_____

33. splenectomy

_____

34. histopathology

_____

35. pleomorphism

_____

36. hemorrhage

_____

37. necrosis

_____

38. hemangiosarcoma

_____

39. chemotherapy

_____

40. protocol

_____

Critical Thinking Exercise

The following story and questions contain words and concepts from this chapter and are designed to stimulate critical thinking through class discussion, Internet searches, or brief writing responses.

Junior is a 5-year-old pug that has been lethargic for 5 days. He has vomited once and had some bloody diarrhea. He has been anorexic for 2 days. On PE, his HR = 120 bpm, RR = 28 breaths/min, T = 102°F, CRT = normal but gums were pale. His chest ausculted normally; abdominal palpation = enlarged spleen. The veterinarian recommended that the dog have some bloodwork done and an abdominal radiograph taken. While the dog was in the treatment room, he had hematuria and melena. Following are the results of Junior's CBC:

COMPONENT	CANINE NORMAL VALUES	FELINE NORMAL VALUES	UNITS	JUNIOR
Hgb	12.0–18.0	8.0–15.0	g/dL	12.0
hct	37–55	24–45	%	26
retic	0–1.5%	0–1%	%	2%
WBC	6–17.1	5.5–19.5	×1,000/ul	25.5
segs	3.6–11.5	2.5–12.5	×1,000/ul	19.8
bands	0–0.3	0–0.3	×1,000/ul	1.3
eos	0.01–1.25	0.0–1.5	×1,000/ul	0.2
lymphos	1.0–4.8	1.5–7.0	×1,000/ul	3.2
monos	0.15–1.35	0.0–0.85	×1,000/ul	1.0
basos	0–0.2	0–0.15	×1,000/ul	0.0
platelets	2–9	3–7	×100,000/ul	5

1. In common terms and medical terms, list the abnormalities on Junior's CBC.

2. Based on the CBC results, what might be occurring in this dog?

Chapter 16
TESTING TESTING

Objectives

Upon completion of this chapter, the reader should be able to:

- Describe terms and equipment used for the basic physical examination

- Recognize, define, spell, and pronounce terms associated with physical examinations

- Recognize, define, spell, and pronounce terms associated with laboratory analysis

- Recognize, define, spell, and pronounce basic medical and disease terms

- Recognize, define, spell, and pronounce terms associated with pathogenic organisms

- Recognize, define, spell, and pronounce terms associated with different types of disease and terms used to describe them

- Recognize, define, spell, and pronounce terms associated with endoscopic and centesis procedures

- Describe positioning for radiographic and imaging procedures

- Recognize, define, spell, and pronounce terms associated with imaging procedures

- Interpret the meaning of abbreviations associated with physical examination, laboratory analysis, pathogenic organisms, disease, endoscopy, centesis, and imaging techniques

BASIC PHYSICAL EXAMINATION TERMINOLOGY

Physical examinations are performed to assess a patient's condition. **Assessment** is the term used to describe the evaluation of a condition. After a patient is assessed, the information is written in a medical record. The animal's signalment should always be included. A **signalment** is a description of the animal with information about the animal, including the species, breed, age, and sexual status (intact or neutered).

Vital Signs

Vital signs are parameters taken from the animal to assess its health.

Temperature is the vital sign that provides an indication of an animal's core body temperature. An animal's temperature is recorded in degrees Fahrenheit or Celsius. Different species have different normal temperature ranges. An elevated body temperature is called a **fever**. **Febrile** (fē-brīl or feh-brīl) is the medical term for fever; **afebrile** (ā-fē-brīl or ā-feh-brīl) means without a fever. **Pyrexia** (pī-rehck-sē-ah) is another medical term for fever. **Pyr/o** means fire. A decrease in body temperature is known as **hypothermia**.

Pulse, another vital sign, tells the number of times the heart beats per minute. The pulse is also called the **pulse rate**. Pulse is taken by palpating an artery. Heart rate may also be considered a vital sign and is taken by auscultating the heart with a stethoscope (Figures 16–1 and 16–2). **Heart rate** is the number of times the heart contracts and relaxes per minute. Heart rate is abbreviated HR.

Respiration is the number of respirations per minute. Respiration is one total inhale and one total exhale. Respiration is also called the **respiration rate** and is abbreviated RR (Figure 16–3).

Blood pressure is another vital sign that may be taken on veterinary patients. A **sphygmomanometer** (sfihg-mō-mah-nohm-eh-tər) is an instrument used to measure blood pressure (Figure 16–4). A Doppler is an instrument used to listen to blood sounds during the measurement of blood pressure in animals. BP is the abbreviation for blood pressure.

Listening

Auscultation (aws-kuhl-tā-shuhn) is the act of listening, which usually involves the use of a stethoscope to listen to body sounds. Auscultation can be used to assess the condition of the heart, lungs, pleura, and abdomen.

Copyright © 2015 Cengage Learning®.

FIGURE 16–1 Parts of a stethoscope. (a) Diaphragm. The diaphragm is the flat, circular portion of the chestpiece covered with a thin membrane. The diaphragm transmits high-pitched sounds, such as those produced by the bowel, lungs, and heart. (b) Bell. The bell is not covered by a membrane. The bell facilitates auscultation of lower-frequency sounds, such as the third and fourth sounds of the heart.

Courtesy of Teri Raffel, CVT

FIGURE 16–2 A stethoscope is used to listen to body sounds.

Touching

Palpation (pahl-pā-shuhn) is examination by feeling. During palpation, one can feel the texture, size, consistency, and location of body parts or masses.

	Heart rate (beats/min)	Respiratory rate (breaths/min)	Rectal temperature (°C)	Rectal temperature (°F)
Dogs	70–160	8–20	38–39	100.5–102.5
Cats	150–210	8–30	38–39	100.5–102.5
Hamsters	250–500	35–135	37–38	99–100.5
Guinea pigs	230–280	42–104	37–39.5	99–103
Rabbits	130–325	30–60	38.5–40	101.5–104
Horses	28–50	8–16	37.5–38.5	99.5–101.5
Cattle	40–80	12–36	38–39	100.5–102.5
Sheep	60–120	12–50	39–40	102.5–104
Goats	70–135	12–50	38.5–40.5	101.5–105
Pigs	58–100	8–18	39–40	102.5–104
Llamas	60–90	10–30	37–39	99–102.5
Ferrets	230–250	33–36	38–40	100.5–104

Copyright © 2015 Cengage Learning®.

FIGURE 16–3 Normal vital sign ranges.

Courtesy of Teri Raffel, CV

(a)

Copyright © 2015 Cengage Learning®.

(b)

FIGURE 16–4 (a) Sphygmomanometer. (b) A technician restrains a dog for a blood pressure measurement. Note the pressure cuff around the front leg.

Percussion (pər kuhsh uhn) is examination by tapping the surface to determine density of a body area (see Figure 9–10). Sound may be produced when the surface is tapped with a finger or an instrument. The sound produced by percussion varies depending on the amount of fluid, solid, or air present in the area being examined.

Looking

Various instruments can be used during the physical examination to obtain a better view of a body system. Examples include an ophthalmoscope and otoscope. A **speculum** (spehck-yoo-luhm) is an instrument used to enlarge the opening of a canal or cavity. A speculum is

Courtesy of Teri Raffel, CVT

(a)

Courtesy of Ron Fabrizius, DVM, Diplomate ACT

(b)

FIGURE 16–5 Specula. (a) A speculum is attached to an otoscope to allow better visualization of a canal or cavity. (b) A mouth speculum is used to enhance visualization of the oral cavity.

BIOLOGY BONUS

Most people do not consider a physical examination a procedure; however, our hands, eyes, and ears are great tools for determining the cause of disease as well as monitoring disease progression and evaluating the success or failure of treatment. Taking good histories and preforming thorough physical exams are vital to successful animal care. For example, animals with cardiac disease should be evaluated through listening (auscultation of the thorax), touching (palpation for the presence of thrills or alterations in intensity or location of the pulse), and seeing (mucous membrane color and capillary refill time). Concurrent auscultation and palpation of pulses use two senses to determine the status of cardiac function. Before rushing to conduct a slew of tests, it is important to gather as much information from a thorough history and extensive physical exam. Developing these skills will greatly benefit the animals in your care.

attached to an otoscope (or another scope) to provide a better view of a canal or cavity. A mouth speculum is used to better visualize the oral cavity (Figure 16–5a and b).

LABORATORY TERMINOLOGY

Specialized terminology has been developed to describe tests and results of laboratory tests. Blood for laboratory tests usually is collected via venipuncture.

Venipuncture (vehn-ih-puhnk-tər) is withdrawing blood from a vein (usually with a needle and syringe).

Terminology of laboratory tests is given in Table 16–1.

Special equipment is needed to prepare the samples for many laboratory tests. A **refractometer** (rē-frahck-tah-mē-tər) is an instrument used to determine the deviation of light through objects. Refractometers are used to measure solute (particle)

TABLE 16–1 Blood Test Terminology

Term	Pronunciation	Definition
agglutination	(ah-gloo-tih-nā-shuhn)	clumping together of cells or particles
assay	(ahs-ā)	assessment or test to determine the number of organisms, cells, or amount of a chemical substance found in a sample
complete blood count		diagnostic evaluation of blood to determine the number of erythrocytes, leukocytes, and thrombocytes per cubic millimeter of blood; abbreviated CBC
differential	(dihf-ər-ehn-shahl)	diagnostic evaluation of the number and types of white blood cells per cubic millimeter of blood
diluent	(dihl-yoo-ehnt)	liquid used to make a dilution
enzyme linked immunosorbent assay	(ehn-zīm lihnkehd ihm-yoo-nō-sohr-behnt ahs-ā)	laboratory test to determine either antigen or antibody in blood; abbreviated ELISA (Figure 16–6)
hematocrit	(hē-maht-ō-kriht)	percentage of erythrocytes in blood; "to separate blood"; also called **crit**, **PCV**, or **packed cell volume** (Figure 16–7)
hemogram	(hē-mō-grahm)	record of the findings in examination of blood especially with reference to the numbers, proportions, and morphology of the blood cells
immunofluorescence	(ihm-yoo-nō-floo-rehs ehns)	method of tagging antibodies with a luminating dye to detect antigen–antibody complexes
leukogram	(loo-kō-grahm)	numeric and descriptive data in the distribution and appearance of WBCs; used to identify a pathologic process
profile	(prō-fīl)	group of laboratory tests performed on serum; also called **screen** or **panel**; includes tests that measure levels of glucose, liver enzymes, and kidney enzymes
prothrombin time	(prō-throhm-bihn)	diagnostic evaluation of the number of seconds needed for thromboplastin to coagulate plasma
radioimmunoassay	(rā-dē-ō-ihm-yoo-nō-ahs-ā)	laboratory technique in which a radioactively labeled substance is mixed with a blood specimen to determine the amount of a particular substance in the mixture; also called **radioassay**
red cell count		number of erythrocytes per cubic millimeter of blood
serology	(sē-rohl-ō-jē)	laboratory study of serum and the reactions of antigens and antibodies
white cell count		number of leukocytes per cubic millimeter of blood

FIGURE 16–6 A heartworm ELISA test. Three tests are shown: The first test is in process (left); the second test is positive showing two red lines representing the positive result and validation (control) result (upper right); the third test is negative showing only the red line for the validation (control) result (lower right).

FIGURE 16–7 Microhematocrit tubes are used to determine a patient's hematocrit.

concentration of serum, urine, and other body fluids (Figure 16–9).

A **centrifuge** (sehn-trih-fūj) is a machine that spins samples very rapidly to separate elements based on density. A centrifuge is used to separate the formed elements of blood from the liquid portion of blood. A centrifuge also separates the liquid portion of urine from the heavier solids (Figure 16–10a, b, and c).

The pH of a sample may also provide information about a patient's status. The terms *acid* and *alkaline* are used to describe a patient's pH status. **Acid** (ah-sihd) is the property of low pH, or high number of hydrogen ions. **Alkaline** (ahl-kah-lihn) is the property of high pH,

Records are kept under many different methods. One method is the SOAP method. **SOAP** is an acronym for subjective, objective, assessment, and plan analysis (Figure 16–8). An acronym is a word formed by the initial letters of the major parts of the name. Some diseases, structures, and procedures are derived from a person's name, and the name is known as an **eponym**.

> **Animal Medical Hospital**
>
> Ⓢ Dog presented BAR c̄ hx of persistent cough, anorexia, and wt loss.
> Ⓞ T = 103.4 °F HR = 120 bpm RR = 30 breaths/min CRT = 1 sec mm = pink. Lungs auscultated harshly, Heart sounds appear Ⓝ. Rest of PE - WNL.
> Ⓐ Respiratory disease — R/O Kennel cough, pneumonia.
> Ⓟ Radiograph chest CBC Medication pending results.

FIGURE 16–8 SOAP method of record keeping.

or low number of hydrogen ions. Alkaline is also called **basic.** The pH of a variety of body fluids may be analyzed to determine the health status of an animal. For example, urine pH can be detected with colorimetric reagent strips (Figure 16–11). The pH of urine is affected by diet (herbivores have more alkaline (higher pH) urine than carnivores), medications, kidney disease, and metabolic disease such as diabetes mellitus. Indicator dyes in the pH reagent pad form colors from yellow-orange for acid urine to green-blue for alkaline urine.

Urinalysis terminology is discussed in Table 7–3.

BASIC MEDICAL AND DISEASE TERMS

Basic medical and disease terms are listed in Table 16–2.

PATHOGENIC ORGANISMS

A **pathogen** (pahth-ō-jehn) is a microorganism that produces disease. A **microorganism** (mī-krō-ōr-gahn-ihzm) is a living organism of microscopic dimensions (Table 16–3). Not all microorganisms are pathogens; some microorganisms are called *normal flora* and live in an animal's body without causing disease. The term **virulence** (vihr-yoo-lehns) is used to describe the ability of an organism to cause disease.

(c)

FIGURE 16–9 Refractometer. (a) A refractometer measures specific gravity optically. A drop of urine is placed on the glass plate. (b) The technician looks through the eyepiece to view the scale. (c) This refractometer scale shows a urine specific gravity of 1.034 (lower left scale) and serum or plasma protein of 5.6 (middle scale).

FIGURE 16–10 Centrifuge. (a) Microhematocrit centrifuge. (b) Proper placement of sealed capillary tubes in a microhematocrit centrifuge. (c) Proper placement of tubes in a centrifuge rotor; bottom A–C diagrams show correct placement of the tubes (small dark blue circles) across from each other in rotor and, D, incorrect placement of tubes.

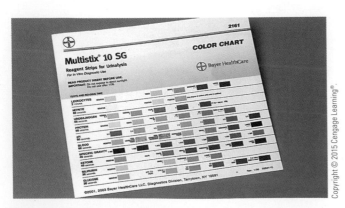

FIGURE 16–11 Reagent strip color chart for chemical testing of urine. One test available on the reagent strip is pH.

TYPES OF DISEASES

Disease is deviation from normal. There are different types of diseases, depending on how they are spread and what causes them. Types of diseases include the following:

- **contagious** (kohn-tā-juhs) **disease** = disease that can be spread from one animal to another by direct or indirect contact. Direct contact is spread from animal to animal, whereas indirect contact is spread through contact with contaminated objects or a healthy animal/insect intermediary. Contagious disease may also be referred to as **communicable** (kuh-mū-nih-kuh-buhl).

TABLE 16-2 Basic Medical Terms

Diagnosis (dī-ahg-nō-sihs) is the determination of the cause of disease; to "know completely"; plural is **diagnoses** (dī-ahg-nō-sēz). To **diagnose** (dī-ahg-nōs) is to determine the cause of disease.	**Differential diagnosis** (dihf-ər-ehn-shahl dī-ahg-nō-sihs) is the determination of possible causes of diseases; a list of possible causes of disease.	**Prognosis** (prohg-nō-sihs) is the prediction of the outcome of disease; to "know before."
Sign (sīn) is a characteristic of disease that can be observed by others.	**Symptom** (sihmp-tuhm) is a characteristic of disease that can be sensed only by the patient; incorrect term in veterinary medicine.	**Syndrome** (sihn-drōm) is a set of signs that occur together.
Acute (ah-kūt) means having a short course with a sudden onset; implies severe. **Peracute** (pər-ah-kyūt) means having an excessively acute onset.	**Chronic** (krohn-ihck) means having a long course with a progressive onset; persisting for a long time.	**Remission** (rē-mih-shuhn) is partial or complete disappearance of disease signs.
Endemic (ehn-dehm-ihck) is the ongoing presence of disease in a group; also called **enzootic** (ehn-zō-oh-tihck) if the disease is always present in an animal community.	**Epidemic** (ehp-ih-dehm-ihck) is the sudden and widespread outbreak of disease in a group; also called **epizootic** (ehp-ih-zō- oh-tihck) if the outbreak attacks many animals in a group; called **episodic** (ehp-ih-sohd-ihk) if the outbreak occurs sporadically.	**Pandemic** (pahn-dehm-ihck) is disease outbreak occurring over a large geographic area; also called **panzootic** (pahn-zō-oh-tihck) if the widespread outbreak affects many animals over a large geographic area.

- **iatrogenic** (ī-aht-rō-jehn-ihck) **disease** = disorder caused by physicians or veterinarians (and the treatment ordered).
- **idiopathic** (ihd-ē-ō-pahth-ihck) **disease** = disorder of unknown cause. Idiopathic disease is a disease peculiar to an individual and not likely to be seen in others.
- **infectious** (ihn-fehck-shuhs) **disease** = disorder caused by pathogenic organisms.
- **noncontagious** (nohn-kohn-tā-juhs) **disease** = disorder that cannot be spread from one animal to another animal or by contact with a contaminated object.
- **noninfectious** (nohn-ihn-fehck-shuhs) **disease** = disorder not caused by organisms (e.g., genetic, traumatic, and iatrogenic).
- **nosocomial infection** (nōs-ō-kō-mē-ahl ihn-fehck-shuhn) = disorder caused by pathogenic organisms contracted in a facility or clinic.

DISEASE TERMINOLOGY

- **asymptomatic** (ā-sihmp-tō-mah-tihck) = without signs of disease.

- **atraumatic** (ā-traw-mah-tihck) = pertaining to, resulting from, or caused by a noninjurious route.
- **carrier** (kahr-ē-ər) = animal that harbors an infectious agent without displaying clinical signs and who may transmit the infectious agent to others.
- **clinical** (klihn-ih-kahl) = visible, readily observed, pertaining to treatment.
- **contract** (kohn-trahckt) = to catch a disease.
- **disease** (dih-zēz) = deviation from normal health.
- **epidemiology** (ehp-ih-dē-mē-ohl-ō-jē) = study of relationships determining frequency and distribution of diseases.
- **etiology** (ē-tē-ohl-ō-jē) = study of disease causes.
- **excessive** (ehck-sehs-ihv) = more than normal.
- **focus** (fō-kuhs) = localized region.
- **germ** (jərm) = common term for any pathogenic microorganism, but especially bacterial and viral organisms.
- **incidence** (ihn-sih-dehns) = number of new cases of disease occurring during a given time.
- **labile** (lā-bīl) = unstable.

TABLE 16–3 Types of Organisms

Organism	Examples
bacterium (bahck-tē-rē-uhm) is a microscopic, prokaryotic unicellular organism; plural is **bacteria** (bahck-tē-rē-ah). A prokaryote (prō-kahr-ē-ōt) is an organism without a membrane-bound nucleus.	**staphylococci** (stahf-ih-lō-kohck-sī) are grapelike clusters of spherical bacteria; **coccus** (kohck-uhs) means sphere; (cocci is plural); **staphyl/o** means grapelike clusters; when the "s" in staphylococcus is lower case, it is used to describe morphology; if the "s" in Staphylococcus is upper case, it is used as a genus name. Staphylococcus
	streptococci (strehp-tō-kohck-sī) are spherical bacteria that form chains; **strept/o** means chain; when the "s" in streptococcus is lower case, it is used to describe morphology; if the "s" in Streptococcus is upper case, it is used as a genus name. Streptococcus
	bacilli (bah-sihl-ī) are rod-shaped bacteria; (bacillus is singular); when the "b" in bacillus is lower case, it is used to describe morphology; if the "b" in Bacillus is upper case, it is used as a genus name. Bacillus
	spirochete (spī-rō-kēt) are spiral-shaped bacterium that are tightly coiled; spirochetes is plural. Spirochete
	endospore (ehn-dō-spōr) is a resistant, oval body formed in some bacteria. Endospore in bacterium — Free endospore
	rickettsia (rih-keht-sē-ah) is a small rod-shaped bacterium transmitted by lice, fleas, ticks, or mites; rickettsiae (rih-keht-sē-ā) is plural.
fungus (fuhng-guhs) is a eukaryotic organism without chlorophyll; plural is **fungi** (fuh-jī). A eukaryote (yoo-kahr-ē-ōt) is an organism with a membrane-bound nucleus.	**yeast** (yēst) is a budding form of fungus. **mold** (mohld) is a filamentous form of fungi. Yeast — Mold
parasite (pahr-ah-sīt) is an organism that lives on or in another living organism.	Parasite
virus (vī-ruhs) is a small organism that is not visualized via light microscopy; viruses live only by invading cells.	Virus

- **lethal** (lē-thahl) = causing death.
- **morbid** (mōr-bihd) = afflicted with disease.
- **morbidity** (mōr-bihd-ih-tē) = ratio of diseased animals to well animals in a population.
- **moribund** (mōr-ih-buhnd) = near death.
- **mortality** (mōr-tahl-ih-tē) = ratio of diseased animals that die to diseased animals.
- **palliative** (pah-lē-ah-tihv) = able to relieve but not cure a condition.
- **phobia** (fō-bē-ah) = extreme fear.
- **prevalence** (preh-vah-lehns) = number of cases of disease in a population at a certain time.
- **prophylaxis** (prō-fihl-ahcks-sihs) = prevention.
- **sequela** (sē-kwehl-ah) = condition occurring as a consequence of another condition.
- **subclinical** (suhb-klihn-ih-kahl) = without showing signs of disease.
- **susceptible** (sah-sehp-tih-buhl) = lacking resistance.
- **swollen** (swohl-ehn) = enlarged by fluid retention.
- **transmissible** (trahnz-mihs-ih-buhl) = ability to transfer from one animal to the next.
- **transmission** (trahnz-mih-shuhn) = transfer from one animal to the next. There are different types of disease transmission, including **blood-borne transmission** (spread of disease via blood or body fluids), **sexual transmission** (spread of disease via contact with reproductive areas or through copulation), **airborne transmission** (spread of disease via respiratory droplets), and **fecal–oral transmission** (spread of disease via eating, drinking, or licking food, water, or objects contaminated by feces).
- **traumatic** (traw-mah-tihck) = pertaining to, resulting from, or causing injury.
- **zoonosis** (zō-ō-nō-sihs) = disease that can be transmitted between animals and humans.

ENDOSCOPY TERMINOLOGY

Endoscopy is the visual examination of the interior of any cavity of the body by means of an endoscope. The procedures and instruments are named for the body parts involved.

Specific endoscopic procedures are covered in the chapter of the body system on which they are used.

Endoscopic surgery is a procedure using an endoscope to aid in surgical procedures so that only very small incisions are made.

FIGURE 16–12 Metal and indwelling plastic trocar. Both trocars have a solid pointed spike in the center that is removed once the trocar is in place. With the spike removed, the tube portion of the trocar allows gas to escape.

CENTESIS TERMINOLOGY

Centesis is the surgical puncture to remove fluid or gas for diagnostic purposes or for treatment. A **trocar** (trō-kahr) is a sharp, needlelike instrument that has a cannula (tube) that is used to puncture the wall of a body cavity to withdraw fluid or gas (Figure 16–12). A **cannula** (kahn-yoo-lah) is a hollow tube. Specific centesis procedures are covered in the chapter of the body system in which they are used.

IMAGING TECHNIQUES TERMINOLOGY

Imaging techniques are used to visualize and examine internal structures of the body.

Radiology

The first imaging technique developed involves the use of ionizing radiation, or X-rays, to produce an image. **Radiography** (rā-dē-ohg-rah-fē) is the procedure in which film is exposed as ionizing radiation passes through the patient and shows the internal body structures in profile. A **radiograph** (rā-dē-ō-grahf), or X-ray, is the record of ionizing radiation used to visualize internal body structures. Note that **graph** (as opposed to gram) is used to mean *record* in this case.

Radiographs are composed of shades of gray and are useful for showing abnormalities such as bone fractures, internal organ enlargement, detecting some types of foreign objects or neoplasms, and dental abnormalities. Hard tissues such as bone are called radiopaque. **Radiopaque** (rā-dē-ō-pāk) means appearing white or light gray on a radiograph because ionizing radiation does not penetrate these structures well. Air and soft tissues are called radiolucent. **Radiolucent** (rā-dē-ō-loo-sehnt) means appearing black or dark gray on a radiograph because ionizing radiation passes quickly through these structures. Substances in between radiopaque and radiolucent appear in various shades of grey on the radiograph.

Radiology (rā-dē-ohl-ō-jē) is the study of internal body structures after exposure to ionizing radiation. A radiologist (rā-dē-ohl-ō-jihst) is a specialist who studies internal body structures after their exposure to ionizing radiation (Figure 16–13).

There are different types of radiographs depending on how they are taken and whether any additional diagnostic tool is used. A **scout film** is a plain radiograph made without the use of a contrast medium. A scout radiograph indicates whether abnormalities exist in the patient or whether there is need for further testing or for more specialized radiographic studies.

X-rays were discovered in 1895 by Wilhelm Konrad Roentgen, a German physicist. **Radiology** comes from the Latin word **radius**, meaning a rod, and the suffix **-logy**, meaning the science of or the study of. Radiology also is known as roentgenology as a tribute to its discoverer. A **roentgen** (rehnt-gehn) is the international unit of radiation. Another term, **rad** (an acronym for *radiation absorbed dose*), is a unit by which absorption of ionizing radiation is measured.

Two abbreviations commonly used in radiography are kVp and MAS. **kVp** stands for kilovoltage peak and represents the strength of the X-ray beam. **MAS** stands for milliamperes per second and represents the number of X-ray beams (because it is based on time).

(a)

(b)

Courtesy of Teri Raffel, CVT

FIGURE 16–13 Radiographic machine. (a) Radiographic table and X-ray tube. (b) Radiographic control panel.

Some body structures, such as the intestinal lumen, are difficult to assess using X-rays alone. In those cases, a radiographic contrast medium may be used. **Radiographic contrast** (rā-dē-ō-grahf-ihck kohn-trahst) **medium** or material is a substance used to show structures on X-ray that are otherwise difficult to see. **Barium sulfate** (bār-ē-uhm suhl-fāt), or **barium,** is one example of contrast material.

The type of contrast radiograph taken depends on which structures are to be visualized (see Figure 6-34). The route of administration of a contrast medium also varies depending on which structures are to be visualized.

A **lower GI** is a type of contrast radiograph used to visualize the structures of the lower gastrointestinal tract. In a lower GI, an enema is used to introduce contrast material into the colon; therefore, it also is called a **barium enema** (bār-ē-uhm ehn-ah-mah). An **upper GI** is a type of contrast radiograph used to visualize the structures of the upper gastrointestinal tract. In an upper GI, contrast material is swallowed; therefore, it is also called a **barium swallow** (bār-ē-uhm swahl-ō) (see Figure 6–20).

Contrast material can also be injected intravenously. An intravenous contrast medium is injected into the vein to make visible the flow of blood through the blood vessels and organs.

Radiographic techniques are also named for the vessels or organs involved. An example is a **lymphangiography** (limh-fahn-jē-ohg-rah-fē), which is a radiographic examination of the lymphatic vessels after injection of contrast material.

Projection Positioning

Two terms frequently used in radiography are *projection* and *positioning*. **Projection** (prō-jehck-shuhn) is the path of the X-ray beam (Table 16–4). **Positioning** (pō-sih-shuhn-ihng) is the specified body position and the part of the body closest to the film (Figure 16–14a and b). **Recumbency** is used in reference to positioning. (Refer to Chapter 2 for positional terms.) **Anatomical position** (ahn-ah-tohm-ihck-ahl pō-sih-shuhn) refers to the animal in its normal standing position.

TABLE 16–4 Types of Projection

Projection	Pronunciation	Definition
craniocaudal projection	(krā-nē-ō-kaw-dahl prō-jehck-shuhn)	X-ray beam passes from cranial to caudal; formerly called **anteroposterior projection** (ahn-tēr-ō-pohs tēr-ē-ər prō-jehck-shuhn), or A/P projection.
caudocranial projection	(kaw-dō-krā-ne-ahl prō-jehck-shuhn)	X-ray beam passes from caudal to cranial; formerly called **posteroanterior projection** (pohs-tēr-ō-ahn-tēr-ih-ər prō-jehck-shuhn), or P/A projection.
dorsoventral projection	(dōr-sō-vehn-trahl prō-jehck-shuhn)	X-ray beam passes dorsally to ventrally; abbreviated D/V.
ventrodorsal projection	(vehn-trō-dōr-sahl prō-jehck-shuhn)	X-ray beam passes ventrally to dorsally; abbreviated V/D.
lateral projection	(laht-ər-ahl prō-jehck-shuhn)	X-ray beam passes from side to side, with the patient at right angles to the film; when used for the skull, cervical spine, thorax, and abdomen, they are described as right or left lateral projections; when used for extremities, they are described as lateromedial or mediolateral projections.
oblique projection	(ō-blēk prō-jehck-shuhn)	X-ray beam passes through the body on an angle. May be further described by the direction of the beam and the degree of the angle (i.e., 60 degrees DP oblique of the hoof) or simply the direction (i.e., dorsomedial-palmarolateral oblique [DMPaLO] of the carpus).

(a)

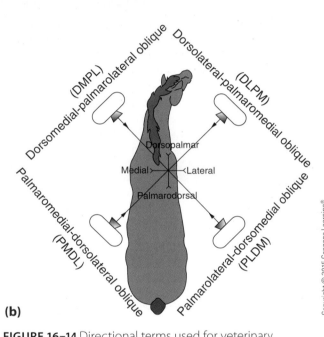

(b)

FIGURE 16–14 Directional terms used for veterinary radiography with their standard abbreviations (based on American College of Veterinary Radiologists) are shown in these figures. Cr = cranial, D = dorsal, V = ventral, R = rostral, Pa = palmer, Pl = plantar, L = lateral, and M = medial. (a) When radiographic views of given structures are written, medial and lateral should be subservient when used in combination with other terms (e.g., dorsomedial); (b) on the limbs, dorsal, palmar, plantar, cranial, and caudal should take precedence when used in combination with other terms (e.g., dorsoproximal).

Computed Tomography

Computed tomography (kohm-puh-tehd tō-moh-grah-fē) is the procedure in which ionizing radiation with computer assistance passes through the patient and shows the internal body structures in cross-sectional views. It also is called **CT scan** or **CAT** (computed axial tomography) **scan**. **Tomography** (tō-moh-grah-fē) is a recording of the internal body structures at predetermined planes that produces

three-dimensional images. Information obtained by radiation detectors is downloaded to a computer, analyzed, and converted to grayscale images corresponding to anatomical body slices. Those images are viewed on a monitor or as a printed hard copy (Figure 16–15).

Magnetic Resonance Imaging

Magnetic resonance imaging (mahg-neh-tihck reh-sohn-ahns ih-mah-jihng) is the procedure in which radio waves and a strong magnetic field pass through the patient and show the internal body structures in three-dimensional views. Magnetic resonance imaging, abbreviated MRI, creates images by tracking the magnetic properties in the nucleus of cells. As the cells move, some atoms respond to magnetic fields and emit radio waves that produce an image. MRI is used for imaging the brain, spine, and joints (Figure 16–16). MRIs do not use X-rays and are considered safe and effective. Most MRIs do not require a contrast medium, but one may be used to enhance a scan in certain areas such as blood vessels. MRI equipment consists of a tube into which the patient is placed and while the patient is lying absolutely still (anesthetized), the magnet in the equipment obtains the scan.

Fluoroscopy

Fluoroscopy (floor-ohs-kō-pē) is the procedure used to visually examine internal body structures in motion using radiation to project images onto a fluorescent screen as light rays are emitted when the X-rays pass through the patient. The combining form **fluor/o** means luminous. **Luminous** (loo-mih-nuhs) means giving off a soft light.

FIGURE 16–15 CT scan.

FIGURE 16–16 MRI with contrast material in a dog with a cerebellar tumor.

Ultrasound

Ultrasound (uhl-trah-sownd), or **ultrasonography** (uhl-trah-soh-noh-grah-fē), is the imaging of internal body structures by recording echoes of high-frequency sound waves. The sound waves are received by a detector that converts them to electrical impulses, which can then be seen on a video monitor. The equipment used for an ultrasound consists of a wand, which is attached to a monitor on which the image is seen, that is moved back and forth over the area being observed. Ultrasound is most effective for viewing solid organs or soft tissues not blocked by bone or air. Ultrasound is also effective for viewing body parts through fluid, as in an ultrasound of a gravid uterus. A special type of ultrasound unit called a Doppler is used on blood vessels and was described at the beginning of this chapter and in Chapter 8. A **sonogram** (soh-nō-grahm) or ultrasound image shows the internal body structures by recording echoes of pulses of sound waves above the range of human hearing (Figure 16–17).

The following terms are related to ultrasound techniques:

amplitude (ahm-plih-tood) = intensity of an ultrasound wave.

anechoic (ahn-eh-kō-ihck) = ultrasonic term for when waves are transmitted to deeper tissue and none are reflected back.

attenuation (ah-tehn-yoo-ā-shuhn) = loss of intensity of the ultrasound beam as it travels through tissue.

echoic (eh-kō-ihck) = ultrasound property of producing adequate levels of reflections (echoes) when sound waves are returned to the transducer and displayed.

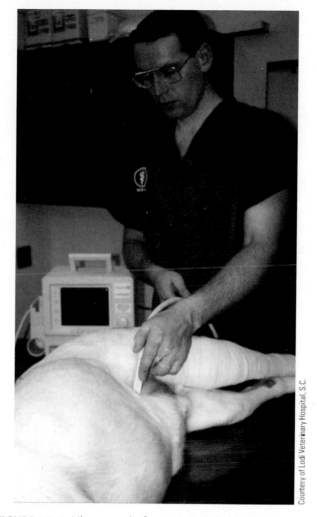

FIGURE 16–17 Ultrasound of a canine abdomen.

frequency (frē-kwehn-sē) = number of cycles per unit of time; abbreviated f.

hyperechoic (hī-pər-eh-kō-ihck) = tissue that reflects more sound back to the transucer than the surrounding tissues; appears bright (Figure 16–18).

hypoechoic (hī-pō-eh-kō-ihck) = tissue that reflects less sound back to the transducer than the surrounding tissues; appears dark.

isoechoic (ī-sō-eh-kō-ihck) = tissue that has the same ultrasonic appearance as that of the surrounding tissue.

resolution (rehs-ō-loo-shuhn) = ability to separately identify different structures on radiograph or ultrasound.

velocity (vehl-oh-sih-tē) = speed at which something travels through an object; abbreviated v.

wavelength (wāv-lehngth) = length that a wave must travel in one cycle; abbreviated λ.

Courtesy of Fern Delany, RDMS, VMTH University of Wisconsin-Madison

FIGURE 16–18 Ultrasound image of spleen and kidney showing hyperechoic and hypoechoic areas. The kidney is hypoechoic to the spleen.

ABBREVIATIONS RELATED TO TESTING

Radiography Abbreviations

- a or amp = ampere
- A/P = anterior/posterior
- APMLO = anterior to posterior and medial to lateral oblique
- BE = barium enema
- Ba = barium
- CT or CAT = computed (axial) tomography
- DL-PaMO = dorsolateral-palmaromedial oblique
- DL-PlMO = dorsolateral-plantaromedial oblique
- DM-PaLO = dorsomedial-palmarolateral oblique
- DP = dorsopalmar
- DPl = dorsoplantar
- DPr DDiO = dorsoproximal-dorsodistal oblique (often called the skyline)*
- D/V = dorsal/ventral
- f = frequency
- Hz; hertz = (a unit of frequency)
- λ = wavelength
- IVP = intravenous pyelogram
- kV = kilovolt
- kVp = kilovolt peak

*Abbreviation may be preceded with *flexed* and may include a number referring to the deviation from perpendicular for a beam; for example, flexed D30Pr-DdiO is a flexed dorsoproximal-dorsodistal oblique taken at a 30-degree angle to the cassette.

- kW = kilowatt
- LM = lateral to medial
- LMO = lateral to medial oblique
- mA = milliamperage
- mAs = milliamperage in seconds
- mc = millicurie
- ML = medial to lateral
- MLO = medial to lateral oblique
- MRI = magnetic resonance imaging
- O = oblique
- P/A = posterior/anterior
- Pa45Pr-PaDiO = palmaroproximal-palmarodistal oblique (often called the skyline navicular)
- PALMO = posterior to anterior and lateral to medial oblique
- PAMLO = posterior to anterior and medial to lateral oblique
- rad = radiation absorbed dose (unit of measurement of absorbed dose of ionizing radiation)
- RT = radiation therapy
- T = period of time (used in ultrasound)
- v = velocity
- V = volt
- V/D = ventral/dorsal
- W = watt
- X-ray = roentgen ray

Physical Examination Abbreviations

- ASAP = as soon as possible
- BAR = bright, alert, responsive
- BP = blood pressure
- CC = chief complaint
- DD = differential diagnosis
- DDN= dull, depressed, nonresponsive
- ddx = differential diagnosis
- DOA = dead on arrival
- dx = diagnosis
- ET = endotracheal
- GSW = gunshot wound
- HBC = hit by car
- HR = heart rate
- hx = history
- mm Hg = millimeters of mercury
- NA or N/A = not applicable
- POVMR = problem-oriented veterinary medical records
- RR = respiration rate
- SOAP = subjective, objective, assessment, plan (record-keeping method)
- stat = immediately (*statim*)

- T = temperature
- TLC = tender loving care
- TPR = temperature, pulse, and respiration
- tx = treatment
- WNL = within normal limits
- wt = weight

Laboratory Abbreviations

- Ab = antibody
- ABG = arterial blood gas
- ABO = human blood groups
- ag = antigen
- alb = albumin
- alk phos = alkaline phosphatase
- ALT = alanine aminotransferase (formerly SGPT)
- amyl = amylase
- AST = aspartate aminotransferase (formerly SGOT)
- BUN = blood urea nitrogen
- CBC = complete blood count
- CFT = complement fixation test
- CHOL = cholesterol
- CK – creatine kinase
- CPR = cardiopulmonary resuscitation
- CREA = creatine
- diff = differential white blood count
- EDTA = ethylenediaminctetraacetic acid; (type of anticoagulant)
- ESR = erythrocyte sedimentation rate
- GGT = gamma glutamyl transpeptidase
- GLU = glucose
- GTT = glucose tolerance test
- H&E = hematoxylin and eosin stain
- Hb or Hgb = hemoglobin

- Hct or crit = hematocrit
- HDL = high-density lipoprotein
- HPF = high-power field
- HW = heartworm
- LDH = lactate dehydrogenase
- LDL = low-density lipoprotein
- LPF = low-power field
- MCH = mean corpuscular hemoglobin
- MCHC = mean corpuscular hemoglobin concentration
- MCV = mean corpuscular volume
- ME = myeloid-erythroid ratio
- NRBC = nucleated red blood cell
- PCV = packed cell volume
- PMN = polymorphonuclear neutrophil leukocyte
- PT = prothrombin time
- PTT = partial thromboplastin time
- qns = quantity not sufficient
- qs = quantity sufficient
- RBC = red blood cell
- rpm = revolutions per minute
- SAP = serum alkaline phosphatase
- sed or SR = sedimentation rate
- SGOT = serum glutamic oxaloacetic transaminase; now abbreviated AST
- SGPT = serum glutamic pyruvic transaminase; now abbreviated ALT
- sp. gr. = specific gravity
- T_3 = triiodothyronine (one type of thyroid hormone)
- T_4 = thyroxine (one type of thyroid hormone)
- TBIL = total bilirubin
- TNTC = too numerous to count
- TP = total protein
- WBC = white blood cell

REVIEW EXERCISES

Multiple Choice

Choose the correct answer.

1. Examination by tapping the surface to determine density of a body area is called
   a. palpation
   b. auscultation
   c. percussion
   d. tapping

2. The percentage of RBCs in blood is called a
   a. hematocrit
   b. PCV
   c. crit
   d. all of the above

3. A machine that spins samples very rapidly to separate elements based on weight is a
   a. counter
   b. centrifuge
   c. refractometer
   d. cannula

4. The quality of appearing white or light gray on a radiograph is called
   a. radiopaque
   b. radiolucent
   c. radiodense
   d. radiopale

5. The determination of the cause of disease is the
   a. prognosis
   b. diagnosis
   c. symptom
   d. sign

6. A set of signs that occur together is called a(n)
   a. sign
   b. symptom
   c. endemic
   d. syndrome

7. The medical term for sphere is
   a. bacilli
   b. spirochete
   c. coccus
   d. strept/o

8. The term for relief of condition but not a cure is
   a. prognosis
   b. diagnosis
   c. prevalence
   d. palliative

9. Pertaining to fever is
   a. febrile
   b. friable
   c. lethal
   d. morbid

10. A disease that can be transmitted between animals and humans is said to be
   a. animalistic
   b. zoonotic
   c. humanistic
   d. sequela

11. What is the term for the ultrasound property of producing adequate levels of reflections (echoes) when sound waves are returned to the transducer and displayed?
   a. echoic
   b. anechoic
   c. hyperechoic
   d. hypoechoic

12. The acronym used to describe a unit by which absorption of ionizing radiation is measured is
   a. MAS
   b. kVp
   c. CT
   d. rad

13. Surgical puncture to remove fluid or gas for diagnostic purposes or for treatment is known as
   a. cannula
   b. centesis
   c. contagious
   d. communicable

14. Examination by feeling is
   a. percussion
   b. auscultation
   c. assessment
   d. palpation

15. Clumping of cells is known as
   a. an assay
   b. agglutination
   c. serology
   d. a profile

16. The medical term for prevention is
   a. prevalence
   b. palliative
   c. prophylaxis
   d. phobia

17. The description of an animal with information about the animal, including the species, breed, age, and sexual status, is called the animal's
   a. symptoms
   b. signalment
   c. assessment
   d. auscultation

18. A decrease in body temperature is known as
   a. afebrile
   b. febrile
   c. hypothermia
   d. prognosis

19. The term for having a short course is
   a. remission
   b. acute
   c. chronic
   d. endemia

20. The study of disease causes is
   a. etiology
   b. toxicology
   c. biology
   d. pathology

21. An outbreak that occurs sporadically in animals is called

    a. epizootic

    b. enzootic

    c. panzootic

    d. episodic

22. The "M" in MRI stands for

    a. magnitude

    b. middle

    c. magnetic

    d. medical

23. The "k" in kVp stands for

    a. killed

    b. kilo

    c. kennel

    d. ketones

24. Coccus is to cocci as bacillus is to

    a. bacillae

    b. bacilla

    c. bacilli

    d. bacillum

25. The abbreviation for wavelength is

    a. w

    b. wl

    c. W

    d. λ

Matching

Match the testing term in Column I with the definition in Column II.

Column I

1. _____ infectious disease

2. _____ contagious disease

3. _____ noncontagious disease

4. _____ noninfectious disease

5. _____ communicable disease

6. _____ iatrogenic disease

7. _____ nosocomial infection

8. _____ zoonotic disease

9. _____ diagnosis

10. _____ susceptible

11. _____ trocar

12. _____ projection

13. _____ roentgen

14. _____ bacillus

15. _____ coccus

16. _____ chronic

Column II

a. disorder caused by physicians or veterinarians and the treatment ordered

b. disorder caused by pathogenic organisms contracted in a facility or clinic

c. disease spread from one animal to another by direct or indirect contact

d. disorder caused by pathogenic organisms

e. disease that can be transmitted between animals and humans

f. disorder transmitted from animal to animal or through contact with contaminated objects; another name for contagious disease

g. disorder not caused by organisms (e.g., genetic, traumatic, and iatrogenic)

h. disease that cannot be spread to another animal by contact or contact with a contaminated object

i. international unit of radiation

j. path of the X-ray beam

k. spherical bacterium

l. determination of the cause of disease

m. having a long course

n. sharp, needlelike instrument that has a tube that is used to puncture the wall of a body cavity to withdraw fluid or gas

o. rod-shaped bacterium

p. lacking resistance

Fill in the Blanks

Write the correct term in the space provided.

1. Pyr/o means _____.

2. Fluor/o means _____.

3. The prefix a- means _____.

4. HR is the abbreviation for _____.

5. RR is the abbreviation for _____.

6. V/D is the abbreviation for _____.

7. MLO is the abbreviation for _____.

8. mAs is the abbreviation for _____.

9. kVp is the abbreviation for _____.

10. DP is the abbreviation for _____.

Spell Check

Circle the term that is spelled correctly.

1. ability of an organism to cause disease: virilence   virulence   virulense

2. able to be spread from one animal to another by direct or indirect contact: contagious   kontagious   contagous

3. laboratory study of serum: sirology   sarology   serology

4. fever: pyrexia   pyrexa   pirexa

5. instrument used to enlarge the opening of a canal or cavity: speckulum   speculum   speckulum

Crossword Puzzle

Testing Terms

Supply the correct term in the appropriate space for the definition listed.

Across

3 tissue that reflects less sound back to the transducer than the surrounding tissues; appears dark

8 number of leukocytes per cubic millimeter of blood

10 ultrasound property of producing adequate levels of reflections (echoes) when sound waves are returned to the transducer and displayed

11 number of cycles per unit of time

17 high pH property

18 spiral-shaped bacteria

20 concentration of hydrogen ions

22 diagnostic evaluation of blood to determine the number of rbc, wbc, and thrombocytes per cubic millimeter of blood (abbr)

23 rod-shaped bacteria

25 combining form for chain

26 loss of intensity of the ultrasound beam as it travels through tissue

27 tissue that reflects more sound back to the transducer than the surrounding tissues; appears bright

Down

1 ability to separately identify different structures on radiograph or ultrasound

2 international unit of radiation

5 path of the X-ray beam

6 ultrasonic term for when waves are transmitted to deeper tissue and none are reflected back

7 microscopic living organism

9 plant-like eukaryotic organism without chlorophyll

12 budding form of fungus

13 combining form for grapelike clusters

14 speed at which something travels through an object

15 tissue that has the same ultrasonic appearance as that of the surrounding tissue

16 low pH property

19 specified body position and the part of the body closest to the film

21 length that a wave must travel in one cycle

24 intensity of an ultrasound wave

Case Studies

Supply the proper definitions for the underlined terms. Leptospirosis is <u>zoonotic</u> disease of humans and animals caused by the <u>spirochete bacterium</u> *Leptospira*. The dog is the reservoir host for the <u>pathogenic</u> *Leptospira interrogans*. The potentially <u>lethal</u> disease presents in some dogs as <u>acute renal</u> failure characterized by <u>icterus</u> and <u>uremia</u> while other dogs may be <u>asymptomatic carriers</u>. The bacterium can be <u>transmitted</u> directly between hosts in close contact through urine, <u>placental</u> transfer, bites, broken skin, or ingestion of infected tissues because the organism penetrates mucosa. The organism is typically shed through urine. Indirect transmission may occur through exposure of susceptible animals or humans to a contaminated environment. Water contact is the most common means of spread, especially stagnant or slow-moving warm water. The organism is identified with <u>serologic</u> methods such as an <u>agglutination</u> test.

1. zoonotic   _____

2. spirochete   _____

3. bacterium   _____

4. pathogenic   _____

5. lethal   _____

6. acute   _____

7. renal   _____

8. icterus   _____

9. uremia   _____

10. asymptomatic   _____

11. carrier   _____

12. transmitted   _____

13. placental   _____

14. serologic   _____

15. agglutination   _____

A 3-yr-old thoroughbred mare had become <u>acutely</u> lame during her normal training session and presented to the clinic with a <u>tendon</u> bow in the left front leg (when the fibers in tendons tear, inflammation and swelling cause the tendon to bulge or bow outward). PE revealed pain on <u>palpation</u> of the affected region and mild <u>edema</u> in the distal limb. The veterinarian performed <u>ultrasound</u> evaluation of the right front <u>flexor</u> tendon and identified a mostly <u>anechoic</u> core lesion of the <u>superficial digital flexor</u> tendon (a core lesion is a hole). The patient was discharged with instructions to limit activity to stall rest, to continue anti-inflammatory medication, and to be reevaluated in 30 days.

On the recheck appointment, there was <u>focal</u> swelling of the mid-<u>cannon</u> flexor region. Ultrasound revealed no changes in the <u>lesion</u> size. The horse was sedated and given a <u>local anesthetic</u> to collect <u>subcutaneous adipose</u> tissue for stem cell recovery. Two days later, <u>viable</u> cells were delivered to the core lesion by ultrasound-guided injection and the horse was discharged to stall rest for 1 week followed by a physical rehabilitation program. One month later the horse's tendon was reevaluated by ultrasound and there was normal <u>echoic</u> tissue and decreased swelling near the affected tendon.

16. acutely _____

17. tendon _____

18. palpation _____

19. edema _____

20. ultrasound _____

21. flexor _____

22. anechoic _____

23. superficial _____

24. digital _____

25. flexor _____

26. focal _____

27. cannon _____

28. lesion _____

29. local anesthetic _____

30. subcutaneous _____

31. adipose _____

32. viable _____

33. echoic _____

Critical Thinking Exercise

The following story and questions contain words and concepts from this chapter and are designed to stimulate critical thinking through class discussion, Internet searches, or brief writing responses.

You are reading an article about radiographic technique because you want to make sure that you are doing everything possible to take good X-rays while limiting exposure to your patients and coworkers. The following is an excerpt from the article you read. Fill in the blanks with the proper medical terminology.

Positioning

It is important to properly position veterinary patients for radiographs. A radiographic study should always consist of two views. The two views should be taken so that the radiographic beams cross at right angles. For example, when radiographing the abdomen, take one view with the animal lying on its side (_____ view) and a second view with the animal lying on either its back (_____ view) or belly (_____ view) For some areas of the body, three or more views are required. For example, to examine the thorax, take one view with the animal lying on its right side (_____), a second view with the animal lying on its left side (_____), and a third view with the animal positioned on its back (_____ view).

Exposure Technique

*Exposure techniques are determined using calipers to measure the thickness of the body part to be radiographed. Using the measurement, the technician should then consult the technique chart for the specific X-ray machine to determine the settings. Machine settings consist of the length of time of exposure (the _____), which represents the number of X-ray beams, and the (_____), which represents the strength of the X-ray beam. The mAs is responsible for the density or blackness of the film while the kVp controls the contrast of the X-ray film. The technique chart will also indicate whether the X-ray film should be placed directly on the table top or in the bucky tray located under the table top. Once the machine is ready, the radiographic beam should be narrowed **(collimated)** to the area being radiographed. This is done by using the light on the tube and the corresponding dials, which control the size of the lighted area on the table. The field of light represents the area of the radiographic beam during radiographic exposure. Position the animal so that the area of interest is in the center of the film and expose the radiograph by using the hand switch or foot pedal. Hold the switch in or pedal down until the machine has finished beeping.*

Chapter 17
DRUGS AND DISSECTION

Objectives

Upon completion of this chapter, the reader should be able to:

- Recognize, define, spell, and pronounce the terms associated with pharmacology and drugs used in various treatments

- Recognize, define, spell, and pronounce the terms associated with different types of surgery and the instruments used in surgery

- Interpret the meaning of abbreviations associated with pharmacology and surgery

PHARMACOLOGIC TERMS

Pharmacology (fahrm-ah-kohl-lō-jē) is the study of the nature, uses, and effects of drugs. Some drugs are dispensed only by a licensed professional, and other drugs are not. A **prescription** (prē-skrihp-shuhn) **drug** is a medication that may be purchased by prescription or from a licensed professional. An **over-the-counter** (ō-vər theh kount-ər) **drug** is a medication that may be purchased without a prescription. A **generic** (jehn-ār-ihck) **drug** is a medication not protected by a brand name or trademark. (It also is called a **nonproprietary drug**.)

Pharmacology

- **agonist** (ā-gohn-ihst) = substance that produces effect by binding to an appropriate receptor.
- **antagonist** (ahn-tā-gohn-ihst) = substance that inhibits a specific action by binding with a particular receptor instead of allowing the agonist to bind to the receptor.
- **antiserum** (ahn-tih-sēr-uhm) = serum containing specific antibodies extracted from a hyperimmunized animal, or an animal that has been infected with the microorganisms containing antigen.
- **antitoxin** (ahn-tih-tohcks-sihn) = specific antiserum aimed at a poison that contains a concentration of antibodies extracted from the serum or plasma of a healthy animal.
- **bacterin** (bahck-tər-ihn) = bacterial vaccine.
- **chelated** (kē-lā-tehd) = bound to and precipitated out of solution.
- **contraindication** (kohn-trah-ihn-dih-kā-shuhn) = recommendation not to use.
- **diffusion** (dih-fū-shuhn) = movement of solutes from an area of high concentration of particles to one of low concentration of particles (Figure 17–1).

Molecule

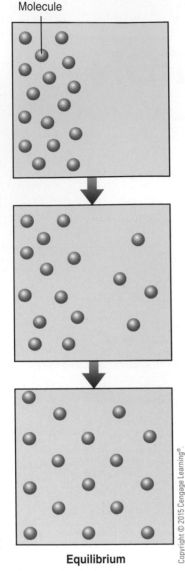

Equilibrium

FIGURE 17-1 Diffusion is the random movement of molecules from an area of high concentration of particles to an area of low concentration of particles. Diffusion can occur in the presence or absence of a membrane.

- **dosage** (dō-sahj) = amount of medication based on units per weight of animal (such as 10 mg/lb and 2 mg/kg).
- **dosage interval** (dō-sahj ihn-tər-vahl) = time between administrations of a drug (such as bid or q12h).
- **dose** (dōs) = amount of medication measured (such as milligrams, milliliters, units, and grams).
- **drug** (druhg) = agent used to diagnose, prevent, or treat a disease.
- **efficacy** (ehf-ih-kah-sē) = extent to which a drug causes the intended effects; effectiveness.
- **endogenous** (ehn-dah-jehn-uhs) = originating within the body.
- **exogenous** (ehcks-ah-jehn-uhs) = originating outside the body.
- **hydrophilic** (hī-drō-fihl-ihck) = water-loving; ionized form.
- **hyperkalemia** (hī-pər-kā-lē-mē-ah) = excessive level of blood potassium.
- **hypernatremia** (hī-pər-nā-trē-mē-ah) = excessive level of blood sodium.
- **hypertonic** (hī-pər-tohn-ihck) **solution** = solution that has more particles than the solution or cell to which it is being compared. The tonicity of solutions usually is compared to blood cells (Figure 17–2).
- **hypokalemia** (hī-pō-kā-lē-mē-ah) = deficiency of blood potassium.
- **hyponatremia** (hī-pō-nā-trē-mē-ah) = deficiency of blood sodium.
- **hypotonic** (hī-pō-tohn-ihck) **solution** = solution that has fewer particles than the solution or cell to which it is being compared (Figure 17–2).
- **hypovolemia** (hī-pō-vō-lē-mē-ah) = low circulating blood volume.
- **ionized** (ī-ohn-īzd) = electrically charged.
- **isotonic** (ī-sō-tohn-ihck) **solution** = solution that has equal particles to the solution or cell to which it is being compared (Figure 17–2).

Hypertonic solution
(more particles
outside cell)

Hypotonic solution
(fewer particles
outside cell)

Isotonic solution
(equal particles inside
and outside cell)

FIGURE 17-2 Solution tonicity.

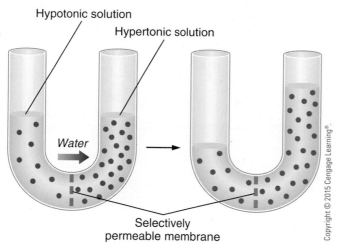

Hypotonic solution

Hypertonic solution

Water

Selectively
permeable membrane

FIGURE 17-3 Osmosis is the movement of water across a selectively permeable membrane along its concentration gradient. In this example, water moves across the membrane to equalize the concentration of particles.

FIGURE 17-4 Examples of vaccines (clockwise from lower left). (a) Modified live vaccines are typically freeze-dried powders that must be reconstituted with liquid before using; (b) an oral vaccine (bolus) may be used to provide passive immunity to newborn calves; (c) a modified-live vaccine may be administered intranasally. The white nasal cannula in front of the bottles is applied to the end of a syringe to administer the vaccine into the nasal passages; (d) a killed vaccine is ready to be administered.

- **lipophilic** (lihp-ō-fihl-ihck) = fat-loving; nonionized form.
- **monovalent** (mohn-ō-vā-lehnt) = vaccine, antiserum, or antitoxin developed specifically for a single antigen or organism.
- **nonionized** (nohn-ī-ohn-īzd) = not charged electrically.
- **osmosis** (ohz-mō-sihs) = movement of water across a selectively permeable membrane along its concentration gradient (Figure 17–3).
- **pharmacodynamics** (fahrm-ah-kō-dī-nahm-ihcks) = physiological effects of drugs and their mechanisms of action.
- **pharmacogenomics** (fahrm-ah-kō-jehn-oh-mihcks) = study of the impact of genetic variation on drug effects with hopes of achieving "personalized medicine."
- **pharmacokinetics** (fahrm-ah-kō-kihn-eht-ihcks) = movement of drugs or chemicals; consists of absorption, distribution, biotransformation, and elimination.
- **placebo** (plah-sē-bō) = inactive substance that is given for its suggestive effects or substance used as a control in an experimental setting.
- **polyvalent** (poh-lē-vā-lehnt) = vaccine, antiserum, or antitoxin that is active against multiple antigens or organisms; mixed vaccine.
- **prevention** (prē-vehn-shuhn) = avoidance; also called **prophylaxis** (prō-fih-lahck-sihs).
- **regimen** (reh-jeh-mehn) = course of treatment.
- **turgor** (tər-gər) = degree of fullness or rigidity caused by fluid content.
- **vaccine** (vahck-sēn) = product containing a disease-causing organism or antigen administered

to stimulate the production of antibodies to fight infection; may be **live**, **modified live** (disease-causing organism has been chemically or thermally altered to make it less pathogenic), or **killed** (disease-causing organism is dead and cannot reproduce in an animal's body) (Figure 17–4).

Routes of Administration

- **inhalation** (ihn-hah-lā-shuhn) = vapors and gases taken in through the nose and mouth and absorbed into the bloodstream through the lungs (Figure 17–5).
- **intra-arterial** (ihn-trah-ahr-tehr-ē-ahl) = within the artery; abbreviated IA.
- **intradermal** (ihn-trah-dər-mahl) = within the skin; abbreviated ID.
- **intramuscular** (ihn-trah-muhs-kyū-lahr) = within the muscle; abbreviated IM.
- **intraocular** (ihn-trah-ohck-yoo-lahr) = within the eye.
- **intraosseous** (ihn-trah-ohs-ē-uhs) = within the bone (medullary cavity of a long bone).
- **intraperitoneal** (ihn-trah-pehr-ih-tohn-ē-ahl) = within the peritoneal cavity; abbreviated IP.
- **intrathecal** (ihn-trah-thē-kahl) = within a sheath; injection of a substance through the spinal cord and into the subarachnoid space; **thec/o** means sheath; abbreviated IT.

FIGURE 17–5 Inhalation anesthesia is given to the patient via an anesthetic machine.

- **intratracheal** (ihn-trah-trā-kē-ahl) = within the trachea, or windpipe.
- **intravenous** (ihn-trah-vehn-uhs) = within the vein; abbreviated IV.
- **nebulization** (nehb-yoo-lih-zā-shuhn) = process of making a fine mist; a method of drug administration.
- **nonparenteral** (nohn-pah-rehn-tər-ahl) = administration via the gastrointestinal tract.
- **oral** (ōr-ahl) = by mouth; abbreviated PO or p.o. *Nothing orally* is abbreviated NPO or n.p.o.
- **parenteral** (pah-rehn-tər-ahl) = through routes other than the gastrointestinal tract (Figure 17–6).
- **percutaneous** (pehr-kyoo-tā-nē-uhs) = through the skin.
- **rectal** (rehck-tahl) = by rectum.
- **subcutaneous** (suhb-kyoo-tā-nē-uhs) = under the skin, or dermal layer; abbreviated SQ, SC, or subq.
- **sublingual** (suhb-lihng-wahl) = under the tongue.
- **transdermal** (trahnz-dər-mahl) = across the skin. Medication is stored in a patch placed on the skin, and the medication is absorbed through the skin.

FIGURE 17–6 Examples of parenteral routes of drug administration.

Drug Categories

- **analgesic** (ahn-ahl-jē-zihck) = substance that relieves pain without affecting consciousness.
- **anesthetic** (ahn-ehs-theht-ihck) = substance that produces a lack of sensation (Figure 17–7).
- **antacid** (ahnt-ahs-ihd) = substance that neutralizes stomach acid.
- **anthelmintic** (ahn-thehl-mihn-tihck) = substance that works against intestinal worms.
- **antibiotic** (ahn-tih-bī-ah-tihck) = substance that inhibits the growth of or kills bacteria. Antibiotics can be **bacteriostatic** (bahck-tē-rē-ō-stah-tihck), which means controlling bacterial growth (inhibiting bacterial replication), or **bactericidal** (bahck-tē-rē-sī-dahl), which means killing bacteria.
- **anticoagulant** (ahn-tih-kō-ahg-yoo-lahnt) = substance that inhibits clot formation.
- **anticonvulsant** (ahn-tih-kohn-vuhl-sahnt) = substance that prevents seizures.
- **antidiarrheal** (ahn-tih-dī-ər-rē-ahl) = substance that prevents watery, frequent bowel movements.
- **antiemetic** (ahn-tih-ē-meh-tihck) = substance that prevents vomiting.

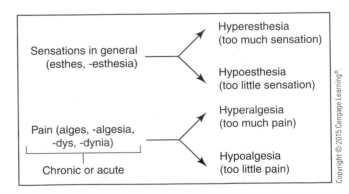

FIGURE 17–7 Word parts for sensation and pain.

- **antineoplastic agent** (ahn-tih-nē-ō-plah-stihck ā-jehnt) = substance that treats neoplasms; usually used against malignancies.
- **antipruritic agent** (ahn-tih-prər-ih-tihck ā-jehnt) = substance that controls itching.
- **antipyretic** (ahn-tih-pī-reh-tihck) = substance that reduces fever.
- **antiseptic** (ahn-tih-sehp-tihck) = chemical agent that kills or prevents the growth of microorganisms on living tissue.
- **antitussive** (ahn-tih-tuhs-ihv) = substance that reduces coughing.
- **asepsis** (ā-sehp-sihs) = state without infection.
- **chronotrope** (krohn-ō-trōp) = substance that changes heart rate. Positive chronotropes increase heart rate, while negative chronotropes decrease heart rate.
- **cytotoxic agent** (sī-tō-tohcks-ihck ā-jehnt) = substance that kills or damages cells.
- **disinfectant** (dihs-ehn-fehck-tahnt) = chemical agent that kills or prevents the growth of microorganisms on inanimate objects.
- **emetic** (ē-meh-tihck) = substance that induces vomiting.
- **endectocide** (ehnd-ehck-tō-sīd) = agent that kills both internal and external parasites.
- **immunosuppressant** (ihm-yoo-nō-suhp-prehs-ahnt) = substance that prevents or decreases the body's reaction to invasion by disease or foreign material.
- **inotrope** (ihn-ō-trōp) = substance affecting muscle contraction. Positive inotropes increase myocardial contractility, while negative inotropes decrease myocardial contractility.
- **miotic agent** (mī-ah-tihck ā-jehnt) = substance used to constrict the pupils.
- **mucolytic** (mū-kō-lih-tihck) = substance that breaks up mucus and reduces its viscosity.
- **mydriatic agent** (mihd-rē-ah-tihck ā-jehnt) = substance used to dilate the pupils.

- **sterilize** (stehr-ih-līz) = to destroy all organisms including bacterial endospores.

Weights and Measures

- **centimeter** (sehn-tih-mē-tər) = metric unit of length equal to one one-hundredth of a meter; abbreviated cm.
- **dram** (drahm) = apothecary unit of measure used for prescription vials. One dram equals 1.8 ounce (by weight), and 1 fluid dram equals 4 mL.
- **gram** (grahm) = metric base unit of weight equal to 0.035 ounce; abbreviated g.
- **kilogram** (kihl-ō-grahm) = metric unit of weight that is 1000 grams; 1 kilogram is approximately 2.2 pounds; abbreviated kg.
- **liter** (lē-tər) = metric base unit of volume equal to 0.2642 gallons; abbreviated L.
- **meter** (mē-tər) = metric base unit of length equal to 1.09 yards; abbreviated m.
- **milligram** (mihl-ih-grahm) = metric unit of weight equal to one one-thousandth of a gram; abbreviated mg.
- **milliliter** (mihl-ih-lē-tər) = metric unit of volume equal to 0.034 of an ounce or one one-thousandth of a liter; abbreviated mL; equivalent to 1 cubic centimeter (cc).
- **millimeter** (mihl-ih-mē-tər) = metric unit of length equal to one one-thousandth of a meter; abbreviated mm.
- **percent** (pər-sehnt) = part per 100 parts; represented by %.

SURGICAL TERMS

Surgery is the branch of science that treats diseases, injuries, and deformities by manual or operative methods. Most surgeries are described based on the location of the body (thoracic surgery is performed on the thorax),

BIOLOGY BONUS

When physicians prescribe drugs to people, pharmacists provide instructions regarding side effects of drugs as well as what to avoid while taking the medication (e.g., some drugs should not be taken before operating a motor vehicle because they cause drowsiness). These issues may not be as important in animals; however, giving animals drugs with food may affect their action or a treatment plan. For example, consider a veterinarian prescribing an antibiotic for a dog with pyoderma secondary to food allergies. It would be important to advise the client not to give the pill in a hot dog or piece of cheese if the dog is only to be fed a hypoallergenic diet.

obstruction being removed (cholecystectomy is surgical removal of the gall bladder and commonly called gall bladder surgery), machine (laparoscopic surgeries are performed with a laparoscope) or technique being used (a Fredet-Ramstedt pyloromyotomy is a technique for obtaining a biopsy of the pyloric mucosa), or where it is performed (outpatient surgery means the patient does not stay overnight in the clinic). There are many types of surgery including **preventative** (designed to prevent further disease), **manipulative** (changed without an incision such as alignment of a fracture; also known as closed), **diagnostic** (designed to help finalize a diagnosis), **minimally invasive** (designed to use the smallest incision as possible such as those using an endoscope), **reconstructive** (designed to improve on or return a part of the body to its original appearance or function), **cryogenic** (involving the use of freezing to destroy tissue), and **cauterizing** (involving the use of heat to destroy tissue). Surgical terms were developed to describe concisely many surgical procedures. Some surgical terms include the following:

- **appositional** (ahp-ō-sih-shuhn-ahl) = placed side to side.
- **aseptic technique** (ā-sehp-tihck tehck-nēk) = precautions taken to prevent contamination of a surgical wound.
- **avulsion** (ə-vuhl-shuhn) = tearing away of a part.
- **coaptation** (kō-ahp-tā-shuhn) = act of approximating.
- **curettage** (kyoo-reh-tahj) = removal of material or growths from the surface of a cavity.
- **debridement** (deh-brēd-mehnt) = removal of foreign material and devitalized or contaminated tissue.
- **dehiscence** (dē-hihs-ehns) = disruption or opening of the surgical wound (Figure 17–8).

- **dissect** (dī-sehckt) = separation or cutting apart; **dissecare** is Latin for "to cut up."
- **enucleation** (ē-nū-klē-ā-shuhn) = removal of an organ in whole; usually used for removal of the eyeball (Figure 17–9).
- **epithelialization** (ehp-ih-thē-lē-ahl-ih-zā-shuhn) = healing by growth of epithelium over an incomplete surface.
- **eversion** (ē-vər-shuhn) = turning outward (Figure 17–10).
- **eviscerate** (ē-vihs-ər-āt) = removal or exposure of internal organs.
- **excise** (ehck-sīz) = to surgically remove.
- **exteriorize** (ehcks-tēr-ē-ōr-īz) = to move an internal organ to the outside of the body (Figure 17–11); also Figure 12–28, which shows exteriorization of the female reproductive tract.
- **fenestration** (fehn-ih-strā-shuhn) = perforation.

(a)

(b)

FIGURE 17–9 (a) A cat with a corneal injury. (b) The eye was enucleated due to the severity of the injury.

FIGURE 17–8 Wound dehiscence in a cat.

Appositional suture pattern (simple interrupted)

Everting suture pattern (horizontal mattress)

Inverting suture pattern (vertical mattress)

FIGURE 17–10 Appositional versus everting versus inverting suture patterns.

FIGURE 17–12 Skin graft in a dog.

FIGURE 17–11 Intraoperative photograph of an exteriorized urinary bladder. A cystotomy was performed on this dog to remove urinary calculi.

- **flap** (flahp) = mass of tissue for grafting in which part of the tissue still adheres to the body; used to repair defects adjacent to the mass site.
- **fracture** (frahck-shər) = breaking of a part, especially a bone.
- **friable** (frī-ah-buhl) = easily crumbled.
- **fulguration** (fuhl-gər-ā-shuhn) = destruction of living tissue by electric sparks generated by a high-frequency current.
- **graft** (grahft) = tissue or organ for transplantation or implantation (Figure 17–12). There are different types of grafts. An **allograft** (ah-lō-grahft) is a graft from another individual of the same species. An **autograft** (awt-ō-grahft) is a graft from the same individual. An **isograft** (ī-sō-grahft) is a graft from genetically identical animals, such as twins or inbred strains.

- **imbrication** (ihm-brih-kā-shuhn) = overlapping of opposing surfaces to realign organs and provide extra support.
- **implant** (ihm-plahnt) = material inserted or grafted into the body.
- **incise** (ihn-sīz) = to surgically cut into.
- **intraop** (ihn-trah-ohp) = common term for *during surgery*; intraoperatively.
- **inversion** (ihn-vər-shuhn) = turning inward.
- **involucrum** (ihn-voh-loo-kruhm) = covering or sheath that contains a sequestrum of bone.
- **laceration** (lah-sihr-ā-shuhn) = act of tearing.
- **lavage** (lah-vahj) = irrigation of tissue with fluid.
- **ligate** (lī-gāt) = to tie or strangulate. A **ligature** (lihg-ah-chūr) is any substance used to tie or strangulate. Ligatures usually are made of suture material.
- **lumpectomy** (luhmp-ehck-tō-mē) = general term for surgical removal of a mass.
- **pinning** (pihn-ihng) = insertion of a metal rod into the medullary cavity of a long bone.
- **postop** (pōst-ohp) = common term for after surgery; postoperatively.
- **preop** (prē-ohp) = common term for before surgery; preoperatively.
- **resect** (rē-sehckt) = to remove an organ or tissue. Resect is used in reference to holding tissue or an organ out of the surgical field.
- **rupture** (ruhp-chuhr) = forcible tearing.
- **sacculectomy** (sahk-yoo-lehck-tō-mē) = surgical removal of a saclike part; usually refers to surgical removal of the anal sacs.
- **seroma** (sehr-ō-mah) = accumulation of serum beneath the surgical incision.

- **suction** (suhck-shuhn) = aspiration of gas or fluid by mechanical means (Figure 17–13a and b).
- **suture** (soo-chuhr) = to stitch or close an area; also refers to the material used in closing a surgical or traumatic wound with stitches. Suture material may be absorbable or nonabsorbable. (Suture also is a type of joint.)
- **transect** (trahn-sehckt) = to cut across; a cross section or section made across a long axis. **Sect** means to cut.
- **transfix** (trahnz-fihcks) = pierce through and through. A transfixion suture pierces through an organ before ligation (Figure 17–14).
- **transplant** (trahnz-plahnt) = to transfer tissue from one part to another part.
- **wicking** (wihck-ihng) = applying material to move liquid from one area to another.

(a)

(b)

FIGURE 17–13 (a) Suction is used to remove fluid (or gas) during surgery. (b) Suction unit showing the collection chamber.

Transfixion suture

Copyright © 2015 Cengage Learning®

FIGURE 17–14 Transfixion suture. Transfixion sutures are used for large, isolated vessels and organs to prevent slippage of the ligature.

Surgical Equipment

- **autoclave** (aw-tō-klāv) = apparatus for sterilizing by steam under pressure (Figure 17–15).
- **bandage** (bahn-dahj) = to cover by wrapping or the material used to cover by wrapping (Figure 17–16).
- **belly band** (behl-ē bahnd) = common term for abdominal wrap; circumferentially wrapping the abdomen with bandages to apply pressure to the area.
- **bone plate** = flat metal bar with screw holes that is used in bone fracture repair (Figure 17–17a and b).
- **bone screw** = screw that holds bone fragments together to repair bone fractures.
- **boxlock** (bohcks-lohck) = movable joint of any ringed instrument (Figure 17–18).
- **cast** (kahst) = stiff dressing used to immobilize various body parts.
- **cautery** (caw-tər-ē) = application of a burning substance, a hot instrument, an electric current, or another agent to destroy tissue.
- **cerclage** (sihr-klahj) **wire** = band of metal that completely (cerclage) or partially (hemicerclage) goes around the circumference of bone that is used in conjunction with other stabilization techniques to repair bone fractures (Figure 17–19).
- **chuck** (chuhck) = clamping device for holding a drill bit.
- **clamp** (klahmp) = instrument used to secure or occlude things.
- **curette** (kyoor-reht) = instrument with cupped head to scrape material from cavity walls.
- **drain** (drān) = device by which a channel may be established for the exit of fluids from a wound (Figure 17–20a and b).

(a)

(b)

FIGURE 17–15 (a) An autoclave provides high-pressure steam heat to sterilize surgical instruments. (b) In a surgery pack (a collection of instruments used for one procedure) that has been properly autoclaved, the indicator tape will change from light-colored lines (roll on right) to dark-colored lines (tape on pack).

Orthopedic wire

Plate and screws

(a)

(b)

FIGURE 17–17 (a) Examples of orthopedic equipment. (b) Radiograph showing the repair of a tibial fracture using a bone plate and screws.

FIGURE 17–16 Puppy with a Robert Jones bandage.

Tip

Jaw

Boxlock

Boxlock pin

Shank

Ratchet

Ring handle

Ratchet

FIGURE 17–18 Parts of surgical instruments.

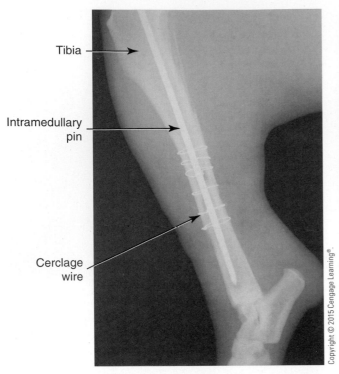

Tibia

Intramedullary pin

Cerclage wire

Copyright © 2015 Cengage Learning®.

FIGURE 17–19 Radiograph showing the repair of a fractured tibia. The bone was repaired with an intramedullary pin and six cerclage wires.

- **drape** (drāp) = cloth arranged over a patient's body to provide a sterile field around the area to be examined, treated, or incised.
- **dressing** (drehs-sihng) = various materials used to cover and protect a wound.
- **elastrator** (ē-lahs-trā-tər) = bloodless castration device using small elastic bands.
- **elevator** (ehl-eh-vā-tər) = instrument used to reflect tissue from bone.
- **emasculatome** (ē-mahs-kyoo-lah-tōm) = instrument used to crush and sever the spermatic cord through intact skin.
- **emasculator** (ē-mahs-kyoo-lā-tər) = instrument used in closed castrations to crush and sever the spermatic cord.
- **hemostatic forcep** (hē-mō-stah-tihck fōr-sehp) = locking instrument used to grasp and ligate vessels and tissues to control bleeding; also called **hemostat** (hē-mō-staht).
- **intramedullary pins** (ihn-trah-mehd-yoo-lahr-ē pihnz) = metal rods that are inserted into the medullary cavity of long bones to repair stable fractures (Figure 17–19).
- **prosthesis** (prohs-thē-sihs) = artificial substitute for a diseased or missing part of the body.
- **ratchet** (rah-cheht) = graded locking portion of an instrument located near the finger rings (Figure 17–18).

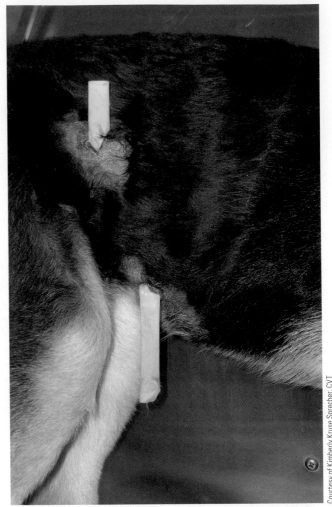

Courtesy of Kimberly Kruse Sprecher, CVT

(a)

Courtesy of Kimberly Kruse Sprecher, CVT

(b)

FIGURE 17–20 Surgical drain. (a) A Penrose drain was placed in the flank area of this dog to maintain an opening in the skin that allows accumulating fluid to drain to the exterior. The fluid does not drain through the center of the tubing, but is allowed to leak from the skin openings. (b) Dog with an active drain and reservoir unit that allows measurement of drainage volume and description of the fluid appearance.

- **retractor** (rē-trahck-tər) = instrument used to hold back tissue (Figure 17–21a).
- **rongeurs** (rohn-jūrz) = forceps with cupped jaws used to break large bone pieces into smaller ones or to cut through tough tissue (Figure 17–21a).

(a)

(b)

(c)

FIGURE 17–21 (a) and (b) Examples of surgical instruments. (c) Cautery unit for hemostasis.

(a)

(b)

FIGURE 17–22 (a) Selection of bandage materials and splints. (b) A splint being applied to a dog's leg.

- **scalpel** (skahl-puhl) = small, straight knife with a thin, sharp blade used for surgery and dissection.
- **serration** (sihr-ā-shuhn) = sawlike edge or border.
- **sling** (slihng) = bandage for supporting part of the body.
- **splint** (splihnt) = rigid or flexible appliance for fixation of movable or displaced parts (Figure 17–22a and b).
- **tissue forceps** (tihs-yoo fōr-sehps) = tweezer-like, nonlocking instruments used to grasp tissue (Figure 17–21b).

Surgical Approaches

In surgery, the specific procedure by which an organ or a part is exposed is called the **approach** (ah-prōch). Different approaches allow the best exposure to different parts of the body. Following are examples of different surgical approaches (Figure 17–23):

- **flank incision** (flahnk ihn-sihz-shuhn) = surgical cut perpendicular to the long axis of the body, caudal to the last rib.

Ventral midline incision

Paramedian incision

Flank incision

Paracostal incision

FIGURE 17–23 Incision types.

- **paracostal incision** (pahr-ah-kah-stahl ihn-sihz-shuhn) = surgical cut oriented parallel to the last rib.
- **paramedian incision** (pahr-ah-mē-dē-ahn ihn-sihz-shuhn) = surgical cut lateral and parallel to the ventral midline but not on the midline.
- **ventral midline incision** (vehn-trahl mihd-līn ihn-sihz-shuhn) = surgical cut along the midsagittal plane of the abdomen along the linea alba.

Biopsies

The term **biopsy** (bī-ohp-sē) means removing living tissue to examine. Biopsy is also used for the specimen removed during the procedure. The first definition is more correct; however, the term *biopsy* is commonly used both ways.

Types of biopsies include the following:

- **excisional biopsy** (ehcks-sih-shuhn-ahl bī-ohp-sē) = removing entire mass, tissue, or organ to examine.
- **incisional biopsy** (ihn-sih-shuhn-ahl bī-ohp-sē) = cutting into and removing part of a mass, a tissue, or an organ to examine.

- **needle biopsy** (nē-dahl bī-ohp-sē) = insertion of a sharp instrument (needle) into a tissue for extraction of tissue to be examined.

A **necropsy** (nē-krohp-sē) is a postmortem examination that consists of a thorough examination of a dead animal to determine the cause and manner of death and to evaluate any disease or injury that may be present.

Needles and Sutures

Suture material and needles are used by surgeons to close wounds or to tie things (Figure 17–24). Terms used in reference to suture material and needles include the following:

- **blunt** (bluhnt) = dull, not sharp; used to describe needles or instrument ends.
- **cutting** (kuht-ihng) **needle** = needle that has two or three opposing cutting edges.
- **ligation** (lī-gā-shuhn) = act of tying.
- **ligature** (lihg-ah-chūr) = substance used to tie a vessel or strangulate a part.
- **monofilament** (mohn-ō-fihl-ah-mehnt) = single strand of material; used to describe suture.
- **multifilament** (muhl-tī-fihl-ah-mehnt) = several strands that are twisted together; used to describe suture.
- **stapling** (stā-plihng) = method of suturing that involves the use of stainless steel staples to close a wound.
- **surgical clip** (sihr-jīh-kahl klihp) = metal staple-like device used for vessel ligation.
- **swaged** (swehgd) **needle** = needle joined with suture material in a continuous unit; eyeless needle.
- **taper** (tā-pər) **needle** = needle with a rounded tip that is sharp to allow piercing but not cutting of tissue.

FIGURE 17–24 Needles and staples.

Copyright © 2015 Cengage Learning®.

Courtesy of Teri Raffel, CVT

ABBREVIATIONS RELATED TO PHARMACOLOGY AND SURGERY
Pharmacology Abbreviations

- ac = before meals (ante cibum)
- ad lib = as much as desired (ad libitum)
- bid = twice daily (bis in die)
- BSA = body surface area
- cal = calorie
- cap = capsule
- cc = cubic centimeter (same as mL)
- cm = centimeter
- conc = concentration
- dr = dram; equal to ⅛ oz or 4 mL
- D_5W = 5% dextrose in water
- ED = effective dose
- ED_{50} = median effective dose
- fl oz = fluid ounce
- g = gram
- gal = gallon
- gr = grain; unit of weight approximately 65 mg
- gt = drop (gutta)
- gtt = drops (guttae)
- hr = hour
- IA = intra-arterial
- IC = intracardiac
- ID = intradermal
- IM = intramuscular
- IP = intraperitoneal
- IT = intrathecal
- IU = International Unit or intrauterine
- IV = intravenous
- kg = kilogram
- km = kilometer
- L or l = liter
- lb or # = pound
- LD = lethal dose
- LRS = lactated Ringer's solution
- m = meter
- MBC = minimum bactericidal concentration
- mcg or μg = microgram
- MED = minimal effective dose
- mEq = milliequivalent
- mg = milligram
- MIC = minimum inhibitory concentration
- MID = minimum infective dose
- Ml = milliliter (same as cc)
- MLD = minimum lethal dose
- mm = millimeter (also used for muscles)
- NPO or npo = nothing by mouth (non per os)
- NS = normal saline
- NSAID = nonsteroidal anti-inflammatory drug

- OTC = over the counter
- oz = ounce
- pc = after meals (post cibum)
- PDR = *Physician's Desk Reference*
- % = percent
- pH = hydrogen ion concentration (acidity and alkalinity measurement)
- PO or po = orally (per os)
- ppm = parts per million
- PR = per rectum
- prn = as needed
- pt = pint
- PZI = protamine zinc insulin
- q = every
- qd = every day
- qh = every hour
- q4h = every 4 hours
- q6h = every 6 hours
- q8h = every 8 hours
- q12h = every 12 hours
- q24h = every 24 hours
- qid = four times daily (quater in die)
- qn = every night
- qod or eod = every other day
- qp = as much as desired
- qt = quart
- ® = registered trade name (superscript next to drug name)
- rx = prescription
- sid = once daily (q24h is more common abbreviation for once daily)
- sig = let it be written as (used when writing prescriptions)
- sol'n or soln = solution
- SQ, SC, subq, or subcu = subcutaneous
- T = tablespoon or tablet (or temperature)
- tab = tablet
- tid = three times daily (ter in die)
- TM = registered trade mark (superscript next to drug name)
- tsp or t = teaspoon
- vol = volume
- VPB = *Veterinary Pharmaceuticals and Biologicals*

Surgery Abbreviations

- bx = biopsy
- fx = fracture
- ICU = intensive care unit
- OR = operating room
- sx = surgery

REVIEW EXERCISES

Multiple Choice

Choose the correct answer.

1. A monovalent vaccine, antiserum, or antitoxin is one that is developed for
   a. many organisms
   b. one organism
   c. monocyte activation
   d. subcutaneous injection

2. Hydrophilic substances are
   a. solid substances
   b. liquid substances
   c. water-loving substances
   d. fat-loving substances

3. Movement of water across a selectively permeable membrane along its concentration gradient is
   a. diffusion
   b. efficacy
   c. chelation
   d. osmosis

4. Substances that control itching are called
   a. antipruritic
   b. antimiotic
   c. antitussive
   d. antiseptic

5. To surgically cut out is to
   a. incise
   b. excise
   c. ligate
   d. evert

6. The abbreviation for nothing orally is
   a. PO
   b. qid
   c. prn
   d. NPO

7. The surgical term for side-by-side placement is
   a. inversion
   b. eversion
   c. fulguration
   d. apposition

8. A device by which a channel may be established for the exit of fluids from a wound is a
   a. sling
   b. cast
   c. drain
   d. dressing

9. A sawlike edge or border is a
   a. serration
   b. cautery
   c. chuck
   d. cerclage

10. Another name for an eyeless needle is
    a. monofilament
    b. multifilament
    c. swaged
    d. taper

11. The term for an excessive level of sodium in the blood is
    a. hyperkalemia
    b. hypercalcemia
    c. hypernatremia
    d. hypersodemia

12. A solution that is less concentrated than what it is being compared with is known as
    a. hypotonic
    b. isotonic
    c. hypertonic
    d. equivalent

13. The route of drug administration via the gastro-intestinal tract is
    a. parenteral
    b. nonparenteral
    c. rectal
    d. transdermal

14. An emetic is a substance that
    a. induces vomiting
    b. controls vomiting
    c. eases constipation
    d. controls diarrhea

15. The term meaning to tie or strangulate is
    a. lavage
    b. gavage
    c. ligate
    d. excise

16. The graded locking portion of an instrument located near the finger rings is the
    a. boxlock
    b. ratchet
    c. chuck
    d. clamp

17. A needle that has a rounded tip is called a
    a. swaged needle
    b. cutting needle
    c. stapling needle
    d. taper needle

18. An instrument that controls bleeding is a
    a. cerclage
    b. sling
    c. hemostat
    d. prosthesis

19. Irrigation of tissue with fluid is called
    a. lavage
    b. gavage
    c. ligate
    d. excise

20. The abbreviation *bid* stands for
    a. once daily
    b. twice daily
    c. three times daily
    d. four times daily

21. A product containing a disease-causing organism or antigen administered to stimulate the production of antibodies to fight infection is a
    a. vaccine
    b. polyvalent
    c. monovalent
    d. ligature

22. An artificial substitute for a diseased or missing part of the body is a
    a. serration
    b. prosthesis
    c. dressing
    d. clamp

23. An instrument used to reflect tissue from bone is an
    a. emasculatome
    b. elastrator
    c. emasculator
    d. elevator

24. When an animal is being prepped for surgery, a cloth called a _____ is arranged over its body to provide a sterile field around the area.
    a. clamp
    b. dressing
    c. drape
    d. coaptation

25. There are _____ milliliters in 1 liter.
    a. 10
    b. 100
    c. 1000
    d. 1,000,000

Matching

Match the abbreviation in Column I with its meaning in Column II.

Column I	Column II
1. _____ tid	a. drop
2. _____ gt	b. as needed
3. _____ c̄	c. nothing orally
4. _____ IV	d. subcutaneously
5. _____ IM	e. three times daily
6. _____ SQ	f. every
7. _____ prn	g. intravenously
8. _____ NPO	h. cubic centimeter (same as milliliter)
9. _____ cc	i. intramuscularly
10. _____ q	j. with

Fill in the Blanks

Write the correct term in the space provided.

1. A term meaning to remove an organ or tissue is _____.

2. An accumulation of serum beneath the surgical incision is _____.

3. The removal or exposure of internal organs is _____.

4. Another word for perforation is _____.

5. The degree of fullness or rigidity caused by fluid content is _____.

6. The abbreviation for twice daily is _____.

7. The abbreviation for intradermal is _____.

8. The abbreviation for tablet is _____.

9. The abbreviation for every 8 hours is _____.

10. The abbreviation for orally is _____.

True or False

If the statement is true, write T on the line. If the statement is false, write F on the line.

1. _____ All medications require a prescription.

2. _____ IM medications are administered through an anesthetic machine.

3. _____ An IV injection is an example of a parenteral route of drug administration.

4. _____ An analgesic produces unconsciousness in animals.

5. _____ A disinfectant kills or inhibits the growth of microorganisms on inanimate objects.

6. _____ An emetic is given to animals to prevent vomiting.

7. _____ A mydriatic agent may be used to dilate an animal's eye for a retinal examination.

8. _____ Puppies are often prescribed anthelmintics to treat roundworm infections.

9. _____ A RBC will burst if placed in a hypotonic solution.

10. _____ A cat with hyperkalemia may be treated with potassium supplements.

Spell Check

Circle the term that is spelled correctly.

1. metric unit of volume equivalent to 1 cc: mililiter   milliliter   mililitir
2. excessive levels of potassium in the blood: hypercalemia   hyperkallemia   hyperkalemia
3. administration of drug through routes other than the gastrointestinal tract: parental   parentral   parenteral
4. amount of medication based on units per weight of animal: doseage   doesage   dosage
5. water-loving: hydrofillic   hydrophillic   hydrophilic

Word Scramble

Use the definitions to unscramble the terms.

1. act of tying   ntoiaigl   _____
2. blunt point   rtpae   _____
3. cut into   eiincs   _____
4. stiff dressing   satc   _____
5. bandage for supporting a body part   nligs   _____
6. graded locking portion of an instrument   tthecar   _____
7. movable joint of any ringed instrument   ooxkclb   _____
8. forcible tearing   uutprer   _____
9. removal of an organ in whole   eeucnlaiont   _____
10. act of approximating   aattoinpco   _____
11. exposure of internal organs   eevtiarsce   _____
12. pierce through and through   xtriansf   _____

Crossword Puzzle

Pharmacology Terms

Supply the correct term in the appropriate space for the definition listed.

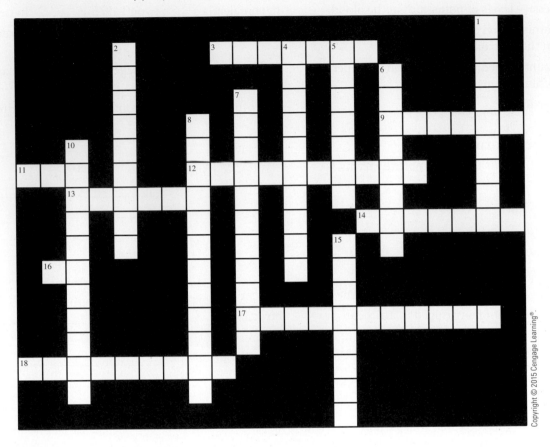

Across

3 without infection

9 substance that induces vomiting

11 as needed (abbr)

12 non-GI route of administration

13 amount of medication based on units/weight

14 course of treatment

16 drop (abbr)

17 substance that reduces coughing

18 substance that dilates pupils

Down

1 to destroy all organisms

2 substance that relieves pain

4 to avoid

5 electrically charged

6 bound to and precipitated out of solution

7 within the skin

8 excessive blood potassium levels

10 agent that kills internal and external parasites

15 bacterial vaccine

Case Studies

Define the underlined terms in each case study.

A 5-yr-old M/N Dalmation HBC was presented to the veterinary clinic with <u>tachypnea</u> and <u>bradycardia</u>. PE revealed T = 99°F, mm = pale gray, CRT = 2.0 sec. The veterinarian was concerned that the dog might be in <u>hypovolemic</u> shock so she requested that the technician start an <u>IV</u> <u>isotonic</u> solution in this patient. The veterinarian pulled up a <u>dose</u> of atropine (a <u>positive chronotrope</u>) for this dog and an <u>antibiotic</u> to treat any infection secondary to skin <u>lacerations</u> from the accident. The dog had a radiograph taken of the thorax and it was determined he had a <u>diaphragmatic hernia</u>. The dog was given an <u>anesthetic</u> and taken to surgery to repair the hernia.

1. tachypnea _____

2. bradycardia _____

3. hypovolemic _____

4. IV _____

5. isotonic _____

6. dose _____

7. positive chronotrope _____

8. antibiotic _____

9. lacerations _____

10. diaphragmatic hernia _____

11. anesthetic _____

A 1-yr-old, 30 <u>kg</u> M Chow presented to the veterinary clinic with a 1 week history of weakness and vomiting. For the past 2 days, the vomit has a coffee ground appearance and the dog has become <u>anorexic</u> and has lost a lot weight. On PE, T = 102°F, HR = 150 bpm, RR = 20 breaths per minute, and the dog was estimated to be 10% dehydrated. MM were pale and tacky; heart and lungs ausculted normally. The dog was painful on abdominal palpation and <u>melena</u> was seen on his hind quarters. Radiographs were taken of this dog's abdomen to determine if he had an <u>obstruction</u>. Abdominal radiographs revealed a <u>dilated</u> duodenum, <u>peritoneal</u> <u>effusion</u>, and a gas <u>opacity</u> ventral to the liver and <u>cranial</u> to the stomach. The veterinarian diagnosed a <u>pyloric</u> obstruction. The dog was prepped for surgery and an abdominal surgery pack was opened. The veterinarian used a <u>scalpel</u> to make a <u>ventral midline incision</u> and used <u>retractors</u> to hold the abdominal walls open so she could check the small intestines for obstructions. There was evidence of small gastric obstructions and <u>incisional biopsies</u> of the stomach were taken to determine whether the dog has <u>gastric carcinoma</u>. Excess fluid from the abdomen was suctioned, the incision was <u>sutured</u>, and the dog returned to the surgical ward. The biopsies were sent to a reference laboratory and the <u>histopathology</u> report was pending. The veterinarian told the technician to watch the incision for signs of <u>dehiscence</u> prior to calling the owner with a progress report.

12. kg _____

13. anorexic _____

14. melena _____

15. obstruction _____

16. dilated _____

17. peritoneal _____

18. effusion _____

19. opacity _____

20. cranial _____

21. pyloric _____

22. scalpel _____

23. ventral midline incision _____

24. retractors _____

25. incisional biopsies _____

26. gastric carcinoma _____

27. suctioned _____

28. sutured _____

29. histopathology _____

30. dehiscence _____

Critical Thinking Exercise

The following story and questions contain words and concepts from this chapter and are designed to stimulate critical thinking through class discussion, Internet searches, or brief writing responses.

A dog is being prepped for a spay surgery and you are asked to help get the animal ready. The dog has been clipped. The following list of terms represents some of the instruments/equipment/chemicals used for a spay surgery. Based on your knowledge of these terms, list the order in which they would be used.

- Tissue forceps

- Hemostatic forceps

- Scalpel

- Drape

- Swaged needle

- Clamp (towel clamp)

- Clamp (vessel, tissue, or organ clamp)

- Antiseptic

- Disinfectant

- Ligature

Chapter 18
LIKE CATS AND DOGS

OVERVIEW

Dogs and Cats	Brief description of canines and felines
Anatomy and Physiology Terms	Describes unique anatomy and physiology of canines and felines
Breed-Related Terms	Defines terms used to describe breed and color terms of canines and felines
Descriptive and Management Terms	Lists terms used to describe canines and felines
Canine and Feline Vaccinations	Lists vaccines used for canines and felines
Abbreviations Related to Canines and Felines	Lists abbreviations for this chapter

Objectives

Upon completion of this chapter, the reader should be able to:

- Recognize, define, spell, and pronounce terms related to dogs and cats

- Interpret the meaning of abbreviations related to dogs and cats

- Analyze case studies and apply medical terminology in a practical setting

DOGS AND CATS

For many years, people have used dogs and cats for different purposes. Dogs were one of the early species man domesticated for work such as herding and protection. Cats were domesticated to control vermin such as rodents. Vital differences in cat and dog personalities resulted in a different relationship between people and these two popular companion species. A dog's willingness to be trained and work resulted in the development of a wide range of breeds suited to do a variety of tasks. The cat's independent nature made it less suited as a working animal, but rather to serve as a hunter. Many believe that dogs came into close contact with people due to refuse from humans serving as a canine food source, while cats came into close contact with people due to the ready source of mice and rats that ate grains

people stored. Although dogs and cats still may be used for work, they are more commonly kept as pets.

The American Kennel Club (AKC) registers dog breeds in the United States, currently recognizes 150 breeds, and divides them into seven groups and a miscellaneous class. The seven groups are Hound Group (e.g., basset hounds, beagles, coonhounds, bloodhounds, dachshunds, Afghan hounds, basenjis, and greyhounds); Herding Group (e.g., Australian shepherds, border collies, collies, German shepherds, Old English sheepdogs, and Welsh corgis); Nonsporting Group (e.g., American Eskimo dog, Boston terrier, Chinese shar-pei, Dalmation, poodle, and schipperke); Sporting Group (e.g., cocker spaniel, English setter, golden retriever, Labrador retriever, and vizsla); Terrier Group (e.g., airedale, cairn, and Scottish terriers); Toy Group (e.g., Chihuahua, papillon, Pekingese, Pomeranian, and shih tzu); Working Group (e.g., boxer, Doberman pinscher, Great Dane, Great Pyrenees, Newfoundland, Rottweiler, and Siberian husky); and Miscellaneous Group (include breeds that are waiting to be accepted to the AKC.

The Cat Fanciers Association (CFA) registers cat breeds in the United States and currently recognizes 41 breeds of cats. Most cats in the United States are not registered, but are mixed breed cats. Examples of cat breeds include Abyssinian, Burmese, Maine Coon, Manx, Persian, and Siamese.

Many of the anatomy and physiology concepts and medical terms related to dogs and cats have been covered in previous chapters. The lists in this chapter apply more specifically to the care and treatment of dogs and cats.

ANATOMY AND PHYSIOLOGY TERMS

- **anal** (ā-nahl) **glands** = secretory tissues composed of apocrine and sebaceous glands located in the anal sac; secretion of the anal glands is stored in the anal sacs and may play a role in territorial marking, as a defense mechanism, or as a pheromone for sexual behavior.
- **anal sacs** (ā-nahl sahcks) = pair of pouches that store an oily, foul-smelling fluid secreted by the anal glands located in the skin between the internal and external anal sphincters (located at the five o'clock and seven o'clock positions); each sac has a duct that opens to the skin at the anal orifice, and fluid is expressed during defecation, excitement, and social interaction (Figures 18–1 and 18–2).
- **carnassial** (kahr-nā-zē-ahl) **tooth** = large, shearing cheek tooth; the upper fourth premolar and lower first molar in dogs (Figure 18–3a) and the upper third premolar and lower first molar in cats; carnassial teeth may develop abscesses secondary to trauma or disease (Figure 18–3b).

FIGURE 18–2 Digital expression of fluid from anal sac.

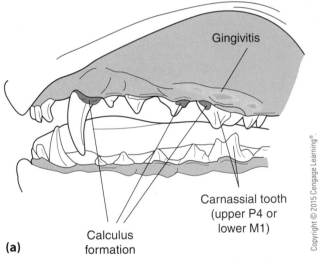

Gingivitis

Carnassial tooth (upper P4 or lower M1)

Calculus formation

(a)

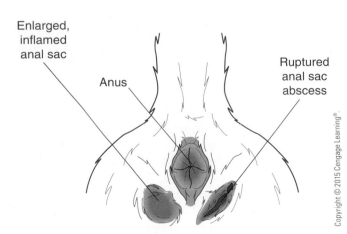

Enlarged, inflamed anal sac

Anus

Ruptured anal sac abscess

FIGURE 18–1 Line drawing of anal sac location.

(b)

FIGURE 18–3 (a) Carnassial teeth of dogs. (b) Carnassial tooth abscess in a dog.

FIGURE 18–4 Liver of a cat with hepatic lipidosis.

- **hepatic lipidosis** (heh-pah-tihck lihp-ih-dō-sihs) = syndrome characterized by excess fat accumulation in the liver of cats that typically occurs after a period of anorexia (Figure 18–4).
- **induced ovulator** (ihn-doosd ohv-yoo-lā-tər) = species that ovulates only as a result of sexual activity (cats, rabbits, ferrets, llamas, camels, and mink).
- **polydactyly** (poh-lē-dahck-tih-lē) = more than the normal number of digits (Figure 18–5a and b).

BREED-RELATED TERMS

- **angora** (ahn-gōr-ah) = type of long fur on cats (and other species).
- **brindle** (brihn-duhl) = mixture of black and other colored hair; the black is uniformly mixed throughout the hair coat.

- **calico** (kahl-ih-kō) = cat with three colors of fur (black, orange, and white); usually female; if male, they have a genotype of XXY (Figure 18–6a).
- **domestic** (dō-mehs-tihck) **longhair** = cat breed that has long guard hairs; abbreviated DLH.
- **domestic** (dō-mehs-tihck) **shorthair** = cat breed that has short guard hairs; abbreviated DSH.
- **mackerel tabby** (mahck-ər-ehl tah-bē) = two-toned feline fur with stripes.
- **mongrel** (mohn-grehl) = mixed breed of any animal.
- **purebred** (pər-brehd) = member of a recognized breed.
- **ruddy** (ruhd-dē) = orange-brown color with ticking of dark brown or black.
- **self** (sehlf) = one-color fur.
- **tabby** (tah-bē) = feline fur with two colors that may be in stripes or spots (Figure 18–6b).
- **ticked coat** (tihckd kōt) = fur in which darker colors are found on the tips of each guard hair.
- **ticking** (tihck-ihng) = fur coat that has guard hairs with darker tips mixed in.
- **tortoiseshell** (tōr-tihs-shehl) = feline fur with two colors (orange and black) producing a spotted or blotched pattern (Figure 18–6c).

DESCRIPTIVE AND MANAGEMENT TERMS

- *ad libitum* **feeding** (ahd lihb-ih-tuhm) = food available at all times so it can be consumed when the animal wants and how much they want; commonly called *ad lib* **feeding**; also called **free choice**.

(a)   **(b)**

FIGURE 18–5 Polydactyly in a cat. (a) Paw of a polydactyl cat. (b) Radiograph of a polydactyl cat.

(a)

(b)

(c)

FIGURE 18–6 (a) Calico cat. (b) Tabby cat.
(c) Tortoiseshell cat.

- **allogrooming** (ahl-ō-groom-ihng) = practice of grooming other members of the same social groups.
- **Association of American Feed Control Officials** = organization that oversees the animal feed industry and ensures that regulations governing the industry are followed; abbreviated AAFCO.
- **bloom** (bloom) = sheen of a coat in top condition.
- **body condition** = amount of fat on an animal evaluated on a scale of 1–10. A score of 1 indicates an animal is extremely emaciated; a score of 5 is assigned to a dog in good body condition; and a score of 10 indicates a dog is grossly obese (Figure 18–7a and b).
- **cattery** (caht-ər-ē) = place where cats are raised or kept.
- **champion** = designation a dog earns by meeting AKC qualifying requirements. Champion or the abbreviation Ch permanently becomes part of the dog's name.
- **choke chain** (chōk) = chain collar that operates like a noose; the tighter the leash is pulled, the tighter the collar becomes around the dog's neck.
- **collar** (kohl-ahr) = fabric, nylon, or leather strap that fits around a dog's neck to provide control.
- **constitution** (kohn-stih-too-shuhn) = physical makeup of an animal.
- **coprophagy** (kohp-rohf-ah-jē or kō-prō-fā-jē) = ingestion of feces.
- **debarking** (dē-bahrk-ihng) = surgical procedure that cuts vocal folds to soften a dog's bark; also called **devocalization** (dē-vō-kahl-ih-zā-shuhn).
- **docile** (doh-sīl) = tame and easygoing.
- **brush** (bruhsh) = fluffy tail of a long-haired cat.
- **digest** (dī-jehst) = flavoring ingredient in pet foods made from digesting animal tissue with enzymes.
- **dull** (duhl) = lacking shine to haircoat; also used to describe behavior more lethargic than normal.
- **Elizabethan** (ē-lihz-ah-bēth-ahn) **collar** = device placed around the neck and head of dogs or cats to prevent them from traumatizing an area; commonly called an **E-collar** (Figure 18–8).
- **euthanasia** (yoo-thehn-ā-zhah) = inducing death of an animal quickly and painlessly; "putting an animal to sleep."
- **fancier** (fahn-sē-uhr) = person who is actively involved in breeding purebred dogs or cats.
- **feral** (fehr-ahl) = wild; not domesticated.
- **food hopper** = food-storing device that drops food into a dish as it is consumed; especially convenient way to feed cats when people are gone for a few days and works best with dry foods.

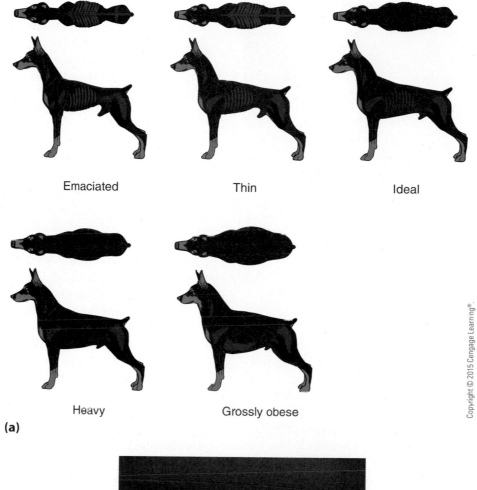

Emaciated   Thin   Ideal

Heavy   Grossly obese

(a)

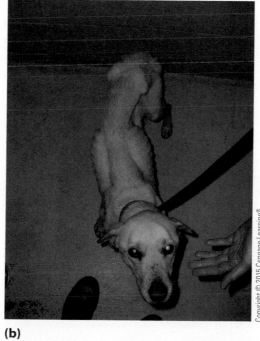

(b)

FIGURE 18–7 (a) Body condition scoring in dogs.
(b) A dog with a body condition score of 1.

FIGURE 18–8 Elizabethan collar on a dog.

Courtesy of Kimberly Kruse Sprecher, CVT

- **gait** (gāt) = way an animal moves.
- **gloves** (gluhvz) = white paws.
- **guaranteed analysis** = information on a commercial bag of food that describes the nutrients found in that food (Figure 18–9).
- **harness** (hahr-nehs) = series of straps that fit around the shoulders of a dog to provide control (Figure 18–10).
- **head halter** = nylon piece of equipment that fits around the muzzle and behind the ears of a dog, giving the handler control of the dog, especially large breeds that are difficult to control; does not apply pressure to the throat like a choke chain and is considered more humane (Figure 18–11).
- **kennel** (kehn-uhl) = facility where dogs are housed.
- **kneading** (nē-dihng) = repetitive massaging motion of the paws; kittens will knead the mammary gland while nursing and some cats knead a location before lying down.
- **leash** (lēsh) = strap connecting a dog's collar to the handler's hand.
- **libido** (lih-bē-dō) = sex drive of an animal.
- **limit feeding** = feeding only a certain amount of food in a given time (not necessarily all the animal can consume).
- **litter pan (box)** = container filled with an absorbent material (litter) in which a cat defecates or urinates.
- **luster** (luhs-tər) = shine.
- **mask** (mahsk) = coloration around the eyes and face of a cat (especially Siamese) or darker color face especially around the eyes in dogs.
- **natural breed** = breed developed in nature, not by human selection.

- **net quality statement** = amount of food present in a feed container; required by the United States Food and Drug Administration (FDA) to be on all feed packages.
- **nutritional adequacy statement** = statement on a feed package indicting whether the feed meets nutritional requirements from a feed trial using AAFCO nutrient profiles for the life stage indicated.
- **obesity** (ō-bē-siht-ē) = excessive fat accumulation in the body.
- **obligate carnivore** = animal that must have a meat-based diet; cats are obligate carnivores.
- **pedigreed cat** (pehd-ih-grēd) = cat registered with the CFA or other breed association.
- **pet quality** = animal that shows the characteristics of the breed, but lacks the characteristics to be sold as a show or breeding animal.
- **points** (poyntz) = color of nose (mask), ears, tail, and feet of an animal (Figure 18–12); also called **colorpoints**. Includes blue point (bluish points), chocolate point (brown points), lilac point (light grayish points), and seal point (dark brownish black points).
- **quarantine** (kwahr-ehn-tēn) = isolation of animals to determine whether they have or carry a disease.
- **retractile** (rē-trahck-tīl) = ability to draw back; feline claws can be drawn back.
- **ruff** (ruhf) = hair around the neck that in some breeds is brushed up to frame the face.
- **sex-linked trait** = inherited trait associated with the sex of the animal (e.g., calico coat color is seen in female cats).
- **sexing** = determining the sex of young animals by identifying the shape of the genital opening and the anogenital distance (distance between the anus and genitalia) (Figure 18–13).
- **sheen** (shēn) = shininess or luster.
- **socialization** (sō-sihl-ih-zā-shuhn) = exposure of a young animal to a wide variety of stimuli, locations, people, and animals to familiarize them with such experience.
- **spraying** (sprā-ihng) = urination on objects to mark territory.
- **staunch** (stawnch) = strong and steady while on point.
- **steatitis** (stē-aht-ī-tihs) = inflammation of fat, usually caused by feeding cats too much oily fish; also called **yellow fat disease**.
- **stray** (strā) = animal with no home or owner.
- **stud** (stuhd) = male animal used for breeding purposes.

100% Complete and Balanced

Holistic Select® Lamb Meal & Rice Formula provides complete and balanced nutrition for maintenance and is comparable in nutritional adequacy to a product which has been substantiated using AAFCO feeding tests.

ingredients: lamb meal, ground brown rice, oatmeal, chicken fat (preserved with natural mixed tocopherols and citric acid), dried beet pulp, flaxseed, brewers dried yeast, rice bran, dried egg product, salt, potassium chloride, air dried peas, ground carrot cubes, inulin, glucosamine hydrochloride, dl-methionine, vitamin a acetate, vitamin d3 supplement, vitamin e supplement, riboflavin supplement, vitamin b12 supplement, d-pantothenic acid, niacin supplement, choline chloride, pyridoxine hydrochloride, thiamine mononitrate, folic acid, ascorbic acid, biolin, rosemary extract, inositol, dehydrated kelp, polysaccharide complexes of zinc, iron, manganese, copper and cobalt, potassium Iodate, sodium selenite, *Yucca Schidigera Extract, Lactobacillus acidophilus, Lactobacillus casei, Enterococcus faecium, B. Subtillus, Bacillus lichenformis, Bacillus coagulins, Aspergillus oryzae, and Aspergillus niger.*

ingredients: Farine d' agneau, riz brun moulu, flacon d' avoine, gras de poulet (conservateurs: mélange de tocophérols et acide citrique), pulpe de betterave déshydraté, lin, levure de biére déshydraté son de riz, oeuf entier déshydraté, sel, chlorure de potassium, pois sèche a l'air, cube de carottes broyées, inuline, hydrochlorure de glucosamine, DL méthionine, acélate de vitamine A, supplement de vitamine D3, supplément de vitamine E, supplément de riboflavine, supplément de vitamin B12, d-pantothenic acide, supplément de niacine, chlorure de choline, chlohydrate de pyridoxine, mononitrate de thiamine, acide folique, acide ascorbique, biotine, extrait de romarin, inositol, varech déshydraté, polysaccharide complexe de zinc, fer, manganèse, cuivre et cobalt, potassium iodate, sodium sélénite, extrait de yucca schidigera, lactobacilles acidophiles, lactobacilles casai, faecium d'entérocoque, subtillus-b, bacilles de lichenformis, bacilles de coagulins, d'oryzae et Niger d'aspergille.

Guaranteed Analysis / Analyse Garantie:

Crude protein/protéines brutes	Min.	22%
Crude fat/matières grasses brutes	Min.	15%
Crude fiber/fibres brutes	Mix.	4.0%
Moisture/humidité	Mix.	10%
Calcium/calcium	Min.	1.3%
Phosphorus/phosphore	Min.	0.85%
Vitamin A/vitamine A	Min.	22,000 IU/kg.
Vitamin E/vitamine E	Min.	125 IU/kg.
Omega 6 fatty acids*/Oméga 6*	Min.	1.6%
Omega 3 fatty acids*/Oméga 3*	Min.	0.17%
Glucosamine hydrochloride*	Min.	400 ppm

*Total Lactic Acid Producing Live Microorganisms
  (Lactobacillus acidophilus, Lactobacillus casei, Enterococcus faecium)   240 million CFU/LB.
*Total Bacillus Organisms
  (Bacillus Subtillus, Bacillus lichenformis, Bacillus coagulins)   7 million CFU/LB.
*Protease (from Aspergillus oryzae and Aspergillus niger)[1]   280 HUT/LB.
*Cellulase (from Aspergillus oryzae and Aspergillus niger)[2]   100 Cellulase Units/LB.
*Amylase (from Aspergillus oryzae and Aspergillus niger)[3]   5 Dextrin Units/LB.
*Not recognized as an essential nutrient by the AAFCO Dog Food Nutrient Profiles.
*Ne figure pas à la liste des éléments nutritifs essentiels reconnus par l'AAFCO en matière d'alimentation pour chiens. Enzyme Functionality Statement: Product contains enzymes. Protease for protein hydrolysis; Cellulase for cellulose hydrolysis; Amylase for starch hydrolysis.

(1) One HUT unit of proteolytic (protease) activity is defined as that amount of enzyme that produces, in one minute under the specified conditions (40 deg Celsius, pH 4.7), a hydrolysate whose absorbance at 275 nm is the same as a solution containing 1.10 ug per ml of tyrosine in 0.006 N hydrochloric acid.

(2) One Cellulase Unit (CU) is that activity that will produce a relative fluidity change of one in 5 minutes in a defined carboxymethylcellulose substrate under the conditions of an assay (40 deg Celsius, pH 4.5.).

(3) One Dextrinizing Unit (DU), or (SKB), unit of alpha-amylase activity as defined as that amount of enzyme that will dextrinize soluble starch ... at the rate of 1 g per hour at 30 deg Celsius.

FIGURE 18–9 Commercial pet food label.

FIGURE 18–10 A properly fitted harness.

FIGURE 18–11 A head halter.

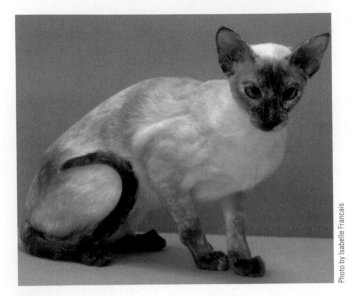

Photo by Isabelle Francais

FIGURE 18–12 Points in a colorpoint shorthair cat.

Copyright © 2015 Cengage Learning®

FIGURE 18–13 Determining the sex of kittens. The female kitten is on the left and the male kitten is on the right.

- **temperament** (tehm-pər-ah-mehnt) = emotional and mental qualities of an individual; also called **disposition** (dihs-pō-sih-shuhn).
- **tie** (tī) = when dogs mate, the swelling near the base of the dog's penis enlarges and temporarily binds the stud to the bitch during copulation. As swelling develops, the bitch's muscles constrict around the penis to hold it in place. This helps ensure proper delivery of sperm and increases the chance for fertilization.
- **thorough** (thər-ō) = working every bit of ground and cover.
- **throwback** (thrō-bahck) = offspring that shows an ancestor's characteristic that has not appeared in previous generations.

- **timid** (tihm-ihd) = showing lack of confidence or shy.
- **topknot** (tohp-noht) = tuft of hair on top of the head.
- **underfur** (uhn-dər-fər) = very dense, soft, short hair that is found beneath the longer, coarser guard hairs; also called **undercoat**.
- **unthrifty** (uhn-thrihf-tē) = not thriving.
- **vigor** (vihg-ər) = healthy growth; also means high energy.
- **wire coat** = hair coat that has stiff hairs.

CANINE AND FELINE VACCINATIONS

- **canine adenovirus** (ahd-nō-vī-ruhs) **2** = adenovirus infection in canines that causes signs of upper respiratory disease; abbreviated CAV-2.
- **canine distemper** (dihs-tehm-pər) **virus** = highly contagious paramyxovirus infection in canines that is associated with respiratory, digestive, muscular, and neurologic signs; abbreviated CDV.
- **canine hepatitis** (hehp-ah-tī-tihs) **virus** = highly contagious adenovirus 1 infection in canines that is associated with ocular ("blue eye"), abdominal, and liver signs; abbreviated ICH or CAV-1.
- **canine herpesvirus** (hər-pēz-vī-ruhs) = herpesvirus infection in canines that primarily affects newborn puppies and is associated with listlessness, nasal discharge, rash, neurologic signs, and death; abbreviated CHV.
- **canine parainfluenza** (pahr-ah-ihn-flū-ehn-zah) **virus** = paramyxovirus infection of canines that contributes to upper respiratory infections and causes subclinical bronchitis.
- **canine parvovirus** (pahr-vō-vī-ruhs) = highly contagious parvovirus infection in canines that is associated with severe diarrhea, vomiting, dehydration, and leukopenia.
- **coronavirus** (kō-rō-nah-vī-ruhs) = coronavirus that causes gastrointestinal disease in dogs and gastrointestinal and systemic disease in cats that is usually spread through contaminated feces; known as feline infectious peritonitis (FIP) in cats; abbreviated CCV in dogs.
- ***Crotalus atrox* toxoid** (krō-tah-luhs ah-trohcks tohcks-oyd) = inactivated toxin from the Western diamondback rattlesnake used in dogs to reduce morbidity and mortality due to envenomation by this snake.
- **feline calicivirus** (kah-lē-sē-vī-ruhs) = picornavirus infection in felines that is associated with upper respiratory and ocular infections.

- **feline chlamydia** (klah-mihd-dē-ah) = bacterial infection in felines caused by the bacterium *Chlamydophila psittaci* (formerly known as *Chlamydia psittaci)* that is associated with upper respiratory and ocular infections.
- **feline immunodeficiency virus** = lentivirus infection in felines that initially presents with fever and lymphadenopathy that over a long period of time progresses to a wide range of clinical signs such as anemia, lethargy, weight loss, and secondary infections; abbreviated FIV.
- **feline infectious peritonitis** = coronavirus infection in felines that is characterized by an insidious onset, fever, and weight loss. The wet form has peritoneal or pleural effusions (or both), whereas the dry form has pyogranulomas in any location; abbreviated FIP.
- **feline leukemia** (loo-kē-mē-ah) **virus** = feline retrovirus that may produce elevated numbers of abnormal leukocytes, immune suppression, cancer, and illness associated with immune suppression; abbreviated FeLV.
- **feline panleukopenia** (pahn-loo-kō-pē-nē-ah) **virus** = parvovirus infection of felines that is associated with fever, vomiting, diarrhea, and a decrease in all types of white blood cells; abbreviated FPV; commonly called feline distemper.
- **feline rhinotracheitis** (rī-nō-trā-kē-ī-tihs) **virus** = herpesvirus infection in felines that is associated with upper respiratory and ocular infections.
- ***Giardia lamblia*** (gē-ahr-dē-ah lahmb-le-ah) = protozoan that may cause asymptomatic disease or cause diarrhea in dogs and cats.
- **infectious tracheobronchitis** (ihn-fehck-shuhs trā-kē-ō-brohng-kī-tihs) = upper respiratory infection caused by the bacterium *Bordetella bronchiseptica* (bōr-dih-tehl-ah brohnk-ō-sehp-tih-kā) that produces a severe hacking cough; also called kennel cough.
- **leptospirosis** (lehp-tō-spī-rō-sihs) = bacterial disease caused by various serotypes of *Leptospira;* signs include renal failure, jaundice, fever, and abortion.
- **Lyme** (līm) **disease** = bacterial disease caused by the bacterium *Borrelia burgdorferi* transported by a tick vector; associated with fever, anorexia, joint disorders, and occasionally neurologic signs; also called Lyme borreliosis.
- **rabies** (rā-bēz) **virus** = fatal zoonotic rhabdovirus infection of all warm-blooded animals that causes neurologic signs; transmitted by a bite or infected body fluid; abbreviated RV. RV is required by public health agencies for licensure of dogs; RV is recommended as a public health measure for cats.

ABBREVIATIONS RELATED TO CANINES AND FELINES

- AAFCO = Association of American Feed Control Officials
- AKC = American Kennel Club
- CAV-2 = canine adenovirus 2
- CCV = canine coronavirus
- CDV = canine distemper virus
- CFA = Cat Fanciers Association
- CHV = canine herpes virus
- DLH = domestic long hair (cat)
- DSH = domestic short hair (cat)
- FDA = Food and Drug Administration
- FeLV = feline leukemia virus
- FIP = feline infectious peritonitis
- FIV = feline immunodeficiency virus
- FPV = feline panleukopenia virus
- ICH = infectious canine hepatitis
- RV = rabies virus
- XXY = genotype of male calico cat

REVIEW EXERCISES

Multiple Choice

Choose the correct answer.

1. The term meaning ingestion of feces is
   a. stoolophagia
   b. fecophagia
   c. coprophagy
   d. bowelophagy

2. The term for more than the normal number of digits is
   a. timid
   b. throwback
   c. polydactyly
   d. mackerel

3. Cats that have white paws are said to have
   a. points
   b. luster
   c. staunch
   d. gloves

4. A cat with three colors of fur is called a(n)
   a. angora
   b. calico
   c. tortoiseshell
   d. ticked coat

5. A male animal used for breeding purposes is known as a
   a. sheen
   b. vigor
   c. stud
   d. feral

6. Urinating on objects to mark territory is called
   a. urination
   b. voiding
   c. micturition
   d. spraying

7. Inflammation of fat is
   a. liposis
   b. lipoma
   c. steatitis
   d. adiposis

8. Devocalization in dogs is commonly called
   a. vocalectomy
   b. cordectomy
   c. silencing
   d. debarking

9. The pouches that store an oily, foul-smelling fluid in dogs and cats are called
   a. anal glands
   b. anal sacs
   c. anal bullae
   d. anal goblets

10. The large, shearing cheek tooth in dogs is commonly called a
    a. premolar tooth
    b. carnassial tooth
    c. polyrooted tooth
    d. deciduous tooth

11. What is the device placed around the neck and head of dogs and cats to prevent them from traumatizing an area?
    a. throwback collar
    b. Elizabethan collar
    c. point collar
    d. gaited collar

12. Inducing death of an animal quickly and painlessly is
    a. quarantine
    b. staunch
    c. euthanasia
    d. unthrifty

13. Feline claws can be drawn back because they are
    a. distal
    b. covered with underfur
    c. unthrifty
    d. retractile

14. A fur coat that has guard hairs with darker tips is a
    a. tortoiseshell
    b. mackerel tabby
    c. ticked coat
    d. calico

15. A mixed breed of any animal is called a
    a. mongrel
    b. purebred
    c. self
    d. domestic

16. A wild animal is called
    a. docile
    b. unthrifty
    c. throwback
    d. feral

17. Cats ovulate as a result of sexual activity and are called
    a. seasonal
    b. polyestrus
    c. induced ovulators
    d. polydactyl

18. Vigor is to timid as
    a. stud is to staunch
    b. luster is to dull
    c. thorough is to bred
    d. tabby is to ruddy

19. The way an animal moves is its
    a. ticking
    b. self
    c. gait
    d. staunch

20. Isolation of animals to determine whether they have or carry a disease is
    a. quarantine
    b. euthanasia
    c. domestic
    d. temperament

21. The "F" in AAFCO stands for
    a. feed
    b. front
    c. feral
    d. fur

22. The "K" in AKC stands for
    a. kitten
    b. kindling
    c. kennel
    d. knee

23. The exposure of a young animal to a wide variety of stimuli, locations, people, and animals to familiarize them with such experience is
    a. familiarization
    b. quarantine
    c. socialization
    d. kneading

24. A series of straps that fit around the shoulders of a dog to provide control is a
    a. head halter
    b. harness
    c. sheen
    d. gloves

25. Having food available at all times so that it can be consumed when the animal wants and how much they want is called
    a. *ad lib* feeding
    b. limit feeding
    c. retractile feeding
    d. vigor feeding

Crossword Puzzle

Dog and Cat Terms

Supply the correct term in the appropriate space for the definition listed.

Across

4 ingestion of feces

6 giving birth to cats

7 fur coat that has guard hairs with darker tips mixed in

8 the way an animal moves

9 tame

10 one color fur

14 not thriving

16 giving birth to dogs

17 urination on objects to mark territory

18 more than the normal number of digits

Down

1 mixed breed of any animal

2 cat having three colors of fur

3 male animal used for breeding

5 color of nose, ears, tail, and feet

11 inducing death of an animal quickly and painlessly

12 wild

13 white paws

15 intact male cat

Case Studies

Define the underlined terms in each case study.

A 3-yr-old F/S black Labrador retriever was presented to the clinic for removal of a round bone from the mandible. Hx: The dog had been chewing on the bone during the day and had gotten the bone stuck on its mandible. The dog had been pawing at the bone for the past 45 min. On PE, the dog was tachycardic, anxious, and tachypnic. The rostral end of the mandible was swollen. The dog was sedated with an IV sedative so that the bone could be removed. The bone was situated caudal to the lower canine teeth. Gigli wire was threaded through the hole in the center of the bone, and the bone was cut in two places to allow its removal. While the bone was being sawed, tissue trauma occurred to the skin of the mandible. The dog was sent home on antibiotics 1 T bid PO 7d.

1. yr _____

2. F/S _____

3. mandible _____

4. Hx _____

5. PE _____

6. tachycardic _____

7. tachypnic _____

8. rostral _____

9. IV _____

10. caudal _____

11. canine teeth _____

12. antibiotics _____

13. T _____

14. bid _____

15. PO _____

16. d _____

A 9-wk-old ♀ DSH kitten was presented to the clinic for inappetence. On PE, it was noted that the kitten had bilateral yellow-green mucopurulent ocular and nasal discharge. T = 103.8°F, HR = 170 BPM, RR = 40 breaths/min, MMs = pink, CRT = 2 sec. The kitten was alert. An audible wheeze was heard on thoracic auscultation; lungs had increased bronchial sounds and referred URT sounds. The conjunctiva was reddened and edematous. The abdomen palpated normally. Dx: URI; DDx: (1) rhinotracheitis virus, (2) calicivirus, (3) chlamydia.

17. wk _____

18. ♀ _____

19. DSH _____

20. inappetence _____

21. PE _____

22. bilateral _____

23. mucopurulent _____

24. ocular _____

25. nasal _____

26. T _____

27. °F _____

28. HR _____

29. BPM _____

30. RR _____

31. min _____

32. MMs _____

33. CRT _____

34. sec _____

35. wheeze _____

36. thoracic _____

37. auscultation _____

38. bronchial _____

39. URT _____

40. conjunctiva _____

41. edematous _____

42. abdomen _____

43. palpated _____

44. Dx _____

45. URI _____

46. DDx _____

47. rhinotracheitis _____

A 5-yr-old F/S DSH cat was presented to the clinic with stranguria and hematuria. T = 102.4°F, HR = 180 BPM, RR = 35 breaths/min, MMs = pink and moist, CRT = 1 sec. Heart and lungs auscultated normally. Oral exam revealed mild tartar with grade II gingivitis. Abdominal palpation yielded normal kidneys, normal intestinal loops, a tense and painful caudal abdomen, and a turgid urinary bladder. Dx: cystitis; DDx: (1) FUS, (2) crystalluria.

48. yr _____

49. F/S _____

50. DSH _____

51. stranguria _____

52. hematuria _____

53. auscultated _____

54. oral _____

55. tartar _____

56. gingivitis _____

57. abdominal _____

58. palpation _____

59. caudal _____

60. turgid _____

61. Dx _____

62. cystitis _____

63. DDx _____

64. FUS _____

65. crystalluria _____

A 2-yr-old <u>intact</u> male golden retriever was presented with a <u>4"</u> <u>laceration</u> with extensive <u>hemorrhage</u> on his right <u>carpus</u>. Pressure bandages were immediately applied for <u>hemostasis</u>. When the bleeding was under control, the dog was <u>anesthetized</u> so that the blood vessels could be <u>ligated</u> and the wound <u>sutured</u>.

66. intact _____

67. 4" _____

68. laceration _____

69. hemorrhage _____

70. carpus _____

71. hemostasis _____

72. anesthetized _____

73. ligated _____

74. sutured _____

A 6-mo-old F black Labrador retriever was presented to the clinic for <u>OHE</u>. A <u>preanesthetic</u> <u>blood screen</u> (<u>PCV</u>, <u>ALT, BUN, GLU</u>) and <u>IV</u> fluid line were done before surgery. The animal was <u>anesthetized</u>, clipped, and prepped for surgery. A <u>ventral midline incision</u> was made, and the reproductive tract was identified. The ovaries, uterine horns, and uterus were removed after proper <u>ligation</u>. When the abdominal incision was being closed, the veterinarian noted pooling of blood in the abdomen. The ligatures were rechecked and still in place. A large amount of blood was coming from the abdominal incision, and the veterinarian had the technician reassess the animal.

The CRT was prolonged, the MMs were pale, and the animal was tachycardic and hypothermic. Blood was taken for another PCV, and the fluid rate was increased. The PCV was low normal. The owner was called to see whether the dog had been sick recently, and the owner stated that the dog was seen eating rat bait about 3 days earlier. Additional blood was collected in a heparin tube for assessment of clotting times, and the dog was given vitamin K1. The incision was closed, and the dog was closely monitored during recovery. The dog made a slow recovery and was hospitalized an additional night for observation. Clotting times from the lab demonstrated prolonged clotting times.

75. OHE _____

76. preanesthetic _____

77. blood screen _____

78. PCV _____

79. ALT _____

80. BUN _____

81. GLU _____

82. IV _____

83. anesthetized _____

84. ventral midline incision _____

85. ligation _____

86. CRT _____

87. MMs _____

88. tachycardic _____

89. hypothermic _____

90. heparin _____

A 10-yr-old M/N cockapoo was presented to the clinic for scooting (dog assumes a sitting position and drags the anal region along the ground) and licking the perianal region. On PE, it was noted that the dog was obese and had dermatitis of the tail head region and oily skin. The TPR were normal. The dog had an hx of tenesmus and reluctance to stand. Rectal palpation of the anal sacs revealed moderately enlarged sacs. Both anal sacs were expressed, and inspissated material was expressed. Both anal glands were flushed with an antiseptic. The dog was discharged with antibiotics, and an appointment was made to reassess the anal sacs in 7 days.

91. perianal _____

92. dermatitis _____

93. TPR _____

94. hx _____

95. tenesmus _____

96. rectal palpation _____

97. anal sacs   _____

98. inspissated   _____

99. antiseptic   _____

Critical Thinking Exercise

The following story and questions contain words and concepts from this chapter and are designed to stimulate critical thinking through class discussion, Internet searches, or brief writing responses.

A client who owns a cat comes into the veterinary clinic to ask about cat food. The owner gives you a label from the food she is currently feed her cat (Figure 18–14). After reviewing the information on the cat food label, you begin teaching your client about pet foods. The following is a summary of what you tell this client:

1. Pet foods are available in three basic formulations: moist or canned, semi-moist, and dry. Dry foods typically contain 10–14% water, are convenient for owners to feed their pets, contain more cereal grains than meat because using meat adds to the feed's moisture content, and provide abrasive action on the teeth to limit tartar formation. Semi-moist foods contain 15–35% water, have a soft texture, contain sugar to increase palatability, and are coated with a carbohydrate and treated with an acid to prevent spoilage. Canned foods contain about 75% water, are more palatable and digestible than dry food, have higher protein and fat levels, and are more odorous, which may increase an animal's interest in the food. How might feeding a cat canned food affect its water consumption?

FIGURE 18–14 A pet food label.

2. The government establishes rules for pet food labeling such as requiring the words "dog food" or "cat food," the name and address of the manufacturer or distributor, the guaranteed analysis, ingredients listed in descending order based on weight, and a description of the usage or purpose of the food. Some companies may describe the food for a specific life stage or condition (puppy or kitten, adult, inactive, overweight, or senior). Many foods are also labeled as "complete and balanced for all life stages." Does this label meet all of labeling requirements?

3. There are two ways the AAFCO monitors pet foods and establishes national standards. The cheapest and least expensive way is based on calculations that are performed in a laboratory to determine the nutrient content. This analysis is then compared to the nutrient requirements of a pet at that given life stage. What are disadvantages to this method of evaluating pet foods?

4. The second method AFFCO uses to evaluate pet food is performing feeding trials. What are advantages and disadvantages to conducting feeding trials?

5. By what method has AFFCO evaluated this cat food?

Chapter 19
HORSE SENSE

OVERVIEW

Horses, Donkeys, Mules, and Ponies	Brief description of equine
Mode of Movement	Defines terms used to describe movement in equine
Anatomy, Physiology, Disease, and Coat Color Terms	Describes anatomy, physiology, and disease terms used in equine
Markings	Defines terms used to describe horse markings
Management Terms	Defines terms used for equine management
Equipment	Describes equipment used for equine
Types of Horses	Describes horse types
Terms for Unsoundness in Horses	Describes terms used to describe unsoundness in equine
Equine Vaccinations	Lists vaccines used for equine
Abbreviations Related to Equine	Lists abbreviations for this chapter

Objectives

Upon completion of this chapter, the reader should be able to:

- Recognize, define, spell, and pronounce terms related to horses, donkeys, mules, and ponies

- Interpret the meaning of abbreviations related to equine animals

- Analyze case studies and apply medical terminology in a practical setting

HORSES, DONKEYS, MULES, AND PONIES

Throughout history, equine have occupied a unique place in society. Horses, donkeys, mules, and ponies have served as food sources to power sources for field work to modes of transportation to sources of recreation and companionship. Horses are from the genus and species *Equus caballus* and developed from prehistoric types of horses. Donkeys are in the same genus as horses (*Equus*), but their species is *asinus*. Mules are a cross of a female horse and male donkey and its genus and species is *Equus caballus* × *Equus asinus*. Ponies are small equine with the scientific name *Equus caballus* var. and are defined based on size (most are less than 58 inches at the withers).

There are hundreds of breeds of horses with cross-breeding being less common in the equine industry than in other livestock industries. Grade horses, which are unregistered horses of mixed or pure breeding, comprise a significant number of animals; however, most breeding programs focus on developing pure-bred animals. Many of the anatomy and physiology concepts and medical terms related to equine species

459

have been covered in previous chapters. The lists in this chapter apply more specifically to the care and treatment of equine species.

MODE OF MOVEMENT

See Figure 19–1.

- **amble** (ahm-buhl) = lateral gait that is different from the pace by being slower and more broken in cadence.
- **back** = slowly trotting in reverse.
- **barrel racing** = racing in which horses race against a timer to complete a cloverleaf pattern around three barrels.

- **beat** (bēt) = time when the foot (or feet if simultaneous) touches the ground.
- **canter** (kahn-tər) = slow, restrained three-beat gait in which the two diagonal legs are paired.
- **dressage** (druh-sahzh) = method of riding in which a rider guides (rather than uses hands, feet, or legs) a trained horse through natural maneuvers.
- **English** = method of riding in a saddle without a horn; the rider rises and sits in rhythm with each stride, and the rider uses both hands on the reins.
- **equitation** (ehk-wih-tā-shuhn) = act and practice of riding a horse.

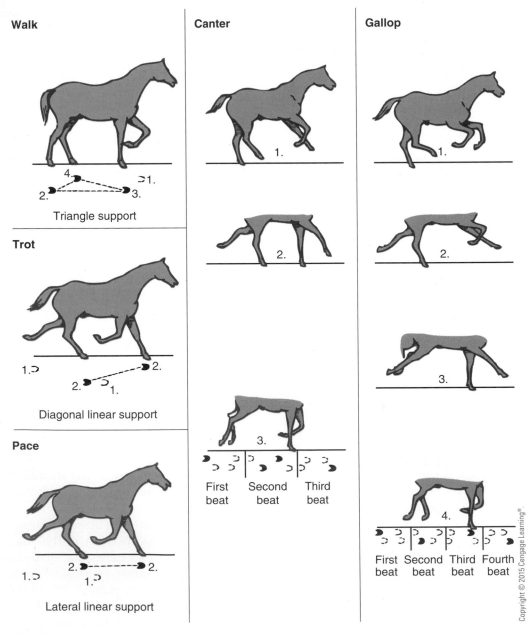

FIGURE 19–1 Basic gaits of equine.

- **fox trot** (fohcks troht) = slow, short, broken type of gait in which the head usually nods.
- **gait** (gāt) = pattern of footfalls resulting in movement. There are **natural gaits** that animals instinctively perform and **artificial gaits** which animals need to learn.
- **gallop** (gahl-ohp) = fast four-beat gait in which the feet strike the ground separately (first = one hind foot, second = other hind foot, third = front foot on the same side as first step, and fourth = other front foot on the same side as second step); also called **run**.
- **gymkhana** (jihm-kah-nah) = racing events on horseback; for example, barrel racing.
- **jog** (johg) = slow trot.
- **pace** (pās) = fast two-beat gait in which the front and hind feet on the same side start and stop at the same time.
- **pointing** (poyn-tihng) = stride in which extension is more pronounced than flexion.
- **racing** = activity in which horses compete directly against one another to complete a distance in the shortest time possible; examples include **endurance racing** (races up to 100 miles), **sprint racing** (races 1/4 mile or less), and **thoroughbred racing** (races from 1/2 to 2 miles).
- **rack** (rahck) = fast, flashy unnatural four-beat gait in which each foot meets the ground separately at equal intervals; also called **single-foot**.
- **rolling** (rō-lihng) = excessive side-to-side shoulder motion.
- **running** (ruhn-ihng) **walk** = slow four-beat gait intermediate in speed between a walk and rack.
- **showing** = any of a wide range of activities in which a horse is exhibited, either ridden, driven, or based on its conformation; common show classes are Western Pleasure (Western riding style), English Pleasure (English riding style), and Showmanship (competes by how well the animal exhibits).
- **stride** (strīd) = distance between successive imprints of the same foot.
- **suspension** (suh-spehn-shuhn) = time in which none of the feet are in contact with the ground.
- **swing** (swihng) = non–weight-bearing phase of a stride.
- **trappy** (trahp-pē) = short, quick, choppy stride.
- **trot** (troht) = natural, rapid two-beat diagonal gait in which the front foot and the opposite hind foot take off and hit the ground at the same time.
- **walk** = natural, slow flat-footed four-beat gait in which each foot takes off and strikes the ground at separate intervals.

- **Western** (wehs-tərn) = method of riding in which the stirrup length is long, the rider rides in an upright posture, and the rider has a one-handed hold on the reins.

ANATOMY, PHYSIOLOGY, DISEASE, AND COAT COLOR TERMS

- **bad mouth** = malocclusion in which the top and bottom teeth do not meet (Figure 19–2).
- **bag up** = development of mammary glands or udder near parturition; also called **bagging up**.
- **bars** (bahrz) = support structure that angles forward from the hoof wall to keep it from overexpanding; also the gap between a horse's incisors and molars; also the side points on the tree of a saddle.
- **bay** = coat color that is light brown to reddish-brown with a black mane, tail, and legs.
- **bishoping** (bihsh-ohp-ihng) = artificial altering of teeth of an older horse to sell it as a younger horse.
- **buckskin** = coat color that is dark golden color with a black mane, tail, and legs.
- **check ligament** (chehck lihg-ah-mehnt) = one of two ligaments to the digital flexors of equine; maintains the limbs in extended position during standing.
- **chestnut** (chehst-nuht) = horny growths on the medial surface of the equine leg either above the knee in the front limb or toward the caudal area of the hock in the rear limb; also a coat color of horses that is reddish-to-brown with a mane and tail the same or lighter color than the coat.
- **cracks** (krahkz) = hoof wall defects that form because the hoof is too long and not trimmed frequently enough.
- **croup** (krūp) = top part of equine rump.

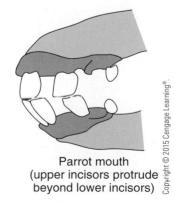

Monkey mouth
(lower incisors protrude
beyond upper incisors)

Parrot mouth
(upper incisors protrude
beyond lower incisors)

FIGURE 19–2 Types of bad mouth (malocclusions).

- **cups** (kuhpz) = deep indentations of the incisors in the center of the occlusal surface in young permanent teeth.
- **curb** (kərb) = enlargement on the caudal aspect of the hind leg below the hock.
- **dental stars** (dehn-tahl stahrz) = marks on the occlusal surface of the incisor teeth appearing first as narrow, yellow lines, then as dark circles near the center of the tooth.
- **dun** = coat color that is brownish golden color with the same colored or slightly darker mane; often has a darker stripe down the back and may have striping on the medial surface of the legs.
- **feather** = long hair below the knee (front limb) and hock (rear limb) that may cover the hoof in some breeds.
- **flehmen** (fleh-mehn) **reaction** = response of a stallion to the scent of a female horse's urine in which he extends his head and curls his upper lip (Figure 19–3).
- **flexor tendon** (flehck-sər tehn-dohn) = tendon that causes the fetlock joint to flex.
- **foal heat** (fōl hēt) = first estrus that occurs shortly after parturition (usually not fertile).
- **full-mouthed** (fuhl mouthd) = having all of the permanent teeth and cups present.
- **Galvayne's groove** (gahl-vānz groov) = mark on labial surface of the equine tooth; used to determine age; usually appears around 10 years of age (Figure 19–4).
- **guttural pouch** (guht-ər-ahl powch) = large, air-filled ventral outpouching of the eustachian tube in equine (Figure 19–5).
- **hindgut** (hihnd-guht) = collective term for the cecum, small colon, and large colon.
- **in wear** = condition in which a tooth has risen to the masticatory level; when opposing teeth have reached sufficient height above the gum line to grind against one another.
- **lamina** (lah-mih-nah) = tissue that attaches hoof to the underlying foot structures.
- **laminitis** (lahm-ihn-ī-tihs) = inflammation of the sensitive laminae under the horny wall of the hoof; a sequela of laminitis is **founder** (fownd-ər) (Figure 19–6).
- **milk teeth** = first teeth that the animal develops.
- **monkey mouth** = condition in which the mandible is longer than the maxilla, causing the lower incisors to protrude beyond the upper incisors (Figure 19–2).
- **nippers** (nihp-pərz) = central incisors of equine; also a tool to remove excess hoof wall.
- **overo** (ō-vār-ō) = marking pattern for Paint or pinto horses that are irregular and scattered across the body; the markings usually have uneven edges, with one or more legs often dark.
- **parrot mouth** = condition in which the maxilla is longer than the mandible, causing the upper incisors to protrude beyond the lower incisors (Figure 19–2).
- **periople** (pehr-ē-ō-puhl) = varnishlike coating that holds moisture in the hoof and protects the hoof wall.
- **pinto** (pihn-tō) = coat pattern of white and any other color in large patches.
- **quidding** (kwihd-ihng) = condition in which a horse drops food from the mouth while chewing.
- **quittor** (kwihd-ər) = festering of the foot anywhere along the border of the coronet.
- **roan** (rōn) = coat color with intermixed solid-colored hairs and white hairs. A red or strawberry roan is a combination of chestnut and white hairs; a blue roan is a combination of black and white hairs.
- **scratches** (skrahch-ehz) = low-grade infection or scab in the skin follicles around the fetlock; also called **grease heel**.
- **smooth mouth** = condition in which no cups are present in the permanent teeth.
- **stay apparatus** (stā ahp-ahr-ah-tuhs) = anatomical mechanism of the equine limb that allows the animal to stand with little muscular effort; includes many muscles, ligaments, and tendons.
- **stock type** (stohck tīp) = muscular body type that is consistent with that desired for working on a ranch; a Quarter Horse body type is a stock type.
- **tobiano** (tō-bē-ah-nō) = color pattern for Paint or pinto horses that has large, instead of scattered, patches. The ends of the pattern tend to be uniform and white often crosses over the back.

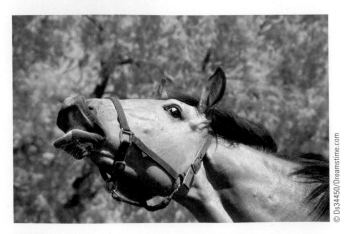

FIGURE 19–3 Flehmen reaction in a stallion.

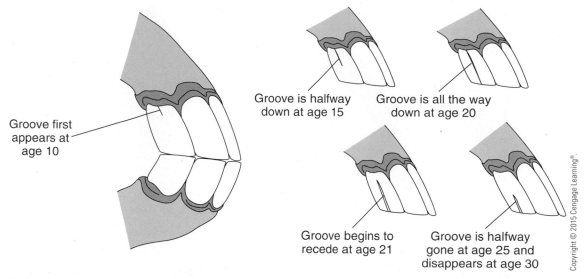

FIGURE 19–4 Galvayne's groove.

Groove first appears at age 10

Groove is halfway down at age 15

Groove is all the way down at age 20

Groove begins to recede at age 21

Groove is halfway gone at age 25 and disappears at age 30

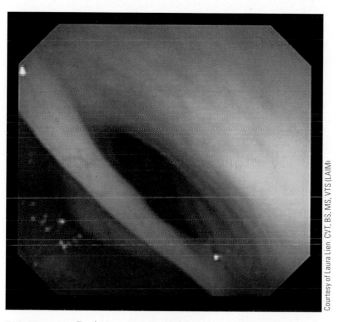

FIGURE 19–5 Endoscopic view of the entrance to the guttural pouch.

- **tush** (tuhsh) = canine tooth in a horse (usually found only in males).
- **waxed teats** (wahcksd tētz) = accumulation of sticky, clear to yellow-colored dried milk at the nipple openings that may occur before parturition.
- **winking** (wihnk-ihng) = opening of the labia to expose the clitoris while the female assumes a mating position.
- **wolf teeth** = rudimentary first upper premolars in equine that are usually shed in maturity.

MARKINGS

See Figure 19–7.

- **ankle** = white marking from the coronet to the fetlock on a horse's leg; also called **sock**.
- **bald face** = wide white marking that extends beyond both eyes and nostrils; also called **apron**.
- **banding** (bahn-dihng) = style of mane that is sectioned and fastened with rubber bands; seen in Western show horses.
- **blaze** (blāz) = broad white stripe on the face of a horse.
- **blemish** (blehm-ihsh) = unattractive defect that does not interfere with performance.
- **bloom** (bloom) = shiny coat for show horses.
- **chin spot** (chihn spoht) = white marking on the chin of a horse's face.
- **coronet** (kohr-ō-neht) = white marking covering the coronary band on a horse's leg.
- **distal** (dihs-tahl) **spots** = dark circles on a white coronet band.
- **half heel** (hahf hēl) = white marking on the medial or lateral aspect of the caudal region of the hoof.
- **half pastern** (hahf pahs-tərn) = white marking from the coronet to the middle of the pastern on a horse's leg.
- **half stocking** (hahlf stohk-ihng) = white marking from the coronet to the middle of the cannon.
- **heel** (hēl) = white marking across the entire heel.
- **pastern** (pahs-tərn) = white marking from the coronet to the pastern on the horse's leg.

FIGURE 19-6 Laminitis. The normal relationship of the coffin bone to the laminae and hoof wall. In chronic founder, the coffin bone is detached and rotated toward the sole.

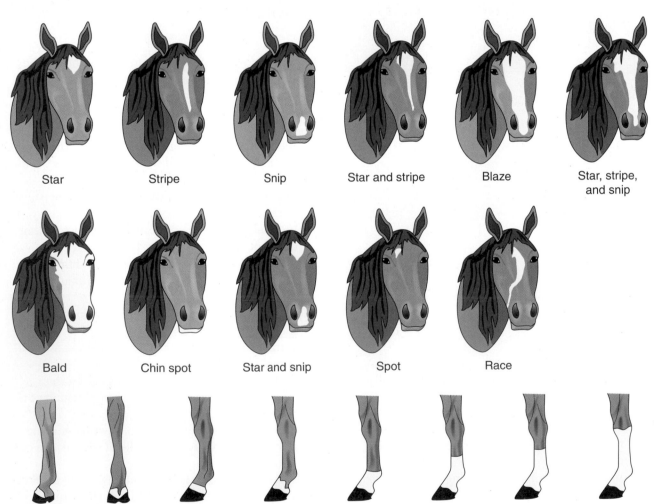

FIGURE 19-7 Natural face and leg markings.

- **points** (poyntz) = black coloration from the knees and hocks down in bays and browns (may include the ear tips).
- **race** (rās) = long wave or irregular stripe down a horse's face.
- **snip** (snihp) = small white marking on a horse's muzzle.
- **spot** (spoht) = small white marking on a horse's face.
- **star** (stahr) = white mark, often in a diamond shape, found between the eyes on the face of the horse.
- **stocking** (stohck-ihng) or **full stocking** = white marking from the coronet to the knee.
- **stripe** (strīp) = long, straight marking down a horse's nose; sometimes called **strip**.

EQUIPMENT

- **aids** (ādz) = means by which a rider communicates with a horse (e.g., voice, hands, legs, seat).
- **bit** (biht) = part of the bridle that is put in the horse's mouth to control the animal. **Curb bits** have a shank extending from the mouthpiece to the reins (creates leverage when pressure is applied to the bit through the reins) while **snaffle bits** have reins that connect directly to a ring attached to the mouthpiece of the bit (the direct connection does not provide leverage such as that with the curb bit).
- **breeching** (brē-chihng) = part of a harness that passes around the rump of a harnessed horse.
- **bridle** (brī-duhl) = part of a harness that includes the bit, reins, and headstall (Figure 19–8).
- **calks** (kawkz) = grips on the heels and the outside of the front shoes of horses.
- **cinch** (sihnch) = part of a saddle used to hold it onto the horse; placed around the girth area.

- **clinch cutter** (klihnch kuht-tər) = tool used to remove horseshoes.
- **cradle** (krā-duhl) = device used to prevent an animal from licking or biting an injured area.
- **halter** (hahl-tər) = device used to lead and tie a horse; also called a **head collar** (Figure 19–9).
- **hobble** (hohb-uhl) = type of restraint in which the front feet or hind feet are placed in straps to keep them from moving.
- **hoof pick** (huhf pihck) = instrument used to clean the sole, frog, and hoof wall.
- **hoof testers** (huhf tehs-tərs) = instrument used to test sensitivity in the equine foot.
- **pincher** (pihn-shər) = tool used to remove horseshoes; pincers are also central incisors.
- **puller** (puhl-ər) = tool used to remove worn horseshoes.
- **rasp** (rahsp) = tool used for leveling a horse's foot (Figure 19–10).
- **reins** (rānz) = strap extending from the bit to the rider's hand.
- **Scotch hobble** (skohtch hohb-uhl) = type of restraint in which all four feet are tied in an *X* pattern to keep them from moving or kicking.
- **shoe** (shū) = plate or rim of metal nailed to the palmar or plantar surface of an equine hoof to protect the hoof from injury or to aid in hoof disease management.
- **tack** (tahck) = equipment used in riding and driving horses.
- **throatlatch** (thrōt-lahtch) = bridle part that connects the bridle to the head located under the horse's throat; also area under throat where the head and neck are joined and where the harness throatlatch fits (Figure 4–1).

MANAGEMENT TERMS

- **as-fed basis** (ahs-fehd bā-sihs) = amount of nutrients in a diet expressed in the form in which it is fed.
- **bedding** (behd-ihng) = material used to cushion the animal's shelter.
- **birth date** (bərth dāt) = for racing or showing, a foal's birthday is considered January 1 regardless of the actual month it was born.
- **blistering** (blihs-tər-ihng) = application of an irritating substance to treat a blemish.
- **board** (bōrd) = to house.
- **bolt** (bōlt) = to eat rapidly or to startle.
- **bots** (bohtz) = larvae of the bot fly, *Gastrophilus*; occur in the stomach (Figure 19–11).
- **bowline knot** (bō-līn noht) = type of nonslippable knot.

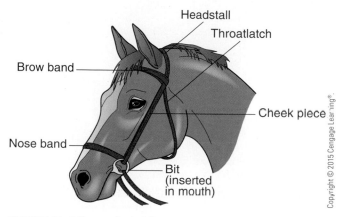

Headstall

Throatlatch

Brow band

Cheek piece

Nose band

Bit (inserted in mouth)

FIGURE 19–8 Parts of a bridle.

- **box stall** (bohcks stahl) = enclosure where a horse can move freely.
- **cast** (kahst) = to be caught in a recumbent position and unable to rise.
- **casting** (kahs-tihng) = restraint method using ropes to place animals in lateral recumbency.
- **covered by a stallion** = mare who was bred by a stallion.

(a)

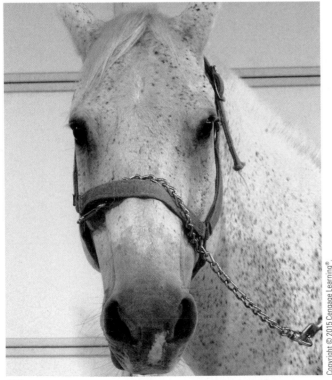

(b)

FIGURE 19–9 (a) Haltering a horse will allow the handler to properly lead the horse and control its movement. (b) A chain can be attached to a lead rope and placed on the halter to apply pressure over the muzzle, under the chin, under the lip, or in the mouth.

(a)

(b)

(c)

FIGURE 19–10 Trimming a horse's hoof. (a) Use of nippers to cut the horny wall to a proper length and angle. (b) Use of hoof knife to pare dead and flaky tissue from the sole. (c) Use of rasp to eliminate jagged edges and sharp corners.

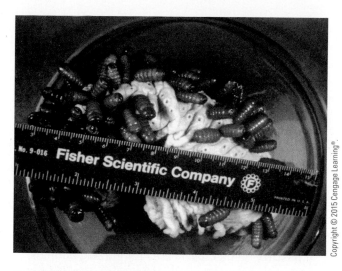

FIGURE 19–11 Larvae of the horse bot fly (*Gastrophilus*) attached to the stomach lining of a horse.

- **cribbing** (krihb-ihng) = vice in which an equine grasps an object between the teeth, applies pressure, and swallows air (Figure 19–12).
- **cross-tying** (krohs-tī-ihng) = method of using two ropes to secure a horse so that the head is level.
- **diluter** (dī-loo-tēr) = type of fluid used to increase the volume of semen (thus diluting the sample).
- **driving** (drī-vihng) = harnessing and controlling horses from behind.
- **dry matter basis** (drī mah-tər bā-sihs) = method of expressing concentration of a nutrient based on absence of water in the feed.

- **extender** (ehcks-tehn-dər) = additive used to extend the lifespan of sperm cells.
- **farrier** (făr-ē-ər) = person who cares for equine feet, including trimming and shoeing.
- **feathering** (feh-thər-ihng) = fringe of hair around an equine foot just above the hoof; also used to describe the fringe of hair on caudal aspects of canine limbs.
- **firing** (fihr-ihng) = making a series of skin blisters with a hot needle over an area of lameness.
- **flighty** (flī-tē) = nervous.
- **floating** (flō-tihng) = filing off the sharp edges of equine teeth (Figure 19–13).
- **get** (geht) = sire's offspring.
- **grade** (grād) = animal that is not registered with a specific breed registry.
- **groom** (gruhm) = to clean a horse with brushes; also a person who is responsible for the care of the horse.
- **grooming equipment** = brushes used to groom a horse; examples include curry comb (rubber disk with blunted teeth that works loose dirt and hair), stiff brush (bristled brush with stiff bristles that removes loosened hair and dirt), soft brush (brush with soft bristles used to brush sensitive areas such as the face and lower legs), comb (untangles mane and tail without pulling out hairs), and hoof pick (used to clean debris from the hoof and is used from heel to toe to ensure that any sharp debris is not pushed into the sensitive parts of the hoof) (Figure 19–14).

FIGURE 19–12 Cribbing.

FIGURE 19–13 Twitch applied to a horse's upper lip to allow teeth to be floated.

- **hand** = unit used to measure height of an equine at the withers; equal to 4 inches.
- **heaving** (hē-vihng) = extra contraction of the flank muscles during respiration; caused by loss of lung elasticity.
- **hunters** (huhn-tərz) = horses that are judged while jumping fences or chasing fox.
- **in foal** = pregnant female horse.

- **jumpers** (juhm-pərz) = horses that compete at shows by jumping and are judged on height, time, and faults.
- **lather** (lah-thər) = accumulation of sweat on a horse's body.
- **leg cues** (lehg kūz) = signals given to the horse through movement of the rider's legs.
- **longe** (luhng) = act of exercising a horse on the end of a long rope, usually in a circle (Figure 19–15); also spelled *lunge.*
- **near side** = left side of horse.
- **off side** = right side of horse; also called **far side**.
- **paddock** (pah-dohck) = small fenced-in area; also called **corral**.
- **pasture** (pahs-chər) = area for grazing animals; also means grass or other forage that grazing animals eat.
- **pasture mating** (pahs-chər mā-tihng) = natural breeding; also called **natural cover**.
- **plumb** (pluhm) **line** = line formed when a weight is placed on the end of a string to measure the perpendicularity of something (used to detect straightness of a horse leg).
- **quick-release knot** = knot that breaks loose easily.
- **saddle** (sahd-uhl) = piece of tack placed over the back of an equine for riding, draft, or pack (Figure 19–16).

(a)

(b)

(c)

(d)   (e)

FIGURE 19–14 Grooming tools. (a) Curry comb, (b) stiff brush, (c) soft brush, (d) comb, and (e) hoof pick.

FIGURE 19–15 Longeing a horse. In longeing, a horse on a long strap or line travels in a large circle around the handler. Longeing helps train young horses, exercises horses, and improves balance and development of stride. Longeing also is spelled *lungeing*.

- **saddle pad** = thick piece of cloth placed between the horse and saddle that absorbs sweat and cushions the horse's back from the saddle.
- **settle** (seht-uhl) = to breed successfully; said of a mare when she becomes pregnant.

- **shod** (shohd) = equine with horseshoes.
- **strike** (strīk) = defensive or aggressive movement of a horse in which the front leg is moved quickly and cranially.
- **stock** (stohck) = device used to restrain equine by inhibiting movement and confining the animal's body (Figure 19–17).
- **tease** (tēz) = act of determining whether a mare is in heat (estrus) by presenting a stallion to her.
- **teaser** (tē-zər) = stallion used to determine which mares are in heat (estrus).
- **twitch** (twihtch) = mode of restraint in which a device is twisted on the upper lip or muzzle (see Figure 19–13).
- **waxing** (wahcks-ihng) = accumulation of colostrum on the nipples of mares usually before foaling; also called **waxed teats**.

TYPES OF HORSES

- **draft** (drahft) **horse** = large breed of working horse (usually over 17 hands) (Figure 19–18). Examples of draft horses include Belgian, Clydesdale, Percheron, and Shire.
- **light horse** = breed of horse that is intermediate in size and stature (usually greater than 14.2 hands) (Figure 19–19). Examples of light horses include Paint, Quarter Horse, Saddlebred,

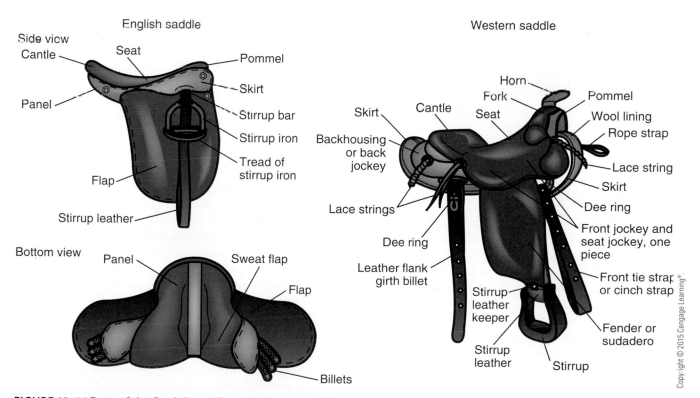

FIGURE 19–16 Parts of the English saddle and Western saddle.

FIGURE 19–17 Equine stock.

FIGURE 19–19 Arabian horses are examples of light horses.

FIGURE 19–18 Belgian horses are examples of draft horses.

Appaloosa, Arabian, Morgan, Standardbred, Tennessee Walking Horse, and Thoroughbred.

- **miniature horse** = small breed of horse (usually less than 8.2 hands). Minature horses must look identical to a full-sized horse except they are smaller.
- **pony** (pō-nē) = small breed of equine (usually about 14 hands). Examples of pony breeds include Pony of the Americas, Shetland, and Welsh.
- **pony of Americas** = breed of pony (usually between 11.5 and 14 hands) that originated from a cross between a Shetland pony stallion and an Arab/Appaloosa mare.

TERMS FOR UNSOUNDNESS IN HORSES

See Figure 19–20.

- **bog spavin** (bohg spah-vihn) = enlargement of proximal hock caused by distention of the joint capsule; **spavin** means swelling.
- **bone spavin** (bōn spah-vihn) = bony enlargement at the base and medial surface of the hock.
- **bowed tendons** (bōd tehn-dohnz) = thickening of the caudal surface of the leg proximal to the fetlock.
- **capped hock** (kahpd hohck) = thickening of the skin or large callus at the point of the hock.
- **fistulous withers** (fihs-tyoo-luhs wih-thərz) = inflammation of the withers.
- **grease heel** (grēs hēl) = infection or scab in the skin around the fetlock; also called **scratches** (skrahch ehz).
- **osselets** (ohs-eh-lehts) = soft swellings on the cranial (and sometimes sides) of the fetlock joint.
- **poll evil** (pōl ē-vihl) = fistula on the poll that does not heal easily.
- **quarter crack** (kwahr-tər krahck), **toe crack** (tō krahck), or **heel crack** (hēl krahck) = cracks in quarters, toe, or heel, respectively, of hoof wall caused by poor management.
- **quittor** (kwihd-ər) = festering of the foot along the border of the coronet.
- **ringbone** (rihng-bōn) = bony enlargement on the pastern bones; high ringbone occurs at the pastern joint; low ringbone occurs at the coffin joint.
- **splints** (splihntz) = inflammation of the interosseous ligament that holds the splint bones to the cannon bone.

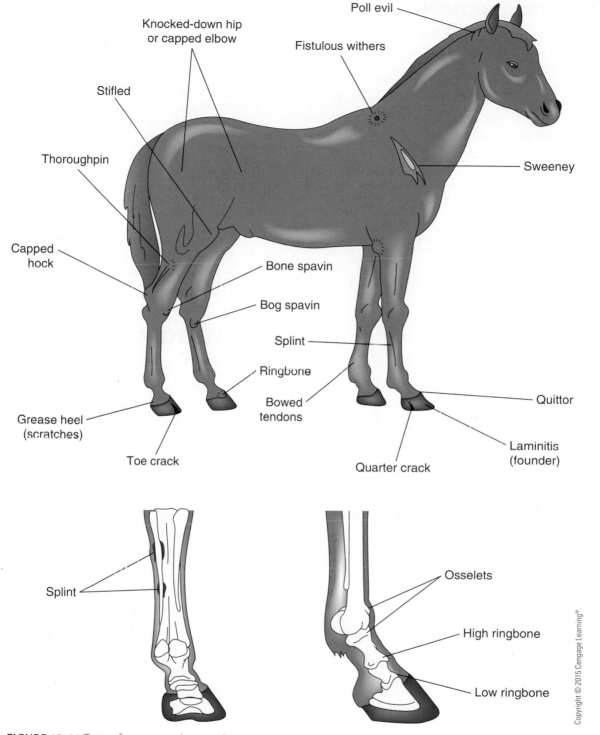

FIGURE 19–20 Terms for unsoundness in horses.

- **stifled** (stī-flehd) = displaced patella.
- **sweeney** (swē-nē) = atrophy of the shoulder muscles.
- **thoroughpin** (thər-ə-pihn) = fluctuating enlargement located in the hollows proximal to the hock;

thoroughpins can be pressed from side to side, hence the name (inflammation of deep digital flexor tendon sheath causes swelling on the medial and lateral sides proximal to the hock).

EQUINE VACCINATIONS

- **equine encephalomyelitis** (ehn-sehf-ah-lō-mī-ih-lī-tihs) = mosquito-transmitted infectious alphaviral disease of horses that is associated with motor irritation, paralysis, and altered consciousness; there are three types: Eastern (EEE), Western (WEE), and Venezuelan (VEE); also known as sleeping sickness.
- **equine influenza** (ihn-flū-ehn-zah) = myxovirus infection of horses that is associated with mild fever, watery eyes, and persistent cough; commonly called **flu**.
- **equine protozoal myelitis** (mī-eh-lī-tihs) = protozoal infection caused by *Sarcocystis neurona* (and perhaps other protozoa) that causes weakness, ataxia (especially in the hindquarters), weight loss, seizures, and other CNS signs; abbreviated EPM.
- **equine viral arteritis** (ahr-tər-ī-tihs) = togavirus infection of horses that is associated with upper respiratory disease signs, abortion, and lesions in small arteries; abbreviated EVA.
- **equine viral rhinopneumonitis** (rī-nō-nū-mohn-ī-tihs) = herpesvirus infection of horses that is associated with signs of upper respiratory disease and abortion; abbreviated EVR; also called equine herpesvirus.
- **Potomac** (pō-tō-mihck) **horse fever** = rickettsial bacterial disease of horses that is associated with fever, anorexia, incoordination, diarrhea, and edema of the extremities; abbreviated PHF; also called **equine ehrlichiosis** (ehr-lihck-ē-ō-sihs).

- **rabies** (rā-bēz) **virus** = fatal zoonotic rhabdovirus infection of all warm-blooded animals that causes neurologic signs; transmitted by a bite or infected body fluid; abbreviated RV.
- **strangles** (strān-guhlz) = contagious bacterial disease of horses caused by the bacteria *Streptococcus equi*; signs include high fever, nasal discharge, anorexia, and swollen and abscessed mandibular lymph nodes.
- **tetanus** (teht-ah-nuhs) = highly fatal bacterial disease caused by the toxin of *Clostridium tetani* that is associated with hyperesthesia, tetany, and convulsions; also called **lockjaw**.
- **West Nile virus** = mosquito-borne viral encephalitis that entered the United States (probably from the Middle East) in 1999. Affected horses may exhibit ataxia, paralysis, depression, head pressing, head tilt, seizures, and death; abbreviated WNV.

ABBREVIATIONS RELATED TO EQUINE

- EEE = Eastern equine encephalitis
- EIA = equine infectious anemia
- EPM = equine protozoal myelitis
- EVA = equine viral arteritis
- EVR = equine viral rhinopneumonitis
- PHF = Potomac horse fever
- RV = rabies virus
- WEE = Western equine encephalitis
- WNV = West Nile Virus
- VEE = Venezuelan equine encephalitis

REVIEW EXERCISES

Multiple Choice

Choose the correct answer.

1. A person who cares for equine feet, including trimming and shoeing, is called a(n)
   a. equestrian
   b. farrier
   c. hobble
   d. quittor

2. A natural, rapid two-beat diagonal gait of horses is called the
   a. pace
   b. trot
   c. canter
   d. gallop

3. In horses, a rapid two-beat gait in which the front and hind feet on the same side start and stop at the same time is called the
   a. pace
   b. trot
   c. canter
   d. gallop

4. What is the term for the large, air-filled out-pouching of the eustachian tube in equine?
   a. Galvayne's groove
   b. guttural pouch
   c. nippers
   d. periople

5. A common sequela of laminitis is
   a. scratches
   b. stay apparatus
   c. breeching
   d. founder

6. The term given to the broad white stripe on the face of a horse is its
   a. star
   b. stripe
   c. race
   d. blaze

7. For racing or showing, a foal's birthday is considered to be
   a. January 1
   b. April 1
   c. June 1
   d. December 1

8. To house a horse is called
   a. blistering
   b. bolting
   c. boarding
   d. bedding

9. The part of the bridle that is located under the horse's throat is called the
   a. tack
   b. throatlatch
   c. chinstrap
   d. halter

10. The tool used for leveling the foot of an equine is the
    a. pincher
    b. hoof pick
    c. rasp
    d. nippers

11. The act of exercising a horse on the end of a long rope is
    a. longeing
    b. driving
    c. handing
    d. plumbing

12. The term for the white marking from the coronet to the knee of a horse is
    a. pastern
    b. stocking
    c. sock
    d. coronet

13. The vice in horses in which they grasp an object between the teeth, apply pressure, and swallow air is
    a. flighting
    b. lathering
    c. cribbing
    d. cross-typing

14. The unit used to measure the height of an equine at the withers is
    a. grade
    b. get
    c. cast
    d. hand

15. The part of the bridle that is put in the horse's mouth is the
    a. bit
    b. cinch
    c. cradle
    d. rasp

16. The canine tooth in a horse that is usually found only in males also is called the
    a. wolf tooth
    b. tush
    c. Galvayne's groove
    d. nipper

17. The act and practice of riding a horse is
    a. equitation
    b. Western
    c. dressage
    d. ambling

18. Teeth that have risen to the masticatory (chewing) level are said to be
    a. cracked
    b. curbed
    c. in swing
    d. in wear

19. The development of the mammary glands or udder near parturition is called
    a. bishoping
    b. cupping
    c. bagging up
    d. curbing

20. The first teeth that an animal develops are its
    a. nippers
    b. milk teeth
    c. wolf teeth
    d. cups

21. The coat color that is light brown to reddish-brown with a black mane, tail, and legs is
    a. bay
    b. buckskin
    c. pinto
    d. roan

22. The coat color that is dark golden color with a black mane, tail, and legs is
    a. bay
    b. buckskin
    c. pinto
    d. roan

23. The coat color with intermixed solid-colored hairs and white hairs is
    a. bay
    b. buckskin
    c. pinto
    d. roan

24. The coat pattern of white and any other color in large patches is
    a. bay
    b. buckskin
    c. pinto
    d. roan

25. To clean a horse with brushes is called
    a. casting
    b. grooming
    c. feathering
    d. firing

Crossword Puzzle

Horse Terms

Supply the correct term in the appropriate space for the definition listed.

Across

3 fast, four-beat gait

6 distance between successive imprints of the same foot

7 nervous

10 left side of horse

11 top part of the equine rump

13 tool used for leveling the foot

16 sire's offspring

17 equine with horseshoes

19 broad white stripe on the face of a horse

20 large, air-filled ventral outpouching of the eustachian tube in equine

Down

1 first teeth that an animal develops

2 founder is a sequela of this condition

4 time when the foot touches the ground

5 condition in which a horse drops food from the mouth while chewing

7 person who cares for equine feet

8 equipment used in riding and driving horses

9 to house

12 area for grazing animals

14 small fenced-in area

15 white marking from the coronet to the knee

18 to cat rapidly or startle

19 shiny coat

Case Studies

Define the underlined terms in each case study.

A 6-yr-old quarter horse <u>mare</u> was presented for signs of pawing at the abdomen, <u>flank</u>-watching, <u>anorexia</u>, and lack of <u>stool</u> production. On PE, she was <u>tachycardic</u> and <u>hyperpnic</u>, <u>MMs</u> were tacky and discolored, and the <u>CRT</u> was 2 seconds. Her ears and limbs were cold to the touch. The gut was <u>auscultated</u>, and <u>borborygmus</u> was noted. The veterinarian did a <u>rectal palpation</u> of the horse and noted gas-filled segments of bowel that felt <u>proximal (oral)</u> to an <u>impaction.</u> <u>Ventral midline abdominocentesis</u> was performed, and blood was collected for a <u>CBC</u>. Laboratory results revealed few <u>peritoneal</u> fluid changes on the abdominocentesis sample, and the CBC was unremarkable. The veterinarian thought the impaction was mild and opted for medical treatment. A <u>naso-gastric tube</u> was passed to relieve the gas distention, and mineral oil was given via the <u>NG tube</u> (Figure 19–21). An <u>analgesic</u> was given <u>IV</u> for pain relief. The horse recovered uneventfully.

1. mare _____

2. flank _____

3. anorexia _____

4. stool _____

5. tachycardic _____

6. hyperpnic _____

7. MMs _____

8. CRT _____

9. auscultated _____

10. borborygmus _____

11. rectal palpation _____

12. proximal (oral) _____

13. impaction _____

14. ventral midline abdominocentesis _____

15. CBC _____

16. peritoneal _____

17. nasogastric tube _____

18. NG tube _____

19. analgesic _____

20. IV _____

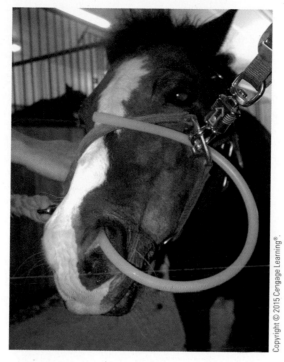

FIGURE 19–21 Nasogastric tube placement in a horse.

A 2-yr-old Arabian <u>colt</u> was anorectic and reluctant to stand. The owner called the veterinarian for an examination. The veterinarian walked the horse, and it showed a short-striding gait. The horse was <u>febrile</u> and <u>tachycardic</u>. Upon <u>palpation</u> of the hoof, the area near the <u>coronary band</u> was warm and a <u>pulse</u> could be palpated. The veterinarian suspected <u>laminitis</u> and recommended <u>radiographs</u> of the hoof. Radiographs revealed rotation of the <u>coffin bone</u> from the hoof wall. The <u>Dx</u> of laminitis (which had progressed to <u>founder</u>) was made. Treatment consisted of <u>NSAIDs</u> and hoof trimming and reshoeing by a <u>farrier.</u>

21. colt

22. febrile

23. tachycardic

24. palpation

25. coronary band

26. pulse

27. laminitis

28. radiographs

29. coffin bone

30. Dx

31. founder

32. NSAIDs

33. farrier

Critical Thinking Exercise

The following story and questions contain words and concepts from this chapter and are designed to stimulate critical thinking through class discussion, Internet searches, or brief writing responses.

A client who owns two dogs is interested in purchasing a horse to ride. She is familiar with the feeding and care of small animals, but has limited experience with large animals. In particular, she would like to know how feeding horses is different from feeding dogs.

1. The first thing you decide to teach your client is the difference between the GI systems of dogs and horses. Briefly describe these differences.

2. You then decide to describe the ways horses are fed. Describe each of the following:
   a. grazing
   b. roughage
   c. concentrates
   d. moldy feed

3. Some horses develop habits that may potentially be harmful to their gastrointestinal tracts. Describe some of those habits.

Chapter 20
MAKE ROOM FOR THE RUMINANTS

OVERVIEW

Objectives

Upon completion of this chapter, the reader should be able to:

- Recognize, define, spell, and pronounce terms related to cattle, sheep, goats, and camelids

- Interpret the meaning of abbreviations related to ruminants

- Analyze case studies and apply medical terminology in a practical setting

RUMINANTS

A **ruminant** (roo-mihn-ehnt) is a cud-chewing animal with a forestomach that allows fermentation of ingesta. Cattle, sheep, and goats have four stomach compartments. The first three—the rumen, reticulum, and omasum—are actually outpouchings of the esophagus. The abomasum is considered the true, or glandular, stomach. Llamas have three stomach compartments and are called pseudoruminants.

CATTLE

Cattle were domesticated around 6,500 B.C. and have been integral to human survival since then. Cattle provide humans with meat, milk, hides, bones, and other by-products used for food, tools, and clothing. Cattle are not indigenous to the United States; they were brought here by Europeans when they were colonizing different parts of the world. American cattle breeds developed in the United States were founded on breeds from other parts of the world.

There are basically two types of cattle: dairy and beef. Dairy cattle are bred for their milk-producing qualities, whereas beef cattle are bred for meat. Some breeds are considered **dual purpose**, which means they have both dairy and beef traits. Examples of dairy breeds include Ayrshire, Brown Swiss, Guernsey, Holstein-Friesian (commonly called Holstein), Jersey, and Milking Shorthorn. Examples of beef breeds include Angus, Brahman, Charolais, Chianina, Galloway, Herefords, Shorthorns, Santa Gertrudis, Scotch Highland, Simmental, and Texas Longhorns.

Many of the anatomy and physiology concepts and medical terms related to cattle have been covered in previous chapters. The lists in this chapter apply more specifically to the care and treatment of cattle.

EQUIPMENT AND INDUSTRY TERMS

- **artificial insemination** (ahr-tih-fih-shahl ihn-sehm-ihn-ā-shuhn) = breeding method in which semen is collected, stored, and deposited in the uterus or vagina without copulation taking place (Figure 20–1); abbreviated AI.
- **auction** (awck-shuhn) = method of selling cattle by bidding in a public venue; used for selling small numbers of animals.
- **backgrounding** = time between weaning and going to the feedlot when a calf is fed primarily a roughage diet.
- **balling** (bahl-ihng) **gun** = tool used to administer pills, boluses, or magnets to livestock (Figure 20–2); also called **bolus gun**.
- **barren** (bār-ehn) = animal that was not bred or did not conceive.
- **beef checkoff** = federal program where one dollar from every beef animal sold goes to state and national organizations to promote the beef industry and fund research related to the beef industry.

FIGURE 20–1 During artificial insemination, semen is placed in the female reproductive tract, allowing the process of fertilization to take place just as in natural mating.

Courtesy of Ron Fabrizius, DVM, Diplomate ACT

FIGURE 20–2 Balling gun used to administer a magnet to a Holstein cow.

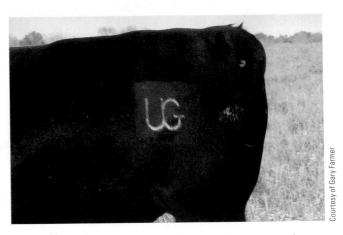

Courtesy of Gary Farmer

FIGURE 20–3 Freeze branding kills the hair pigment, thus marking the animal.

- **body capacity** (boh-dē kah-pah-siht-ē) = heart girth and barrel.
- **brand** (brahnd) = method of permanently identifying an animal by scarring the skin with heat, extreme cold, or chemicals (Figure 20–3).
- **bred** (brehd) = said of an animal that has mated and is pregnant.
- **breed** (brēd) = group of animals that are genetically similar in color and conformation so that when mated to each other, they produce young identical to themselves; also the act of breeding.
- **breed character** = characteristics unique to a breed.
- **bulk tank** = holding tank on a dairy farm used to store milk until it is picked up by a milk cooperative; the bulk tank cools the milk to 40°F and maintains that temperature until transport arrives (Figure 20–4).
- **California mastitis test** = a test that is a cow-side indicator of somatic cell count to determine the presence of mastitis (see Figure 12–31).
- **calving interval** (kahv-ihng ihn-tər-vahl) = amount of time between the birth of a calf and birth of the next calf from the same cow.

Copyright © 2015 Cengage Learning®.

FIGURE 20–4 A bulk tank stores milk until it is collected by a milk cooperative.

- **carcass** (kahr-kuhs) = body of animal after it has been slaughtered; usually has head, hide, blood, and offal removed.
- **casting** (kahs-tihng) = restraint method using ropes to place an animal in lateral recumbency.
- **cattle** (kah-tuhl) = more than one member of the genus *Bos*.
- **chute** (shoot) = mechanical device that is used to restrain cattle (Figure 20–5).
- **cleaning a cow** = common term for removal of a retained placenta; also called **cleansing a cow**.
- **cod** (kohd) = remnants of steer scrotum.
- **colostrometer** (kō-lah-stohm-ih-tər) = instrument designed to float in a sample of colostrum to measure the specific gravity (density) of colostrum (Figure 20–6); higher-quality colostrum will have more immunoglobulins (increased density)

(a)

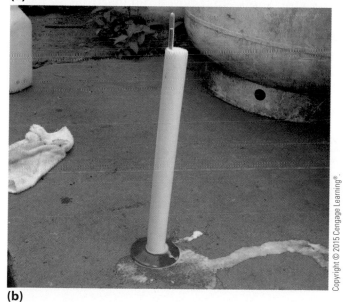

(b)

FIGURE 20–6 A colostrometer is used to measure the quality of colostrum. (a) The weighted float is placed into a column of milk. Higher-quality colostrum has elevated levels of immunoglobulin, which increases its density. As a result the float within the colostrometer is higher. When the float is high, the top of the column of milk is seen on the green region of the scale. (b) Lower-quality colostrum has depressed levels of immunoglobulin, which decrease its density. As a result the float within the colostrometer is lower. The float settles deeper into the sample and the scale is read in the red range. This milk sample shows the float in the red zone and would not be a good source of immunoglobulins.

FIGURE 20–5 A squeeze chute with headgate provides immobilization via a head catch and squeezable sides. The side panels can be dropped to examine feet and legs, while the side bars can be dropped to examine the animal's dorsum.

and will cause the colostrometer to float higher in the sample than colostrum with less immunoglobulins (decreased density).

- **colostrum** (kō-lah-struhm) = first milklike substance produced by the female after parturition; it is thick, yellow, and high in protein and antibodies.
- **composite breed** = a new breed developed from combining established breeds.
- **condition** = amount of fat cover on a breeding animal.
- **Confined Animal Feeding Operations** = large operations that raise animals in a limited amount of space; abbreviated CAFO. Facilities identified as CAFOs must comply with Environmental Protection Agency (EPA) guidelines regarding waste removal.
- **conformation** (kohn-fōr-mā-shuhn) = shape and body type of an animal.
- **corium** (kōr-ē-uhm) = specialized, highly vascular cells that nourish the hoof and horn.
- **crossbred** (krohs-brehd) = having resulted from the mating of two different breeds within the same species.
- **cull** (kuhl) = to remove an animal from the rest because it does not meet a specific standard or is unproductive.
- **dairy character** = term used to describe how closely an animal fits a dairy type; animals with dairy character should be lean without excess fat because their energy should be focused on production of milk (not meat and fat). Dairy character is a visual representation of energy distribution.
- **dehorn** (dē-hōrn) = to remove horns or horn buds by mechanical, thermal, or chemical means.
- **dental pad** (dehn-tahl pahd) = hard surface of the upper mouth of ruminants that serves in place of upper teeth (Figure 20–7).
- **downer** = a recumbent animal that has failed to respond to treatment; downer cows could have a variety of conditions including milk fever (parturient paresis), fractured bone, or other disease (Figure 20–8).
- **dual purpose** (dool pər-puhs) = bred and used for both meat and milk production (cattle) or meat and wool (sheep).
- **ear tagging** (ēr tahg-ihng) = placement of identification device in the ear (Figure 20–9). Ear tags for identification in cattle are placed in the left ear (the right ear is reserved for the brucellosis vaccination tattoo).
- **embryo transfer** (ehm-brē-ō trahnz-fər) = removal of an embryo from a female of superior genetics and placing it in the reproductive tract of another female.

© Margo/Dreamstime.com

FIGURE 20–7 Sheep and cattle have a dental pad instead of upper teeth.

Copyright © 2015 Cengage Learning®.

FIGURE 20–8 A cow with listeriosis (a bacterial infection) that is unable to rise is an example of a downer cow.

- **estrus synchronization** (ehs-truhs sihnk-rihn-ih-zā-shuhn) = practice of using hormones to have multiple animals come into estrus at the same time to increase efficiency of artificial insemination and produce calves of a more uniform age.
- **F_1 generation** (F-1 jehn-ər-ā-shuhn) = first offspring from purebred parents of different breeds or varieties; F1 stands for first filial.
- **feeders** (fē-dərz) = beef cattle that are placed in a feedlot based on age and weight.
- **feedlot** (fēd-loht) = confined area where an animal is fed until it is slaughtered (Figure 20–10a).

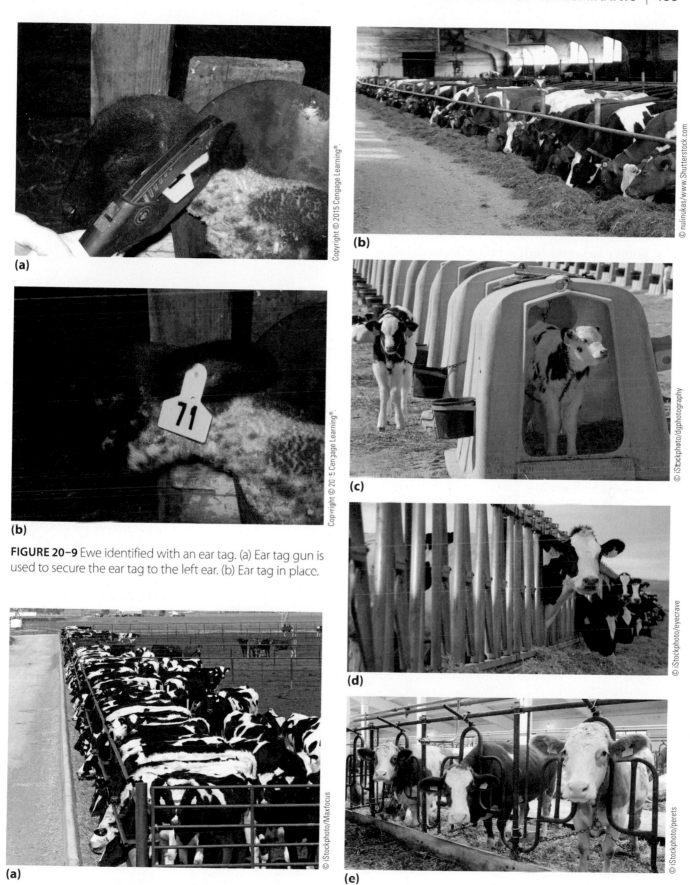

FIGURE 20–9 Ewe identified with an ear tag. (a) Ear tag gun is used to secure the ear tag to the left ear. (b) Ear tag in place.

FIGURE 20–10 Cattle housing. (a) Feedlot; (b) Free stall housing for dairy cattle; (c) Hutch; (d) Stanchion; (e) Tie stall.

- **finish** = degree of fat on an animal that is ready for slaughter.
- **flushing** (fluhsh-ihng) = act of increasing feed before breeding or embryo transfer to increase the number of ova released; typically includes feeding a grain-based diet.
- **fly strike** (flī strīk) = infestation with maggots.
- **frame** (frām) = skeleton of the animal; animals are judged to be large-framed or small-framed depending on the size of their skeletons.
- **free stall barn** (frē stahl) = barn for cattle in which each animal is free to lie down, feed, move, or seek out other animals (Figure 20–10b).
- **gomer** (gō-mər) **bull** = bull used to detect female bovines in heat; bull may have penis surgically deviated to the side, may be treated with androgens, or may be vasectomized so as not to impregnate females; also called **teaser bull**.
- **grade** = an animal not registered with a breed association; the animal can either be a purebred (e.g., a grade Holstein) or of an unknown breed (e.g., a grade cow).
- **gutter** (guh- tər) = channel behind cows in stanchions that collects manure deposited during the milking process. The **gutter cleaner** is a mechanical device that moves the manure out of the barn to a manure collection area (Figure 20–11a). Another type of barn cleaner is an **alley scrapper**, which uses a dragged blade to move manure to the end of the alley (Figure 20–11b).
- **halter** (hahl-tər) = head harness worn by animals for restraint that extends behind the head and over the nose (see Figure 20–12).
- **herd health program** = health care plan to maintain the health of a herd of animals; consists of all aspects of herd management such as vaccination, nutrition, reproduction, and facility management.
- **heart girth** (hahrt gərth) = circumference around the thoracic cavity used to estimate an animal's weight and capacity of the heart and lung.
- **hutch** (huhch) = individual housing pen for calves (and other small animals such as rabbits) (Figure 20–10c).
- **hybrid** (hī-brihd) = offspring resulting from mating of two different species.
- **hybrid vigor** (hī-brihd vihg-ohr) = mating of dissimilar breeds to increase productivity and performance in the F1 generation of crossbred animals over that shown by either parent; also called **heterosis** (heh-tər-ō-sihs).

(a)

(b)

FIGURE 20–11 (a) A gutter cleaner, which is a chain link system of paddles that remove manure from gutters, up a chute, and into a manure spreader. These are often seen in tie-stall or stanchion barns. (b) An alley scraper has a v-shaped blade that is dragged over the alley by a chain to pull manure to a collection area at the end of the alley. Then the blade is collapsed and drawn back to the opposite end of the alley.

- **implant** (ihm-plahnt) = device inserted in the ear that provides a slow release of hormone to the animal to increase feed efficiency.
- **inbred** (ihn-brehd) = resulting from the mating of two closely related animals (e.g., son to dam, sire to daughter); also called **close breeding**.

FIGURE 20–12 Nose tongs are used as a form of restraint in bovines. When applying the nose tong, one prong is placed in the nostril and the device is rotated across the nose to insert the other tong in the other nostril. Tension is applied to the rope to keep the tongs in place. This cow's head is secured with a halter.

- **lead rope** (lēd rōp) = piece of rope, leather, or nylon that is attached via a clasp to a halter.
- **magnet** (māg-neht) = charged metal device that is used to prevent hardware disease (traumatic reticuloperitonitis); it is given orally and placed in the reticulum.
- **malpresentation** (mahl-prē-sehn-tā-shuhn) = abnormal position of a fetus just before parturition.
- **marbling** (mahr-blihng) = streaks of fat interdispersed throughout meat that increase its tenderness.
- **market animal** = animal raised for sale to a slaughter market.
- **muscling** = amount of muscle throughout an animal's body.
- **nose tongs** (nōz tohngz) = clamp used to grasp the nasal septum to restrict movement; the nasal septum is very sensitive and applying pressure to it will distract the animal (Figure 20–12).
- **offal** (aw-fuhl) = inedible visceral organs and unusable tissues removed from the carcass of a slaughtered animal.
- **parturient paresis** (pahr-too-rē-ahnt pahr-ē-sihs) = hypocalcemic metabolic disorder of ruminants seen in late pregnancy or early lactation; also called **milk fever**.
- **pinch** (pihnch) = common term for a bloodless castration using an emasculatome.

- **preconditioning** = preparing a calf for moving to a feedlot; includes vaccination, castration, tattooing or branding, deworming, and weaning.
- **proven** (proov-ehn) = said of an animal whose ability to pass on specific traits is known and predictable.
- **range** (rānj) = large tracts of land for raising animals; most often used to describe land in the Western or Plains states.
- **rectal palpation** (rehck-tahl pahl-pā-shuhn) = method of determining pregnancy, phase of the estrous cycle, or disease process by inserting a gloved arm into the rectum of the animal and feeling for a specific structure (Figure 20–13).
- **registered** = animal whose parentage is documented with an established association.
- **render** (rehn-dər) = to melt down fat by heat.
- **replacement** (rē-plās-mehnt) = animal that is raised for addition to the herd (one that replaces a less desirable animal).
- **ruminating** (roo-mihn-ā-tihng) = cud-chewing process (see page 140).
- **scurs** (skərz) = underdeveloped horns that are not attached to the skull.
- **somatic** (sō-mah-tihck) **cell count** = determination of number of cells (e.g., leukocytes and epithelial cells) in milk to test for mastitis; abbreviated SCC.
- **spotter** (spoh-tər) **bull** = vasectomized male bovine used to find and mark female bovines in estrus.
- **springing** (sprihng-ihng) = anatomic changes in a ruminant that indicate parturition is near.
- **stall** (stahl) = small compartment to house an animal.

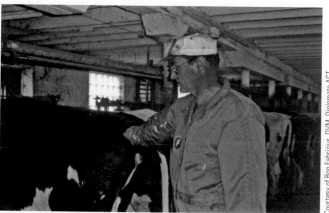

FIGURE 20–13 Rectal palpation in a cow.

- **stanchion** (stahn-chuhn) = restraint device that secures cattle around the neck to allow accessibility for milking, feeding, and examining (Figure 20–10d).
- **standing heat** = phase of estrus in which a female bovine will stand to be mounted.
- **switch** (swihtch) = distal part of a bovine tail that consists of long, coarse hairs (Figure 20–14).
- **tailing** (tā-lihng) = restraint technique used in cattle in which the tailhead is grasped and raised vertically; also called a **tail jack** (Figure 20–15).
- **tankage** (tahnk-ahj) = animal residues left after fat is rendered in a slaughterhouse; used for feed or fertilizer.
- **tattoo** (taht-too) = permanent identification of an animal using indelible ink that is injected under the skin.
- **tie stall** (tī stahl) = stall large enough for only one animal, which is usually tied by a neck chain (cattle) or halter (horse) (Figure 20–10e).

- **tilt table** = restraint table that can be tilted, which is helpful when treating or trimming an animal's feet.
- **traumatic reticuloperitonitis** (traw-mah-tihck reh-tihck-yoo-loh-pehr-ih-tō-nī-tihs) = relatively common disease in adult cattle caused by ingestion and migration of a foreign body into the reticulum; occasionally the foreign body penetrates the diaphragm and pericardium causing pericarditis; abbreviated TRP; commonly called **hardware disease**.
- **veal** (vēl) = confined young dairy calf that is fed only milk or milk replacer to produce pale, soft, and tender meat.
- **wean** (wēn) = to remove young from their mother so they can no longer nurse.
- **windbreak** (wihnd-brāk) = shelter in which an animal can stand and be protected from the wind.

MILK-RELATED TERMS

- **alveoli** (ahl-vē-ō-lī) = milk-secreting sacs of mammary gland; also used to describe gas exchange sac of respiratory system.
- **dry** (drī) = not lactating.
- **drying off** = ending the production of milk when milk yield is low or before freshening.
- **gland cistern** (sihs-tərn) = area of udder where milk collects before entering the teat cistern (Figure 20–16).
- **milk grade** = farms are certified to produce grades of milk based on standards of facility maintenance and herd health. Grades of milk are Grade A milk (highest-quality milk, which is the only kind sold in the United States as fluid milk) and Grade B milk (milk produced at facilities that do not meet Grade A standards and is used to make butter or cheese, which are further processed to ensure safety for consumption).
- **milk letdown** = release of milk from the alveoli to allow removal by the milking machine or calf; cattle can be conditioned to let down milk when they enter the milking environment or have their udders washed.
- **milk solids** = portion of milk that is left after water is removed; includes protein and fat.
- **milk veins** = veins found near the ventral midline of a cow (Figure 20–17); also called **mammary veins**.
- **milk well** = depression in the cow's ventral underline where milk veins enter the body.
- **milk yield** = amount of milk produced in a given period.

FIGURE 20–14 The distal part of a bovine tail is the switch.

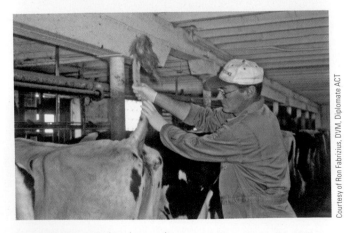

FIGURE 20–15 Tail jack or tailing a cow.

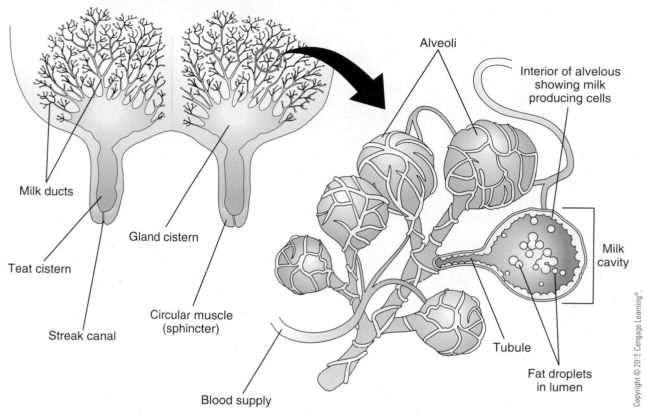

Milk ducts

Gland cistern

Teat cistern

Circular muscle
(sphincter)

Streak canal

Blood supply

Alveoli

Interior of alvelous
showing milk
producing cells

Milk
cavity

Tubule

Fat droplets
in lumen

FIGURE 20–16 Parts of the udder.

FIGURE 20–17 The veins found on a cow's ventral midline are the milk veins.

- **milking** = process of drawing milk from the mammary glands (Figures 20–18). Milk can be collected in containers next to the cow (pail milkers), into a pipeline that moves milk through the barn to the bulk tank (pipeline milkers; see Figure 20–18a), or into containers that are suspended off the ground yet around the cow's body (suspension milkers).

- **milking parlor** = system of milking in which cattle enter in groups, and a person in a recessed pit attaches the milking machine (Figure 20–19).
- **streak canal** (strēk kah-nahl) = passageway that takes milk from the teat cistern to the outside of the body (Figure 20–16); also called the **papillary duct** or **teat canal.**
- **strip cup** = metal cup with a lid that is used for detecting mastitis.
- **supernumerary teats** (soo-pər-nū-mahr-ehr-ē tētz) = more than the normal number of teats.
- **teat** (tēt) = nipple, especially the large nipples of ruminants and equine.
- **teat cannula** (tēt kahn-yoo-lah) = short and narrow, round-pointed metal or plastic tube used to pass from the exterior through the teat canal and into the teat cistern to relieve teat obstructions (Figure 20–20).
- **teat cistern** (sihs-tərn) = cavity in the udder where milk is secreted before leaving the teat (Figure 20–16).
- **teat dipping** = submerging or spraying the nipple with antiseptic to prevent the development of mastitis.

(a)

Milk

Milk

Metal teat
cup shell

Rubber liner
or inflation

Outer
chamber

Vacuum

Outer
chamber

Air at
atmospheric
pressure

Pulsator
vaccum
stroke

Pulsator
air stroke

Milking phase

Rest phase

(b)

FIGURE 20–18 (a) Teat cups are placed on the cow's teats to collect milk. (b) Milking is accomplished using a vacuum system that pulsates on the teats.

FIGURE 20–19 Milking parlor.

FIGURE 20–20 Teat cannula.

- **teat stripping** = removing milk from the teat before or after milking by occluding the teat at the top between the thumb and forefinger and then pulling downward to express all of the milk; also called **stripping**.
- **three quartered** (thrē kwar-tərd) = condition in cows in which one quarter is no longer used due to previous damage or infection.
- **udder** (uh-dər) = milk production organ in ruminants and equine species.

FEEDING-RELATED TERMS

- **bypass protein** (bī-pahs prō-tēn) = protein that is heat or chemically treated so that it is not altered in the ruminant stomach.
- **concentrate** (kohn-sehn-trāt) = type of feed that is high in total digestible nutrients and low in fiber.
- **creep feed** (krēp fēd) = high-energy feed that is fed to young animals in special feed devices so that adult animals cannot gain access to the feed (Figure 20–21).
- **cudding** (kuhd-ihng) = act of chewing cud; **cud** (kuhd) is regurgitated food particles, fiber, rumen fluid, and rumen microorganisms.
- **energy feed** = type of concentrate with less than 20% protein and provides energy through carbohydrates and fats (e.g., corn, oats, and barley).
- **ensiling** (ehn-sī-lihng) = process in which a forage is chopped, placed in a storage unit that excludes oxygen, and fermented to allow longer preservation of feed.
- **feed efficiency** = effectiveness with which an animal turns feed into growth and meat/milk. Increased feed efficiency results in higher profit margins for producers.
- **feedstuff** (fēd-stuhf) = any dietary component that provides a nutrient; also called **feed**.
- **finishing** (fihn-ihsh-ihng) = act of feeding beef cattle high-quality feed before slaughter to increase carcass quality and yield.
- **grass** = roughage that is relatively low in protein; examples include timothy, orchard grass, and Bermuda grass.
- **graze** (grāz) = to eat grasses and plants that grow close to the ground.
- **hand-fed** = use of a bottle to feed calves; after 24 hours, calves are removed from the mother and hand-fed milk collected from the cow for about 3 days. Gradually, calves are hand-fed milk replacer for 2–3 weeks (Figure 20–22).

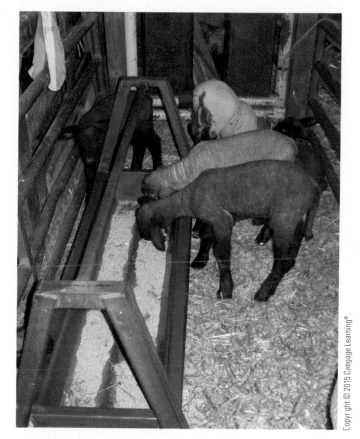

Copyright © 2015 Cengage Learning®

FIGURE 20–21 Creep feed is fed to young animals in special feeders so adult animals cannot gain access to the feed.

- **legumes** (lehg-yooms) = roughage plants that have nitrogen-fixing nodules on their roots; examples include alfalfa, clover, and soybeans.
- **milk replacer** = substitute for milk that has increased amounts of fat and protein for growth.
- **premix** (prē-mihx) = ration mixed with various feedstuffs at the feedmill.
- **protein supplement** = feeds high in protein, typically more than 20% (e.g., meat, milk products, soybean oil meal, and cottonseed oil meal).
- **ration** (rah-shuhn) = amount of food consumed by an animal in a 24-hour period.
- **roughage** (ruhf-ahj) = type of feed that is high in fiber and low in total digestible nutrients; examples include pasture and hay; also called **forage**.
- **silage** (sī-lahj) = type of roughage feed that is produced by fermenting chopped corn, grasses, or plant parts under specific moisture conditions to ensure preservation of feed without spoilage (Figure 20–23).
- **supplement** (suhp-lah-mehnt) = additional feed product that improves and balances a poorer ration; also called **additive**.

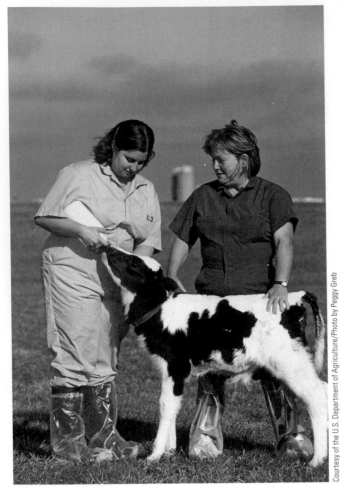

FIGURE 20–22 Calves are hand-fed by bottle following the first 24 hours of nursing from the cow.

- **sweetfeed** (swēt-fēd) = food that consists of grains and pellets mixed with molasses to increase palatability.
- **total mixed ration** = method of combining ration ingredients so each bite contains all necessary nutrients for the animal; abbreviated TPR.

SHEEP

Sheep and goats were among the first animals domesticated and are a major part of the animal industry in countries other than the United States. Historically, sheep and goats frequently accompanied explorers and settlers, as their small size and lower feed needs made them easier to transport than cattle.

Sheep are raised for wool, meat, and research models. As in cattle, some breeds are better known for their wool production and others are better known for their meat quality. Fine wool breeds are bred primarily for their wool production and produce a lower quality

(a)

(b)

FIGURE 20–23 Silage is a type of roughage feed produced by fermenting chopped plant material to ensure preservation of feed without spoilage. (a) Corn silage is fermented chopped corn. (b) Haylage is fermented chopped hay.

meat carcass; examples include Debouillet, Merino, and Rambouillet. Medium wool breeds are raised primarily for meat production; examples include Cheviot, Dorset, Finnish Landrace, Hampshire, Montadale, and Suffolk. Long wool breeds have coarse wool and low-quality meat carcasses; examples include Cotswold and Lincoln. Crossbred wool breeds produce high-quality meat and good wool; examples include Columbia and Targhee. Hair sheep differ from other sheep in that they have more hair follicles and fewer wool follicles, and do not need to be sheared. Hair sheep do well in hot climates and include Dorper and Katahdin. Dual-purpose sheep were developed for both meat and wool production.

Many of the anatomy and physiology concepts and medical terms related to sheep have been covered in previous chapters. The following list of words is used more specifically for sheep and sheep production. Additional terms pertaining to sheep are listed under the cattle section because they apply to both animals.

- **accelerated lambing** = management practice of lambing three times in two years (spring, fall, spring); this requires the use of ewes that will breed out of season as well as excellent record keeping and management. Sheep usually give birth to twins rather than single lambs (Figure 20–24).
- **band** (bahnd) = large group of range sheep.
- **broken mouth** = loss of teeth that indicates the animal is at least 5 years old.
- **browse** = woody twigs and leaves consumed by sheep and goats; generally they are the younger parts of the plants that are higher in nutritional value.
- **carding** (kahr-dihng) = process of separating, straightening, and aligning wool fibers.
- **clip** (klihp) = one season's wool yield.
- **combing** (kō-mihng) = separating long fibers from short fibers and then arranging and laying fibers parallel by length prior to further treatment (Figure 20–25).
- **crimp** (krihmp) = amount of wave in wool.
- **crutching** (kruhtch-ihng) = process of clipping wool from dock, udder, and vulva of sheep before lambing; also called **tagging**.
- **docking** (dohck-ihng) = removal of the distal portion of the tail; also means reducing in value (Figure 20–26).

FIGURE 20–25 In the combing process, the fibers are untangled and smoothed.

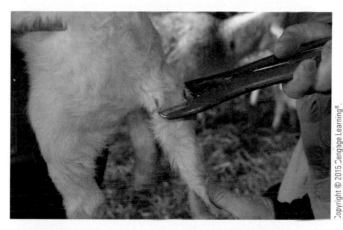

FIGURE 20–26 Docking. Tail docking of a lamb with an elastrator.

- **early spring lambs** = lambs born in January and February, from ewes that breed early in the breeding season.
- **fall lambs** = lambs born before December 25, from breeds that produce "out of season."
- **farm flock** = sheep raised in a pasture and farm situation.
- **fed lambs** = lambs raised on grain.
- **feeder lambs** = lambs purchased after weaning and are raised and fed to market weight.
- **felting** (fehl-tihng) = property of wool fibers interlocking when rubbed together under heat, moisture, or pressure.
- **fleece** (flēs) = another term for wool.
- **flocking instinct** = desire of sheep to stay together in a flock, which is a valuable quality in sheep maintained on rangeland.
- **grease wool** = wool that has been shorn from a sheep and has not been cleaned.
- **lamb** (lahm) = young sheep meat.

FIGURE 20–24 Twins are common in sheep.

- **lambing pen** = enclosure that separates the ewe from the rest of the flock for lambing (Figure 20–27).
- **lanolin** (lahn-ō-lihn) = fatlike substance secreted by the sebaceous glands of sheep.
- **late spring lambs** = lambs born between March and May; these lambs were conceived at the height of the normal breeding season.
- **mixed grazing** = practice of grazing sheep or goats, with cattle, on the same land. Cattle consume the grass while the sheep and goats consume the browse.
- **mutton** (muh-tihn) = adult sheep meat.
- **open face** = sheep that have no wool on the face.
- **rumping** (ruhm-pihng) = method of restraining sheep by placing them in a sitting position with the front legs elevated; also called **tipping**.
- **scouring** (skow-ər-ihng) = cleaning wool (Figure 20–28); also diarrhea in livestock.
- **shear** (shēr) = to shave off wool, hair, or fur (Figure 20–29).
- **singleton** (sihn-guhl-tohn) = one offspring born.
- **yearling** = sheep between one and two years old.

FIGURE 20–28 Scouring removes material from the wool.

FIGURE 20–29 Shearing sheep.

GOATS

Goats have been domesticated for thousands of years. When explorers travelled around the world, goats often accompanied them as they were smaller and were able to produce milk or meat on a wider variety of forage than cattle. Goats are raised for milk, meat, wool, and research models and as pets. Examples of goats raised for meat or milk include Boer, French Alpine, LaMancha, Nubian, Pygmy (also used as pets), Saanen, and Toggenburg. An example of a wool breed is the Angora.

Many of the anatomy and physiology concepts and medical terms related to goats have been covered in previous chapters. The following list applies more specifically to the care and treatment of goats. Additional terms pertaining to goats are listed under the cattle or sheep section because they apply to both animals.

- **cabrito** (kah-brē-tō) = meat of young goats under 50 pounds.
- **cashmere** (kahzh-mər) = fine undercoat of goats (can be harvested from any goat).
- **chevon** (shehv-ehn) = adult goat meat.

FIGURE 20–27 Lambing pen.

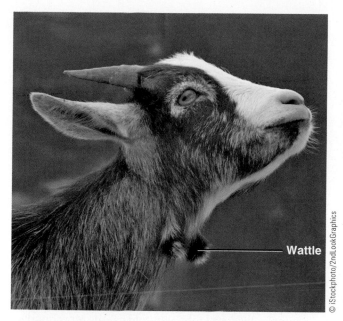

FIGURE 20–30 Wattle in a goat.

- **clip** (klihp) = hair harvested from one animal in one shearing.
- **disbud** (dihs-buhd) = to remove horn growth in kids or calves by use of a hot iron or caustic substance; also called **debudding**.
- **wattle** (waht-tuhl) = appendage suspended from the head (usually the chin) in chickens, turkeys, and goats (Figure 20–30).

CAMELIDS

Camelids are large, herbivorous animals that have slender necks and long legs. Camelids do not have horns or antlers. Camelids are considered **pseudoruminants** because they have three stomach compartments (reticulum, omasum, and abomasum) instead of four. Camelids are induced ovulators. There are three genera of camelids: the Old World camelids (genus *Camelus*), which includes the Bactrian camel (two humps; from Asia) and Dromedary camel (one hump; from Africa); and the New World camelids (smaller without humps), which include the wild guanaco and the domesticated alpaca and llama of South America (genus *Lama*) and the vicuña of South America (genus *Vicugna*). Llamas are the larger species with coarser (they have coarse guard hairs and a fine undercoat), less desirable wool, banana-shaped ears, and typically weigh between 280 and 450 pounds. Llamas are becoming popular pets and are used as pack animals and for fiber production (Figure 20–31). Alpacas are smaller than llamas and have high-quality wool, shorter, more pointed ears, and weigh between 100 and 175 pounds. Alpacas are

(a)

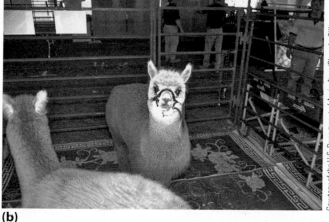

(b)

FIGURE 20–31 (a) Llama. (b) Alpaca.

typically raised for their fleece, for showing, or for pets (Figure 20–31). There are two types of alpacas: Huacaya, which have crimped fleece, and Suri, which have long, wavy fleece. Vicunas are undomesticated and are the smallest and rarest with fine, high-quality wool; and guanacos are slightly smaller than llamas and are undomesticated.

Many of the anatomy and physiology concepts and medical terms related to llamas have been covered in previous chapters. The following list applies more specifically to the care and treatment of llamas.

- **Berserk Male Syndrome** (bər-sərk māl sihn-drōm) = undesirable behaviors sometimes seen in male llamas, especially those that are hand-raised; these animals may be overly aggressive and frequently need to be euthanized because they are unsafe; abbreviated BMS.
- **cushing** (kuhsh-ihng) = common term for copulation of llamas; also the position camelids take when lying down.

FIGURE 20–32 Fighting teeth of a llama.

- **dust pile** = an area of the ground that camelids use for rolling.
- **fighting teeth** = set of six teeth in llamas that include an upper vestigial incisor and an upper and lower canine on each side (Figure 20–32).
- **fleece** (flēs) = hair of llama or alpaca; also known as **fiber**.
- **guard llama** = animal that lives with livestock (mainly sheep) to protect them from predatory animals.
- **harem breeding** (hār-uhm brē-dihng) = style of breeding where a male is left with females most of the year.
- **humming** = muted sound produced by camelids to communicate with each other.
- **kush** (kuhsh) = act of a camelid lying down sternally or the actual position when it is lying down; it is also the command to get a llama to attain this position. Kush is achieved by initially restraining a hind limb, and then placing a rope loop around the abdomen, grasping the opposite hind limb and placing it in the loop.
- **orgle** (ohr-gehl) = call of a male llama or alpaca.
- **spitting** (spiht-ihng) = behavior in which an animal spits saliva at an animal or a person who is perceived to be a threat or an annoyance; spitting also may be used to establish hierarchy. Spitting starts as a gurgle, and then stomach contents are brought up and sprayed from the mouth.

RUMINANT VACCINATIONS

- **bovine respiratory syncytial** (sihn-sihsh-ahl) **virus** = paramyxovirus infection of bovine that is associated with fatal pneumonia; abbreviated BRSV or RSV.
- **bovine viral diarrhea** = togavirus infection of bovine that is associated with acute stomatitis, gastroenteritis, and diarrhea; abbreviated BVD.

- **brucellosis** (broo-sehl-ō-sihs) = bacterial infection caused by *Brucella abortus* that causes abortion in cattle.
- **clostridial** (klohs-trihd-ē-ahl) **disease** = group of bacterial infectious conditions of ruminants caused by various species of *Clostridium*, which includes blackleg, malignant edema, pulpy kidney, enterotoxemia, and overeating disease.
- **coronavirus** (kō-rohn-ah-vī-ruhs) = corona virus infection that causes diarrhea in calves.
- *E. coli* (ē kō-lī) = bacterial infection that typically causes scours in calves; also known as colisepticemia and septicemic colibacillosis.
- **enzootic abortion** = bacterial infection caused by *Chlamydophila abortus* (formerly known as *Chlamydia psittaci*) that causes abortion in sheep.
- *Haemophilus somnus* (hē-moh-fihl-uhs sohm-nuhs) = bacterial infection that causes reproductive, urinary, respiratory, and septicemic disease in cattle.
- **infectious bovine rhinotracheitis** (rī-nō-trā-kē-ī-tihs) = herpesvirus infection of bovine that is associated with fever, anorexia, tachypnea, and cough; abbreviated IBR.
- **leptospirosis** (lehp-tō-spī-rō-sihs) = bacterial disease caused by various serotypes of *Leptospira*; signs include renal failure, jaundice, fever, and abortion.
- **parainfluenza** (pār-ah-ihn-flū-ehn-zah) = paramyxovirus infection of ruminants that is associated with fever, cough, and diarrhea; one part of the shipping fever complex; abbreviated PI-3.
- **pasteurellosis** (pahs-tər-ehl-ō-sihs) = bacterial infection caused by *Pasteurella multocida* and *Mannheimia hemolytica* (formerly known as *Pasteurella hemolytica*) that causes respiratory disease in cattle, sheep, and goats.
- **rotovirus** (rō-tō-vī-ruhs) = reovirus that causes scours in calves.
- **soremouth** (sōr-mowth) = poxvirus infection of sheep, goats, and camelids that causes mouth lesions (Figure 10–17); also called **contagious ecthyma** and **orf**.
- **tetanus** (teht-ah-nuhs) = highly fatal bacterial disease caused by the toxin of *Clostridium tetani* that is associated with hyperesthesia, tetany, and convulsions; also called **lockjaw**; abbreviated TE (Figure 20–33).
- **vibriosis** (vihb-rē-ō-sihs) = *Campylobacter fetus* bacterial infection that is associated with infertility and irregular estrous cycles; bulls are vaccinated; also called **campylobacteriosis**.

FIGURE 20–33 A cow with tetanus displaying a wide stance and stilted gait.

ABBREVIATIONS RELATED TO RUMINANTS

- AI = artificial insemination
- BMS = Berserk Male Syndrome
- BRSV = bovine respiratory syncytial virus
- BVD = bovine viral diarrhea
- CAFO = Confined Animal Feeding Operation
- CMT = California mastitis test
- F1 = first offspring
- IBR = infectious bovine rhinotracheitis
- PI-3 = parainfluenza virus
- RSV = respiratory syncytial virus
- SCC = somatic cell count
- TE = tetanus
- TRP = traumatic reticuloperitonitis
- WMT = Wisconsin mastitis test

REVIEW EXERCISES

Multiple Choice

Choose the correct answer.

1. The common term for removing a retained placenta in a cow is
   a. coding
   b. culling
   c. cleaning a cow
   d. chuting a cow

2. The inedible visceral organs and unusable tissues removed from the carcass of a slaughtered animal are the
   a. marbling
   b. scurs
   c. render
   d. offal

3. What is the name of the passageway that takes milk from the teat cistern to the outside of the udder?
   a. milk duct
   b. gland cistern
   c. alveoli
   d. streak canal

4. What is the term for the restraint device that secures cattle around the neck to allow accessibility for milking, feeding, and examining?
   a. switch
   b. stanchion
   c. corium
   d. lead rope

5. An individual housing pen for calves is called a
   a. halter
   b. stanchion
   c. hutch
   d. barren

6. The term for shaving off wool, hair, or fur is
   a. cutting
   b. scouring
   c. carding
   d. shearing

7. Another term for wool is
   a. fur
   b. curl
   c. fleece
   d. crimp

8. The common term for llama copulation is
   a. tupping
   b. cushing
   c. felting
   d. finishing

9. The amount of food consumed by livestock in a 24-hour period is its
    a. graze
    b. ration
    c. roughage
    d. silage

10. The appendage suspended from the head of goats is the
    a. wattle
    b. comb
    c. crimp
    d. clip

11. The behavior in which a camelid propels saliva at an animal or a person who is perceived to be a threat is called
    a. regurgitation
    b. vomiting
    c. spitting
    d. drooling

12. The term for adult goat meat is
    a. mutton
    b. veal
    c. lamb
    d. chevon

13. The milk production organ in ruminants is called the
    a. nipple
    b. udder
    c. milk well
    d. gland cistern

14. The type of roughage feed produced by fermenting chopped plant parts is
    a. concentrate
    b. sweetfeed
    c. silage
    d. hay

15. The hypocalcemic metabolic disorder of ruminants that is seen in late pregnancy or early lactation is
    a. parturient paresis
    b. malpresentation
    c. rendering
    d. brucellosis

16. The mechanical device that is used to restrain cattle is a
    a. cast
    b. chute
    c. cod
    d. cull

17. Removing an animal from the rest because it does not meet a specific standard is known as
    a. coding
    b. culling
    c. cleaning
    d. cross-breeding

18. The mating of dissimilar breeds to increase productivity and performance in the next generation is known as
    a. heart girth
    b. inbred
    c. casting
    d. hybrid vigor

19. A large group of range sheep is a
    a. band
    b. card
    c. clip
    d. crimp

20. Roughage plants that have nitrogen-fixing nodules on their roots are
    a. feedstuff
    b. legumes
    c. roughages
    d. sweetfeeds

21. The amount of fat cover on a breeding animal is
    a. condition
    b. scurs
    c. offal
    d. switch

22. The practice of grazing sheep or goats, with cattle, on the same land is
    a. browsing
    b. pasturing
    c. late grazing
    d. mixed grazing

23. The "M" in TMR stands for
    a. meat
    b. milk
    c. mixed
    d. marbling

24. The release of milk from the alveoli to allow removal by the milking machine or calf is

    a. milking interval
    b. milk letdown
    c. milk release
    d. milk solids

25. An animal whose parentage is documented with an established association is

    a. rendered
    b. registered
    c. ruminated
    d. rationed

Crossword Puzzle

Ruminant Terms

Supply the correct term in the appropriate space for the definition listed.

Copyright © 2015 Cengage Learning®

Across

3 remove an animal from the rest

5 another term for wool

6 metal cup with a lid that is used for detecting mastitis

7 large group of range sheep

8 milk production organ in ruminants and equine

10 chewing cud

12 distal part of the bovine tail with long, coarse hairs

13 young llama

16 adult sheep meat

17 animal that is not lactating

18 milk producing organ of ruminants and equine

19 not able to reproduce

Down

1 copulation in llamas

2 one season's wool yield

3 common name for removal of a retained placenta

4 appendages suspended from the chin of goats

9 removal of horn growth in kids

11 mechanical device used to restrain cattle

14 amount of food consumed in 24-hour period

15 California mastitis test

Case Studies

Define the underlined terms in each case study.

A 2-yr-old French alpine <u>doe</u> was presented with recurrent abdominal distention, decreased milk production, and <u>dyspnea</u>. On PE, the <u>MMs</u> were pink, <u>CRT</u> was normal, <u>TPR</u> was normal, breathing was labored but no abnormal respiratory sounds were <u>auscultated</u>, and the mammary glands were normal. The doe is in a closed herd, and the rest of the animals were normal. Urine was tested and revealed a trace amount of glucose and negative ketones. Rumen fluid had a <u>pH</u> of 6.9 and a healthy population of small and medium <u>protozoal</u> organisms. The majority of the large protozoal population was dead. A <u>rumenostomy</u> was performed, and administration of 0.5 <u>L</u> warm water was initiated.

1. doe _____

2. dyspnea _____

3. MMs _____

4. CRT _____

5. TPR _____

6. auscultated _____

7. pH _____

8. protozoal _____

9. rumenostomy _____

10. L _____

A 2-yr-old Suffolk <u>ram</u> was examined for <u>lethargy, anorexia</u>, and weight loss. The <u>PE</u> was <u>WNL</u> except for pale <u>mucous membranes</u> and a distended abdominal area. The owner was questioned about grazing, nutritional, and deworming practices on the farm. The veterinarian took a blood sample for a <u>hematocrit</u> and a stool sample for parasite testing. The hematocrit revealed that the ram was mildly <u>anemic</u>, and the fecal exam revealed *Haemonchus* eggs. The owner was advised to administer a broad-spectrum <u>anthelmintic</u>, to rotate pastures, and to monitor animals for signs of anemia.

11. ram _____

12. lethargy _____

13. anorexia _____

14. PE _____

15. WNL _____

16. mucous membranes _____

17. hematocrit _____

18. anemic _____

19. anthelmintic _____

A <u>herd</u> of Hereford cattle were showing signs of <u>alopecia</u> and crusty skin <u>lesions</u> on the head and neck. Some of the more severely affected cattle had lesions over most of their bodies. The <u>PE</u> was unremarkable except for the skin lesions. The veterinarian took several <u>skin scrapings</u> to identify possible mite infestation. <u>Microscopic examination</u> revealed *Psoroptes* <u>mites</u>. The veterinarian diagnosed <u>scabies</u>, recommended pour-on medication, and advised the owner to observe withdrawal times. The cattle recovered uneventfully.

20. herd _____

21. alopecia _____

22. lesions _____

23. PE _____

24. skin scrapings _____

25. microscopic examination _____

26. mites _____

27. scabies _____

A 3-year-old <u>bull</u> llama presented with signs of neurologic disease including hindlimb <u>paralysis</u> and a head tilt. The owner stated that initially the llama had walked with a slight limp and had weakness in his hind limbs. Since the animal was not <u>anorexic</u>, the owner decided to monitor the animal with the suspicion that the llama had injured its leg. In a few days, the condition progressed to <u>ataxia</u>. Routine laboratory tests were performed on the llama, including a <u>CBC, chem panel</u>, and <u>fecal examination</u>. The CBC showed <u>eosinophilia</u>; however, the other tests were WNL. Since the llama is pastured with other llamas in an area of the Midwest that has white-tailed deer and had eosinophilia, the veterinarian suspected that the llama had <u>meningeal</u> worm (*Parelaphostrongylus tenuis*) and treated him with an <u>antiparasitic</u> drug. The owner was given a poor <u>prognosis</u> due to the fact that the llama had already progressed to the paralytic stage.

28. bull _____

29. paralysis _____

30. anorexic _____

31. ataxia _____

32. CBC _____

33. chem panel _____

34. fecal examination _____

35. eosinophilia _____

36. meningeal _____

37. antiparasitic _____

38. prognosis _____

Critical Thinking Exercise

The following story and questions contain words and concepts from this chapter and are designed to stimulate critical thinking through class discussion, Internet searches, or brief writing responses.

A 3-yr-old female Holstein has been producing 70 pounds of milk per day; however, as her milk production started to decrease, the farmer noticed she had mastitis in one teat. He would like the veterinarian to come to the farm to check on this cow. After you schedule the appointment, you begin to take a history on this cow so the veterinarian has this information prior to arriving on the farm.

Based on your knowledge of industry terms, what would you like to ask this farmer about his management practices and type of farm he manages? Things to address include the process of milking and how animals are prepared for milking, the type of facility and how it is managed, and the care each animal receives.

Chapter 21
HOG HEAVEN

OVERVIEW

Pigs	Brief history of swine
Equipment and Industry Terms	Describes terms used in the swine industry
Swine Management Terms	Describes terms used in raising swine and types of swine operations
Swine Vaccinations	Lists vaccines used for pigs
Abbreviations Related to Swine	Lists abbreviations for this chapter

Objectives

Upon completion of this chapter, the reader should be able to:

- Recognize, define, spell, and pronounce terms related to swine

- Interpret the meaning of abbreviations related to swine

- Analyze case studies and apply medical terminology in a practical setting

PIGS

Pigs have been domesticated since 6,500 B.C. and have been used as livestock both as a forager in developing countries and as confinement-raised animals in commercial operations in industrialized nations. Today pigs are used for their meat and hides, as research models, and for pharmaceutical production. Some pigs, such as potbellied pigs, have more recently been housed as pets.

Swine are monogastric animals and have organ systems similar to dogs and cats. They have cloven hooves, meaning they have a split in the toe of the hoof that separates the hoof into two parts. Most swine breeds weigh between 220 and 240 pounds, but some select bacon breeds can top 400 pounds. Swine used for food products are usually young pigs because the older hogs, especially boars, develop a strong flavor in the meat. Most hogs raised for meat yield a dressing of 65 to 80 percent of their body weight.

There are two types of swine—the meat-type hog and the bacon-type hog. Meat-type hogs have a large amount of meat (ham) on the body, while bacon-type hogs have a large amount of fat (bacon) on the body. The meat-type hog is muscular and produces cuts of meat such as ham, loins, and roasts. The bacon-type hog is fatter and produces meat used for bacon and sausage. Most producers select hogs that have a verified pedigree. Sows are selected from gilts born to large litters that have no genetic flaws and have adequate backfat measurement. Sows are reevaluated between 150 and 200 pounds of growth to ensure that they are growing properly and will be excellent breeders. Boar selection is based on temperament and body structure. Boars are typically selected from large litters, have a medium to large body structure, should reach adult size by 155 days or less, and have less than 1 inch of backfat. Ideal weight gain for breeding boars is 2 pounds of body weight gain per day. Pigs typically farrow between 7 and 12 piglets

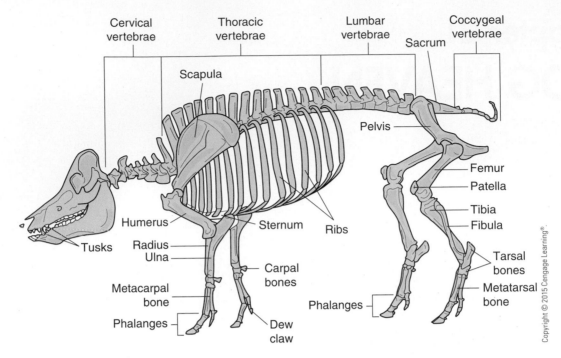

FIGURE 21–1 Pig skeleton.

per year and have a gestation length of 114 days. Piglets are born with enough iron reserves in their livers to last 7–10 days, making anemia a concern in piglets who do not receive an injection of iron 1–3 days after birth. Baby pig anemia has become prevalent since swine producers began farrowing litters in confinement, denying the nursing pig access to iron in the soil.

Many of the anatomy and physiology concepts and medical terms related to pigs have been covered in previous chapters. The lists in this chapter and in Figure 21–1 apply more specifically to the care and treatment of pigs.

EQUIPMENT AND INDUSTRY TERMS

- **backfat** (bahk-faht) = thickness of fat along the dorsum of the pig.
- **bacon** (bā-kohn) = meat product from the sides of the pig.
- **boar taint** (bōr tānt) = odor of pork that is harvested from an adult boar; as boar ages, boar taint becomes more pronounced and results in an unpalatable product.
- **casting** (kahs-tihng) = restraint method using ropes to place animal in lateral recumbency.
- **checkoff** (chehck-ohf) = system where a portion of the sale price of every pig goes to the National Pork Board to promote and improve the pork industry.

- **cross-fostering** (krohs-fohs-tər-ihng) = moving piglets from one litter to another to balance litter size (done in the first day or two of life).
- **dressing** (drehs-ihng) = amount of meat produced by one pig.
- **dunging pattern** (duhn-gihng pah-tərn) = tendency for animals to eliminate wastes in a particular location.
- **ear notching** (ēr nohtch-ihng) = identification method used in swine in which notches of various patterns are cut in the ear (Figure 21–2).
- **farrowing crate** (fār-ō-ihng krāt) = holding pen that limits sow movement before and during parturition (Figure 21–3a).
- **farrowing house** (fār-ō-ihng hows) = building dedicated to the delivery and raising of piglets to weaning.
- **farrowing pen** (fār-ō-ihng pehn) = sow holding area that has guardrails and floor junctures that allow young pigs to escape; used before and during parturition; pen is larger than a crate (Figure 21–3).
- **finish** (fihn-ihsh) = degree of fat on an animal that is ready for slaughter.
- **free-access gestation housing** = holding area that allows sows to move about and socialize and have access to an individual stall when they want to eat or rest (Figure 21–3).
- **hide** (hīd) = pig skin.
- **hog hurdle** (hohg hər-duhl) = portable partition used to move swine by blocking the area in which the pig should not go (Figure 21–4).

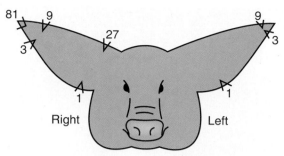

Key to standard ear notching system

Examples of litter numbers

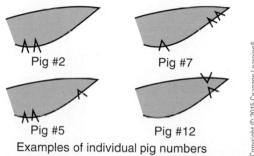

Examples of individual pig numbers

Copyright © 2015 Cengage Learning®

(a)

© gudak/www.Shutterstock.com

(b)

FIGURE 21–2 (a) Ear notching identification in swine. (b) Pig with notched ears.

- **hog snare** (hohg snār) = restraint method in which the pig's snout is secured by a loop tie that is attached to a long handle; also called **snare** (Figure 21–4).
- **hog-tight** (hohg-tīt) = fencing that prevents animal escape.

- **lard** (lahrd) = soft, white fat that is the product of rendering pig fat.
- **needle teeth** = eight temporary incisors and canine teeth of young swine (Figure 21–5).
- **piles** (pīlz) = common term for a prolapsed rectum in swine (Figure 21–6).
- **ringing** = act of implanting a wire ring through a pig's nose to discourage rooting.
- **sling** (slihng) = swine restraint device with four leg holes and an additional hole under the neck for blood collection; device looks like a hammock.
- **snout** (snowt) = upper lip and apex of nose of swine.
- **specific pathogen free** (speh-sihf-ihck pahth-ō-jehn frē) = national system of accrediting an animal free of specific diseases; animals are obtained by cesarean section and raised in isolation to prevent certain infectious diseases; abbreviated SPF; not disease-free.
- **tail docking** = removal of the distal part of the tail to prevent fecal contamination and/or tail biting (Figure 21–7).
- **tusk** (tuhsk) = overgrown canine tooth of boar.
- **wallow** (wahl-ō) = natural or artificial wading area for swine to cool themselves or avoid biting insects.

SWINE MANAGEMENT TERMS

- **closed herd** = group of animals that restricts entrance of new animals.
- **farrow-to-finish** (făr-ō too fihn-ihsh) = full-service swine operation that houses breeders, newborns, weanlings, and feeder stock.
- **farrow-to-wean** (făr-ō too wēn) = swine operation that houses breeding sows and offspring until they reach weaning age or weight.
- **feeder-to-finish** (fē-dər too fihn-ihsh) = swine operation that raises weanling pigs to market weight.
- **finisher pig** (fihn-ihsh-ər pihg) = swine over 100 pounds to slaughter.
- **grower pig** (grō-ər pihg) = swine from about 40 to 100 pounds.
- **open herd** = group of animals in which animals from other groups are allowed to join the existing group.
- **starter pig** (stahr-tər pihg) = swine from about 10 to 40 pounds; also called **feeder pig**.

SWINE VACCINATIONS

- *Bordetella* (bōr-dih-tehl-ah) = bacteria causing respiratory disease; atrophic rhinitis pathogen.
- *Clostridium perfringens type C* (klohs-trih-dē-uhm pər-frihn jehns tīp C) = bacteria causing enterotoxemia that results in diarrhea and high mortality.

FIGURE 21–3 Housing of swine. (a) Farrowing or gestation crate restrains sows during farrowing. (b) Farrowing pen allows baby pigs to move away from the sow to prevent being crushed. (c) Larger pigs are housed away from the sow. (d) Free-access gestation stalls allow sows to gather in a socialization area and enter and exit individual stalls as desired.

FIGURE 21–4 Restraint of pigs. (a) Hog hurdle. (b) Hog snare.

FIGURE 21–5 Needle teeth. Needle teeth of baby pigs are clipped to prevent injury to the sow during nursing and to minimize wounds from sibling pigs.

Copyright © 2015 Cengage Learning®.

FIGURE 21–6 Rectal prolapse in a pig may be caused by stress or constipation or may be secondary to other illness.

Courtesy of Eric Klaphake, DVM, Diplomate ABVP

FIGURE 21–7 Tail docking. Tail docking is done shortly after piglets are born to prevent feces from collecting on the tail and to prevent piglets from biting each other's tails.

Courtesy of Dr. Frank B. Flanders

- *E. coli* (ē kō-lī) = bacteria causing colibacillosis that causes edema, ataxia, and death.
- *Erysipelas* (ehr-ih-sihp-eh-lahs) = bacteria causing acute septicemia, skin lesions, chronic arthritis, and endocarditis.
- *Haemophilus* (hē-mohf-ih-luhs) = bacteria causing respiratory disease associated with acute onset, pyrexia, and reluctance to move.
- **leptospirosis** (lehp-tō-spī-rō-sihs) = bacterial (*Leptospira*) disease of swine associated with pyrexia, anorexia, neurologic signs, and abortion.
- *Mycoplasma* (mī-kō-plahz-mah) = bacteria causing respiratory disease seen largely in young pigs with a severe cough.
- **parvovirus** (pahr-vō-vī-ruhs) = parvovirus infection affecting mainly gilts and associated with abortion.
- *Pasteurella* (pahs-too-rehl-ah) = bacteria causing respiratory disease that sometimes leads to pericarditis and pleuritis.
- **porcine proliferative enteropathy** (por-sīn prō-lihf-ər-ah-tihv ehn tər-oh-pah-thē) = bacterial disease (caused by *Lawsonia intracellularis*) that causes ileitis.
- **porcine** (por-sīn) **respiratory and reproductive syndrome** = arteriviral disease of swine characterized by reproductive failure in sows and respiratory disease in young and growing pigs; abbreviated PRRS.
- **pseudorabies** (soo-dō-rā-bēz) = herpesvirus infection associated with pyrexia and neurologic signs.
- **rotavirus** (rō-tō-vī-ruhs) = rotavirus associated with villous destruction in the intestine, malabsorption, and diarrhea.
- *Streptococcus suis* (strehp-tō-kohk-kuhs sū-hihs) = bacterial infection that causes meningitis.
- **swine influenza** (swīn ihn-flū-ehn-zah) = orthomyxoviral infection that causes respiratory disease.
- **transmissible gastroenteritis** (trahnz-mihs-ih-buhl gahs-trō-ehn-tehr-ī-tihs) = coronaviral disease of swine characterized by villous destruction of jejunum and ileum, malabsorption, diarrhea, and dehydration; abbreviated TGE.

ABBREVIATIONS RELATED TO SWINE

- PRRS = porcine respiratory and reproductive syndrome
- SPF = specific pathogen free
- TGE = transmissible gastroenteritis

REVIEW EXERCISES

Multiple Choice

Choose the correct answer.

1. The eight temporary incisors and canine teeth of young swine are known as
   a. piles
   b. needle teeth
   c. finishing teeth
   d. tusks

2. A swine over 100 pounds to slaughter is known as a(n)
   a. finisher pig
   b. grower pig
   c. starter pig
   d. open pig

3. Swine weighing 10 to 40 pounds are called
   a. finisher pigs
   b. grower pigs
   c. starter pigs
   d. open pigs

4. What identification method is used for swine?
   a. tattooing
   b. branding
   c. ear notching
   d. tail notching

5. What is the name of the restraint method that uses ropes to place swine in lateral recumbency?
   a. hog-tighting
   b. hog-tying
   c. slinging
   d. casting

6. The overgrown canine tooth of the boar is its
   a. needle tooth
   b. molar
   c. tusk
   d. snout

7. A swine operation that raises weanling pigs to market weight is what type of management system?
   a. farrow-to-finish
   b. farrow-to-wean
   c. feeder-to-finish
   d. feeder-to-farrow

8. The upper lip and apex of the nose of swine is its
   a. tusk
   b. wallow
   c. snout
   d. lard

9. What type of restraint method has the pig's snout secured via a loop tie that is attached to a long handle?
   a. hog hurdle
   b. snare
   c. sling
   d. crate

10. A holding pen that limits sow movement before and during parturition is known as a
   a. farrowing crate
   b. farrowing pen
   c. finishing pen
   d. starter pen

11. The common term for a prolapsed rectum in swine is
   a. ringing
   b. sling
   c. piles
   d. backfat

12. A group of animals that restricts entrance of new animals is a(n)
   a. open herd
   b. closed herd
   c. farrowing herd
   d. finisher herd

13. Moving piglets from one litter to another to balance litter size is
   a. checkoff
   b. boar taint
   c. piglet transfer
   d. cross-fostering

14. A portable partition to move swine by blocking the area is a
   a. hog snare
   b. hog hurdle
   c. hog ring
   d. hog sling

15. Pig skin also is called
    a. hide
    b. dung
    c. piles
    d. lard

16. The tendency for animals to eliminate wastes in a particular location is known as a
    a. notching pattern
    b. farrowing pattern
    c. slinging pattern
    d. dunging pattern

17. The odor of pork that is harvested from an adult boar is called
    a. pork taste
    b. boar taint
    c. boar odor
    d. pork taint

18. White fat that is the product of rendering pig fat is
    a. wallow
    b. piles
    c. lard
    d. bacon

19. Swine from about 40 to 100 pounds in weight are
    a. finisher pigs
    b. grower pigs
    c. starter pigs
    d. farrowing pigs

20. SPF pigs are
    a. completely disease-free
    b. obtained by cesarean section
    c. raised in large intermingling groups
    d. part of a nonaccredited system

21. The "S" in SPF stands for
    a. sensitive
    b. species
    c. swine
    d. specific

22. The "T" in TGE stands for
    a. teeth
    b. transmissible
    c. transfer
    d. tusk

23. The "P" in PRRS stands for
    a. pig
    b. piglet
    c. porcine
    d. parvovirus

24. Needle teeth are
    a. retained teeth that help piglets eat solid food
    b. helpful because they help pigs root in the dirt
    c. permanent teeth that allow pigs to defend themselves
    d. clipped in baby pigs to prevent injury to the sow

25. To avoid being bitten by insects, a pig may
    a. stay near its dunging pattern
    b. wallow in the mud
    c. move to a farrowing crate
    d. move to a farrowing pen

Crossword Puzzle

Swine Terms

Supply the correct term in the appropriate space for the definition listed.

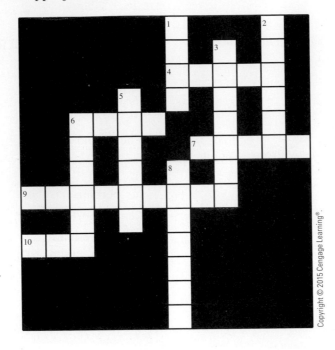

Copyright © 2015 Cengage Learning®.

Across

4 restraint method in which the pig's snout is secured via a loop tie

6 intact male pig

7 common term for prolapsed rectum in swine

9 giving birth to swine

10 female pig

Down

1 overgrown canine tooth of boar

2 eight temporary incisors and canine teeth of young swine

3 restraint method using ropes to place animal in lateral recumbancy

5 wading area to cool swine

6 castrated male pig

8 implanting a wire ring through a pig's nose

Case Studies

Define the underlined terms in each case study.

A group of 4-<u>mo</u>-old <u>barrows</u> was presented with <u>clinical</u> <u>signs</u> of sneezing, <u>purulent</u> <u>nasal</u> discharge, and decreased weight gain. Most of the barrows had a mild to moderate <u>deviation</u> of the <u>snout</u>. The farmer sacrificed one pig for <u>necropsy</u> and found that the nasal <u>turbinates</u> were <u>atrophied</u> and <u>asymmetrical</u>. <u>Dx</u> was atrophic <u>rhinitis</u>, which is a common disease of pigs caused by two types of bacteria. Control measures such as better ventilation and improved hygiene were discussed with the farmer (Figure 21–8).

1. mo _____

2. barrows _____

3. clinical _____

4. signs _____

5. purulent _____

6. nasal

_____

7. deviation

_____

8. snout

_____

9. necropsy

_____

10. turbinates

_____

11. atrophied

_____

12. asymmetrical

_____

13. Dx

_____

14. rhinitis

_____

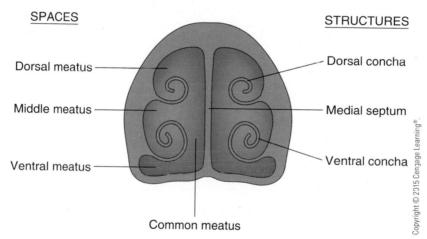

SPACES   STRUCTURES

Dorsal meatus —   — Dorsal concha

Middle meatus —   — Medial septum

Ventral meatus —   — Ventral concha

Common meatus

FIGURE 21–8 Atrophic rhinitis. Atrophic rhinitis causes atrophy of the turbinates and distortion of the nasal septum. The turbinates soften and erode due to infection by bacteria.

A 3-yr-old sow was seen by the area veterinarian for dystocia. She had previously farrowed two litters normally but had developed problems with mastitis and agalactia. Because the pig had uterine inertia, the veterinarian had to perform an emergency C-section to retrieve the remaining pigs.

15. yr

_____

16. sow

_____

17. dystocia

_____

18. farrowed

_____

19. mastitis

_____

20. agalactia

_____

21. uterine inertia

_____

22. C-section

_____

Provide the medical term for the underlined definitions. A litter of 2-wk-old pigs was presented with <u>vomiting</u>, <u>abnormal frequency and liquidity of fecal material</u>, <u>incoordination</u>, and <u>elevated body temperature</u>. A blood sample was taken, and the pigs tested positive for pseudorabies, also called Aujeszky's disease, which is a herpesvirus infection. In adult pigs, pseudorabies can cause respiratory disease and <u>termination of pregnancy</u>. All positive animals were isolated and kept under observation. Pseudorabies control measures, such as adding only serologically negative animals to the herd, avoiding visiting infected premises, keeping <u>wild</u> animals away from swine, and providing separate equipment for each group of animals, were discussed with the owner.

23. vomiting _____

24. abnormal frequency and liquidity of fecal material _____

25. incoordination _____

26. elevated body temperature _____

27. termination of pregnancy _____

28. wild _____

Critical Thinking Exercise

The following story and questions contain words and concepts from this chapter and are designed to stimulate critical thinking through class discussion, Internet searches, or brief writing responses.

Mr. Meyer, a local dairy farmer, called the veterinary clinic to say that he has purchased a Hampshire sow and 7 piglets. He is comfortable raising cattle, but does not really know anything about raising swine. He knows that his cattle are checked by the veterinarian every few months, but asks when he should have the veterinarian look at his pigs. You schedule a veterinary call to the Meyer farm, but he would like advice on how to monitor the animals until the veterinarian arrives.

1. What should you warn Mr. Meyer about regarding piglet and sow safety?

2. What types of management practices may you want to mention to Mr. Meyer so that he is prepared when the veterinarian arrives on the farm?

Chapter 22
BIRDS OF A FEATHER

Objectives

Upon completion of this chapter, the reader should be able to:

- Recognize, define, spell, and pronounce terms related to birds

- Interpret the meaning of abbreviations related to birds

- Analyze case studies and apply medical terminology in a practical setting

BIRDS

Birds are two-legged, egg-laying, warm-blooded animals that have a unique anatomy because of their ability to fly. To allow flight, their respiratory and skeletal systems differ greatly from those of other vertebrates (Figure 22–1). Birds have thin skin that consists of two layers: the epidermis (superficial layer) and dermis (deep layer). From the dermis emerge feathers that may be a variety of different types. Birds have beaks that are varied in shape to accommodate their various dietary preferences. Birds that are seed eaters have short, strong bills that can crack open a hard-shelled seed easily, while birds that eat flying insects have flat bills with a wide base so that they can catch a small moving target. Raptors such as hawks and eagles have hooked bills that are sharp, making it easy to rip apart their prey. The skeletal system of birds is modified to enable the bird to fly. These modifications include fusion of bones (provides a rigid structure) and the development of wing bones that are hollow, lack bone marrow, and contain air (reduce weight to enhance flight). Birds have a right jugular vein, while most birds have a very small left jugular vein or no vein at all. This is especially important to know when collecting a blood sample from the jugular vein of a bird.

Birds and the terms associated with them vary depending on the types of birds people keep. Some birds, such as psittacines, are popular as pets. Other birds, such as poultry and ratites, are used as livestock. The terms related to birds typically refer to one or the other of these uses for birds.

Some anatomy and physiology concepts and medical terms related to avian species have been covered in previous chapters. Additional anatomy and physiology concepts are covered in this chapter.

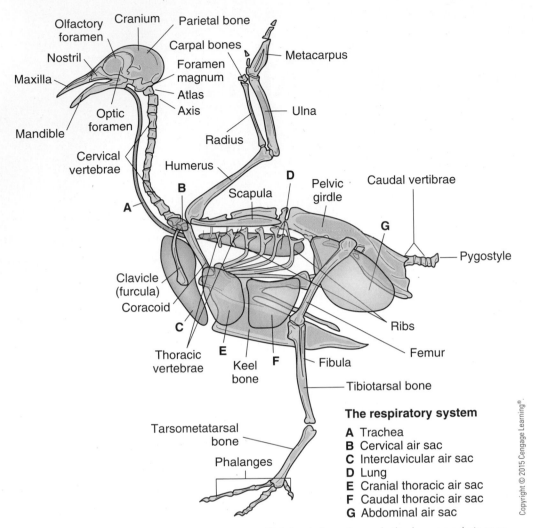

The respiratory system

A Trachea
B Cervical air sac
C Interclavicular air sac
D Lung
E Cranial thoracic air sac
F Caudal thoracic air sac
G Abdominal air sac

FIGURE 22–1 Skeletal and respiratory system of birds; air flow through the lungs and air sacs. (1) First inspiration: Air flows into the trachea and through the primary bronchi and goes to the caudal and abdominal air sacs. Air already in the caudal air sacs moves to the cranial air sacs. (2) First expiration: Air travels back to the parabronchi and gas exchange occurs. (3) Second inspiration: Air moves from the parabronchi to the cranial air sacs. (4) Second expiration: Air moves out of the cranial air sacs, into the parabronchi, and out of the trachea. Birds do not have a diaphragm to assist in respiration.

ANATOMY AND PHYSIOLOGY TERMS
Respiratory System

- **air sacs** = thin-walled sacs in the respiratory tract of birds that store air and provide buoyancy for flight.
- **choana** (kō-ā-nah) (or **choanal space**) = caudal naris; the cleft in the hard palate of birds (Figure 22–2).
- **nasal gland** = gland that allows sea birds to drink saltwater; found in the rostral portion of the beak.
- **parabronchi** (pahr-ah-brohnck-ī) = tiny passages in bird lungs that are the primary sites of gas exchange between air and blood; birds do not have alveoli.
- **syrinx** (sehr-ihncks) = voice organ of birds located at the tracheal bifurcation (where the trachea splits into bronchi).

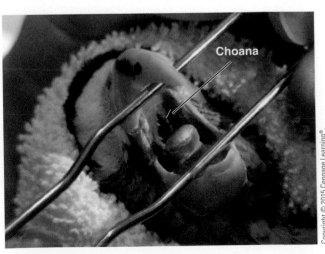

FIGURE 22–2 Choana of a parrot.

Integumentary System

- **apterium** (ahp-tehr-ē-uhm) = area or tract of skin without feathers or down; plural is **apteria** (ahp-tehr-ē-ah).
- **barb** (bahrb) = one of the parallel filaments projecting from the main feather shaft (rachis); forms the feather vane.
- **barbule** (bahr-byool) = one of the small projections fringing the edges of the barbs of feathers; attach to adjacent barbules to give the vane rigidity.
- **beak** (bēk) = hard mouth part of bird that is a modified epidermal structure that covers the rostral part of the maxilla and mandible.
- **calamus** (kahl-ah-muhs) = hollow shaft at the proximal end of the feather shaft; also called the **quill**.
- **central tail feathers** = primary feathers at the center of the tail.
- **cere** (sēr) = thickened skin at base of *external nares* of birds; may be different colors in some birds to denote sex (Figure 22–3).

(a)

(b)

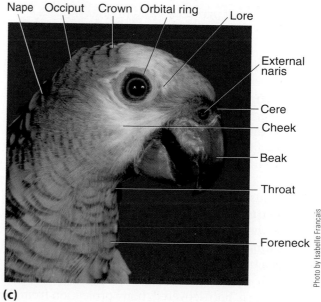

(c)

Copyright © 2015 Cengage Learning®.

Photo by Isabelle Francais

FIGURE 22–3 (a) External parts of a bird. (b) External parts of a bird on a blue and gold macaw. (c) External parts of a bird's head on a blue-fronted Amazon parrot.

- **cheek** (chēk) = area of the face below the eye of a bird.
- **contour feathers** (kohn-tər fehth-ərz) = body or flight feathers arranged in rows.
- **coverts** (kuh-vərhtz or kō-vehrtz) = small feathers that cover the bases of large feathers on the body; named based on their location.
- **feather** (fehth-ər) = epidermal structure analogous to hair; used for insulation and thermoregulation, in courtship displays, and for flight (Figure 22–4).
- **filoplume** (fihl-ō-ploom) = type of small, hairlike feather with barbs along the shaft that resembles a bristle feather topped by a down feather; filoplumes help guide the bird in flight and reposition the feathers.
- **lateral tail feathers** = large tail feathers to each side of the central tail feathers.
- **mantle** (mahn-tuhl) = feathers across the dorsum (back) (Figure 22–3).
- **molt** (mōlt) = process of casting off feathers before replacement feathers appear.
- **orbital ring** (ohr-bih-tahl rihng) = ring of unfeathered skin around the eye (Figure 22–3).
- **pin feathers** = developing feathers that have blood flowing through them and can grow as new feathers or to replace molted feathers; also called **blood feathers**.
- **plume** (ploom) = feathers that are close to the body and provide warmth; also known as **down feathers**.
- **primary feathers** = longer, thinner remiges connected to the carpometacarpus and phalanges that project along the outer edge of a bird's wing; also called **primaries** and **primary quills**.
- **pteryla** (tehr-ih-lah) = feather tract of birds; plural is **pterylae** (tehr-ih-lā).
- **rachis** (rā-kuhs) = distal end of the feather shaft.
- **rectrices** (rehck-trih-sehz) = tail flight feathers.
- **remiges** (rehm-ih-jehz) = primary wing feathers; singular is **remix**.
- **secondary feathers** = smaller, lighter remiges than the primary feathers that are connected to the ulna; also called **secondaries**.
- **setae** (sē-tā) = sensitive bristles that grow on the heads of many birds; also called **bristles**.
- **shaft** (shahft) = quill or central part of a contour feather; also called **scapus** (skā-puhs).
- **snood** (snood) = long, fleshy extension at the base of a turkey's beak.
- **spur** (spər) = projecting body (as from a bone) or a sharp, horn-covered, bony projection from the shank of some species of male birds.

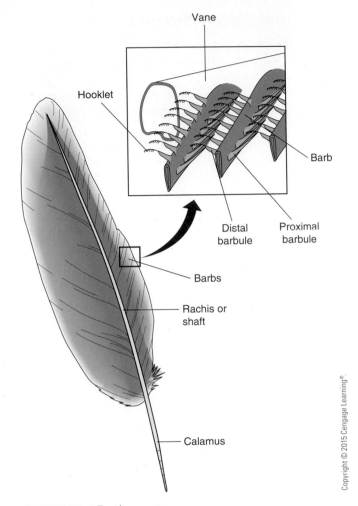

FIGURE 22–4 Feather parts.

- **tertiary feathers** = shorter remiges that are connected to the humerus in some bird species; also called **tertiaries** and **tertials**.
- **uropygial** (yoor-ō-pihj-ē-ahl) **gland** = gland located laterally to the tail feather attachment that secretes oil used to waterproof or preen feathers; also called the **preen** (prēn) **gland**. **Preening** is the process by which birds groom and care for their feathers. Birds preen by picking through their feathers (removing any debris and arranging out-of-place feathers) and distributing the oil that is secreted from the uropygial gland.
- **wattle** (waht-tuhl) = appendage suspended from the head (usually the chin) in chickens, turkeys, and goats.

External Anatomy Terms

- **breast** (brehst) = chest area of birds.
- **crown** (krown) = area caudal to the forehead to the cranial portion of the neck; the top of the head.

- **forehead** (fōr-hehd) = part of the head rostral to the eyes.
- **foreneck** (fōr-nehck) = area cranial to the breast where the clavicle (wishbone) is located.
- **lore** (lōr) = lateral area between the cranial portion of the eyes and bill of birds.
- **nape** (nāp) = caudodorsal portion of the neck of birds.
- **occiput** (ohck-sihp-uht) = sloping caudal part of the head of birds.
- **rump** (ruhmp) = area that overlies the pelvis cranial to the tail.
- **throat** (thrōt) = area caudal to the head and cranial to the chest.

Gastrointestinal System

- **cloaca** (klō-ā-kah) = common passage for fecal, urinary, and reproductive systems in birds and lower vertebrates (Figure 22–5).
- **coprodeum** (kōp-rō-dē-uhm) = rectal opening into the cloaca (Figure 22–6).
- **crop** (krohp) = esophageal enlargement that stores, moistens, and softens food in some birds (Figure 22–7).
- **droppings** = composite of feces and urine in birds.

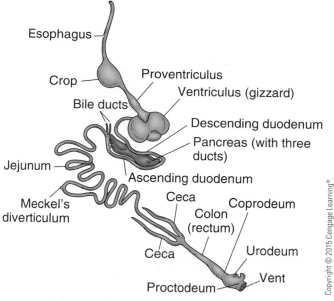

FIGURE 22–6 Gastrointestinal tract of the bird.

FIGURE 22–7 A young, parent-fed cockatiel chick with a full crop.

FIGURE 22–5 Cloacal swab.

- **Meckel's diverticulum** (mehck-ehlz dī-vər-tihck-yoo-luhm) = structure at the terminal end of the jejunum that functions as a lymphatic organ.
- **proventriculus** (prō-vehn-trihck-yoo-luhs) = elongated, spindle-shaped glandular stomach of birds.
- **vent** (vehnt) = external opening of the cloaca of birds.
- **ventriculus** (vehn-trihck-yoo-luhs) = muscular stomach of birds; also called the **gizzard**.

Musculoskeletal System

- **columella** (kohl-uhm-eh-lah) = bony structure that replaces the malleus, incus, and stapes in the middle ear.
- **furcula** (fuhr-kuh-lah) = fused clavicle in birds; the **wishbone**.

FIGURE 22–8 Keel bone of a bird.

- **keel** (kēl) = sternum, or breastbone, of birds (excluding ratites) (Figure 22–8).
- **pygostyle** (pihg-ō-stīl) = bony termination of the vertebral column in birds where tail feathers attach; also called **rump post**.
- **scleral** (skleh-rahl) **ring** = overlapping bony plate encircling the eye at the corneal–scleral junction.

Urogenital System

See Figure 22–9.

- **infundibulum** (ihn-fuhn-dihb-yoo-luhm) = portion of the oviduct located closest to the ovary that captures ovulated eggs.
- **isthmus** (ihs-muhs) = portion of the oviduct located farthest from the ovary that adds the shell membranes.
- **magnum** (mahg-nuhm) = middle portion of the oviduct that separates the albumin and chalaza over the egg yolk and sperm.
- **sperm nests** = clusters of spermatozoa held in readiness in the infundibulum to fertilize the egg as it comes from the ovary.
- **urodeum** (yoo-rō-dē-uhm) = area of the cloaca in which the ureters and vagina open.
- **uterus** (yoo-tər-uhs) = portion of the female reproductive tract in birds that produces the shell and shell pigments; also the area in which the egg turns around so that it is laid blunt end first.
- **vagina** (vah-jī-nah) = portion of the female reproductive tract in birds that directs the egg to the cloaca.

POULTRY TERMS

- **average daily gain** = amount of weight an animal gains over time; calculated by taking a weight at the beginning and end of a period, and then dividing the difference by the number of days in the period.

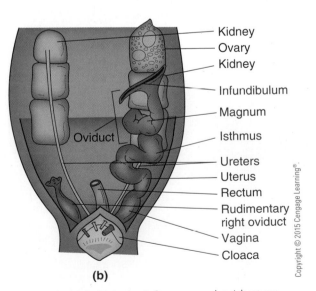

FIGURE 22–9 Urogenital system of the bird. (a) Male bird; (b) Female bird. Only the left ovary and oviduct are functional in female birds.

- **battery cage** = system of housing poultry in rows of cages in an environmentally controlled building (cage size and number of birds per cage is regulated by various federal organizations).
- **bleaching** = loss of pigmentation from the legs and beaks of laying hens that occurs naturally as hens move through the production cycle.
- **breed** (bred) = set of birds with similar characteristics.
- **broiler** (broy-lər) = young chicken approximately 8 weeks old weighing 1.5 kg or more; also called **fryer** or **young chicken**.
- **brood** (brūd) = young that are hatched or cared for at the same time.
- **brooder** (brū-dər) = housing unit for rearing birds after hatching; includes a heating device to keep young chicks warm.
- **brooding** (brūd-ihg) = mothering behavior in hens when they are choosing to sit on eggs and incubate them.
- **cage operation** = method of raising chickens in which the hens are kept in confinement as they produce eggs.
- **candling** (kahn-dlihng) = process of shining a light through an egg to check embryo development (Figure 22–10).
- **cannibalism** (kahn-ih-bahl-ihz-uhm) = practice of one bird attacking another and causing injury or death. Cannibalism may be one bird against another bird, or groups of birds attacking an individual.
- **chalaza** (kahl-ā-zah) = ropelike structure that holds the yolk to the center of the egg (Figure 22–11).
- **class** = the region of a bird's origin.

FIGURE 22–10 Candling is a process by which eggs pass over an intense light to reveal cracks in the shell or to check on embryo development.

- **comb** (kōm) = in domestic fowl, the vascular, red cutaneous structure attached in a sagittal plane to the dorsum of the skull (Figure 22–12).
- **confinement** (kohn-fīn-mehnt) = practice of raising birds in a restricted, climate-controlled environment.
- **contract growers** = people hired by a company to raise a certain number of birds or produce a certain number of eggs for a contracted price.
- **coop** (koop) = enclosure to keep birds.
- **debeaking** (dē-bēk-ihng) = removing about one-half of the upper beak and a small portion of the lower beak in poultry to prevent feather picking, cannibalism, and fighting; also called **beak trimming** in poultry (Figure 22–13).
- **dressed bird** = bird that has had its feathers removed and is cleaned and ready for food preparation.

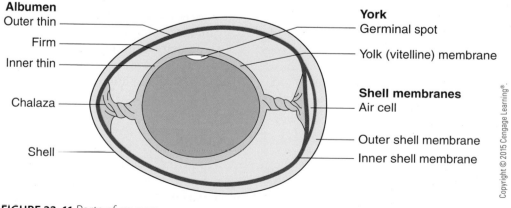

Albumen
Outer thin
Firm
Inner thin
Chalaza
Shell

York
Germinal spot
Yolk (vitelline) membrane

Shell membranes
Air cell
Outer shell membrane
Inner shell membrane

FIGURE 22–11 Parts of an egg.

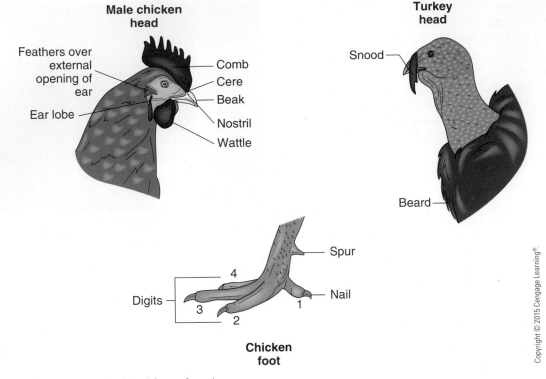

Male chicken head

Feathers over external opening of ear

Ear lobe

Comb
Cere
Beak
Nostril
Wattle

Turkey head

Snood

Beard

Spur

Nail

Digits

4
3
2
1

Chicken foot

FIGURE 22–12 Head and feet of poultry.

FIGURE 22–13 Debeaking a chicken.

- **feed efficiency** = calculation of how much feed an animal is consuming to gain a pound of weight; calculated by dividing the weight gained over a period of time by the weight of the feed consumed over that period of time.
- **force-feeding** = practice of administering food to ducks and geese through a tube to encourage development of the liver for the product foie gras (fwah-grah).
- **free-range** = practice of raising birds in enclosed yards with shelters.

- **hatch** (hahtch) = emergence of a baby bird from the shell.
- **hatchability** (hahtch-ah-bihl-ih-tē) = percentage of fertilized eggs that hatch.
- **hatchery** (hahtch-eh-rē) = business that sells fertilized eggs ready to hatch, or hatches the eggs and sells the chicks.
- **hen-day production** = daily rate of egg production reported as a percentage.
- **hierarchy** (hī-ər-ahr-kē) = social ranking from most dominant to least dominant that can be linear or nonlinear; also called **pecking order**.
- **incubation** (ihn-kyoo-bā-shuhn) = process of a fertilized poultry egg developing into a newly hatched bird.
- **incubator** (ihn-kyoo-bā-tohr) = mechanical device that is maintained at the optimal temperature and humidity for the development of embryos and hatching birds.
- **layer** (lā-ər) = chicken raised for egg production.
- **litter** = combination of avian waste and bedding; the bedding before use is known as "**fresh litter**."
- **poultry** (pōl-trē) = any domesticated fowl raised for meat, eggs, or feathers.
- **variety** = a subgroup of birds within a breed, usually based on color or markings.
- **yolk** (yōk) = yellow part of the egg that contains the germinal cells (Figure 22–11).

Feather Patterns and Comb Types

The American Poultry Association has divided modern chicken breeds into different classes, breeds, and varieties, typically based on color or markings. Feather pattern and comb type are two features that may identify varieties within a breed.

Feather Patterns

- **barred** (bahrd) = feather with alternating strips of different colors across the width (Figure 22–14).
- **laced** (lāsd) = feathers with a narrow border of a contrasting color.
- **mottled** (moht-uhld) = some feathers are tipped in white.
- **penciled** (pehn-sihld) = feathers with narrow uniform lines of contrasting color (may be single or multiple).
- **spangled** (spahng-gehld) = feathers with a contrasting color at the tip that may be black or white.

Comb Types

- **buttercup** (buht-tər-kuhp) = cup-shaped comb centered on the head, with evenly spaced points all around (Figure 22–15).

FIGURE 22–14 The barred feather pattern on chickens.

- **cushion** = comb is small and close to the head with no spike, bumps, or other irregularities.
- **pea** = comb is fairly close to the head with three ridges along its length.
- **rose** = wide, fleshy comb that is low to the head with a flat top and ends in a spike near the back of the head.
- **silkie** = round, lumpy comb that may be wider than it is long (most often seen on birds with a crest).
- **single** = single, blade-shaped comb centered on the top of the head, with five or six clearly defined points.
- **strawberry** = egg-shaped comb set close to the head.
- **V-shaped** = comb with two hornlike sections that join in the center of the head and form a definite V shape.

Buttercup   Cushion   Pea   Rose

Silkie   Single   Strawberry   V-Shaped

FIGURE 22–15 Comb patterns in chickens.

PET BIRD TERMS

- **aviary** (āv-ē-ahr-ē) = outdoor enclosure for birds that is large enough for birds to fly freely.
- **aviculture** (ā-vih-kuhl-chər) = the keeping of companion birds.
- **beak trimming** = trimming the tip of the beak to keep the beak properly aligned (Figure 22–16).
- **cage** = portable enclosure for birds. A properly sized cage should allow birds to fully extend their wings.
- **clipping** (klihp-ihng) = trimming wings of birds to alter their flight; also called **pinioning** (pihn-yehn-ihng) (Figures 22–17 and 22–18).

Courtesy of Martin G. Orr, DVM

FIGURE 22–16 A dremel tool may be used to trim a bird's beak.

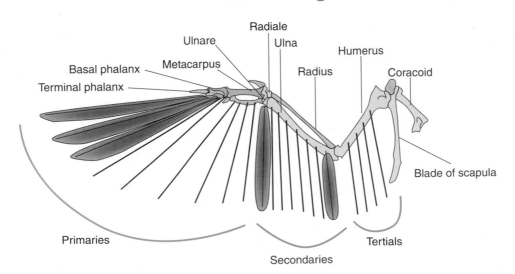

FIGURE 22–17 Parts and types of feathers of a wing.

Copyright © 2015 Cengage Learning®.

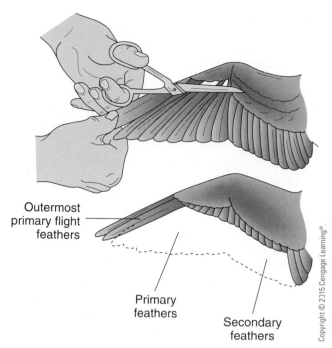

Outermost primary flight feathers

Primary feathers

Secondary feathers

FIGURE 22–18 Wing clipping in a pet bird.

FIGURE 22–20 Feather picking is a common behavioral problem in pet birds.

- **columbiformes** (kō-luhm-bih-fōrmz) = group of dovelike birds that includes doves and partridges that have short beaks and two forward-facing toes and two rear-facing toes.
- **cuttlebone** (kuht-uhl bōn) = shell (endoskeleton) of a cuttlefish that is typically provided in the cage for the bird to use in wearing down its beak (Figure 22–19).
- **feather picking** = undesired behavior in which birds remove their own feathers mainly due to stress or disease; also called **feather plucking** (Figure 22–20).
- **fledgling** (flehdj-lihng) = young bird that has recently acquired its flight feathers and typically is out of the nest but not eating on its own (Figure 22–21).

FIGURE 22–21 Fledgling budgerigar.

- **hand-raised** = refers to a bird that has been raised by humans (Figure 22–22).
- **passeriformes** (pahs-ər-ih-fōrmz) = group of perching birds that includes most songbirds (such as finches, sparrows, mynahs, and canaries) that have three forward-facing toes and one rear-facing toe that are adapted for perching (see Figure 22–23).
- **perch** (pərch) = stick or dowel provided for the bird to sit on; some perches have rough surfaces that assist the bird in wearing down its toenails (Figures 22–23).
- **psittacine** (siht-ah-sēn) = group of parrotlike birds that includes parrots, macaws, budgerigars, cockatiels, cockatoos, conures, lovebirds, and

FIGURE 22–19 Cuttlebones are given to small birds as a source of calcium and trace amounts of iodine. A cuttlebone is the endoskeleton of this saltwater cuttlefish.

FIGURE 22–22 This young conure is being hand raised. Hand feeding is one way to ensure that hand-raised birds have enough nutrition.

parakeets that have a strong curved beak and two forward-facing toes and two rear-facing toes that are adapted for grasping (Figure 22–23).

- **self-mutilation** = any behavior in which a bird causes injury to itself.

RATITE TERMS

- **breeding animals** = animals purchased, often as pairs, to produce offspring.
- **emu** (ē-mū) = large, fast flightless bird native to Australia that has been imported to zoos and used in animal agriculture; males are slightly smaller than females and have gray, black, and brown feathers (Figure 22–24).
- **emu oil** = oil produced from emu fat that is used in pharmaceuticals and cosmetics.
- **meat** = ratite muscle that is a red meat yet is lower in fat and cholesterol than beef.
- **ostrich** (ow-strihch) = large, flightless bird native to Africa that is the largest ratite (can weigh over 400 pounds) and may be black, blue, or red (Figure 22–25).
- **ostrich feathers** = feathers of ostriches that are used commercially for feather dusters and decoration.

(a)

(b)

Parrot/2.5 cm (1 in) square   Finch/1.25 cm (0.5 in) round   Budgie/1.25 cm (0.5 in) oval

(c)

FIGURE 22–23 (a) African gray parrots on a wooden perch. (b) Gouldian finch (red-headed) on a wooden perch. (c) Ideal perch size and shape for a variety of birds.

FIGURE 22–24 Emu.

FIGURE 22–26 Rhea.

FIGURE 22–25 Ostrich.

- **ostrich skin** = skin from ostriches that is tanned into leather and used to make boots and other leather products.
- **ratite** (rah-tīt) = class of large, flightless birds that are raised for their meat and hides.

- **rhea** (rē-ah) = large, flightless bird native to South America that may be pale gray to brown in color and lacks tail feathers; females lay their eggs in the same nest, and the male incubates the eggs (Figure 22–26).
- **rhea skin** = skin of rhea that is tanned into leather and typically used to make boots.

ABBREVIATIONS FOR AVIANS

- ACPV = American College of Poultry Veterinarians

REVIEW EXERCISES

Multiple Choice

Choose the correct answer.

1. What is the term for the spaces in the bird respiratory tract that store air and provide buoyancy for flight?
   a. nasal sacs
   b. air sacs
   c. apteria
   d. ceres

2. The caudal nares (also called the cleft in the hard palate of birds) are called the
   a. choana
   b. cloaca
   c. calamus
   d. crop

3. The sternum, or breastbone, of some birds is the
   a. pygostyle
   b. crop
   c. spur
   d. keel

4. The esophageal enlargement that stores, moistens, and softens food in some birds is the
   a. coprodeum
   b. cloaca
   c. crop
   d. choana

5. A young bird that has recently acquired its flight feathers is known as a
   a. clutch
   b. fledgling
   c. brooder
   d. snood

6. The glandular stomach of birds is the
   a. vent
   b. Meckel's diverticulum
   c. ventriculus
   d. proventriculus

7. A body or flight feather is known as a
   a. down feather
   b. contour feather
   c. lume
   d. filoplume feather

8. The voice organ of birds is the
   a. syrinx
   b. cere
   c. choana
   d. setae

9. The primary wing feathers are known as
   a. rectrices
   b. rachis
   c. pteryla
   d. remiges

10. The process of casting off feathers before replacement feathers appear is known as
    a. pluming
    b. molting
    c. shafting
    d. feathering

11. Large, flightless birds used for their meat and hides are
    a. columbiformes
    b. passeriformes
    c. ratites
    d. psittacines

12. Finches are examples of which type of bird?
    a. columbiformes
    b. passeriformes
    c. ratites
    d. psittacines

13. Cockatiels and cockatoos are examples of which type of bird?
    a. columbiformes
    b. passeriformes
    c. ratites
    d. psittacines

14. Doves are examples of which type of bird?
    a. columbiformes
    b. passeriformes
    c. ratites
    d. psittacines

15. A stick or dowel on which a caged bird sits is called a
    a. scleral ring
    b. barbule
    c. barb
    d. perch

16. Which type of feather has a blood supply running through it that if broken can lead to heavy bleeding?
    a. covert
    b. plume
    c. pin
    d. bristle

17. The common passage for fecal, urinary, and reproductive systems in birds is the
    a. crop
    b. cloaca
    c. choana
    d. coprodeum

18. The ropelike structure that holds the yolk to the center of the egg is the
    a. chalaza
    b. yolk tendon
    c. isthmus
    d. snood

19. The wishbone in birds is the
    a. keel bone
    b. pygostyle
    c. carpus
    d. furcula

20. The external opening of the cloaca in birds is the
    a. ventriculus
    b. vent
    c. diverticulum
    d. infundibulum

21. Which of the group of birds listed would have the characteristic toe arrangement pictured in Figure 22–27?
    a. passerines
    b. psittacines
    c. ratites
    d. poultry

22. How is avian respiration different from that of mammals?
    a. Birds have a syrinx where oxygen and carbon dioxide exchange.
    b. Birds have a cloaca, a cleft in the hard palate of birds, where air enters the lungs.
    c. Birds do not have a diaphragm to assist in respiration.
    d. Birds do not have air sacs, which are found in mammals.

23. Developing feathers that have blood flowing through them and can grow as new feathers or to replace molted feathers are
    a. rachis
    b. primary feathers
    c. pin feathers
    d. contour feathers

24. Chickens raised for egg production are called
    a. layers
    b. hatchers
    c. brooders
    d. broilers

25. In birds, the feces and urine collectively is called
    a. coprodeum
    b. vent
    c. droppings
    d. rump

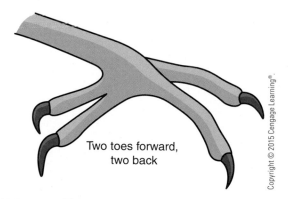

Two toes forward, two back

FIGURE 22–27 Toe arrangement.

Crossword Puzzle

Avian Terms

Supply the correct term in the appropriate space for the definition listed.

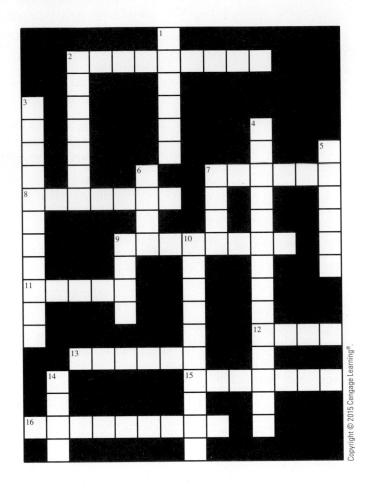

Across

2 preen gland

7 common passage for fecal, urinary, and reproductive systems in birds and lower vertebrates

8 primary wing feathers

9 trimming wings of birds to alter flight

11 chicken raised for egg production

12 red cutaneous structure attached in a sagittal plane to the dorsum of the skull in domestic fowl

13 caudal naris of birds

15 ropelike structure that holds the yolk to the center of the egg

16 young bird that has recently acquired its flight feathers

Down

1 voice organ of birds

2 area of the cloaca in which the ureters and vagina open

3 muscular stomach of birds (gizzard)

4 elongated, glandular stomach of birds

5 distal end of the feather shaft

6 sternum of birds

7 esophageal enlargement in some birds

9 thickened skin at base of external nares of birds

10 group of parrot-like birds

14 process of casting off feathers

Case Studies

Define the underlined terms in each case study.

A <u>neonatal</u> greater sulfur-crested cockatoo was presented with a history of being tube-fed with a red rubber tube. The tube became dislodged from the feeding syringe and was swallowed by the chick. A <u>radiograph</u> was taken that confirmed that the tube was swallowed and was located extending from the <u>crop</u> to the <u>proventriculus</u>. An <u>ingluviotomy</u> was considered. Because it would be a less traumatic procedure, it was decided to retrieve the feeding tube with an <u>endoscope</u>. With the cockatoo under general <u>anesthesia</u> with isoflurane, the tube was successfully removed. The bird recovered uneventfully, and the owner was counseled on hand-feeding with a spoon instead of a feeding tube.

1. neonatal _____

2. radiograph _____

3. crop _____

4. proventriculus _____

5. ingluviotomy (also known as cropotomy) _____

6. endoscope _____

7. anesthesia _____

A 3-yr-old male budgerigar (parakeet) was presented for a growth on its <u>thorax</u> near the <u>thoracic inlet</u>. The bird had been kept singly and was fed a predominantly seed diet. On physical examination, the bird was noted to have a 2.5 <u>cm</u> diameter mass located <u>subcutaneously</u> on the <u>cranial</u> portion of the thorax. Differential <u>diagnoses</u> included <u>neoplasia</u> resulting in <u>lipoma</u> or a <u>xanthoma</u>. A <u>fine needle aspirate (FNA)</u> was performed, which confirmed the mass to be a lipoma. The owner declined surgical removal. The owner did elect to administer Lugol's iodine in the drinking water because this condition may be related to <u>hypothyroidism</u> (caused by iodine deficiency relating to an all-seed diet).

8. thorax _____

9. thoracic inlet _____

10. cm _____

11. subcutaneously _____

12. cranial _____

13. diagnoses _____

14. neoplasia _____

15. lipoma _____

16. xanthoma _____

17. fine needle aspirate (FNA) _____

18. hypothyroidism _____

A commercial <u>flock</u> of chickens developed disease in <u>cockerels</u> and <u>pullets</u> including birds that were pale, <u>anorexic</u>, <u>emaciated</u>, dehydrated, and had bloody <u>droppings</u>. Chicks were <u>asymptomatic</u>. The <u>layers</u> had a reduction in the rate of egg production. A diagnosis of coccidiosis was made from identification of coccidia protozoa by fecal examination. The flock was treated with a coccidiostat added to the feed, and facility management issues such as instituting <u>brooder</u> sanitation and switching to a <u>cage operation</u> were discussed with the owners.

19. flock _____

20. cockerels _____

21. pullets _____

22. anorexic _____

23. emaciated _____

24. droppings _____

25. asymptomatic _____

26. layers _____

27. brooder _____

28. cage operation _____

An <u>ostrich</u> <u>flock</u> had <u>chicks</u> developing neurologic signs including limb <u>paralysis</u>, <u>torticollis</u>, <u>ataxia</u>, and <u>opisthotonus</u>. In five days after initial signs appeared, many of the chicks had died. The outbreak occurred during the summer, producing a high <u>mortality</u> rate. <u>Histological</u> examination of tissues was inconclusive; however, equine encephalitis virus and Newcastle disease virus infection were suspected. <u>Serology</u> was performed on these birds, and the results are pending.

29. ostrich _____

30. flock _____

31. chicks _____

32. paralysis _____

33. torticollis _____

34. ataxia _____

35. opisthotonus _____

36. mortality _____

37. histological _____

38. serology _____

Critical Thinking Exercise

The following story and questions contain words and concepts from this chapter and are designed to stimulate critical thinking through class discussion, Internet searches, or brief writing responses.

You are working in a veterinary practice that sees many birds, and sometimes clients know a lot about birds and other times they do not. As a way to prevent avian husbandry errors, you decide to make a client brochure on bird care and common mistakes of caring for birds. The following is a list of questions that clients have asked over the years. You decide this is a good place to start when making your brochure.

1. Based on what you have observed about how birds fly, what type of cage should an owner purchase based on the size of his/her bird?

2. What type of materials should be avoided when choosing a bird cage?

3. What are some things that should be placed in the bird cage? How should things be arranged?

4. Where should the bird cage be placed? Where should it not be placed?

5. What household items can be dangerous for pet birds?

6. What should an owner feed their pet bird?

7. What routine care does a bird need?

Chapter 23
ALL THE REST

Objectives

Upon completion of this chapter, the reader should be able to:

- Recognize, define, spell, and pronounce terms related to laboratory animals

- Recognize, define, spell, and pronounce terms related to pocket pets

- Recognize, define, spell, and pronounce terms related to reptiles and amphibians

- Interpret the meaning of abbreviations related to laboratory animals, pocket pets, reptiles, and amphibians

- Analyze case studies and apply medical terminology in a practical setting

LABORATORY ANIMALS, POCKET PETS, REPTILES, AND AMPHIBIANS

Laboratory animals include a wide range of species, even species some people consider to be pets and livestock. Pocket pets are small, nontraditional pets that were once thought of solely as laboratory animals or being exotic to many areas of the country.

Reptiles and amphibians are cold-blooded verte-brates that are gaining popularity as nontraditional pets. Many terms used in the laboratory animal field have already been covered in other chapters. New terms pertaining to laboratory animals are related to scientific studies and the facilities in which they occur and are used in the care of pocket pets (Table 23–1).

TABLE 23-1 Common Laboratory and Pocket Pet Terms

Term	Pronunciation	Definition
acclimatization	ahck-lih-mah-tih-zā-shuhn	adjustment of an animal to a new environment
agouti	ah-goo-tē	naturally occurring coat color pattern that consists of dark-colored hair bands at the base of the hair and lighter increments of hair color toward the tip (Figure 23–1)
albino	ahl-bī-nō	animal with a white coat and pink eyes; devoid of melanin (Figure 23–1)
altricial	ahl-trih-shihl	young born or hatched very immature or helpless
ambient	ahm-bē-ahnt	surrounding
analogous structures	ahn-ahl-oh-guhs	refers to structures that differ anatomically but have similar functions (e.g., the jointed legs of an insect and a dog)
anogenital distance	ā-nō-jehn-ih-tahl	area between the anus and genitalia; females have a shorter anogenital distance than males, which is used to determine the sex of animals (Figure 23–2)
antivivisectionist	ahnt-ih-vihv-ih-sehck-shuhn-ihst	person who opposes surgery on live animals for research or educational purposes
arboreal	ahr-bōr-ē-ahl	tree-living
autotomy	aw-toh-tō-mē	voluntary release (of the tail) in reptiles
autosome	aw-tō-zōm	non–sex-determining chromosome
axenic	ā-zehn-ihck	germ-free
barbering	bahr-bər-ihng	behavioral disorder in which dominant animals bite or chew the fur of subordinate animals (Figure 23–3)
barrier sustained	bār-ē-ər suh-stānd	gnotobiotic animals that are maintained under sterile conditions in a barrier unit
biohazard	bī-ō-hahz-ahrd	substance that is dangerous to life
bolt hole		an entrance or exit in a rabbit warren
Bruce effect		abortion of embryos in recently bred mice in order to re-breed with the new male
calvarium	kahl-vahr-ē-uhm	top of the skull
cannibalism	kahn-ih-bahl-ihz-uhm	devouring one's own species
cesarean derived	sē-sā-rē-ahn	animal is delivered via cesarean section into a sterile environment to avoid possible contamination
cheek pouch		space in oral cavity of hamsters that carries food and bedding
chromodachryorrhea	krō-mō-dahck-rē-ō-rē-ah	shedding of colored (blood-colored) tears due to porphyrin release from the **Harderian gland** located behind the eyes of many rodents; may indicate stress or illness
contact bedding		substrate with which animal comes into direct contact; also called **direct bedding**; in general, cedar should not be used as contact bedding because the aromatic fumes are toxic and can lead to respiratory problems or death
control		standard normal against which experimental results are compared; also called **experimental control**
crepuscular	krē-puhs-kuh-lahr	becoming active at twilight or before sunrise

TABLE 23–1 Common Laboratory and Pocket Pet Terms (*Continued*)

Term	Pronunciation	Definition
data	dah-tah	mass of accumulated information or results of an experiment
dusting	duhs-tihng	cleaning method in which chinchillas roll in dust
dwarf	dwohrf	an animal much smaller than the average of its species
ectotherm	ehck-tō-thərm	animals whose internal temperature varies with environmental temperature; commonly called **cold-blooded**; ectotherms obtain heat from the environment
emission	ē-mihsh-uhn	discharge
endotherm	ehn-dō-thərm	animal that generates its own heat to maintain a constant internal body temperature, typically above the temperature of its surroundings; also known as **warm-blooded**
enrichment	ehn-rihch-mehnt	items added to an environment to provide an animal with a visually appealing environment, as well as areas to explore, and opportunity for exercise (see Figure 23–22)
estivate	ehs-tih-vāt	to reduce body temperature, heart and respiration rates, and metabolism to dormancy in summer
exsanguinations	ehcks-sahn-gwih-nā-shuhn	removal of blood or blood loss from the body
fixed formula		diets that are constant; the ingredients and percentages do not change with market availability and price
fomite	fō-mīt	inanimate carrier of disease such as bedding or thermometers
free choice		unhindered access to food at all times
fur-slip	fər-slihp	shedding of hair patches from rough handling of chinchillas
fuzzy	fuhz-ē	mouse or rat pup with a slight coat of hair (Figure 23–4)
genome	jē-nōm	one complete copy of the genetic information in a cell
genotype	jē-nō-tīp	genetic makeup of an individual for a particular trait
gnotobiotic	nōt-ō-bī-ah-tihck	said of germ-free animals that have been introduced to one or two known nonpathogenic mircoorganisms
heliotherm	hē-lē-ō-thərm	animals (like turtles) that actively seek out sunlight for heat
herpetology	hər-peh-toh-lō-jē	study of reptiles and amphibians
heterozygous	heht-ər-ō-zī-guhs	having two different genes for a given genetic trait; usually one gene is dominant over the other; represented with a capital and lowercase letter
hibernate	hī-bər-nāt	to reduce body temperature, heart, and respiration rates and metabolism to dormancy in winter
homeotherm	hō-mē-ō-thərm	animals that maintain a more or less constant body temperature at all times
homologous structures	hō-mohl-ō-guhs	having a common origin but different functions in different species (e.g., the forelimbs of birds and horses)
homozygous	hō-mō-zī-guhs	having two identical genes for a given genetic trait; represented with two capital letters or two lowercase letters
hooded	huhd-ehd	refers to a rat having a white coat with a black "hood" over the head and shoulders and pigmented eyes (Figure 23–5)

(continued)

TABLE 23–1 Common Laboratory and Pocket Pet Terms (*Continued*)

Term	Pronunciation	Definition
hoppers		mice and rats that have a full coat of hair and are fully mobile but are smaller than adult mice
horizontal transmission	hōr-ih-zohn-tahl trahnz-mihs-shuhn	disease transfer from one animal to the other
hutch	huhtch	pen or enclosed coop for small animals or calves
hygrometer	hī-grōm-ih-tər	a humidity gauge
hybrid	hī-brihd	strain resulting from mating two inbred strains
hypothesis	hī-pohth-eh-sihs	statement of research supposition
hystricomorph	hihs-trihck-ō-mōrf	type of rodent that includes guinea pigs, chinchillas, and porcupines
inbred	ihn-brehd	resulting from at least 20 brother–sister or parent–offspring matings
in situ	ihn sih-too	at the normal site
in vitro	ihn vē-trō	outside living organisms; in test tubes or other laboratory glassware
in vivo	ihn vē-vō	inside living organisms
latent infection	lā-tehnt ihn-fehck-shuhn	condition that may not be clinically noticed but under stress or poor health will develop into a recognizable disease state
macroenvironment	mahck-rō-ehn-vī-rən-mehnt	surroundings above the cellular level
metanephric	meht-ah-nehf-rihck	embryonic-like kidney
microenvironment	mīk-rō-ehn-vī-rən-mehnt	surroundings at the cellular level
monogamous	moh-noh-goh-muhs	pairing with one mate
murine	moo-rēn	of mice and rats
outbred	owt-brehd	from unrelated parents; also called **random bred**
parthenogenesis	pahr-thehn-ō-jehn-ih-sihs	development of an egg without fertilization
phenotype	fē-nō-tīp	physical characteristics of an individual
phylogeny	fī-lohj-eh-nē	developmental history of a species
pithing	pihth-ihng	destroying the brain and spinal cord by thrusting a blunt needle into the cranium or vertebral column
pinky	pihnk-ē	mouse or rat hairless neonate (see Figure 23–4)
poikilotherm	poy-kē-loh-thərm	animals who let their body temperature vary to an appreciable extent on a regular basis
polygamous	poh-lihg-ah-muhs	having multiple mates
polytocous	poh-liht-ō-kuhs	giving birth to multiple offspring
precocial	prē-kō-shihl	young born or hatched relatively mature and mobile (Figure 23–6)
prehensile	prē-hehn-sihl	adapted for grasping and seizing
progenitor	prō-jehn-ih-tōr	parent or ancestor
progeny	proh-jehn-ē	offspring or descendants

(continued)

TABLE 23–1 Common Laboratory and Pocket Pet Terms (*Continued*)

Term	Pronunciation	Definition
propagate	proh-pah-gāt	to reproduce
protocol	prō-tō-kawl	written procedure for carrying out experiments
rack	rahck	metal device that supports caging units
reduction	rē-duhck-shuhn	theory of using the minimal number of animals for a project that will yield valid results; one of the three *R's* principles of Russell and Burch
refinement	rē-fīn-mehnt	theory of inflicting minimal stress and pain to animals in research; one of the three *R's* principle of Russell and Burch
replacement	rē-plās-mehnt	theory of using cell or tissue culture or mathematical models instead of animals in research if possible; one of the three *R's* principle of Russell and Burch
ringtail	rihng-tāl	abnormal condition in which annular lesions form on the tails of rats housed in low humidity environments
rodent	rō-dehnt	class of animal that has chisel-shaped incisor teeth
rosette	rō-seht	swirl hair growth pattern in Abyssinian guinea pig; also called **whorl** (wohrl) (see Figure 23–14a)
rudimentary	roo-dih-mehn-tār-ē	incompletely developed
sable	sā-buhl	color pattern that has cream-colored undercoat and black guard hairs on feet, tail, and mask
scruff	skruhf	thick skin of the neck by which an animal can be grasped or lifted (Figure 23–7); scruffing is used as verb
scurvy	skər-vē	common term for vitamin C deficiency
sexual dimorphism	sehcks-yoo-ahl dī-mōrf-ihzm	physical or behavioral differences between females and males of a given species
shoebox	shoo-bohcks	caging that has solid-bottom flooring (Figure 23–8)
suspended cage	suh-spehnd-ehd	caging that hangs from a metal rack and has wire flooring (Figure 23–9)
teratology	tehr-aht-ohl-ō-jē	study of embryo development
test group		collection of animals used for experimental manipulation
transgenic	trahnz-jehn-ihck	refers to removing or synthesizing specific genes from one strain and injecting them into the cells of another strain
urethral cone	yoo-rē-thrahl kōn	organ found in front of the anus of male and female chinchillas
urostyle	yoor-ō-stīl	long caudal vertebra of some amphibians
vector	vehck-tər	organism or vehicle that carries infectious agent from one animal to another
vertical transmission	vər-tih-kahl trahnz-mihs-shuhn	disease transfer from mother to fetus
vestigial	vehs-tih-jē-ahl	said of a structure that has lost a function it previously had (e.g., the pelvis remnant in snakes or the wings of flightless birds)
warren	wahr-ehn	a place where rabbits breed or abound
Whitten effect	whiht-ehn eh-fehct	introduction of a male into a group of females synchronizes their estrous cycle; observed mainly in mice

FIGURE 23–1 Color pattern of animals. (a) Black mouse on left and agouti mouse on right. (b) Abino mouse.

FIGURE 23–2 Anogenital distance. (a) Genital area of a male rat. The testicles are large and pendulous in lightly haired scrotal sacs. (b) Genital area of a female rat. The anogenital distance of a female rat is shorter than the male rat. (c) Genital area of a male mouse. The anogenital distance is much shorter in the male than the female and in mature males the testicles are large and pendulous in lightly haired scrotal sacks. (d) Genital area of a female mouse.

FIGURE 23–3 Barbering in a female mouse.

FIGURE 23–4 Rat and mouse pups may be called pinkies (neonates that are hairless) and fuzzies (pups with a slight coat of hair). There are 4 pinkies and 1 fuzzy in this image.

FIGURE 23–5 Rat with hooded markings.

FIGURE 23–6 Chinchilla kits are born precocial. This kit is one day old.

(a)

(b)

FIGURE 23–7 (a) A mouse is restrained for examination by its scruff. (b) Care must to taken when scruffing a hamster. The cheek pouches should be empty and the scruff sufficient enough to prevent the hamster from turning and being able to bite, yet not too tight as to cause exophthalmos.

FIGURE 23–8 Shoebox cages have solid-bottom flooring.

FIGURE 23–9 Suspended cages hang from a metal rack and have wire flooring.

RODENTS

Rodents are gnawing animals that have continuously growing upper and lower incisors. The most common examples of rodents are hamsters, gerbils, mice, rats, guinea pigs, and chinchillas. The order Rodentia is divided into three suborders based on the musculature of the jaw and other skull structures (Figure 23–10). These suborders are the Sciuromorpha (squirrel-like rodents), the Myomorpha (ratlike rodents), and the Hystricomorpha (porcupine-like rodents). The two suborders of primary importance to the veterinary field are the Hystricomorpha, which includes chinchillas and guinea pigs, and the Myomorpha, which includes rats, mice, hamsters, and gerbils (Figure 23–11).

Mice and Rats

Mice and rats are rodents that are classified in the subfamily Murinae. Rats are larger than mice and have more rows of scales on their tails. Research mice have helped humans decode the human genome and understand countless diseases, while research rats have been used in cancer research and cognition (learning) studies (Figure 23–12). Rats and mice were once used mainly as research animals but are also popular pets. Both are clean and quiet animals to keep and like to be housed with others of their species. Rats and mice have also been bred as feeders (food for carnivores kept in captivity) typically purchased from pet stores or ordered frozen and vacuum packed as pinkies or fuzzies. There are many rat species; however, only some species, such as the black rat (*Rattus rattus*) and the brown rat (*Rattus norvegicus*), have been domesticated and used for research and as pets. The black rat (sometimes called the roof rat) is believed to have originated in Southern Asia and was the major reservoir of the Black Plague in Europe in the 1200s. The brown rat

FIGURE 23–10 Rodent skeleton.

FIGURE 23–11 Classification scheme of rodents.

(sometimes called the Norway rat) is believed to have originated in Eastern Asia and is well established in the United States. Rats have the ability to adapt to many different habitats, environments, and food sources, which explains why rats are found in all parts of the world. Rats are agile climbers and excellent swimmers and are very curious.

Rats are considered to be better pets than mice, are easily socialized as pups, and rarely bite. Male rats are more docile than female rats. They do best when kept with other rats and are primarily nocturnal. Rats

lack a gallbladder or tonsils (Figure 23–13). Rats have 16 teeth, and the dental formula is 2(I 1/1, C 0/0, P 0/0, M 3/3) = 16 open-rooted teeth.

Mice are believed to have originated from Asia and have spread throughout the world. The best-known species of mouse is the common house mouse (*Mus musculus*), which dwells wherever humans live. Mice have a pointed nose and slit upper lip. Mice have 16 open-rooted teeth, and the dental formula of the mouse is 2(I 1/1, C 0/0, P 0/0, M 3/3). Mice have a perfect visual field because of the placement and shape of

Courtesy of Eric Klaphake, DVM

FIGURE 23–12 Tumors are a prevalent problem in rats, making them good research models for cancer research. In pet rats, tumors are a concern because they may become so large as to restrict movement.

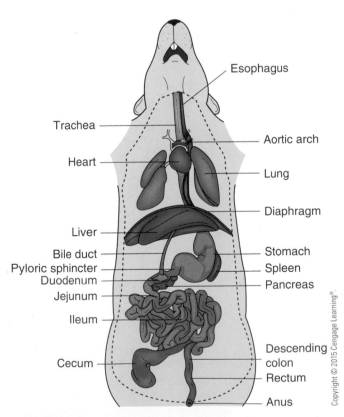

Esophagus

Trachea

Aortic arch

Heart

Lung

Diaphragm

Liver

Bile duct

Stomach

Pyloric sphincter

Spleen

Duodenum

Pancreas

Jejunum

Ileum

Descending colon

Cecum

Rectum

Anus

Copyright © 2015 Cengage Learning®.

FIGURE 23–13 Internal structures of a rat.

their eyes; however, their detailed vision is poor. To compensate, mice have large ears and a highly developed sense of hearing. Mice have a highly developed sense of smell as well.

Mice are more active than rats and will rarely sit quietly long enough to allow for handling or interaction. Mice do best when kept with other mice and are primarily nocturnal. A mouse colony is led by one head male. The head male is the only one allowed to mate with the females. Fighting between males for the head male status is common in mouse colonies. The odor of mice is stronger than rats.

Guinea Pigs

Guinea pigs, also called cavies, are rodents that are believed to have originated in South America. Guinea pigs have short bodies; stocky legs; and short, sharp claws. Guinea pigs tend to be gentle, easy-to-handle rodents that are timid in nature, so they should be kept away from other household pets. Even though guinea pigs are social with each other and respond to human companionship, they can be stressed by other animals, which may result in anorexia. The American Cavy Breeders Association (ACBA) recognizes 13 different guinea pig breeds with several varieties in each category (Figure 23–14a, b, and c). The male guinea pig is larger than the female guinea pig (Figure 23–15a and b). Sows are polyestrus and have many breeding cycles throughout the year. A vaginal plug is seen in guinea pigs and many rodents after copulation but falls away within a few hours after mating. The function of the vaginal plug is unknown, but it consists of ejaculate and is thought to prevent sperm from leaking from the vagina, ensuring enhanced fertility or to prevent

Rosette

(a)

Photo by Isabelle Francais

(b)

Photo by Isabelle Francais

(c)

© Eric Isselee/www.Shutterstock.com

FIGURE 23-14 Examples of guinea pig varieties. (a) Abyssinian; (b) American short-haired; (c) Peruvian.

other boars from mating with the same sow. Guinea pigs are born precocial after a gestation period of 63 days. Guinea pigs have very sensitive senses of hearing and smell.

Guinea pigs produce a wide range of vocal sounds including a whistle that can be a call of greeting or alarm (the exact meaning is determined by the length and pitch of the call), chirps that are low pitched and equally varied in tone and modulation (sound of contentment), and drilling, a low, guttural and rapid d-r-r-r-r-r produced by their teeth (sound of warning or pain).

Guinea pigs cannot synthesize vitamin C as most other mammals can because they lack the liver enzyme L-gulonolactone; therefore, vitamin C must be supplied in their diet. Feeding fresh pellets with supplemental fresh produce can provide the guinea pig with enough vitamin C. Vitamin C has a relatively short half-life and is sensitive to light, heat, and dampness, so it is important to check the expiration date on all bags of pelleted food. Guinea pigs have 20 teeth, and their dental formula is 2(I 1/1, C 0/0, P 1/1, M 3/3). Guinea pigs have a large perianal sebaceous gland they use for scent marking, especially unneutered boars.

Guinea pigs are monogastric and cannot vomit due to the size and position of the stomach relative to the diaphragm. The guinea pig has a large and well-developed cecum with numerous pouches; therefore, feeding adequate roughage is important to maintaining gut motility. Guinea pigs are coprophagic and fecal ingestion by pups is necessary to help establish normal gut flora. Any disruption of normal gastrointestinal flora and/or gut motility can lead to gut stasis, overgrowth of toxin-producing bacteria, and the development of air pockets of gasses trapped within the intestines (Figure 23–16). **Coprophagic therapy**, feeding a suspension of fecal pellets from a healthy animal to an ill animal, has been used to treat Guinea pigs with gastrointestinal disease.

Guinea pig housing should be simple because they do not climb or attempt to jump out of cages. Any enclosed area with sides at least 8 inches high with a floor space of 36 inches in length should provide enough room for exercise and a feeding area. Guinea pigs urinate and defecate without a location preference, resulting in contaminated feeding bowls. A food hopper and sipper bottle can prevent food and water contamination. A hide box should be provided as a place for the guinea pig to escape to and feel safe.

(a)

— Penis

— Anus

(b)

— Penis

— Oil buildup from sebaceous glands

(c)

— Vulva

— Anus

(d)

— Vulva

— Anus

Copyright © 2015 Cengage Learning®

FIGURE 23–15 (a) Genital area of a male guinea pig. Boars have slender testicles on either side of the anus, and the penis may be everted with gentle thumb pressure on the lower abdomen. (b) Large perianal sebaceous glands are used for scent marking and are especially pronounced in unneutered males. (c) Genital area of a female guinea pig. The anogenital area of the sow is more rounded in appearance and has a Y shape. The Y configuration extends from the anus to the vaginal opening. (d) Swollen genital area of a female guinea pig in heat.

Hamsters

Hamsters are rodents that are small (about 5 to 6 inches long), have short stump tails, and originated from the desert areas of Syria. The Syrian Golden, or Golden, hamster (Figure 23–17a and b) was bred from wild hamsters in the 1930s, and all domesticated hamsters sold are descendants of this original breeding. Hamsters are rodents that are nocturnal and solitary (two sexually mature housed together will result in fighting and the death of one of them). Hamsters are meticulous in maintaining their habitats as they choose specific areas for sleeping, stockpiling food, and depositing waste. They do not awaken during the day to defecate, but

FIGURE 23–16 A radiograph showing a large air pocket on the right side of this guinea pig who presented with abdominal distention, decreased amount of fecal droppings, and dyspnea.

(a)

(b)

FIGURE 23–17 Hamsters. (a) Short-haired hamster. (b) Long-haired hamster.

instead upon waking collect fecal material from their sleeping area, place it in their cheek pouches, and move it to the area with other droppings. Hamsters are used as research animals and pets (although not recommended for young children due to their quickness to bite and their nocturnal activities). A few hamster species exist; however, the Golden hamster is most common. Hamsters have almost no tails. Hamster skin is abundant, loose, and pliable (Figure 23–18).

(a)

(b)

FIGURE 23–18 (a) Genital area of a male hamster. At about 4 weeks of age males have a noticeable scrotum with testicles, although the testicles can be retracted. (b) Genital area of a female hamster. The anogenital distance of the female is much shorter than the male. Only female hamsters have nipples.

Hamsters prefer a temperature around 70°F and will go into a deep estivation if the temperature rises above 80°F. Hamsters have prominent scent glands on their flanks that appear as dark brown patches on either side of the back, near the pelvis on Syrian hamsters and on the ventral midline near the umbilicus in dwarf breeds of hamster. They are more prominent in males than females; hamsters use the secretions from these glands to mark territory. When the male is excited, he will lick his scent glands until the entire area is wet. He will also scratch and rub himself as if this area is irritated. Females are seasonally polyestrous, produce a creamy-white vulvar discharge that lasts 4 days during their reproductive cycle; most do not have reproductive cycles during the winter months. Female hamsters need to be left undisturbed with a new litter as they may cannibalize their young. Hamsters also have cheek pouches to store and transport food (Figure 23–19). Hamsters have 16 teeth that continually grow, and the dental formula is 2(I 1/1, C 0/0, P 0/0, M 3/3). Chew blocks provide enrichment and help prevent overgrown teeth.

Gerbils

Gerbils, also known as jirds, are burrowing rodents with bodies that are short and have a hunched appearance (Figure 23–20). The tail is covered with fur, has a bushy tip, and is used for support while standing. Both sexes have an area of alopecia on their abdomens that contains a large sebaceous gland used for scent marking. Regardless of the hair coat color, this area typically has a slightly orange pigment. This sebaceous gland allows adult hamsters to leave their scent by sliding their abdomen across an object (Figure 23–21).

Cheek pouch

FIGURE 23–19 Cheek pouch of a hamster.

Photo by Isabelle Francais

FIGURE 23–20 Gerbil.

The Mongolian gerbil is the most common type of gerbil and is found in parts of China, in the former Soviet Union, and throughout most of Mongolia. Gerbils are kept as pets and are used as research animals. The Mongolian gerbil is the most common type of gerbil kept as pets.

Gerbils are quiet animals; however, they are quite active. Gerbils are monogamous, mate for life, and share in the care of their offspring; therefore, paired gerbils should remain together. Gerbils are hardy and resistant to disease. Like hamsters, gerbils have 16 teeth that continually grow, and the dental formula is 2(I 1/1, C 0/0, P 0/0, M 3/3). The adrenal gland of the gerbil is large and contributes to the gerbil's ability to conserve water.

Chinchillas

Chinchillas are rodents that were originally used for their fur but are becoming increasingly popular as pets and research animals. Chinchillas are believed to have originated in South America and were used by the Incas as a source of fur. The native habitat of the chinchilla is the barren areas of the Andes Mountains at elevations up to 20,000 feet. Chinchillas are small, compact animals with dense fur that makes them appear bigger and rounder. They resemble small rabbits because they have long whiskers and large dark eyes, but have shorter ears and bushy tails. Their fur is thick and soft because they have as many as 60 hairs growing from one hair follicle and possess fewer guard hairs than some of the other pocket pets. Chinchillas have four toes on each foot with nailpads that never need trimming. The average life span of a chinchilla is 14 to 15 years (Figure 23–22).

Chinchillas are not strictly nocturnal (they are more active at dusk and during the evening), shelter in crevices and holes along rocks, and live in groups. They are typically very curious animals with very long hind feet allowing them to jump up and over items many

(a)

FIGURE 23–22 Standard chinchilla has a natural coat color that is gray with a lighter underbelly. Now there are many color mutations available. Note the use of enrichment (logs and plastic tube) in this cage.

(b)

FIGURE 23–21 (a) Genital area of a male gerbil. Male gerbils have a longer anogenital distance than females. Males also have a prominent, darkly colored scrotum and a more prominent, elliptical ventral scent gland. (b) Genital area of a female gerbil.

FIGURE 23–23 Dusting in a chinchilla. Dust should be about 1 inch deep and be provided in a shallow pan.

times their own length. Chinchillas are very quick, and their long tail acts as a rudder as they leap from place to place. Chinchillas clean their fur by rolling in dust (Figure 23–23). Chinchillas also can release their fur as a defense mechanism. If grabbed too roughly by a predator or handler, the chinchilla will leave a patch of fur behind.

Chinchillas are herbivores that are hindgut fermenters with a large stomach, jejunum, and cecum. Chinchillas have 20 teeth, and the dental formula is 2(I 1/1, C 0/0, P 1/1, M 3/3). Chinchillas do not eat directly from a food bowl, but pick up and hold individual food items. Chinchilla pellets are made longer so they are easier for them to grasp. Chinchillas are coprophagic, and they produce pellets that are not encased in mucus (like rabbit pellets) but are eaten during the night. Male chinchillas have open inguinal rings. The testes are located in the inguinal canal, without a true scrotal sac (Figure 23–24). Female chinchillas are seasonally polyestrous and—especially those that are obese—are prone to vaginal prolapse, which may be seen during delivery, during the postpartum period, or in nonpregnant animals (Figure 23–25).

— Urethral cone

— Anus

(a)

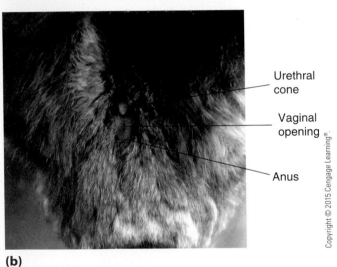

— Urethral cone

— Vaginal opening

— Anus

(b)

FIGURE 23–24 (a) Genital area of a male chinchilla. Note that males do not have a true scrotum and have a longer anogenital distance than females. The anogenital distance is the best way to determine the sex of a chinchilla because their external genitalia look similar. The anogenital distance of the male is about twice that of the female. (b) Genital area of a female chinchilla. Note both sexes have a urethral cone, which may be mistaken in the female for a penis.

FIGURE 23–25 A female chinchilla with a vaginal prolapse. This condition requires immediate veterinary care and may need to be repaired surgically.

FIGURE 23–26 Fitch ferrets. Ferrets may be restrained by scruffing the skin of the dorsal neck. Male ferrets called hobs (right) are larger than female ferrets called jills (left).

FERRETS

Ferrets are mammals that belong to the family Mustelidae, which includes weasels, mink, and polecats (Figure 23–26). Ferrets have been in the United States for more than 300 years and were originally used for rodent control. The American Ferret Association (AFA) recognizes 30 color variations, including sable (the most common and natural color), silver, silver mitt, chocolate, albino, black-eyed whites, and cinnamon. Today ferrets are used primarily as pets (check with state regulations regarding ferrets because they are illegal to have has pets in California); however, they are curious and like to get into small spaces and eat just about anything that is lying around.

Elaborate cages are not needed for ferrets as they spend a fair amount of time sleeping. Ferrets are ground-dwelling and do not need various levels or ramps within their cages because ferrets are nearsighted and are not used to climbing. Cage substrates such as wood shavings, recycled paper products, and corncobs should not be used for ferrets; simply lining the cage with plain newspaper is adequate bedding for a ferret. Ferrets can be trained to use a litter box; however, during play they do not actively seek out the litter box but will run backward to the nearest corner to urinate or defecate. Ferrets elevate their rear quarters to urinate and defecate, so corners should be protected on the ground and up the wall.

Ferrets are also used in research and as working animals because of their ability to reach difficult places due to their long and flexible spine. Ferrets have long, slender bodies and long tails that are about one-half the length of the head and body. Ferrets have short legs and each foot has five toes (with claws that they use to actively dig), and small, rounded ears. Ferrets are primarily nocturnal and can spend as much as 18 to 20 hours sleeping per day; they do not see well in bright light. Ferrets shed their coats twice a year, and their coat color may change from dark to light depending on the season. Ferrets have highly developed senses of hearing, smell, and touch. Ferrets have 34 permanent teeth, and the dental formula is 2(I 3/3, C 1/1, P 3/3, M 1/2).

Female ferrets are induced ovulators, which can cause health problems if the female does not come out of heat. **Hyperestrogenism** (hī-pər-ehs-trō-jehn-ihz-uhm) is elevated blood estrogen levels seen in intact cycling female ferrets if not bred (Figure 23–27). Ferrets are very susceptible to canine distemper and should be vaccinated with USDA-approved vaccines. Ferrets are unusual in that they can become infected with human influenza virus. There are many different strains of the influenza virus, all causing respiratory infections, and clinical signs are similar (pyrexia, coughing, sneezing, ocular and nasal discharge, lethargy, and dehydration). Ferrets are also prone to obstruction due to gastrointestinal foreign bodies (they love to chew on objects), gastrointestinal ulcers due to ingestion of inappropriate substances or non-steroidal anti-inflammatory drug use, adrenal gland disease (see Figure 11–6), and mast cell tumors (Figure 23–28).

(a)

(b)

(c)

FIGURE 23–27 External genitalia of ferrets. (a) Genital area of a male ferret. Male ferrets have a penis on the ventral abdomen and two testicles near the anus. (b) Genital area of a nonestrous female ferret. (c) Genital area of a female ferret in estrus.

(a) **(b)**

FIGURE 23–28 (a) An early stage mast cell tumor in the axilla of a ferret. MCT in ferrets typically first appear as small, hairless areas on the skin surface. (b) A late stage MCT in the mouth of a ferret.

RABBITS

Rabbits are mammals that were once classified as rodents but are now classified as lagomorphs. Rabbits have four upper incisor teeth, whereas rodents have two incisor teeth (Figures 23–29 and 23–30). The American Rabbit Breeders Association (ARBA) recognizes 45 distinct breeds of rabbit and determines breed standards based on body type and acceptable coat colors, patterns, texture, and eye color. Rabbits may be dwarf, mini, standard, or giant in size and may be lop-eared (Figure 23–31). Hares are in the same family as rabbits, usually are larger than rabbits, and have longer ears than rabbits. Hares do not build nests as rabbits do, and young hares are born fully furred with their

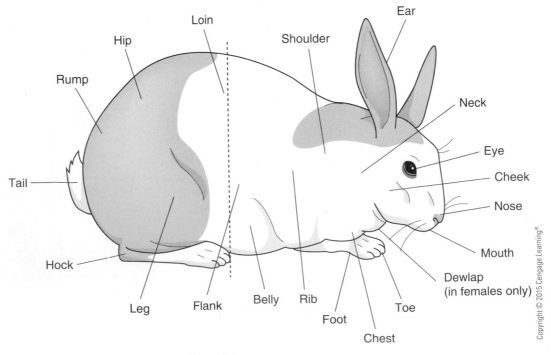

FIGURE 23–29 External anatomy of a rabbit.

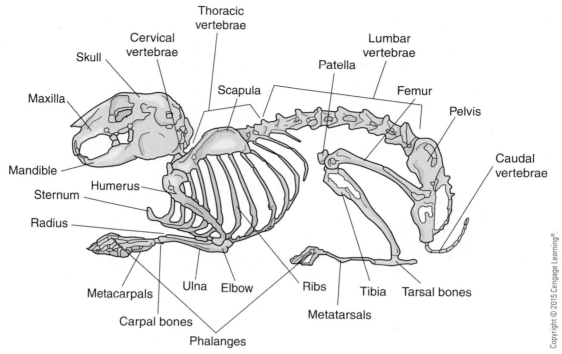

Skull

Maxilla

Mandible

Sternum

Radius

Humerus

Metacarpals

Carpal bones

Phalanges

Cervical vertebrae

Thoracic vertebrae

Scapula

Ulna

Elbow

Patella

Lumbar vertebrae

Femur

Pelvis

Caudal vertebrae

Ribs

Tibia

Tarsal bones

Metatarsals

FIGURE 23–30 Skeleton of a rabbit.

eyes open (Figure 23–32). Hares live above ground and do not dig tunnels as rabbits do. Rabbits are kept indoors or outdoors as pets, are raised for pelts and food, and are used for research.

In general, rabbits are quiet animals and vocalize only when provoked, when in pain, or when extremely frightened. Growling in rabbits is a sign of aggression. A frightened, restrained rabbit may scream and kick with its hind legs, potentially causing injury to the handler (from the rabbit raking its rear claws on the person) or to the rabbit (from its powerful hind legs causing injury during its attempt to escape). Territorial aggression is common in rabbits of both sexes, especially those that are not spayed or neutered. Typically unspayed does are more aggressive than unneutered bucks. Rabbits use scent glands to mark territory. They have submandibular glands, commonly called chin glands, and both males and females have folds of skin on either side of their genitals. These inguinal glands normally produce an unpleasant smelling, thick, waxy substance.

Proper ventilation is important in keeping rabbits disease-free. Rabbits housed outdoors need to have shelter from the heat and cold. Temperatures below 40°F may cause them to be hypothermic or may cause frostbite, while temperatures above 80°F may cause heat stroke. Rabbits do not sweat or pant, so they have limited ability to cool themselves in hot weather. When

selecting a cage or hutch for a rabbit, it should be at least three times the length of the rabbit and should have wire-bottom floors so urine and fecal material fall away from the animal. An area with solid flooring also needs to be provided so the rabbit is not constantly on wire, which can lead to sores on the bottoms of their feet. All cages should have a hide box.

Rabbits have a simple glandular stomach and a large cecum, making them hindgut fermenters, so adequate roughage in the diet is important. Rabbits cannot vomit and are coprophagic. Rabbits produce hard pellets that are passed during the day and consist of undigested waste. **Night stool** (nīt stool), also known as **cecotrophs** (sē-kō-trōfs), is rabbit nocturnal feces that is looser than normal and encased in a mucous membrane. Night stool contains vitamins B and K as well as beneficial bacteria and protein. The rabbit consumes night stool directly from the anus.

Rabbits also need to gnaw to keep the length of their incisors in check by clipping back teeth to the appropriate length. Rabbit teeth are open-rooted and grow continuously (Figure 23–33). Rabbits have 28 permanent teeth, and the dental formula of the rabbit is 2(I 2/1, C 0/0, P 3/2, M 3/3).

Rabbits have powerful hindquarters for jumping. If not restrained properly, a rabbit can kick its rear legs and fracture its spine (see Figure 13–25). In male rabbits, the inguinal canals remain open for life, the

FIGURE 23–31 Types of rabbits. (a) Dwarf. (b) Mini. (c) Standard. (d) Giant. (e) Lop-eared.

(a) **(b)**

FIGURE 23–32 (a) Rabbits build nests for their litters so a hide box should be provided in all cages or hutches. (b) Rabbit kits are born hairless, and their eyes and ears are not open.

FIGURE 23–33 Extremely overgrown lower incisors. The incisor on the left is fractured and a large hairball has formed around both incisors.

FIGURE 23–34 A doe and buck New Zealand White. The doe, on the right, has a pronounced dewlap.

scrotum is hairless, and the testes descend at about 12 weeks of age. Does are induced ovulators and are capable of breeding at any time once they reach sexual maturity (around 4 to 8 months). With maturity, does in many breeds may develop a pronounced dewlap (Figure 23–34). Bucks rarely have dewlaps. Bucks have tiny penile spines that irritate cervical tissue, inducing the ovaries to release eggs. Does do not have a uterine body, but two elongated uterine horns for the development of multiple kits. Copulation is short, and ovulation occurs within 10 hours of breeding. A female should be housed in her own cage after mating because the presence of the male during gestation may be

stressful enough for her to reabsorb the fetuses. Bucks should not be kept with does after does have given birth because they may kill the kits. Does may cannibalize their young and those who do typically continue to cannibalize each litter. The young are born altrical.

Rabbits are prone to some diseases such as the following:

- **snuffles** (snuhf-uhlz) = common term for upper respiratory disease of rabbits caused by *Pasteurella multocida* (Figure 23–35).
- **sore hocks** = ulceration of the foot pads and foot area caused by the animal's body weight pressing down on the foot and the thick covering of fur (instead of foot pads) on rabbit feet; commonly

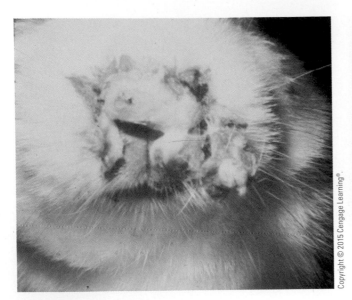

FIGURE 23–35 Snuffles in a rabbit.

FIGURE 23–36 Torticollis in a rabbit.

seen in rabbits housed on wire cage floors; also called **pododermatitis** (see Figure 10–27).

- **torticollis** (tōr-tih-kō-luhs) = contracted state of the cervical muscles producing torsion of the neck; also called **wry neck** (Figure 23–36).
- **urolithiasis** (yoo-rō-lih-thī-ah-sihs) = disorder characterized by the presence of urinary bladder stones (Figure 23–37).

REPTILES

Reptiles are cold-blooded vertebrates that have lungs and breathe air. Reptiles have a body covering (bony skeleton, scales, or horny plates) and a heart that has two atria and in most species one ventricle.

There are four different types of reptiles: Chelonia (turtles, tortoises, and terrapins), Serpentes (snakes,

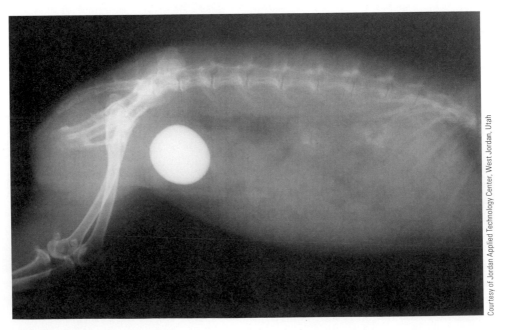

FIGURE 23–37 Uroliths in rabbits may be due to excess protein in the diet. This radiograph shows a large urolith. Rabbits with uroliths may sit with a hunched-over posture, strain to urinate, and grind their teeth when attempting to urinate.

pythons, and boas), Squamata (iguanas and lizards), and Crocodilia (crocodiles and alligators) (Figure 23–38).

Modern-day reptiles have three body types. One type, including lizards, has long bodies and clearly defined tails (Figure 23–39). The second type, including snakes, has long bodies that taper into tails (Figure 23–39). The third type, including turtles, has short, thick bodies encased in shells (Figure 23–39).

Movement in reptiles varies with the type of body structure in each group. Crocodiles and lizards have paired limbs that are attached to the body at right angles. This enables the animal to lift its body off the ground while moving. Crocodiles and alligators have strong limbs, whereas lizards have weaker limbs. Snakelike reptiles do not have limbs, and their movement is due to undulating movements of the body. The scales on the underside of the body project outward as the muscles are contracted and relaxed, allowing pressure to move the animal forward. Some turtles and tortoises have limbs that enable them to move on land using a creeping and crawling motion, yet also have modified limbs that enable them to swim.

Reptile skin has a horny surface layer. In lizards and snakes, this layer forms a hard, continuous covering of scales. These scales lie beneath the superficial layer of the epidermis so that the body can grow; this superficial layer is shed, allowing further growth. Crocodiles, alligators, and some lizards have bony dermal scales covered by a horny epidermal layer that also is shed to allow for body growth. Turtles and tortoises do not molt their thick epidermal skin; each year a new epidermal scale is formed beneath the old one. These epidermal scales (scutes) form rings that can be counted, allowing a person to estimate the animal's age.

Lizards and snakes have teeth that are fused into the jawbones; some snakes have teeth fused to the palate bones. Crocodile teeth are set in sockets. Turtles and tortoises do not have teeth; their jaws form sharp crushing plates. Reptile teeth are **polyphyodontic** (pohl-lē-fī-ō-dohn-tihck), meaning they are reabsorbed or shed and replaced at a rapid rate throughout life.

(a)

Photo by Isabelle Francais

(b)

Photo by Isabelle Francais

(c)

Photo by Isabelle Francais

(d)

Photo by Isabelle Francais

FIGURE 23–38 Types of reptiles. (a) This Western painted turtle is an example of Chelonia. (b) This ball python is an example of Serpentes. (c) This Madagascar day gecko is an example of Squamata. (d) These American alligators are examples of Crocodilia.

Ventral View of a Female Lizard

(b)

FIGURE 23–39 (a) Ventral view of a female lizard. (b) The anatomy of a snake showing elongated internal organs. (c) Turtles and tortoises have an upper and lower shell connected by bony bridges to protect internal organs.

Ventral View of a Turtle

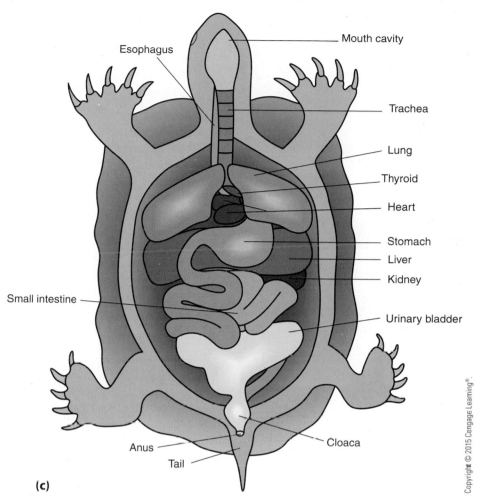

FIGURE 23–39 (continued)

The tongues of reptiles vary greatly. Some have short, fleshy tongues that have little movement, whereas other reptiles have tongues that are long, slender, and forked (some lizards and snakes).

Tortoises, crocodiles, and many lizards and snakes are oviparous (lay eggs that hatch after leaving the female body). Some species are ovoviviparous (eggs hatch inside the body of the female with the live young emerging from the female's body). A common problem occurs when reptiles become egg bound (Figure 23–40). Lack of a suitable site to lay eggs, stress, poor nutrition, or hormonal problems can all lead to a reptile being egg bound.

Terms unique to reptiles include the following:

- **carapace** (kahr-ah-pās) = dorsal region of a turtle shell (Figure 23–41).
- **chin glands** = secretory organs located on the throats of turtles; also called **mental glands**.

- **dysecdysis** (dihs-ehck-dī-sihs) = difficult or abnormal shedding.
- **ecdysis** (ehck-dī-sihs) = shedding or molting.
- **femoral pores** (fehm-ōr-ahl poorz) = sexually dimorphic glands prominent in mature lizards that are located on the ventral surface of lizard thighs; also called **femoral glands**.
- **head gland** = small secretory organ located on the head of snakes.
- **hemipenes** (hehm-ih-pē-nēz) = paired, vascular sacs that open into the caudal cloaca in male snakes and lizards. The two hemipenes are brought together to act as a penis and are inserted into the cloaca of the female during copulation (Figure 23–42).
- **musk glands** = four secretory organs that open lateroventrally near the carapace edge of turtles.

FIGURE 23–40 A radiograph of an egg-bound veiled chameleon.

- **plastron** (plahs-trohn) = ventral region of a turtle shell. The plastron of the male is concave and the plastron of the female is flat.
- **scent gland** = saclike secretory organ located at the base of the tail in snakes.
- **scute** (skoot) = any scalelike structure such as those found on tortoise or turtle shells or on the heads of snakes.
- **spectacle** (spehck-tah-kuhl) = transparent, highly vascular scale that permanently covers the cornea of some reptiles; also called an **eyecap** or a **brille** (brī-uhl) (Figure 23–43). Snakes cannot close their eyes and have spectacles instead of eyelids; the spectacle is fixed and does not open or close. It should be shed when a snake sheds; a retained spectacle is abnormal.
- **tail autotomy** (tā-uhl aw-toh-tō-mē) = lizard's ability to lose its tail.

AMPHIBIANS

Amphibians are animals that live the larval part of their lives in water and their adult lives partially or completely on land. Immature amphibians do not resemble their adult form. All must go through developmental stages of **metamorphosis** (meht-ah-mōr-fih-sihs), a

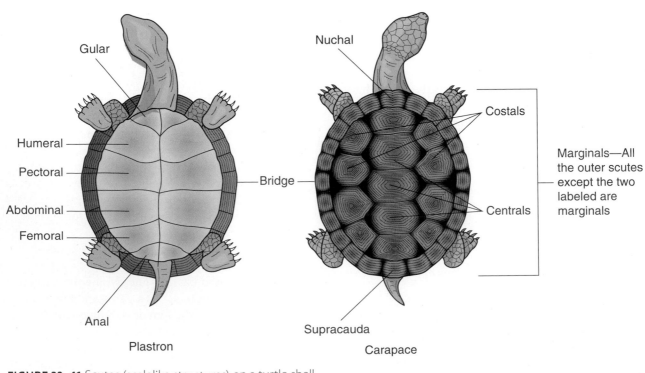

FIGURE 23–41 Scutes (scalelike structures) on a turtle shell.

FIGURE 23–42 (a) The sex of a reptile may be determined by inserting a probe into the vent to observe the hemipenes. (b) The hemipenes of a speckled skink.

FIGURE 23–43 (a) Normal spectacle, or eyecap, in a snake. (b) Spectacle in process of shedding. (c) Shed spectacle.

process by which an animal physically develops after birth or hatching. There are approximately 4,000 species of amphibians in three orders: Gymnophiona (wormlike, limbless amphibians called caecilians), Caudata (newts and salamanders), and Salientia (frogs and toads) (Figure 23–44).

Amphibians have scaleless, thin, moist skin that is used for breathing (cutaneous respiration) and for absorption of water. Amphibian skin is covered with fluid-secreting glands that produce a slimy mucus. This mucus helps conserve moisture; prevents excessive amounts of water from being absorbed into the body; and makes the animals slippery, which aids in their defense. The tongues of amphibians vary greatly from those that do not have tongues to those having long tongues with sticky tips to aid in the capture of small insects. Amphibians typically eat insects;

however, larger species may feed on mice and small rodents.

All amphibians have a life cycle with four stages: egg, aquatic larva with gills, aquatic larva with limb

(a)

(b)

FIGURE 23–44 Types of Salientia amphibians. (a) American toad. Toads have drier, thicker, and nodulated skin. (b) Green tree frog. Frogs have thinner, smooth, mucoid skin.

buds, and a complete metamorphosis into air-breathing adults. Depending on the species and environmental factors, metamorphosis can take two or more years to complete. Some amphibians, such as frogs, perform **amplexus** (ahm-plehck-suhs) during mating in which the male clasps the female's forelegs for breeding purposes. Some male frogs and toads have nuptial pads near their toes which assist in amplexis. Most amphibians are oviparous and lay gelatinous strings of eggs that are fertilized outside the body. The eggs, also known as **spawn** (spahwn), do not have shells and are deposited directly in the water or wet plant beds. Without water, the eggs dry up.

Sexing amphibians can be difficult. Males have internal testes and the sexes may only be able to be differentiated from each other during mating season. Male frogs and toads are more vocal and in some species are half the size of the female or may have nuptial pads near their toes. Some species of newts have crests near their head and along their dorsal surface. Sometimes there are color differences between the sexes of amphibians such as male newts who become brightly colored during breeding season.

All amphibians are **poikilotherms** (poy-kē-loh-thərms) because they allow their body temperature to vary to an appreciable extent on a regular basis and are totally dependent upon their environment to regulate body temperature and metabolic activity. Captive amphibians may die if specific temperature conditions are not met. Basic housing requirements for amphibians include a glass aquarium or plastic tank of an appropriate size, good ventilation and a ventilated lid, water supply, full spectrum lighting, 75–90% humidity, and enrichment such as plants, natural rocks, and logs.

Medical concerns of amphibians include an incomplete or partial shed, gastric overload from feeding too many mealworms, parasitism, fungal and bacterial skin diseases (typically the result of poor husbandry), and corneal lipidosis.

ABBREVIATIONS FOR LABORATORY ANIMALS AND POCKET PETS

- ACBA = American Cavy Breeders Association
- ACLAM = American College of Laboratory Animal Medicine
- AFA = American Ferret Association
- ARBA = American Rabbit Breeders Association
- AZVT = Association of Zoo Veterinary Technicians
- LAT = Laboratory Animal Technician
- LATG = Laboratory Animal Technologist
- 3 R's = reduction, refinement, replacement

REVIEW EXERCISES

Multiple Choice

Choose the correct answer.

1. The ventral region of a turtle shell is the
   a. plastron
   b. carpace
   c. dysecdysis
   d. ecdysis

2. A term for germ-free is
   a. barrier
   b. ambient
   c. axenic
   d. fomite

3. Biting or chewing the fur of subordinate animals is known as
   a. cribbing
   b. barbering
   c. cannibalism
   d. polytocous

4. A term meaning outside living organisms is
   a. in situ
   b. in vitro
   c. in vivo
   d. inbred

5. Rabbit nocturnal feces that is looser than normal is known as
   a. diarrhea
   b. night stool
   c. dark feces
   d. melena

6. The medical term for shedding or molting is
   a. dysecdysis
   b. ecdysis
   c. estivation
   d. transgenic

7. The space in the oral cavity of hamsters that allows them to carry food and bedding is the
   a. oral cavity
   b. cheek cavity
   c. cheek pouch
   d. calvarium

8. The area between the anus and genitalia used to determine the sex of animals is the
   a. analogous space
   b. cesarean space
   c. anogenital distance
   d. crepuscular area

9. Chinchillas clean themselves by
   a. fur-slip
   b. barbering
   c. dusting
   d. shining

10. Incompletely developed parts are known as
    a. homologous
    b. heterozygous
    c. homozygous
    d. rudimentary

11. An animal with a white coat and pink eyes is called a(n)
    a. albino
    b. axenic
    c. cashmere
    d. agouti

12. Something that carries disease from one animal to another is a
    a. test group
    b. vestigial
    c. vector
    d. vertical transmission

13. The standard normal against which experimental results are compared is the
    a. variable
    b. control
    c. fomite
    d. ambient

14. Adjustment of an animal to a new environment is
    a. ambient
    b. acclimatization
    c. calvarium
    d. gnotobiotic

15. Another term for brille is
    a. spectacle
    b. urostyle
    c. scute
    d. plastron

16. Wry neck, or the contracted state of the cervical muscles producing torsion of the neck, also is called
    a. ecdysis
    b. dysecdysis
    c. carapace
    d. torticollis

17. Intact cycling female ferrets may have elevated blood estrogen levels. This condition is called
    a. scurvy
    b. homozygous
    c. homologous
    d. hyperestrogenism

18. To reduce body temperature, heart and respiration rates, and metabolism to dormancy in winter is called
    a. calvariumism
    b. hibernation
    c. estivation
    d. metanephrism

19. To reduce body temperature, heart and respiration rates, and metabolism to dormancy in summer is called
    a. calvariumism
    b. hibernation
    c. estivation
    d. metanephrism

20. The written procedure for carrying out experiments is the
    a. phenotype
    b. genotype
    c. protocol
    d. polytocous

21. The term for surrounding is
    a. analogous
    b. ambient
    c. agouti
    d. autosome

22. The paired, vascular sacs that open into the caudal cloaca in male snakes and lizards is the
    a. hibernate
    b. hooded
    c. hybrid
    d. hemipenes

23. Dewlaps are found more commonly in
    a. female rabbits
    b. male rabbits
    c. female turtles
    d. male turtles

24. The thick skin of the neck by which an animal can be grasped or lifted is the
    a. scute
    b. carapace
    c. scruff
    d. spectacle

25. The long caudal vertebra of some amphibians is the
    a. urostyle
    b. brille
    c. rosette
    d. rack

Crossword Puzzle

Laboratory, Pocket Pet, Repile, and Amphibian Terms
Supply the correct term in the appropriate space for the definition listed.

Across

1 inanimate carrier of disease
3 cleaning method of chinchillas
6 contracted state of the cervical muscles producing torsion of the neck
8 dorsal region of a turtle shell
11 standard "normal" against which experimental results are compared
12 to reproduce
14 inside living organisms
16 study of embryo development
17 transparent, unshed, abnormal covering over the cornea of some reptiles
20 incompletely developed

Down

1 shedding of hair patches from rough handling of chinchillas
2 behavioral disorder in animals where dominant animals bite or chew the fur from subordinate animals
4 organism or vehicle that carries infectious agent from one animal to another
5 offspring or descendants
7 statement of research supposition
9 germ-free
10 dormant state during summer
11 top of the skull
13 ventral region of a turtle shell
15 rabbit noctural feces
18 shedding
19 surrounding
21 of mice and rats

Case Studies

Define the underlined terms in each case study.

A 6-mo-old female green iguana was presented for examination for <u>lethargy</u>, <u>ataxia</u>, and limb twitching. It was kept in an aquarium with food, water, a hot rock, and a white light source. The diet consisted mainly of head lettuce and grapes. Upon <u>PE</u>, the iguana was noted to be underweight and undersized for its age, and limb tremors could be induced during handling. Based on the history, it was likely that this animal had secondary nutritional <u>hyperparathyroidism</u>, otherwise known as <u>metabolic</u> bone disease. <u>Radiographs</u> were taken, and blood was drawn for a <u>serum</u> <u>chemistry profile</u>. Results indicated poor bone density, <u>hypocalcemia</u>, and <u>hypoproteinemia</u>. The iguana was treated with calcium and vitamin D_3 supplementation. The owners were counseled on proper husbandry and nutrition of green iguanas (Figure 23–45).

1. lethargy _____

2. ataxia _____

3. PE _____

4. hyperparathyroidism _____

5. metabolic _____

6. radiographs _____

7. serum _____

8. chemistry profile _____

9. hypocalcemia _____

10. hypoproteinemia _____

FIGURE 23–45 Green iguana.

A 5-yr-old three-toed box turtle was presented with a swelling on the right side of its head. The turtle was housed in an aquarium with dirt substrate, a water dish, and a hide box. On PE, the right <u>tympanic membrane</u> was noted to bulge out <u>laterally</u> from the head. <u>Oral</u> examination revealed <u>stomatitis</u> characterized by yellow <u>plaques</u> on the oral <u>mucosa</u>. A <u>diagnosis</u> of <u>otitis media</u> and potential <u>hypovitaminosis</u> was made. The <u>abscess</u> in the ear was <u>lanced</u> and <u>debrided</u>. The oral plaques were swabbed for <u>cytology</u> and were found to demonstrate <u>squamous</u> <u>metaplasia</u> consistent with hypovitaminosis A. The turtle was treated with antibiotics and vitamin A supplementation. Husbandry and nutrition were discussed with the owner.

11. tympanic membrane _____

12. laterally _____

13. oral _____

14. stomatitis _____

15. plaques _____

16. mucosa _____

17. diagnosis _____

18. otitis media _____

19. hypovitaminosis _____

20. abscess _____

21. lanced _____

22. debrided _____

23. cytology _____

24. squamous _____

25. metaplasia _____

A 5-yr-old M/N ferret was presented for signs of <u>alopecia</u> on its <u>dorsal lumbar</u> and tail regions as well as <u>dysuria</u>. Although differential diagnoses of <u>ectoparasites</u>, <u>dermatophytosis</u>, <u>atopy</u>, <u>cystitis</u>, and <u>urolithiasis</u> were considered, the most likely <u>dx</u> was <u>hyperadrenocorticism</u>. This endocrine disease is caused by adrenal gland <u>hyperplasia</u> or <u>neoplasia</u>. Production of excessive androgen hormones results in <u>cessation</u> of the normal hair growth cycle and may cause <u>prostatic hyperplasia</u>, which in turn produces the clinical signs described. Treatment options include surgical <u>excision</u> of the affected adrenal glands and medical therapy aimed at reducing the functional tissue of the adrenal glands. Many ferrets respond very well to the therapeutic options.

26. alopecia _____

27. dorsal lumbar _____

28. dysuria _____

29. ectoparasites _____

30. dermatophytosis _____

31. atopy _____

32. cystitis _____

33. urolithiasis   _____

34. dx   _____

35. hyperadrenocorticism   _____

36. hyperplasia   _____

37. neoplasia   _____

38. cessation   _____

39. prostatic hyperplasia   _____

40. excision   _____

A 3-yr-old M chinchilla was noted to be limping on its left rear leg. He had taken a fall from the top of a bedroom dresser. Upon presentation to the veterinarian, a <u>midshaft</u>, <u>spiral</u> <u>femoral fx</u> was <u>diagnosed</u> with <u>radiographs</u>. Options included <u>coaptation</u> by <u>splinting</u> the leg, surgical repair of the <u>fx</u>, or restricted cage rest. The owner chose to have the leg splinted. The chinchilla was anesthetized with isoflurane, an <u>inhalant anesthetic</u>, via a face mask. A splint was fashioned with fiberglass casting and bandage materials. One week later it was noted that the splint had shifted, so the animal was returned to the veterinarian. Upon removal of the splint, it was discovered that the leg was no longer <u>viable</u> because <u>vascular</u> and soft tissue damage had occurred. The chinchilla was again anesthetized for surgical <u>amputation</u> of the leg. He recovered well over the next few weeks and <u>ambulated</u> well as a three-legged chinchilla.

41. midshaft   _____

42. spiral   _____

43. femoral fx   _____

44. diagnosed   _____

45. radiographs   _____

46. coaptation   _____

47. splinting   _____

48. fx   _____

49. inhalant anesthetic   _____

50. viable   _____

51. vascular   _____

52. amputation   _____

53. ambulated   _____

A 2-yr-old M guinea pig was noted to have <u>hematuria</u> and <u>pollakiuria</u> from the owner's <u>hx</u>. On PE, the guinea pig was <u>BAR</u>. All systems examined were <u>WNL</u> other than <u>palpation</u> of the urinary bladder, which elicited <u>vocalization</u> as a result of assumed pain. The guinea pig was restrained for <u>cystocentesis</u>, and a <u>UA</u> was performed. The urine contained numerous <u>erythrocytes</u> and <u>leukocytes</u>, as well as calcium oxalate <u>crystals</u>. A <u>urolith</u> was suspected. Whole body <u>radiographs</u> confirmed the presence of a <u>radiopaque</u> density in the urinary bladder. A <u>cystotomy</u> was performed to remove the calcium oxalate stone. The owners were instructed to feed timothy hay instead of alfalfa hay to decrease dietary calcium intake.

54. hematuria _____

55. pollakiuria _____

56. hx _____

57. BAR _____

58. WNL _____

59. palpation _____

60. vocalization _____

61. cystocentesis _____

62. UA _____

63. erythrocytes _____

64. leukocytes _____

65. crystals _____

66. urolith _____

67. radiographs _____

68. radiopaque _____

69. cystotomy _____

A 1-yr-old F domestic mouse was presented for pruritus and a powdery substance on the dorsal aspect of the neck. PE revealed a pruritic mouse with white granular material on the dorsal cervical region. A cellophane tape preparation was performed on the neck debris. Ectoparasitic ascariasis was diagnosed, the most likely agent being *Myobia musculi.* The mouse was treated topically with the parasiticide ivermectin, which was placed on the interscapular skin. The owner was instructed to return for a second treatment in 10 days.

70. pruritus _____

71. dorsal _____

72. dorsal cervical region _____

73. debris _____

74. ectoparasitic _____

75. topically _____

76. parasiticide _____

77. interscapular _____

Socrates, a 1 1/2-yr-old hooded rat, was examined for what the owner described as a cold. Conjunctivitis and sinusitis were noted on PE. The owner was concerned that there was bleeding from the nares. He was informed that when rats become ill, they groom themselves less, which allows nasal discharge to accumulate on the fur. The red tinge that the owner was seeing was not blood, but a porphyrin pigment that is a normal component of the nasal secretions. Various viral and bacterial etiologic agents are responsible for respiratory disease in rats, including *Mycoplasma.* The antibiotic oxytetracycline was prescribed PO for 2 weeks. The owner was informed that this may be a chronic, recurrent disease.

78. conjunctivitis   _____

79. sinusitis   _____

80. nares   _____

81. nasal   _____

82. etiologic   _____

83. antibiotic   _____

84. PO   _____

85. chronic   _____

Critical Thinking Exercise

The following story and questions contain words and concepts from this chapter and are designed to stimulate critical thinking through class discussion, Internet searches, or brief writing responses.

The veterinary clinic you work at occasionally performs examinations on pocket pets. You notice that many of the clients who inquire about pocket pets have children. This is a concern to you because you have seen too many pocket pets adopted into homes that were not appropriate for them. You also feel that there is limited information about zoonotic diseases in pocket pets.

1. What might you want to discuss with clients who have small children and want to adopt pocket pets? Describe specific advantages and disadvantages of each animal you list.

2. What specific concerns do you have regarding each of the pocket pets listed from question 1? Provide a few concerns or special husbandry techniques needed for each animal listed.

Appendix A
ABBREVIATIONS

IMAGING ABBREVIATIONS

a or amp	ampere
A/P	anterior/posterior
APMLO	anterior to posterior and medial to lateral oblique
Ba	barium
BE	barium enema
CT or CAT	computed (axial) tomography
DL-PaMO	dorsolateral-palmaromedial oblique
DL-PlMO	dorsolateral-plantaromedial oblique
DM-PaLO	dorsomedial-palmarolateral oblique
DP	dorsopalmar
DPl	dorsoplantar
DPr DDiO	dorsoproximal-dorsodistal oblique (often called the skyline)*
D/V	dorsal/ventral
f	frequency
Hz	Hertz (a unit of frequency)
λ	wavelength
IVP	intravenous pyelogram
kv	kilovolt
kVp	kilovolt peak
kW	kilowatt
LM	lateral to medial
LMO	lateral to medial oblique
mA	milliamperage
mAs	milliamperage in seconds
mc	millicurie
ML	medial to lateral
MLO	medial to lateral oblique
MRI	magnetic resonance imaging
O	oblique
P/A	posterior/anterior
Pa45Pr-PaDiO	palmaroproximal-palmarodistal oblique (often called the skyline navicular)

* Abbreviation may be preceded with *flexed* and may include a number referring to the deviation from perpendicular for a beam; for example, flexed D30Pr-DdiO is a flexed dorsoproximal dorsodistal oblique taken at a 30-degree angle to the cassette.

PALMO	posterior to anterior and lateral to medial oblique
PAMLO	posterior to anterior and medial to lateral oblique
rad	radiation absorbed dose (unit of measurement of absorbed dose of ionizing radiation)
RT	radiation therapy
T	period of time (used in ultrasound)
v	velocity
V	volt
V/D	ventral/dorsal
W	watt
X-ray	roentgen ray

SPECIALIST OR TITLE ABBREVIATIONS

The letter *D* may appear in front of these specialties, which means that the person is a diplomate of that particular college. For example, DACVIM after someone's name means that the person is a diplomate in the American College of Veterinary Internal Medicine.

AAVSB	American Association of Veterinary State Boards
ABVP	American Board of Veterinary Practitioners
ABVT	American Board of Veterinary Toxicology
ACCC	Animal Chiropractic Certification Commission
ACLAM	American College of Laboratory Animal Medicine
ACPV	American College of Poultry Veterinarians
ACT	American College of Theriogenologists
ACVA	American College of Veterinary Anesthesiologists
ACVB	American College of Veterinary Behaviorists
ACVCP	American College of Veterinary Clinical Pharmacology
ACVD	American College of Veterinary Dermatology
ACVECC	American College of Veterinary Emergency and Critical Care

ACVIM	American College of Veterinary Internal Medicine
ACVM	American College of Veterinary Microbiologists
ACVN	American College of Veterinary Nutrition
ACVO	American College of Veterinary Ophthalmologists
ACVP	American College of Veterinary Pathologists
ACVPM	American College of Veterinary Preventive Medicine
ACVR	American College of Veterinary Radiology
ACVS	American College of Veterinary Surgeons
ACZM	American College of Zoological Medicine
AEVNT	Academy of Equine Veterinary Nursing Technicians
AHT	animal health technician
AIMVT	Academy of Internal Medicine for Veterinary Technicians
ALAT	assistant laboratory animal technician
ASVDT	American Society of Veterinary Dental Technicians
AVBT	Academy of Veterinary Behavior Technicians
AVCA	American Veterinary Chiropractic Association
AVCPT	Academy of Veterinary Clinical Pathology Technicians
AVDC	American Veterinary Dental College
AVECCT	American Veterinary Emergency and Critical Care Technicians
AVST	Academy of Veterinary Surgical Technicians
AVTCP	Academy of Veterinary Technicians in Clinical Practice
AZVT	Association of Zoo Veterinary Technicians
AVZMT	Academy of Veterinary Zoological Medicine Technicians
CH	certified herbalist
CVA	certified veterinary acupuncturist
CVT	certified veterinary technician
DC	doctor of chiropractic
DVM	doctor of veterinary medicine
LAT	laboratory animal technician
LATG	laboratory animal technologist
LVT	licensed veterinary technician
MPH	Master of Public Health
ND	doctor of naturopathy
RVT	registered veterinary technician
RVTG	registered veterinary technologist
SVBT	Society of Veterinary Behavior Technicians
VA	veterinary assistant
VMD	veterinary medical doctor (veterinariae medicinae doctor)
VNCA	Veterinary Nurses Council of Australia
VTNE	Veterinary Technician National Exam

VTS (Internal Medicine—Small Animal)
    Veterinary Technician Specialist in small animal internal medicine

VTS (Internal Medicine—Large Animal)
    Veterinary Technician Specialist in large animal internal medicine

VTS (Internal Medicine—Cardiology)
    Veterinary Technician Specialist in Cardiology

VTS (Internal Medicine—Neurology)
    Veterinary Technician Specialist in Neurology

VTS (Internal Medicine—Oncology)
    Veterinary Technician Specialist in Oncology

ASSOCIATION ABBREVIATIONS

AAHA	American Animal Hospital Association
AALAS	American Association of Laboratory Animal Science
AKC	American Kennel Club
AO	Arbeitsgemeinschaft für Osterosyntesesfragen (Association for the Study of Fracture Treatment in Man, founded by a group of Swiss surgeons); used to describe specialized bone plates and instruments used in orthopedic repair
APA	American Poultry Association
APHIS-VS	Animal & Plant Health Inspector Services—Veterinary Services
ASIF	Association of the Study of Internal Fixation
ASPCA	American Society for the Prevention of Cruelty to Animals
AVMA	American Veterinary Medical Association
CAAHT	Canadian Association of Animal Health Technologists & Technicians
CDC	Centers for Disease Control and Prevention (human)
CVTEA	Committee on Veterinary Technician Education and Activities
DEA	Drug Enforcement Administration
DHIA	Dairy Herd Improvement Association
DOT	Department of Transportation (used for OSHA regulation of hazardous material transfer)
EPA	Environmental Protection Agency
FDA	Food and Drug Administration
FSIS	Food Safety and Inspection Services
IACUC	Institutional Animal Care and Use Committee
NADC	National Animal Disease Center
NAPCC	National Animal Poison Control Center
NAVTA	National Association of Veterinary Technicians in America
NVSL	National Veterinary Services Laboratories
OFA	Orthopedic Foundation for Animals
OSHA	Occupational Safety & Health Administration
USDA	United States Department of Agriculture
USP	United States Pharmacopeia
VTAS	Veterinary Technician Anesthetist Society

PHARMACOLOGY ABBREVIATIONS

ac	before meals (*ante cibum*)
ad lib	as much as desired (*ad libitum*)
bid	twice daily (*bis in die*)
BSA	body surface area
cal	calorie
cap	capsule
cc	cubic centimeter (same as mL)
cm	centimeter
conc	concentration
dr	dram; equal to 1/8 oz or 4 mL
D_5W	5% dextrose in water
ED	effective dose
ED_{50}	median effective dose
fl oz	fluid ounce
g	gram
gal	gallon
gr	grain; unit of weight approximately 65 mg
gt	drop (*gutta*)
gtt	drops (*guttae*)
hr	hour
IA	intra-arterial
IC	intracardiac
ID	intradermal
IM	intramuscular
IP	intraperitoneal
IT	intrathecal
IU	International Unit or intrauterine
IV	intravenous
kg	kilogram
km	kilometer
L or l	liter
lb or #	pound
LD	lethal dose
LRS	lactated Ringer's solution
m	meter
MBC	minimum bactericidal concentration
mcg or µg	microgram
MED	minimal effective dose
mEq	milliequivalent
mg	milligram
MIC	minimum inhibitory concentration
MID	minimum infective dose
Ml	milliliter (same as cc)
MLD	minimum lethal dose
mm	millimeter (also used for muscles)
NPO or npo	nothing by mouth (*non per os*)
NS	normal saline
NSAID	nonsteroidal anti-inflammatory drug
OTC	over the counter
oz	ounce
pc	after meals (*post cibum*)
PDR	*Physician's Desk Reference*
%	percent
pH	hydrogen ion concentration (acidity and alkalinity measurement)
PO or po	orally (*per os*)
ppm	parts per million
PR	per rectum
prn	as needed
pt	pint
PZI	protamine zinc insulin
q	every
qd	every day
qh	every hour
q4h	every 4 hours
q6h	every 6 hours
q8h	every 8 hours
q12h	every 12 hours
q24h	every 24 hours
qid	four times daily (*quater in die*)
qn	every night
qod or eod	every other day
qp	as much as desired
qt	quart
®	registered trade name (superscript next to drug name)
sid	once daily (q24h is more common abbreviation for once daily)
sig	let it be written as (used when writing prescriptions)
sol'n or soln	solution
SQ, SC, subq, or subcu	subcutaneous
T	tablespoon or tablet (or temperature)
tab	tablet
tid	three times daily (*ter in die*)
™	registered trade mark (superscript next to drug name)
tsp or t	teaspoon
vol	volume
VPB	*Veterinary Pharmaceuticals and Biologicals*

LABORATORY ABBREVIATIONS

ab	antibody
ABG	arterial blood gas
ABO	human blood groups
ag	antigen
alb	albumin
alk phos	alkaline phosphatase
ALT	alanine aminotransferase (formerly SGPT)
amyl	amylase
AST	aspartate aminotransferase (formerly SGOT)

BUN	blood urea nitrogen
CBC	complete blood count
CFT	complement fixation test
CHOL	cholesterol
CK	creatine kinase
CMT	California mastitis test
CREA	creatine
diff	differential white blood count
EDTA	ethylenediaminetetraacetic acid; type of anticoagulant
ESR	erythrocyte sedimentation rate
GGT	gamma glutamyl transpeptidase
GLU	glucose
GTT	glucose tolerance test
H&E	hematoxylin and eosin stain
Hb or Hgb	hemoglobin
Hct or crit	hematocrit
HDL	high-density lipoprotein
HPF	high-power field
HW	heartworm
LDH	lactate dehydrogenase
LDL	low-density lipoprotein
LPF	low-power field
MCH	mean corpuscular hemoglobin
MCHC	mean corpuscular hemoglobin concentration
MCV	mean corpuscular volume
ME	myeloid–erythroid ratio
NRBC	nucleated red blood cell
PCV	packed cell volume
PMN	polymorphonuclear neutrophil leukocyte
PT	prothrombin time
PTT	partial thromboplastin time
qns	quantity not sufficient
qs	quantity sufficient
RBC	red blood cell
rpm	revolutions per minute
SAP	serum alkaline phosphatase
SCC	somatic cell count
sed or SR	sedimentation rate
SGOT	serum glutamic oxaloacetic transaminase; now abbreviated AST
SGPT	serum glutamic pyruvic transaminase; now abbreviated ALT
SPF	specific pathogen free
sp. gr.	specific gravity
Staph	*Staphylococcus* bacteria
Strep	*Streptococcus* bacteria
T_3	triiodothyronine (one type of thyroid hormone)
T_4	thyroxine (one type of thyroid hormone)
TB	tuberculin
TBIL	total bilirubin

TNTC	too numerous to count
TP	total protein
UA	urinalysis
WBC	white blood cell
WMT	Wisconsin mastitis test

VACCINATION ABBREVIATIONS

BRSV	bovine respiratory syncytial virus
BVD	bovine viral diarrhea
DHLPP	distemper, hepatitis, leptospirosis, parainfluenza, and parvovirus
DHLPP-CV	distemper, hepatitis, leptospirosis, parainfluenza, parvovirus, and coronavirus
EEE	eastern equine encephalitis
EIA	equine infectious anemia (vaccine not available in North America)
EPM	equine protozoal myelitis
FeLV, FeLeuk, or FeLuk	feline leukemia virus
FIP	feline infectious peritonitis
FIV	feline immunodeficiency virus
FVRCP	feline viral rhinotracheitis, calicivirus, and panleukopenia
FVRCP-C	feline viral rhinotracheitis, calicivirus, panleukopenia, and chlamydia
IBR	infectious bovine rhinotracheitis
MLV	modified live vaccine
PHF	Potomac horse fever
PI-3	parainfluenza 3 virus
PRRS	porcine reproductive and respiratory syndrome
RSV	respiratory syncytial virus
RV	rabies vaccine
TE	tetanus
TGE	transmissible gastroenteritis
WEE	western equine encephalitis
WNV	West Nile virus
VEE	Venezuelan equine encephalitis

PHYSICAL EXAMINATION, PHYSIOLOGY, AND PATHOLOGY ABBREVIATIONS

ACh	acetylcholine
ACH	adrenocortical hormone
AChE	acetylcholinesterase
ACTH	adrenocorticotropic hormone
ADH	antidiuretic hormone
AF	atrial fibrillation
AI	artificial insemination
AKI	acute kidney injury

ANS	autonomic nervous system		FLUTD	feline lower urinary tract disease
ARF	acute renal failure		F/S	female spayed
ASAP	as soon as possible		FSH	follicle-stimulating hormone
ASD	atrial septal defect		FUO	fever of unknown origin
AV	atrioventricular		FUS	feline urological syndrome
BAR	bright, alert, responsive		GERD	gastroesophageal reflux disease
BD/LD	big dog/little dog (used for dogs wounded in a fight)		GFR	glomerular filtration rate
			GH	growth hormone
BM	bowel movement		GI	gastrointestinal
BP	blood pressure		GSW	gunshot wound
BPM	beats or breaths per minute		HBC	hit by car
$\bar{c}$	with		hCG	human chorionic gonadotropin
C	castrated		HD	hemodialysis
C_2, etc.	cervical vertebra (followed by its numeric order)		HR	heart rate
			IBD	inflammatory bowel disease
cath	catheter		ICSH	interstitial cell-stimulating hormone
CC	chief complaint		ICU	intensive care unit
Cd_1, Cd_2, etc.	coccygeal vertebra (followed by its numeric order)		IHD	intermittent hemodialysis
			IVDD	intervertebral disc disease
ChE	cholinesterase		K-9	canine
CHF	congestive heart failure		Ⓛ	left
CNS	central nervous system		L_1, L_2, etc.	lumbar vertebra (followed by its numeric order)
CO	carbon monoxide			
CO_2	carbon dioxide		LA	large animal
COPD	chronic obstructive pulmonary disease		LDA	left displaced abomasum
CP	conscious proprioception		LE	lupus erythematosus
CPR	cardiopulmonary resuscitation		lg	large
CRF	chronic renal failure		LH	luteinizing hormone
CRRT	continuous renal replacement therapy		LOC	level of consciousness
CRT	capillary refill time		LTH	luteotropin hormone
C section	cesarean section		LV	left ventricle
CSF	cerebrospinal fluid		M	male
CSM	carotid sinus massage		M/C	male castrated
CVP	central venous pressure		MDB	minimum database
Cy_1, Cy_2, etc.	coccygeal vertebra (followed by its numeric order)		mm	millimeter
			MM	mucous membrane
DA	displaced abomasum		mm Hg	millimeters of mercury
DD	differential diagnosis		M/N	male neutered
DDN	dull, depressed, nonresponsive		MS	mitral stenosis
DIC	disseminated intravascular coagulation		MSDS	material data safety sheet
DJD	degenerative joint disease		MSH	melanocyte stimulating hormone
DLH	domestic longhair (feline)		N	neutered (or normal on physical examination)
DNA	deoxyribonucleic acid			
DOA	dead on arrival		NA or N/A	not applicable
DSH	domestic shorthair (feline)		NG	nasogastric
ECG or EKG	electrocardiogram or electrocardiograph		NPN	nonprotein nitrogen
EEG	electroencephalogram or electroencephalograph		OA	osteoarthritis
			OB	obstetrics
EMG	electromyogram		OHE or OVH	ovariohysterectomy
ET	endotracheal		OR	operating room
F	Fahrenheit or female		$\bar{p}$	after
FA	fatty acid		P	pulse

P1	phalanx 1
P2	phalanx 2
P3	phalanx 3
PD	polydipsia
PDA	patent ductus arteriosus
PE	physical examination
PEEP	positive end-expiratory pressure
pg	pregnant
PM	postmortem; also abbreviation for evening
PMI	point of maximal intensity
PNS	peripheral nervous system
POVMR	problem-oriented veterinary medical records
PTH	parathyroid hormone
PU	polyuria
PVC	premature ventricular complex
®	respirations or right
RDA	right displaced abomasum
RNA	ribonucleic acid
R/O	rule out
ROM	range of motion
RP	retained placenta
RR	respiration rate
RV	right ventricle
$\bar{s}$	without
S	spayed
S_1, S_2, etc.	sacral vertebra (followed by its numeric order)
SA	sinoatrial or small animal
SOAP	subjective, objective, assessment, plan (record-keeping acronym)
Stat	immediately (*statim*)
STH	somatotropin hormone (also known as growth hormone)
T	temperature (tablespoon or tablet)
T_1, T_2, etc.	thoracic vertebra (followed by its numeric order)
TLC	tender loving care
TPN	total parenteral nutrition
TPO	triple pelvic osteotomy
TPR	temperature, pulse, and respiration
TSH	thyroid-stimulating hormone
TTA	transtracheal aspiration
TTW	transtracheal wash
TVT	transmissible venereal tumor
URI	upper respiratory infection
UTI	urinary tract infection
VM	vagal maneuver
VSD	ventricular septal defect
VT	ventricular tachycardia
WNL	within normal limits
wt	weight

SYMBOLS

≅	approximately equal to
↑	increased
↓	decreased
+	positive (used to describe test results); may have multiple +s to indicate degree
−	negative (used to describe test results); may have multiple −s to indicate degree
.	times or multiplication sign
√	check
=	equal to
#	number (in front of number; for example, #1); pound (following number; for example, 50#)
≠	not equal to
<	less than
≤	equal to or less than
>	greater than
≥	equal to or greater than
@	at or each
%	percent
♂	male
♀	female
°	degree

THE Xs

bx	biopsy
dx	diagnosis
ddx	differential diagnosis
fx	fracture
hx	history
Rx	prescription
sx	surgery
tx	treatment

EYE AND EAR ABBREVIATIONS

AD	right ear
AS	left ear
AU	both ears
IOP	intraocular pressure
OD	right eye (also abbreviation for overdose)
OS	left eye
OU	both eyes
PLR	pupillary light reflex

CHEMICAL ABBREVIATIONS

Ca	calcium
Cl	chlorine
Fe	iron

H	hydrogen
H_2O	water
H_2O_2	hydrogen peroxide
HCl	hydrochloric acid
I	iodine
K	potassium

KCl	potassium chloride
N	nitrogen (also abbreviation for normal)
Na	sodium
NH_3	ammonia
O_2	oxygen

Appendix B
PLURAL FORMS OF MEDICAL TERMS

Many plural word forms are formed by adding *s* to the singular form. This is true for medical terms as well. For example, the plural of *laceration* is *lacerations* and the plural of *bone* is *bones*. However, some rules should be followed when plural forms of medical terms are used. These rules are presented in the following table.

Singular Ending	Change or Deletion from Singular Form	Add Plural Ending	Examples (singular)	Plural Form
a	delete *a*	ae	vertebra	vertebrae
ax	delete *ax*	aces	thorax	thoraces
ch		es	stitch	stitches
en	delete *en*	ina	foramen	foramina
ex	delete *ex*	ices	cortex	cortices
is	delete *is*	es	diagnosis	diagnoses
ix	delete *ix*	ices	cervix	cervices
nx	delete *nx*	nges	phalanx	phalanges
oma#	delete *oma*	omata	stoma	stomata
on†	delete *on*	a	spermatozoon	spermatozoa
s		es	abscess	abscesses
sh		es	brush	brushes
u	delete *u*	ua	cornu	cornua
um	delete *um*	a	bacterium	bacteria
us*	delete *us*	i	alveolus	alveoli
y	delete *y*	ies	capillary	capillaries
yx	delete *yx*	ices	calyx	calices

except plural of *carcinoma* may be *carcinomas*
† except plural of *chorion* is *chorions*
* except plural of *virus* is *viruses* and plural of *sinus* is *sinuses*

Appendix C
PREFIXES, COMBINING FORMS, AND SUFFIXES FOR MEDICAL TERMS

Word Part	Meaning	Word Origin (L = Latin, G = Greek)
A		
a-	away from, no, not, without	L, G
ab-	away from	L
abdomin/o	abdomen	L
able	capable of	L
abort/o	premature expulsion of a nonviable fetus	L
abrad/o or abras/o	scrape off	L
abrupt/o	broken away from	L
abs-	away from	L
absorpt/o	to suck up or in	L
ac-	toward	L
-ac	pertaining to	L, G
acanth/o	spiny	G
acetabul/o	hip socket	L
-acious	characterized by	L
acne/o	point	G
acous/o or acoust/o	hearing	G
-acousia	hearing	G
acr/o	extremities or top	G
acromi/o	point of the shoulder blade	G
acu/o	sharp, severe, sudden, needle	L
acuit/o or acut/o	sharp	L
ad-	toward	L
-ad	toward, to, in the direction of	L
aden/o	gland	G
adhes/o	stickiness or clinginess	L
adip/o	fat	L
adnex/o	bound to	L
adren/o or adrenal/o	adrenal glands	L
aer/o	air or gas	L, G

Word Part	Meaning	Word Origin (L = Latin, G = Greek)
aesthet/o	sensation or feeling	G
af-	toward	L
affect/o	exert influence on	L
ag-	toward	L
agglutin/o	clumping	L
aggress/o	attack or step forward	L
-ago	attack, diseased condition	G
-agra	excessive pain or seizure	G
-aise	ease or comfort	F
al-	similar	L
-al	pertaining to	L, G
alb/i, alb/o, or albin/o	white	L
albumin/o	albumin (one type of protein)	L
alg/e or alg/o	pain	G
algesi/o	pain or suffering	G
-algesia	pain or suffering	G
-algesic	painful	G
algi-	pain or suffering	G
-algia	pain or suffering	G
align/o	to bring in line or the correct position	L
aliment/o	to nourish	L
all-	other or different	G
all/o	other or different	G
alopec/o	baldness	G
alveol/o	small sac	L
amb-	both sides or double	L
amb/i	both sides or double	L
ambly/o	dim or dull	G
ambul/o or ambulat/o	to walk	L
ametr/o	out of proportion	G
-amine	nitrogen compound	L
amni/o	fetal membrane	G

Word Part	Meaning	Word Origin (L = Latin, G = Greek)
amph-	around, on both sides, or doubly	G
amput/o or amputat/o	to cut away or cut off	L
amyl/o	starch	G
an-	no, not, without	G
an/o	anus or ring	L
an/o	upward	G
ana-	up, apart, backward, excessive	G
andr/o	male	G
angi-	vessel	G
angi/o	vessel	G
angin/o	strangling	L
anis/o	unequal	G
ankyl/o	bent or stiff	G
anomal/o	irregularity	G
ante-	before or forward	L
anter/o	front	L
anti-	against	G
anxi/o or anxiet/o	uneasy or distressed	L
aort/o	aorta, or major artery in the body	L, G
ap-	toward	G
aphth/o	small ulcer or eruption	L, G
apic/o	apex or point	L
aplast/o	lack of or defective development	G
apo-	opposed or detached	G
aponeur/o	aponeurosis	G
aqu/i, aqu/o, aque/o	water	L
-ar	pertaining to	L
arachn/o	spider	G
arc/o	bow or arch	L
-arche	beginning	G
arrect/o	upright	L
arteri/o	artery	L, G
arthr/o	joint	G
articul/o	joint	L
-ary	pertaining to	L
as-	toward	L
-ase	enzyme	G
asphyxi/o	absence of a pulse	G
aspir/o or aspirat/o	to breathe in	L
asthen-	weakness	G
-asthenia	weakness	G
asthmat/o	gasping	L, G
at-	toward	L

Word Part	Meaning	Word Origin (L = Latin, G = Greek)
atel/o	incomplete	G
ather/o	plaque or fatty substance	G
athet/o	uncontrolled	G
-atonic	lack of tone or strength	L
atop/o	out of place	G
atres/i	without an opening	G
atri/o	atrium or chamber	L
attenuat/o	diluted or weakened	L
aud-	hear or hearing	L
audi/o, audit/o, aur/i, or aur/o	hear or hearing	L
auscult/o	to listen	L
aut/o	self	G
ax/o	axis or main stem	G
axill/o	armpit	L
azot/o	urea or nitrogen	G

B

Word Part	Meaning	Word Origin (L = Latin, G = Greek)
bacill/o	rod	L
bacteri/o	bacteria	G
bar/o	pressure or weight	G
bas/o	base (not acid)	G
bi-	twice, double, or two	L
bi/o	life	G
bifid/o	split into two parts	L
bifurcat/o	divide into two branches	L
bil/i	bile or gall	L
bilirubin/o	bilirubin, or breakdown product of blood	L
-bin	twice, double, or two	L
bio-	life	G
bis-	twice, double, or two	L
-blast	embryonic	G
blast/i	embryonic	G
blephar/o	eyelid	G
borborygm/o	rumbling sound	L
brachi/o	arm	L
brachy-	short	G
brady-	abnormally slow	G
brev/i or brev/o	short	L
bronch/i, bronch/o, or bronchi/o	bronchial tube	L
brux/o	to grind	G
bucc/o	cheek	L
bucca-	cheek	L
bulla-	blister	L
burs/o	sac of fluid near a joint (bursa)	L, G

C

Word Part	Meaning	Word Origin (L = Latin, G = Greek)
cac-	bad, diseased, or weak	G
cac/o	bad, diseased, or weak	G
cadaver/o	dead body	L
calc/i or calc/o	calcium or the heel	L
calcane/o	heel bone (calcaneus)	L
calcul/o	stone	L
cali/o or calic/o	cup	L
callos/o or call/i	hard	L
calor/i	heat	L
canalicul/o	little duct	L
canth/o	corner of the eye	G
capill/o	hairlike	L
capit/o	head	L
capn/o	carbon dioxide	G
-capnia	carbon dioxide	G
capsul/o	little box	L
carb/o	carbon	L
carcin/o	cancerous	G
cardi/o	heart	G
cari/o	decay	L
carot/o	sleep or stupor	G
carp/o	carpus or joint of front limbs between the radius or ulna and metacarpals	L, G
cartilag/o	cartilage or gristle	L
caruncul/o	bit of flesh	L
cat-, cata-, or cath-	down, lower, under, or downward	G
catabol/o	breaking down	G
cathart/o	cleansing or purging	G
cathet/o	insert or send down	G
caud/o	toward the tail	L
caus/o, caust/o, cauter/o, or caut/o	burning	L, G
cav/i or cav/o	hollow	L
cavern/o	containing hollow spaces	L
cec/o	blind gut, or cecum	L
-cele	hernia, cyst, cavity, or tumor	L
celi/o	abdomen	G
cement/o	cementum	L
cent-	hundred	L
-centesis	surgical puncture to remove fluid or gas	G
cephal-	head	G
cephal/o	head	G

Word Part	Meaning	Word Origin (L = Latin, G = Greek)
-ceps	head	L
cera-	wax	L
cerebell/o	cerebellum	L
cerebr/o	cerebrum, or largest part of brain	L
cerumin/o	cerumen, or earwax	L
cervic/o	neck or necklike	L
chalas/o	relaxation or loosening	G
-chalasis	relaxation or loosening	G
chalaz/o	hailstone, small lump	G
cheil/o	lip	G
chem/i, chem/o, or chemic/o	drug or chemical	G
chol/e	bile or gall	G
cholangi/o	bile duct	G
cholecyst/o	gallbladder	G
choledoch/o	common bile duct	G
cholesterol/o	cholesterol	G
chondr/o or chondri/o	cartilage	G
chord/o	cord or spinal cord	G
chori/o	chorion, or membrane	G
choroid/o	choroid layer of the eye	G
chrom/o or chromat/o	color	G
chym/o	juice	G
cib/o	meal	L
-cidal	death	L
-cide	killing or destroying	L
cili/o	microscopic hairlike projections or eyelashes	L
cine/o	movement	G
circi/o	ring or circle	L
circulat/o	to go around in a circle	L
circum-	around	L
circumscrib/o	confined or limited in space	L
cirrh/o	tawny, orange-yellow	G
cis/o	cut	L
-clasis or -clast	break	G
clav/i	key	L
clavicul/o	collarbone, clavicle	L
clitor/o	female erectile tissue	G
clon/o	violent action	G
-clysis	irrigation	G
co-	together, with	L
coagul/o or coagulat/o	clump or congeal	L
cocc/i or cocc/o	round	L

Word Part	Meaning	Word Origin (L = Latin, G = Greek)
-coccus	round bacterium	L
coccyg/o	tailbone	G
cochle/o	snail-like or spiral	L
-coel	hollow	G
coher/o or cohes/o	to stick together	L
coit/o	coming together	L
col/o or colon/o	colon (part of large intestine)	L, G
coll/a	glue	G
coll/i	neck	L
colp/o	vagina	G
column/o	pillar	L
com-	together	L
comat/o	deep sleep	L, G
comminut/o	to break into pieces	L
communic/o	to share	L
compatibil/o	to sympathize with	L
con-	together or with	L
concav/o	hollow	L
concentr/o	to remove excess water or condense	L
concept/o	to receive or become pregnant	L
conch/o	shell	L, G
concuss/o	shaken together violently	L
condyl/o	knuckle or knob	L, G
confus/o	disorder or confusion	L
conjunctiv/o	pink mucous membrane of the eye or connected	L
consci/o	aware	L
consolid/o	to become solid	L
constipat/o	pressed together	L
constrict/o	to draw tightly together to narrow	L
contact/o	touched	L
contagi/o	touching of something or infection	L
contaminat/o	to pollute	L
contine/o or continent/o	to keep in or restrain	L
contra-	against or opposite	L
contracept/o	prevention of fertilization of egg with sperm	L
contus/o	bruise	L
convalesc/o	to become strong	L
convex/o	arched	L
convolut/o	twisted or coiled	L
convuls/o	to pull together	L
copi/o	plentiful	L
copulat/o	joining together	L

Word Part	Meaning	Word Origin (L = Latin, G = Greek)
cord/o	spinal cord	L, G
cordi/o	heart	L
cor/o or core/o	pupil	G
core-	pupil	G
cori/o	skin or leather	L
corne/o	transparent anterior portion of the sclera; cornea	L
coron/o	crown	L, G
corp/u or corpor/o	body	L
corpuscul/o	little body	L
cort-	covering	L
cortic/o	outer region or cortex	L
cost/o	rib	L
cox/o	hip or hip joint	L
crani/o	skull	L, G
-crasia	mixture	L, G
crepit/o or crepitat/o	crackling	L
crin/o	to secrete or separate	G
-crine	to secrete or separate	G
cris/o or critic/o	turning point	L
-crit	separate	G
cruci-	cross	L
cry/o	cold	G
crypt/o	hidden	G
cubit/o	elbow	L
cuboid/o	cubelike	G
cusp/i	pointed or pointed flap	L
cut-	skin	L
cutane/o	skin	L
cyan/o	blue	G
cycl/o	ciliary body of the eye or cycle	G
cyst-, -cyst	bag or bladder	G
cyst/o	urinary bladder	G
cyt/o	cell	G
-cyte or -cytic	cell	G
-cytosis	condition of cells	L

D

dacry-	tear or tear duct	G
dacry/o	tear or tear duct	G
dacryocyst/o	lacrimal sac	G
dactyl/o	digits	G
dart/o	skinned	G
de-	from, not, down, or lack of	L
dec/i	ten or tenth	L
deca-	ten or tenth	L
decem-	ten	L

Word Part	Meaning	Word Origin (L = Latin, G = Greek)
decidu/o	falling off or shedding	L
decubit/o	lying down	L
defec/o or defecat/o	free from waste	L
defer/o	carrying down or out	L
degenerat/o	breakdown	L
deglutit/o	swallow	L
dehisc/o	to burst open	L
dek- or deka-	ten	G
deliri/o	wandering in the mind	L
delta	triangle	L
dem/o	population or people	G
-dema	swelling (fluid)	G
demi-	half	L
dendr/o	resembling a tree or branching	G
dent-	teeth	L
dent/i or dent/o	teeth	L
depilat/o	hair removal	L
depress/o	pressed or sunken down	L
derm/o or dermat/o	skin	G
derma-	skin	G
-desis	surgical fixation of a bone or joint or to bind	G
deteriorat/o	worsening	L
deut-, deuto-, or deuteron	second	G
dextr/o	right side	L
di-	twice	G
dia-	through, between, apart, or complete	G
diaphor/o	sweat	G
diaphragmat/o	wall across or muscle separating the thoracic and abdominal cavities	G
diastol/o	expansion	G
didym/o	testes or double	G
diffus/o	spread out	L
digest/o	divided or to distribute	L
digit/o	digit	L
dilat/o or dilatat/o	spread out	L
dilut/o	to dissolve or separate	L
diphther/o	membrane	G
dipl/o	double	G
dipla-	double	G
dips/o	thirst	G
-dipsia	thirst	G
dis-	negative, apart, or absence of	L
dis-	duplication or twice	G

Word Part	Meaning	Word Origin (L = Latin, G = Greek)
disc/o	disc (intervertebral)	G
dislocat/o	displacement	L
dissect/o	cutting apart	L
disseminat/o	widely scattered	L
dist/o	far	L
distend/o or distent/o	to stretch apart	L
diur/o or diuret/o	increasing urine output	G
divert/i	turning aside	L
domin/o	controlling	L
don/o	to give	L
dors/i or dors/o	back of body	L
drom/o	running	G
-drome	to run together	G
du/o	two	L
duct/o	to lead	L
-duct	opening	L
duoden/i or duoden/o	duodenum	L
dur/o	dura mater or tough, hard	L
dy- or dyo-	two	L
-dynia	pain	G
dys-	difficult, bad, or painful	G

E

Word Part	Meaning	Word Origin (L = Latin, G = Greek)
e-	out of or from	L
-eal	pertaining to	L
ec-	out or outside	G
ecchym/o	pouring out of juice	G
ech/o	sound	G
eclamps/o or eclampt/o	flashing or shining forth	G
-ectasia or -ectasis	dilation or enlargement	G
ecto-	out or outside	G
-ectomy	surgical removal or excision	G
eczemat/o	eruption	G
edemat/o	swelling, fluid, or tumor	G
edentul/o	toothless	L
ef-	out	L
effect/o	to bring about a response	L
effus/o	pouring out	L
ejaculat/o	to throw out	L
electr/o	electricity	G
eliminat/o	to expel from the body	L
em-	in	G
emaciat/o	wasted away by disease or lean	L
embol/o	something inserted or thrown in	L, G
embry/o	fertilized egg	G

Word Part	Meaning	Word Origin (L = Latin, G = Greek)
-emesis	vomiting	G
emet/o	vomit	G
-emia	blood or blood condition	G
emmetr/o	in proper measure	G
emolli/o	to soften	L
en-	in, into, or within	L, G
encephal/o	brain	G
end/o	within, in, or inside	G
endo-	within, in, or inside	G
endocrin/o	to secrete within	G
enem/o	to inject	G
ennea-	nine	G
enter/o	small intestine	G
ento-	within	G
enzym/o	leaven	G
eosin/o	red or rosy	G
epi-	upon, above, on, or upper	G
epidemi/o	among the people	G
epididym/o	tube at the upper part of each testis (epididymus)	G
epiglott/o	lid covering the larynx (epiglottis)	G
episi/o	vulva	G
epithel/i	outer layer of skin or external surface covering	G
equin/o	horse or horselike	L
erect/o	upright	L
erg/o	work	G
eruct/o or eructat/o	to belch	L
erupt/o	to burst forth	L
erythem/o or erythemat/o	flushed or redness	G
erythr/o	red	G
es-	out of, outside, or away from	L
-esis	abnormal condition or state	G
eso-	inward	G
esophag/o	tube that connects mouth to the stomach (esophagus)	G
esthesi/o	sensation or feeling	G
-esthesia	sensation or feeling	G
esthet/o	feeling or sense of perception	G
estr/o	female	L
ethm/o	sieve	G
eti/o	to cause	G

Word Part	Meaning	Word Origin (L = Latin, G = Greek)
eu-	well, easy, or good	G
evacu/o or evacuat/o	to empty out	L
eviscer/o or eviscerat/o	disembowelment	L
ex-	out of, outside, or away from	L, G
exacerbat/o	to aggravate	L
excis/o	cutting out	L
excori/o or excoriat/o	to scratch	L
excret/o	separate or discharge	L
excruciat/o	intense pain	L
exhal/o or exhalat/o	to breathe out	L
-exia or -exis	condition	G
exo-	out of, outside, or away from	G
exocrin/o	to secrete out of	G
expector/o	to cough up	L
expir/o or expirat/o	to breathe out or die	L
exstroph/o	twisted out	G
extern/o	outside or outer	L
extra-	outside or beyond	L
extrem/o or extremit/o	outermost or extremity	L
extrins/o	contained outside	L
exud/o or exudat/o	to sweat out	L

F

Word Part	Meaning	Word Origin (L = Latin, G = Greek)
faci-	face or form	L
faci/o	face or form	L
-facient	producing	L
fasci/o	fibrous band (fascia)	L
fascicul/o	little bundle	L
fatal/o	death or fate	L
fauc/i	narrow pass	L
febr/i	fever	L
fec/i or fec/o	sediment	L
femor/o	thigh bone (femur)	L
fenestr/o	window	L
fer/o	to carry or bear	L
-ferent or -ferous	carrying	L
fertil/o	productive or fruitful	L
fet/i or fet/o	fetus	L
fibr/o	fiber	L
fibrill/o	muscular twitching	L
fibrin/o	threads of a clot (fibrin)	L
fibros/o	fibrous connective tissue	L
fibul/o	small bone of distal hindlimb (fibula)	L
-fic	producing or forming	L

Word Part	Meaning	Word Origin (L = Latin, G = Greek)
fic/o	producing or forming	L
filtr/o or filtrat/o	to strain through	L
fimbri/o	fringe	L
fiss/o or fissur/o	crack, split, or cleft	L
fistul/o	tube or pipe	L
flamme/o	flame-colored	L
flat/o	rectal gas	L
flex/o	bend	L
fluor/o	luminous	L
foc/o	point	L
foll/i	bag or sac	L
follicul/o	small sac (follicle)	L
foramin/o	opening	L
fore-	before or in front of	L
-form	form, figure, or shape	L
form/o	form, figure, or shape	L
fornic/o	arch or vault	L
foss/o	shallow depression	L
fove/o	pit	L
fract/o	break	L
fren/o	bridle or device that limits movement	L
frigid/o	cold	L
front/o	forehead or front	L
-fuge	to drive away	L
funct/o or function/o	to perform	L
fund/o	bottom, base, or ground	L
fung/i	fungus	L
furc/o	branching or forking	L
furuncul/o	boil or infection	L
-fusion	to pour	L

G

Word Part	Meaning	Word Origin (L = Latin, G = Greek)
galact/o	milk	G
gamet/o	sperm or egg	G
gangli/o or ganglion/o	ganglion	G
gangren/o	gangrene	L, G
gastr/o	stomach	G
gastrocnemi/o	calf muscle of hindlimb	G
gemin/o	twin or double	L
gen/o or genit/o	producing or birth	G
-gene	production, origin, or formation	G
-genesis or -genic	producing	G
genit/o	related to birth or the reproductive organs	L
genous	producing	G
ger/i or geront/o	old age	G

Word Part	Meaning	Word Origin (L = Latin, G = Greek)
germin/o	bud or germ	L
gest/o or gestat/o	to bear or carry offspring	L
gigant/o	giant or very large	G
gingiv/o	gum	L
glauc/o	gray	G
glen/o	socket or pit	G
gli/o	glue	G
globin/o	protein	L
globul/o	little ball	L
-globulin	protein	L
glomerul/o	grapelike cluster or clusters of capillaries in the nephron (glomerulus)	L
gloss/o	tongue	G
glosso-	tongue	G
glott/i or glott/o	back of tongue	G
gluc/o or glyc/o	sugar	G
glute/o	buttocks	G
glyc/o	sweet	G
glycer/o	sweet	L
glycogen/o	animal starch (glycogen)	G
gnath/o	jaw	G
goitr/o	enlargement of the thyroid gland	L
gon/e, gon/o, or goni/o	seed or angle	G
gonad/o	sex glands	L, G
gracil/o	slender	L
grad/i	to move, go, or step	L
-grade	to go	L
-gram	record or picture	G
granul/o	granule(s)	L
-graph	instrument for recording; occasionally used to describe the record or picture	G
-graphy	process of recording	G
grav/i	heavy or severe	L
gravid/o	pregnancy	L
guttur/o	throat	L
gynec/o	woman	L
gyr/o	turning or folding	G

H

Word Part	Meaning	Word Origin (L = Latin, G = Greek)
hal/o	breath	L, G
halit/o	breath	L
hallucin/o	to wander in the mind	L
hect-, hector-, or hecato	hundred	G

Word Part	Meaning	Word Origin (L = Latin, G = Greek)
hem-	blood	G
hem/e	deep red iron-containing pigment	G
hem/o or hemat/o	blood	G
hemangi/o	blood vessel	G
hemi-	half	G
hepa- or hepar-	liver	G
hepat/o	liver	G
hept- or hepta-	seven	G
hered/o or heredit/o	inherited	L
herni/o	protrusion of a part through tissues normally containing it	L
herpet/o	creeping	L, G
heter/o	different	G
hetero-	different	G
hex- or hexa-	six	G
hiat/o	opening	L
hidr/o	sweat	G
hil/o	notch or opening from a body part (hilus)	L
hirsut/o	hairy or rough	L
hist/o or histi/o	tissue	G
holo-	all	G
hom/o	same	G
home/o	unchanging	G
hormon/o	to urge on or excite	G
humer/o	long bone of the proximal forelimb (humerus)	L
hyal/o	glassy, clear	G
hydr/o	water	G
hydra-	water	G
hygien/o	healthful	G
hymen/o	membrane (hymen)	G
hyper-	over, above, increased, excessive, or beyond	G
hypn/o	sleep	G
hypo-	under, decreased, deficient, or below	G
hyster/o	uterus	G

I

Word Part	Meaning	Word Origin (L = Latin, G = Greek)
-ia	state or condition	L, G
-iac	pertaining to	G
-ial	pertaining to	L
-iasis	condition or abnormal condition	G
iatr/o	relationship to treatment or doctor	G
-ible	capable of being	L

Word Part	Meaning	Word Origin (L = Latin, G = Greek)
-ic	pertaining to	L, G
ichthy/o	dry or scaly	G
icter/o	jaundice	L, G
ictero-	jaundice	L, G
idi/o	peculiar to an individual	G
idio-	peculiar to an individual	G
-iferous	bearing, carrying, or producing	L
-ific	producing	L
-iform	resembling or shaped like	L
-igo	diseased condition	L
-ile	capable of being or pertaining to	L
ile/o	ileum	L
ili/o	part of the pelvis (ilium)	L
illusi/o	deception	L
immun/o	protected, safe, or immune	L
impact/o	pushed against	L
impress/o	pressing into	L
impuls/o	pressure or urging on	L
in-	in, into, not, without	L
-in	substance, usually with specialized function	L
incis/o	cutting into	L
incubat/o	hatching or incubation	L
indurat/o	hardened	L
infarct/o	filled in or stuffed	L
infect/o	tainted or infected	L
infer/o	below or beneath	L
infest/o	attack	L
inflammat/o	flame within or to set on fire	L
infra-	beneath, below, or inferior to	L
infundibul/o	funnel	L
ingest/o	to carry or pour in	L
inguin/o	groin	L
inhal/o or inhalat/o	to breathe in	L
inject/o	to force or throw in	L
innominat/o	nameless	L
inocul/o	to introduce or implant	L
insipid/o	tasteless	L
inspir/o or inspirat/o	to breathe in	L
insul/o	island	L
intact/o	whole	L
inter-	between or among	L
intermitt/o	not continuous	L
intern/o	within or inner	L

Word Part	Meaning	Word Origin (L = Latin, G = Greek)	Word Part	Meaning	Word Origin (L = Latin, G = Greek)
interstiti/o	space between things	L	lacrim/o	tear or tear duct	L
intestin/o	intestine	L	lact/i or lact/o	milk	L
intim/o	innermost	L	lactat/o	to secrete milk	L
intoxic/o	to put poison in	L, G	lacun/o	pit	L
intra-	within, into, inside	L	lamin/o	one part of the dorsal portion of the vertebra (lamina)	L
intrins/o	contained within	L			
intro-	within, into, or inside	L			
introit/o	entrance or passage	L	lapar/o	abdomen or flank	G
intussuscept/o	to receive within	L	laps/o	to slip or fall downward	L
involut/o	curled inward	L	laryng/o	voice box	G
-ion	action, process, state or condition	L	lat/i or lat/o	broad	L
			later/o	side	L
ion/o	charged particle	G	lax/o or laxat/o	to loosen or relax	L
ipsi-	same	L	leiomy/o	smooth muscle	G
ir/i or ir/o	iris of eye or rainbow	G	lemm/o	peel	G
isch/o	to hold back	G	lent/i	lens of the eye	L
ischi/o	part of pelvis (ischium)	G	lenticul/o	shaped like a lens	L
-ism	condition or disorder	G	-lepsy	seizure	L, G
iso-	equal	G	lept/o	thin	G
-itis	inflammation	G	letharg/o	drowsiness or indifference	L
-ium	structure	G			
-ive	performs or tends toward	L	leuco-	white	G
			leuk/o	white	G
J			lev/o or levat/o	to lift up	L
jejun/o	jejunum	L	libid/o or libidin/o	sexual drive	L
jugul/o	throat	L	ligament/o	fibrous tissue connecting bone to bone (ligament)	L
juxta-	nearby	L			
			ligat/o	binding or tying off	L
K			lingu/o	tongue	L
kal/i	potassium	L	lip/o	fat	G
kary/o	nucleus	G	-lite or -lith	stone or calculus	L, G
karyo-	nucleus	G	lith/o	stone or calculus	L, G
kata- or kath-	down	G	-lithiasis	presence of stones	L, G
kel/o	growth or tumor	G	lob/i or lob/o	lobe or well-defined part of an organ	L, G
kera-	horn or hardness	G			
kerat/o	hornlike, hard, or cornea	G	loc/o	place	L
ket/o or keton/o	breakdown product of fat (ketones)	German	loch/i	confinement or birthing process	L, G
kinesi/o	movement	G	-logist	specialist	G
-kinesis	motion	G	-logy	study of	G
-kinetics	pertaining to motion	G	longev/o	long-lived	L
koil/o	hollow or concave	G	lord/o	bent backward	G
kraur/o	dry	G	lumb/o	lower back, or loin	L
kyph/o	bent or hump	G	lumin/o	light	L
			lun/o	moon	L
L			lunul/o	crescent	L
labi/o	lip	L	lup/i or lup/o	wolf	L
labyrinth/o	maze (used to describe the inner ear)	G	lute/o	yellow	L
lacer/o or lacerat/o	torn	L	lymph/o	lymphatic tissue	L

Word Part	Meaning	Word Origin (L = Latin, G = Greek)
lymphaden/o	lymph gland	L
lymphangi/o	lymph vessel	L
-lysis	breakdown, destruction, or separation	L, G
-lyst	agent that breaks down	L, G
lytic	to reduce or destroy	L, G

M

Word Part	Meaning	Word Origin (L = Latin, G = Greek)
macr/o	large or abnormal size	G
macro-	large or abnormal size	G
macul/o	spot	L
magn/o	great or large	L
major/o	larger	L
mal-	bad, poor, or evil	L
malac/o	abnormal softening	G
-malacia	abnormal softening	G
malign/o	bad or evil	L
malle/o	hammer	L
malleol/o	little hammer	L
mamm/o	breast or mammary gland	L
man/i	rage or hand	L
man/o	hand	L
mandibul/o	lower jaw	L
-mania	obsessive preoccupation	L, G
manipul/o	handful or use of hands	L
manubri/o	handle	L
mastic/o	to chew	L, G
mast/o	breast or mammary gland	L
mastoid/o	mastoid process of temporal bone	L, G
mastricat/o	bone to chew	L, G
matern/o	of a mother	L
matur/o	ripe	L
maxill/o	upper jaw	L
maxim/o	largest	L
meat/o	opening or passageway	L
med-	middle	L
medi/o	middle	L
mediastin/o	in the middle	L
medic/o	medicine, healing, or doctor	L, G
medicat/o	medication or healing	L
medull/o	inner section, soft, or middle	L
mega-	large	G
megal/o	large	G
-megaly	large or enlargement	G
mei/o	less	L, G
melan/o	black	G
mellit/o	honey	L

Word Part	Meaning	Word Origin (L = Latin, G = Greek)
membran/o	thin skin (membrane)	L
men/o	menstruation	G
mening/o or meningi/o	membrane lining the central nervous system	L, G
menisc/o	crescent	L, G
mens-	menstruation	L
mens/o	menstruation	L
ment/o	mind	L
mes-	middle	L, G
mes/o	middle	G
mesenter/o	mesentery or middle intestine	L, G
mesi/o	middle, or median, plane	G
meta-	beyond, change, behind, after, or next	G
metabol/o	change	G
metacarp/o	beyond the carpus	L
metatars/o	beyond the tarsus	L
-meter	measuring device	G
metr/i, metr/o, or metri/o	uterus	G
mi/o	smaller or less	G
micr/o	small	G
micro-	small; one-millionth of the unit it precedes	G
mictur/o or micturit/o	to urinate	L
midsagitt/o	dividing into equal left and right halves	L
milli-	one-thousandth	L
-mimetic	mimic or copy	L
mineral/o	naturally occurring non-organic solid substance (mineral)	L
minim/o	smallest	L
minor/o	smaller	L
mio-	smaller or less	G
mit/o	thread	G
mitr/o	miter having two points on top	L
mobil/o	capable of moving	L
mon/o	single or one	G
monil/i	string of beads	L
mono-	single or one	G
morbid/o	disease	L
moribund/o	dying	L
morph/o	shape or form	G
mort/i, mort/o, or mort/u	dead	L
mortal/i	death	L
mot/o or motil/o	movement	L

Word Part	Meaning	Word Origin (L = Latin, G = Greek)
muc/o or mucos/o	mucus	L, G
multi-	many	L
muscul/o	muscle	L
mut/a	genetic change	L
my/o	muscle	G
myc/o or myc/e	fungus	G
mydri/o	wide	L, G
myel/o	spinal cord or bone marrow; "white substance"	G
myocardi/o	heart muscle	L
myom/o	muscle tumor	L
myos/o	muscle	L, G
myring/o	tympanic membrane (eardrum)	L
myx or myxo-	mucus or slime	L, G
myx/o	mucus or slime	L, G
myxa-	mucus	L, G

N

Word Part	Meaning	Word Origin (L = Latin, G = Greek)
nar/i	nostril	L
narc/o	numbness or stupor	L, G
nas/i or nas/o	nose	L
nat/i	birth	L
natr/o	sodium	L
nause/o	sick feeling to stomach	L, G
necr/o	death	G
-necrosis	death of tissue	G
ne/o	new	G
neo-	new	G
nephr/o	kidney	G
nephra-	kidney	G
nerv/o	nerve	L, G
neu-	nerves	G
neur/i or neur/o	nerve	G
neutr/o	neither or neutral	L
nigr/o	black	L
niter-	nitrogen	L, G
nitro-	nitrogen	G
noct/i	night	L
nod/o	knot or swelling	L
nodul/o	little knot	L
nom/o	control	G
non-	no	L
non-, nona-, or nonus-	nine	L
nor-	chemical term meaning normal or parent compound	Chemical term
norm/o	usual	L

Word Part	Meaning	Word Origin (L = Latin, G = Greek)
not/o	back	G
novem-	nine	L
nuch/o	nape	L
nucle/o	nucleus	L
nulli-	none	L
numer/o	count or number	L
nunci/o	messenger	L
nutri/o or nutrit/o	nourishment or food	L
nyct/o or nyctal/o	night	G

O

Word Part	Meaning	Word Origin (L = Latin, G = Greek)
o-	egg	G
ob-	against	L
obes/o	extremely fat	L
obliqu/o	slanted or sideways	L
oblongat/o	oblong or elongated	L
obstetr/o or obstetr/i	one who receives	L
occipit/o	back of the skull	L
occlud/o or occlus/o	to close up	L
occult/o	hidden	L
octa-, octo-, or oct-	eight	G
ocul/o	eye	L
oculo-	eye	L
odont/o	tooth	G
-oid	like or resembling	G
olecran/o	proximal projection on the ulna	G
olfact/o	smell	L
oligo-	scanty	G
-ologist	specialist	G
-ology	study of	G
-oma	tumor, mass, or neoplasm (usually benign)	G
oment/o	omentum, or fat	L
onc/o	tumor	G
onych/o	claw or nail	G
oo/o	egg	G
oophor/o	ovary	G
opac/o or opacit/o	shaded or dark	L
oper/o or operat/o	to work or perform	L
opercul/o	lid or cover	L
ophthalm/o	eye	G
-opia, -opsia, -opsis, or -opsy	vision	G
opisth/o	backward	G
opti/o, opt/i, or opt/o	eye	G
optic/o	vision	L, G
or/o	mouth	L
orbit/o	circle, bony cavity, or socket	L

Word Part	Meaning	Word Origin (L = Latin, G = Greek)
orch/o, orchi/o, or orchid/o	testes	G
orex/i or orect/i	appetite	G
organ/o	organ	L, G
orth/o	straight, normal, or correct	G
ortho-	straight, normal, or correct	G
os-	mouth or bone	L
-osis	abnormal condition	G
osm/o	smell or odor; pushing	G
-osmia	smell or odor	G
oss/e or oss/i	bone	L
oste/o	bone	G
-ostomosis or —ostomy	surgically creating an opening	G
ot/o	ear or hearing	G
-otomy	cutting into	G
-ous	pertaining to	L
ov/o or ov/i	egg	L
ovari/o	ovary	L
ox/o, ox/y, or ox/i	oxygen	G
oxid/o	containing oxygen	L
oxy-	sharp, acid, quick, or oxygen	G

P

Word Part	Meaning	Word Origin (L = Latin, G = Greek)
pachy-	thick	G
palat/o	roof of mouth	L
pall/o or pallid/o	pale	L
palliat/o	hidden	L
palm/o	caudal surface of the manus (front paw) including the carpus	L
palpat/o	to touch	L
palpebr/o	eyelid	L
palpit/o	throbbing	L
pan-	all, entire, or every	G
pancreat/o	pancreas	L
papill/o or papill/i	nipplelike	L
papul/o	pimple	L
par/o	apart from, beside, near, or abnormal; also reproductive labor	L, G
para-	apart from, beside, near, rear in quadrupeds, or abnormal	G
-para	to bear or bring forth	L
paralys/o or paralyt/o	to disable	G
parasit/o	near food	G
parathyroid/o	parathyroid glands	G

Word Part	Meaning	Word Origin (L = Latin, G = Greek)
-paresis	weakness	L, G
paret/o or pares/i	to disable	L, G
pariet/o	wall	L
parotid/o	parotid gland	G
paroxysm/o	sudden attack	G
part/o or parturit/o	birth or labor	L
-partum	birth or labor	L
patell/a or patell/o	patella	L
path/o	disease	G
-pathic	pertaining to disease	G
-pathy	disease	G
paus/o	stopping	L
pector/o	chest	L
ped/o	foot	L
ped/o or pedi/a	child	G
pedicul/o	louse	L
pelv/o or pelv/i	hip bone or pelvis	L, G
pen/i	penis	L
-penia	deficiency or reduction in number	G
pent- or penta-	five	G
pept/o or peps/i	digestion	L, G
per-	excessive or through	L
percept/o	to become aware	L
percussi/o	to tap or beat	L
peri-	around or surrounding	G
perine/o	membrane lining the abdominal cavity and its organs (peritoneum)	G
perme/o	to pass through	L
pernici/o	destructive or harmful	L
perone/o	fibular	L
pertuss/i	intense cough	L
petechi/o	skin spot	L
-pexy	surgical fixation	G
phac/o or phak/o	lens of the eye	G
phag/o or -phagia	eating or swallowing	G
-phage	one that eats	G
phalang/o	phalanges	G
phall/o	penis	G
pharmac/o	drug	G
pharyng/o	throat (pharynx)	G
phe/o	dusky	G
pher/o	to bear or carry	G
-pheresis	removal	G
phil/o	love or attraction to	L, G
-philia	attraction for or increase in numbers	L, G

Word Part	Meaning	Word Origin (L = Latin, G = Greek)
phim/o	muzzling or constriction of an orifice	G
phleb/o	vein	G
phlegm/o	thick mucus	L, G
phob/o	fear	G
phon/o	sound or voice	G
-phonia	sound or voice	G
phor/o	carry or movement	G
-phoresis	carrying or transmission	G
-phoria	to bear, feeling, or mental state	G
phot/o	light	G
phren/o	diaphragm	G
-phylactic	protective or preventive	G
-phylaxis	protection or prevention	G
physi/o or physic/o	nature	L
-physis	to grow	G
-phyte	plant	G
pigment/o	color	L
pil/o or pil/i	hair	L
pineal/o	pineal gland	L
pinn/i	external ear, or auricle	L
pituit/o	pituitary gland	L
plac/o	flat patch	G
placent/o	round, flat cake or placenta	L
-plakia	thin, flat layer or scale	G
plan/o	flat	L
plant/o or plant/i	caudal surface of the pes (rear paw) including the tarsus	L
plas/o or plas/i	growth, development, formation, or mold	L
-plasia	formation, development, or growth	G
-plasm	formative material of cells	G
plasm/o	something formed	G
-plast	primitive living cell	G
plast/o	growth, development, or mold	G
-plasty	surgical repair	G
-plegia or -plegic	paralysis	G
pleur/o	membrane lining the lungs (pleura)	G
plex/o	network	L
plic/o	fold or ridge	L
-pnea	breathing	G
-pneic	pertaining to breathing	G
pneu-	relating to air or lungs	G
pneum/o or pneumon/o	lung or air	G

Word Part	Meaning	Word Origin (L = Latin, G = Greek)
pod/o	foot	G
-poiesis	formation	G
poikil/o	irregular	G
poli/o	gray matter of the brain and spinal cord	G
polio-	gray matter of the brain and spinal cord	G
pollaki-	many times	G
poly-	many	G
polyp/o	small growth	G
pont/o	part of the brain (the pons)	L
poplit/o	caudal surface of the patella	L
por/o	small opening (pore)	L, G
porosis	passage or porous condition	G
port/i	gate or door	L
post-	after or behind	L
poster/o	toward the tail	L
potent/o	powerful	L
practic/o or pract/i	practice	G
prandi/o	meal	L
-prandial	meal	L
-praxia	action	G
pre-	before or in front of	L
pregn/o	pregnant or with offspring	L
prematur/o	too early	L
preputi/o	foreskin or prepuce	L
press/o	to press or draw	L
priap/o	penis	L, G
primi-	first	G
pro-	before or on behalf of	L, G
process/o	going forth	L
procreat/o	to reproduce	L
proct/o	anus and rectum	G
prodrom/o	precursor	L, G
product/o	to lead forward or yield	L
prolaps/o	to fall downward or slide forward	L
prolifer/o	to reproduce	L
pron/o or pront/o	bent forward	L
prostat/o	prostate gland	G
prosth/o or prosthet/o	addition or appendage	G
prot- or proto-	first	G
prot/o or prote/o	first, original, or protein	G
proxim/o	near	L
prurit/o	itch	L
pseud/o	false	G

Word Part	Meaning	Word Origin (L = Latin, G = Greek)
psor/o or psor/i	itch	G
psych/o	mind	G
ptomat/o	a fall	G
-ptosis	drooping or sagging	G
-ptyalo	spit or saliva	G
-ptysis	spitting	G
pub/o	part of the hip bone (pubis)	L
pubert/o	adult	L
pudend/o	pudendum	L
puerper/i	labor	L
pulm/o or pulmon/o	lung	L
pulpos/o	fleshy or pulpy	L
puls/o	beating	L
punct/o	sting, puncture, or little hole	L
pupill/o	pupil of the eye	L
pur/o	pus	L
purpur/o	purple	L
pustul/o	blister	L
py/o	pus	G
pyel/o	renal pelvis	G
pylor/o	gatekeeper or pylorus	G
pyr/o or pyret/o	fever or fire	G
pyramid/o	pyramid-shaped	G

Q

Word Part	Meaning	Word Origin
quadr/i or quadr/o	four	L
quart-	fourth	L
quinqu-	five	L
quint-	fifth	L

R

Word Part	Meaning	Word Origin
rabi/o	rage or madness	L
rachi/o	spinal column or vertebrae	G
radi/o	radiation or the radius (bone in distal forelimb)	L
radiat/o	giving off radiant energy	L
radicul/o	root or nerve root	L
ram/o	branch	L
raph/o	seam or suture	G
-raphy	suturing	G
re-	back or again	L
recept/o or recipi/o	receive	L
rect/o	rectum or straight	L
recuperat/o	to recover	L
reduct/o	to bring back together	L
refract/o	to bend back or turn aside	L

Word Part	Meaning	Word Origin (L = Latin, G = Greek)
regurgit/o	to flood or gush back	L
remiss/o	to let go or give up	L
ren/o	kidney	L
restor/o	to rebuild	L
resuscit/o	to revive	L
retent/o	to hold back	L
reticul/o	network	L
retin/o	nervous tissue of the eye (retina) or net	L
retract/o	to draw back	L
retro-	behind or backward	L
rhabd/o	rod	G
rhabdomy/o	striated muscle	G
-rhage or -rhagia	bursting forth	G
-rhaphy	suture	G
-rhea	flow or discharge	G
rheum/o or rheumat/o	watery flow	L, G
-rhexis	rupture	G
rhin/o	nose	G
rhiz/o	root	G
rhonc/o	to snore	L, G
rhythm/o	rhythm	L, G
rhytid/o	wrinkle	G
rigid/o	stiff	L
roentgen/o	X-ray	Named for Wilhelm Conrad Röentgen
rostri	beak	L
rotat/o	to revolve	L
-rrhage or -rrhagia	bursting forth	G
-rrhaphy	suture	G
-rrhea	flow or discharge	G
-rrhexis	rupture	G
rug/o	wrinkle or fold	L

S

Word Part	Meaning	Word Origin
sacc/o or sacc/i	sac	L
sacr/o	sacrum	L
sagitta	arrow	L
saliv/o	spit	L
salping/o	tube (uterine or auditory)	G
-salpinx	uterine tube, oviduct, or fallopian tube	G
sanguin/o or sangu/i	blood	L
sanit/o	soundness or health	L
saphen/o	clear or apparent	L
sapr/o	dead or decaying	G

Word Part	Meaning	Word Origin (L = Latin, G = Greek)
sarc/o or sarcomat/o	flesh (connective tissue) or malignancy of connective tissue	G
sarco-	flesh	G
-sarcoma	malignant tumor	G
scalp/o	to carve or scrape	L
scapul/o	shoulder blade	L
-schisis	cleft or divided	G
schiz/o	cleft or divided	G
scirrh/o	hard	G
scler/o	white of the eye (sclera) or hard	G
-sclerosis	abnormal dryness or hardness	G
scoli/o	curved or crooked	G
-scope	instrument for visual examination	G
-scopic	pertaining to visual examination	G
-scopy	process of visually examining	G
scot/o	darkness	G
scrot/o	bag or pouch	L
seb/o	wax or sebum	L
sect/o	cutting	L
secti/o	to cut	L
secundi-	second	L
segment/o	in pieces	L
sell/o	saddle	L
semi-	half	L
semin/i	sperm, semen, or seed	L
sen/i	old	L
sens/i	feeling or sensation	L
sensitiv/o	affected by or sensitive to	L
seps/o	infection	G
sept- or septi-	seven	L
sept/o	infection or partition	L
ser/o	clear, fluid portion of blood (serum)	L
seros/o	serous	L
sesqui-	one and one-half	L
sex-	six	L
sial/o	saliva or salivary glands	G
sialaden/o	salivary glands	G
sider/o	iron	G
sigmoid/o	sigmoid colon	L, G
sin/o or sin/u	hollow, sinus, or tubelike passage	L
sinistr/o	left	L
sit/u	place	L

Word Part	Meaning	Word Origin (L = Latin, G = Greek)
skelet/o	bony framework of the body (skeleton)	G
soci/o	companion	L
solut/o	dissolved	L
solv/o	dissolved	L
soma-	body	G
somat/o	body	G
somn/o or somn/i	sleep	L
-somnia	sleep	L
son/o	sound	L
sopor/o	sleep	L
-spasm	sudden involuntary contraction or tightening	L, G
spasm/o or spasmod/o	sudden involuntary contraction or tightening	L, G
spec/i	to look at or a sort	L
specul/o	mirror	L
sperm/o or spermat/o	sperm cell or seed	G
sphen/o	wedge	G
spher/o	round, sphere, or ball	G
sphincter/o	tight band	L, G
sphygm/o	pulse	G
spin/o	backbone	L
spir/o	breathing	L
spirill/o	little coil	L
spirochet/o	coiled microorganism	L, G
splen/o	spleen	G
spondyl/o	vertebra	G
spontane/o	unexplained, voluntary	L
spor/o	seed	G
sput/o	spit	L
squam/o	scale	L
-stalsis	contraction	G
staped/o or stapedi/o	one of the middle ear bones (stapes or stirrup)	L
staphyl/o	cluster or bunch of grapes	G
stas/i or stat/i	controlling or stopping	G
-stasis	controlling or stopping	G
-static	controlling or stopping	L, G
steat/o	fat	G
sten/o	narrowing or contracted	G
-stenosis	narrowing, or stricture, of a duct or canal	G
ster/o	solid structure	G
stere/o	solid or three-dimensional	G
steril/i	barren	L
stern/o	breastbone	G

Word Part	Meaning	Word Origin (L = Latin, G = Greek)
stert/o	to snore	L
steth/o	chest	G
sthen/o	strength	G
-sthenia	strength	G
stigmat/o	point or spot	L, G
stimul/o	good or incite	L
stomat/o	mouth	G
-stomosis	making a new opening	G
-stomy	new opening	G
strab/i	to squint	G
strat/i	layer	G
strept/o	twisted chain	G
striat/o	stripe or groove	L
stric-	narrowing	L
strict/o	to draw tightly together or tie	L
strid/o	harsh sound	L
stup/e	stunned	L
styl/o	pointed instrument or pen	L
sub-	under, less, or below	L
subluxat/o	partial dislocation	L
sucr/o	sugar	L
sudor/i	sweat	L
suffoc/o or suffocat/o	to choke	L
sulc/o	furrow or groove	L
super-	above, excessive, or higher than	L
super/o	above, excessive, or higher than	L
superflu/o	overflowing or excessive	L
supin/o	lying on the back	L
supinat/o	bent backward or to place on the back	L
suppress/o	to press down	L
suppur/o or suppurat/o	to form pus	L
supra-	above (cranial) or excessive	L
supraren/o	above (cranial) or on the kidney	L
sutur/o	to stitch	L
sym-	with or together	G
symptomat/o	falling together or symptom	G
syn-	union or association	G
synaps/o or synapt/o	point of contact	G
syncop/o	to cut short or cut off	G

Word Part	Meaning	Word Origin (L = Latin, G = Greek)
-syndesis	surgical fixation of the vertebrae	G
syndesm/o	ligament	G
syndrom/o	running together	G
synovi/o	synovial membrane or synovial fluid	L, G
syring/o	tube	G
system/o or systemat/o	entire body	G
systol/o	contraction	G

T

Word Part	Meaning	Word Origin (L = Latin, G = Greek)
tachy-	abnormally fast	G
tact/i	touch	L
tars/o	tarsus, or edge of the eyelid	G
tax/o	order or coordination	G
techn/o or techni/o	skill	G
tectori/o	covering	L
tele/o	distant or far	G
tempor/o	temple	L
ten/o or tend/o	tendon, strain, or to extend	G
tenac/i	sticky	L
tendin/o	tendon	L
tens/o	to stretch out, strain, or extend	L
terat/o	malformed fetus	G
termin/o	end	L
tert-	third	L
test/o or test/i	testis or testicle	L
tetan/o	rigid	G
tetra-	four	G
thalam/o	inner room or thalamus	G
thanas/o or thanat/o	death	G
the/o	to put or place	G
thec/o	sheath	L, G
thel/o	nipple	G
therap/o or therapeut/o	treatment	G
therm/o	heat	G
thio-	sulfur	G
thorac/o	chest	L, G
-thorax	pleural cavity or chest	G
thromb/o	clot	G
thym/o	thymus gland	G
thyr/o or thyroid/o	shield or thyroid gland	G
tibi/o	tibia	L
-tic	pertaining to	L, G

Word Part	Meaning	Word Origin (L = Latin, G = Greek)
tine/o	gnawing worm or ringworm	L
tinnit/o	ringing	L
toc/o	birth	G
-tocia or -tocin	labor or birth	G
tom/o	to cut or section	G
-tome	instrument to cut	G
-tomy	cutting or incision	G
ton/o	tension or stretching	G
tone/o	to stretch	L, G
tonsill/o	tonsil or throat	L
top/o	place or location	G
tors/o	to twist or rotate	L
tort/i	twisted	L
tox/o or toxic/o	poison	L
trabecul/o	beams or little beam marked with crossbars	L
trache/o or trache/i	windpipe	G
trachel-	neck	G
tract/o	to draw, to pull, path, or bundle of nerve fibers	L
tranquil/o	quiet	L
trans-	across or through	L
transfus/o	to transfer or pour across	L
transit/o	changing	L
transvers/o	across or crosswise	L
traumat/o	injury	G
trem/o	shaking	L
tremul/o	fine tremor or shaking	L
treponem/o	coiled or turning microbe	G
-tresia	opening	G
tri-	three	G
trich/o	hair	G
trigon/o	triangle	L, G
-tripsy	crushing stone	G
trit- or trito-	third	G
-trite	instrument for cutting	L
trochle/o	pulley	G
trop/o	to turn or change	G
troph/o	formation, development, or growth	G
-trophy	formation, development, or growth	G
-tropia	to turn	G
-tropic or -tropin	having an affinity for or turning toward	G
tub/o or tub/i	tube or pipe	L
tubercul/o	little knot or swelling	L
tunic/o	covering or sheath	L

Word Part	Meaning	Word Origin (L = Latin, G = Greek)
turbinat/o	coiled or spiral-shaped	L
tuss/i	cough	L
tympan/o	tympanic membrane, or eardrum	G

U

Word Part	Meaning	Word Origin (L = Latin, G = Greek)
ulcer/o	sore or ulcer	L
uln/o	one of the bones in the distal forelimb (ulna)	L
ultra-	beyond or excess	L
-um	structure	L
umbilic/o	naval	L
un-	not	L, G
ungu/o	nail or hoof	L
uni-	one	L
ur/o	urine or urinary tract	G
-uresis	urination	G
ureter/o	ureter	G
urethr/o	urethra	G
-uria	urination or urine	G
urin/o	urine or urinary organs	L, G
urtic/o	nettle, rash, or hives	L
-us	thing	L, G
uter/o or uter/i	uterus or womb	L
uve/o	vascular layer of the eye, iris, choroid, or ciliary body	L
uvul/o	little grape or uvula	L

V

Word Part	Meaning	Word Origin (L = Latin, G = Greek)
vaccin/o or vaccin/i	vaccine	L
vacu/o	empty	L
vag/o	wandering	L
vagin/o	sheath or vagina	L
valg/o	bent or twisted outward	L
valv/o or valvul/o	valve	L
var/o	bent or twisted inward	L
varic/o	swollen or dilated vein	L
vas/o	vessel, duct, or vas deferens	L
vascul/o	little vessel	L
vaso-	vessel	L
vast/o	great or extensive	L
vect/o	to carry or convey	L
ven/o	vein	L
vener/o	sexual intercourse	L
venter-	abdomen	L
ventilat/o	to expose to air	L
ventr/o	belly side	L

Word Part	Meaning	Word Origin (L = Latin, G = Greek)
ventricul/o	small chamber or ventricle of the brain or heart	L
venul/o	small vein	L
verg/o	to twist or incline	L
verm/i	worm	L
verruc/o	wart	L
vers/o or vert/o	turn	L
-version	to turn	L
vertebr/o	backbone	L
vertig/o or vertigin/o	turning around or revolution	L
vesic/o	urinary bladder	L
vesicul/o	seminal vesicle, blister, or little bladder	L
vestibul/o	entrance	L
vi/o	force	L
vill/i	tuft of hair or thread-like projections from a membrane	L
vir/o	poison or virus	L
viril/o	masculine	L
vis/o	seeing or sight	L
visc/o or viscos/o	sticky	L

Word Part	Meaning	Word Origin (L = Latin, G = Greek)
viscer/o	internal organ	L
vit/a or vit/o	life	L
viti/o	blemish or defect	L
vitre/o	glassy	L
viv/i	alive	L
voc/i	voice	L
vol/o	caudal surface of forelimbs (palm) or hindlimbs (sole)	L
volv/o	to turn or roll	L
vulgar/i	common	L
vulv/o	covering or vulva	L

X

Word Part	Meaning	Word Origin (L = Latin, G = Greek)
xanth/o	yellow	G
xen/o	strange or foreign	G
xer/o	dry	G
xiph/o or xiph/i	sword	G

Z

Word Part	Meaning	Word Origin (L = Latin, G = Greek)
zoo-	life, animal	G
zygomat/o	yoke or cheekbone	G
zygot/o	joined or yoked together	G

INDEX

Note: Page numbers referencing figures are italicized and followed by an "*f*". Page numbers referencing tables are italicized and followed by a "*t*".

A

A- (prefix), *3t*, 29, 201
Ab- (prefix), *3t*
Abbreviations
   cardiovascular system, 208–209
   cats, 449
   digestive system, 156
   dogs, 449
   ears, 359
   endocrine system, 278
   equine, 472
   eyes, 359
   hematologic system, 387
   immune system, 387
   integumentary system, 260
   laboratory, 413
   lymphatic system, 387
   muscular system, 75
   nervous system, 334
   oncology, 387
   pharmacologic terms, 433
   physical examination, 412
   physical examination terminology, 412
   pocket pets, 558
   radiography, 412
   radiology, 412
   reproductive system, 306
   respiratory system, 233
   ruminants, 495
   skeletal system, 75
   surgical terms, 433
   swine, 505
   testing, 412–413
   tumors, 387
   urinary system, 179
Abdomen, 23, 134
Abdomin/o (combining form), 134
Abdominocentesis, 153
Abducent nerve, *325f, 326t*

Abduction, *3t*, 24, *25f*
Abductor, 72
Ablation, 359
Abomasopexy, 153, *154f*
Abomasum, 137, *139f*
Aboral, 140
Abortion, 302
Abort/o (combining form), 302
Abrasion, 252
Abscess, 252
Absorption, 143
-Ac (suffix), 4
Acanthokeratodermia, 256
Acariasis, 256
Accelerated lambing, 491
Accessory nerve, *325f, 326t*
Accessory organs of digestion, 142–143, *142f*
   gallbladder, 143
   liver, 142–143
   pancreas, 143
Accessory sex glands, 292–293, *293t*
Acclimatization, *532t*
Accommodation, 347
Acetabulum, 59, *59f, 64f*
Achalasia, 146
Achilles tendon, 52
Acid, 402
Acidosis, 275
Acne, 252
Acous/o (combining form), *32t*, 353
Acoustic, 353
Acoustic nerve, *325f, 326t*
Acoust/o (combining form), *32t*, 353
Acr/o (combining form), 275
Acromegaly, 275
Acrosome, 290
ACTH stimulation test, 275
Acuity, 347
Acute, *405t*
Acute bronchitis, 230
Acute moist dermatitis, 252
Acute renal failure (ARF), 178
Ad- (prefix), *3t*, 272, 273
Addison's disease, 277